Application of Biomaterials in the Treatment of Cardiovascular Diseases

Yi-Da Tang • Jing Yang • Yufeng Zheng •
Yongjun Li • Yong Zeng
Editors

Application of Biomaterials in the Treatment of Cardiovascular Diseases

Springer

Editors
Yi-Da Tang
Department of Cardiology and Institute of Vascular Medicine
Peking University Third Hospital
Beijing, China

Jing Yang
Laboratory of Biomaterials and Drug Delivery
Institute of Biomedical Engineering, Chinese Academy of Medical Science and Peking Union Medical College
Tianjin, China

Yufeng Zheng
School of Materials Science and Engineering
Peking University
Beijing, China

Yongjun Li
Department of Vascular Surgery
Beijing Hospital
Beijing, China

Yong Zeng
Department of Cardiology
Beijing Anzhen Hospital, Capital Medical University
Beijing, China

ISBN 978-981-19-7711-4 ISBN 978-981-19-7712-1 (eBook)
https://doi.org/10.1007/978-981-19-7712-1

This Springer imprint is published by the registered company Springer Nature Singapore Pte Ltd.
The registered company address is: 152 Beach Road, #21-01/04 Gateway East, Singapore 189721, Singapore

Foreword I

With the Chinese population aging, urbanization progress accelerating, and unhealthy lifestyles increasing, the prevalence and incidence of cardiovascular diseases (CVDs) in China continue to rise. Characterized by high disabling and mortality rates, CVD remains a severe threat to human health, and the leading cause of death in China. Despite many therapeutic methods have been developed over time, the continuously increasing rates of mortality suggest the need for more advanced strategies for the treatment of CVDs. Better understanding of the aetiology of CVDs and implementation of modern techniques improve the outcome of treatment and quality of life.

Biomaterials dominate the field of cardiovascular therapeutics; clinicians are increasingly aware of the benefits in taking a multidisciplinary collaborative approach with researchers from different fields to advance clinical outcomes by improving techniques and biomaterials for CVDs. This boosts the innovation of biomedical materials and its application in cardiovascular conditions. Biomedical materials are biocompatible which allows implantation into human body to repair, replace, or improve the function of defective tissues or organs. On top of traditional oral or intravenous medication, vascular stents and bioprosthetic valves have been deployed in various clinical scenarios for CVDs. This field has evolved beyond merely choosing biocompatible materials and now focuses on the sensible design of controlled structures that integrate with the native cardiovascular tissues. These medical devices are specific applications of biomedical materials in the cardiovascular field. The development of these medical devices has contributed significantly to advances in the treatment of CVD.

Although many different types of biomaterials have been deployed in clinical practice, very limited literature has summarized the experimental studies and clinical evaluations of commonly used biomedical materials in the cardiovascular conditions. Professor Yi-Da Tang from the Department of Cardiovascular Medicine, Peking University Third Hospital, led a joint effort with several experts in the cardiovascular field to compile a book on the application and latest advances of biomedical materials in common CVDs. Based on the normal anatomical structure and the treatment strategies of common CVDs, this book elucidates the application and progress of standard cardiac devices in the cardiovascular field. This book is rich in content, easy to understand, and has significant academic and clinical value. It is believed that the publication of this book will serve as a guide for the clinical practice of cardiovascular physicians and a valuable reference for researchers of all relevant specialties.

Jie Qiao

Chinese Academy of Engineering, Beijing, China

American Academy of Arts and Sciences, Cambridge, MA, USA

Royal College of Obstetricians and Gynaecologists, London, UK

Peking University, Beijing, China

Peking University Health Science Center, Beijing, China

Peking University Third Hospital, Beijing, China

Foreword II

Cardiovascular diseases (CVDs) are the leading cause of death worldwide with an annual toll of 17.7 million deaths which may reach 23.6 million by 2030. CVDs generally can be treated by medication, intravascular intervention, or surgery. The diversification of treatment options for CVDs has been fueled, to a large extent, by evolution in biomaterials, such as drug delivery systems, tissue engineering, and intravascular intervention devices, among others.

Derived from a variety of biomedical materials, such as natural or synthetic polymers, metals, and ceramics, currently available drug delivery systems, including liposomes, micelles, nanoparticles, dendrimers, hydrogels, and viral vectors, are more biocompatible and precise, and can release drugs over extended periods of time.

Intravascular interventions, including stents and balloons, for the management of coronary heart disease have evolved rapidly. Coronary stents have progressed from bare metal to drug-eluting stents and bioabsorbable scaffolds, with each innovation addressing limitations of its predecessor. However, there is room for improvement in terms of stent structural materials and polymeric coatings. As an application of Tissue Engineering and Regenerative Medicine, an emerging cardiac patch technique involving stem cell injection therapy and cellularized or cell-free cardiac patch implants have shown potential in repairing injured cardiac tissue.

Biomaterials are also widely applied in structural valvular heart disease and congenital heart diseases. Both balloon- and self-expanding valves made from bovine or porcine pericardial leaflets with a cobalt-chromium or nickel-titanium alloy frame, are commonly used in transcatheter aortic, mitral, pulmonary, or tricuspid valve replacement or repair. The design of prosthetic heart valves should optimize mechanical properties and biocompatibility, anti-calcification and anti-coagulation characteristics, as well as endothelial cell adhesion, hydrodynamic and anti-fatigue properties. Novel biomaterials are needed to address current procedure-related complications including permanent pacemaker implantation, paravalvular leakage, stroke, and embolism, as well as coronary obstruction. Nondegradable nitinol and polymeric (i.e., polyethylene terephthalate) fabric patches are frequently deployed in occlusion devices for the treatment of congenital heart diseases such as atrial or ventricular septal defect, patent foramen ovale, and patent ductus arteriosus. However, complications such as foreign body reaction, device erosion or fracture, insufficient endothelialization, and thrombus formation may occur. Novel bioabsorbable occlusion devices may further minimize inflammatory response and late complications...

Randomized controlled clinical trials support the widely accepted use of cardiac pacemakers and defibrillators for arrhythmias. Cardiac pacemakers have evolved from traditional (single and dual chamber pacing) to biventricular (cardiac resynchronization therapy, as well as multipoint, His and left bundle pacing) and leadless designs. Similarly, the most recent implantable cardioverter defibrillators notably differ from previous models and are recommended as first-line intervention for primary and secondary prevention of sudden cardiac death. Patients with atrial fibrillation, who have been traditionally managed with oral anti-coagulants, now can be offered an alternative left atrial appendage occlusion strategy to prevent cerebral infarction and systemic thromboembolism. The materials for the latter occlusion devices are similar to those for congenital heart diseases.

Typical examples of biomaterials application in CVDs are prosthetic vascular grafts for vascular repair and total artificial heart implantation for end-stage heart failure. Prosthetic valvular grafts containing polyethylene terephthalate and expanded polytetrafluoroethylene are regularly used in aortic, lower limb arteries, and extra-anatomy bypass reconstructions. Total artificial hearts are used temporarily for bridging heart transplantation; however, a substitute for heart transplantation is under regulatory review. At present, total artificial heart implantation has limited clinical application due to its availability, which may be improved by introduction of better biomaterials and medications, such as antibiotics and anti-coagulants.

Based on the current understanding of applications of biomaterials in CVDs, this book covers the properties and manufacturing methods of available biomaterials in various clinical scenarios, as well as their pros and cons, challenges, and future perspectives. Chapter 1 outlines CVD types and treatments. Chapter 2 focuses on anatomy of the heart and blood vessels. Chapters 3 and 4 describe the qualities of assorted biomaterials in managing CVDs. Chapter 5 introduces material manufacturing methods; and Chapters 7–12 summarize the full spectrum of biomaterials applied in cardiovascular structural and electrophysiological diseases. Chapter 13 reviews current progress and future directions for devices in CVDs. We hope this book will be informative and helpful for students, fellows, experts in the field, and other readers interested in CVDs and biomaterials.

Department of Cardiology,
Cardiovascular Institute and Fuwai
Hospital, National Center for
Cardiovascular Diseases, Chinese
Academy of Medical Sciences and
Peking Union Medical College,
Beijing, China

Runlin Gao MD

Preface

Cardiovascular diseases (CVDs) are the biggest killers of human health, and biomaterials play a significant role in the clinical treatment of CVDs. In order to help researchers and students who are interested in relevant research to understand the progress and needs of this field more quickly, deeply, and comprehensively, an editorial team including clinical and materials experts has been established, led by famous professor Yi-da Tang from the Third Hospital of Beijing Medical College. This book is composed of 13 chapters, the contents including types of CVD, common clinical treatment methods and defects of CVD, anatomy of heart and the blood vessel, ideal cardiovascular material properties, preparation methods, drug delivery system, stent and the infused balloon, cardiac valves, heart occluder, artificial heart and blood vessel, cardiac patch, cardiac pacemaker, and defibrillator.

I would like to thank the team led by the chief editor for their positive contribution to the compilation of this book. I wish that the collaboration of scholars will bridge clinical medicine and engineering, promote the original innovation of related technologies in this field of China, and build a healthy China.

This is also a gift from Chinese Academy of Medical Science and Peking Union Medical College to all interesting in this research field.

Beijing, China	Yi-Da Tang
Tianjin, China	Jing Yang
Beijing, China	Yufeng Zheng
Beijing, China	Yongjun Li
Beijing, China	Yong Zeng

Contents

Chapter 1
Summary (Types of Cardiovascular Diseases, Common Clinical Treatment Methods, and Defects of Cardiovascular Diseases)

Yida Tang

1.1 Introduction

Cardiovascular diseases include congenital heart disease, adult structural heart disease, atherosclerosis-based ischemic heart disease, arrhythmias and electrophysiological diseases of the heart, cardiomyopathies (hereditary, myocardial inflammation, myocardial damage in systemic diseases), pericardial disease, and the end-stage manifestations of heart disease—heart failure. Along with an ever-deepening understanding of disease and the emergence of new treatments, the medical problems we are currently facing require a more intensive interdisciplinary approach and the establishment of a holistic management mindset.

1.2 Congenital Heart Disease

The incidence of congenital heart disease (CHD) has gradually decreased in developed countries in recent years, but remains a serious health problem in developing countries. Surgical procedures and transcatheter treatment are the main treatments for CHD, and the outcomes of CHD have improved significantly due to the progressive development of equipment. In recent years, there are still many impressive achievements in the field of CHD, and aspects of adult CHD are even more of a focus for clinicians, and considerable results have been achieved in the study of

Y. Tang (✉)
Department of Cardiology and Institute of Vascular Medicine, Key Laboratory of Molecular Cardiovascular Science, Ministry of Education, Peking University Third Hospital, Beijing, China
e-mail: tangyida@bjmu.edu.cn

Y.-D. Tang et al. (eds.), *Application of Biomaterials in the Treatment of Cardiovascular Diseases*, https://doi.org/10.1007/978-981-19-7712-1_1

pathogenesis, thus allowing management nodes to move forward and achieve more precise treatment whenever possible.

1.2.1 CHD Pathogenesis Research Is Progressing Rapidly

With the advancement of WES, scRNA-seq, and other technologies, the research on CHD mechanism has progressed rapidly, mainly focusing on two aspects: on the one hand, through high-throughput sequencing of large samples and cohorts to find CHD susceptibility genes and clarify the genetic basis of CHD, on the other hand, through scRNA-seq, cell lineage tracing technology, and gene editing technology, we can find the transcription factors regulating CHD from the perspective of developmental biology and the means of intervention.

Many CHD susceptibility genes have been reported. Sevim Bayrak et al. [1] from the Icahn School of Medicine at Mount Sinai, USA, found that CHD patients carry novel variants in 23 genes, including HSP90AA1, ROCK2, IQGAP1, and CHD4, which have similar biological functions, molecular pathways, and interactions with known CHD causative genes. Lahm et al. [2] from the Munich Heart Center in Germany performed a genome-wide association study (GWAS) of 4034 CHD patients in Europe and identified 20 novel SNP loci associated with CHD. Izarzugaza et al. [3] from the University of Copenhagen, Denmark, identified 218 candidate genes in 32 CHD families by WES sequencing, and enrichment analysis revealed that the functions of these genes are closely related to the calcium pathway, and later functional experiments in zebrafish confirmed the involvement of ITPR1, PLCB2, and ADCY2 genes in cardiac development. Chromosomal effects on CHD are not limited to the coding regions, as Richter et al. from the Icahn School of Medicine at Mount Sinai [4] found that patients with CHD may have more pathogenic de novo variants in noncoding regions than thought, at least as many as de novo variants in coding regions. In addition to germline mutations, Hsieh et al. [5] from Columbia University found that somatic chimeric mutations that may cause cardiac malformations were detected by EM-mosaic technology in the peripheral blood of 1% of CHD patients, especially those with relatively high allele ratios of pathogenic variants, while cardiac tissue sequencing detected up to 5% of patients carrying somatic chimeric variants. Morton et al. [6] of Harvard Medical School found that a higher proportion of loss-of-function mutations (LoF) in tumor-associated genes were found in patients with CHD compared to healthy controls, often combined with abnormal manifestations outside the heart and an increased risk of tumor. Miao et al. [7] from Cincinnati Children's Hospital, USA, illustrated the important role of endocardium in the pathogenesis of HLHS by using scRNA-seq technique and patient-specific iPSC from left ventricular dysplasia syndrome (HLHS) and heart samples. We also found that exogenous addition of Fibronectin 1 protein significantly ameliorated functional impairment of HLHS iPSC cardiomyocytes, suggesting that Fibronectin 1 may be an important therapeutic target for HLHS.

Current analytical protocols may underestimate the proportion of copy number variants (CNV) and single-nucleotide variants (SNV) because of limitations in detection and difficulty in predicting whether the identified variants are pathogenic. In addition, multiple disease patterns are involved in CHD, including polygenic inheritance, somatic mutations, and mutations in noncoding region DNA. As more patients with CHD undergo more comprehensive sequencing evaluations, we will be able to determine the genetic basis of most patients and ultimately achieve precision medicine for patients with CHD.

1.2.2 CHD Imaging Assessment Is Becoming More Sophisticated

Multimodal cardiovascular imaging plays an important role in the treatment of patients with CHD. In 2020, the American Heart Association (AHA), the American College of Cardiology (ACC), and nine other cardiovascular societies [8] introduced the appropriate use criteria (AUC) for multimodality imaging follow-up assessment of ACHD, covering transthoracic/transesophageal echocardiography (TTE, TEE), cardiac magnetic resonance (CMR)/computerized tomography (CT), stress imaging, lung scan, and other noninvasive. Imaging is used to clarify the indications of multimodality imaging for the evaluation of various ACHD, for preoperative and postoperative evaluation of conventional or catheter-based interventions/surgeries. The aim is to provide a rational selection of imaging methods for different CHDs and to reduce the choice of "almost inapplicable" assessment methods. A new technique and analytical tool based on traditional methodology, CMR/CT 4D flow imaging for cardiovascular hemodynamic assessment, will also be introduced into the CHD assessment arena and will be validated in continuous application and practice. Integrating imaging assessment into the lifelong management proposed in the 2020 ESC ACHD Guidelines, the establishment of a follow-up system for ACHD patients will gradually enter a standardized mode, thus benefiting CHD patients for life.

1.2.3 The Concept of CHD Treatment Is Being Upgraded

Mortality from CHD has decreased significantly in the last 20 years. Data from the National Center for Health Statistics [9] show a 16.4% decrease in all-cause mortality and a 39.4% decrease in CHD-related mortality in the 19 years from 1999 to 2017. There has been a tremendous upgrade in CHD treatment concepts, especially in adult congenital heart disease (ACHD), leading to the publishment of new 2020 edition of the European Adult Congenital Heart Disease guidelines [10]. The new edition of the guidelines introduced a standardized nomenclature for "ACHD," classified for the first time by mild, moderate, or severe, and emphasized a new concept of lifelong management of ACHD. The guidelines focus on recommendations for the management of ACHD combined with pulmonary arterial hypertension (PAH). PAH, caused by prolonged left-to-right shunting, vasoconstriction, inflammatory response, cellular proliferation, and vascular remodeling, ultimately leads to Eisenmenger syndrome, which accounts for 4%–10% of CHD and is closely associated with treatment decisions, outcomes, and long-term prognosis. When PAH is suggested by noninvasive methods in patients with shunted CHD, pulmonary vascular resistance (PVR) must be measured by cardiac catheterization, and the decision to close the defect must be based on PVR. The risk of heart failure in patients with ASD combined with left heart remodeling due to a rapid increase in left heart blood volume and pressure load after closure of the ASD was also emphasized. The new version of the guidelines revises the criteria for identifying PAH by right heart catheterization. Risk assessment should be performed in all patients with PAH. Initial combination therapy with oral targeted agents or sequential combination therapy is recommended, and initial combination therapy including prostacyclin should be used in high-risk patients. Guidance on pregnancy, contraception, and genetic counseling for patients with ACHD has been added to the guidelines, and high-risk pregnancies have been categorized and are not recommended for patients with PAH.

Arrhythmias such as atrial fibrillation, atrial flutter, ventricular premature, ventricular tachycardia, and atrioventricular block are the most common complications of ACHD and can lead to heart failure and sudden cardiac death. The new guidelines recommend that patients with ACHD combined with arrhythmias should be seen promptly at a specialized center to understand the causes and mechanisms of arrhythmias and the importance of cardiovascular anatomy of malformations. Arrhythmias treated with transcatheter or surgical intervention are the best treatment options. For symptomatic supraventricular and ventricular tachycardia, early catheter ablation may be an alternative to long-term pharmacological therapy, and patients with bradycardia and intra-atrial folding tachycardia benefit more from pacemaker implantation. Studies of ICD implantation in patients with tetralogy of Fallot (TOF) have shown a high burden of long-term ICD complications, and therefore the conditions for ICD implantation in patients with TOF should be carefully controlled [11].

There is a growing emphasis on the role of patent foramen ovale (PFO) in the occurrence of stroke. A meta-analysis [12] that pooled the results of 1032 studies worldwide showed that approximately 25% of people have a PFO (with slightly different proportions depending on the means of examination); in patients with cryptogenic stroke (CE), the incidence of PFO was three times higher than in healthy controls and twice as high as in non-CE patients, and paradoxical embolism was an important mechanism for stroke in patients with PFO. PFO occlusion can significantly reduce the risk of stroke recurrence compared with pharmacological treatment [13].

In recent years, many noteworthy achievements have emerged in the field of CHD, but the underlying mechanisms of how CHD develops, key issues in the developmental biology of the heart, and how to prevent birth defects are unanswered questions that require increased input and in-depth research. While it is important to treat patients with CHD that has already occurred, more importantly, we need to find the key pathways to prevent CHD from occurring.

1.3 Coronary Heart Disease

Coronary heart disease is the most prevalent disease of the cardiovascular system. Current treatments for coronary heart disease include pharmacological, interventional, and surgical approaches, among which the development of interventional cardiology is always in the limelight. With the advancement of various assessment techniques and interventional devices, as well as the presentation of a large amount of evidence-based medical evidence, coronary interventional therapy has undergone a long development of more than half a century and has moved from how to complete the procedure to precision intervention.

Thanks to the development of intravascular imaging and physiological assessment techniques, clinical interventional strategies have been significantly optimized, complementing and correcting the deficiencies of QCA-based PCI and improving the long-term prognosis of patients. Intravascular imaging techniques can assess the composition, severity, and extent of involvement of coronary lesions, as well as the diameter of the reference vessel, which can guide the operator in choosing the right strategy, as well as help evaluate the immediate results after stent placement and possible problems for optimal interventional outcomes. The PROSPECT II trial published in 2020 found that NIRS can help identify lipid-rich of non-target lesions that are at significantly increased risk for future coronary events [14]. Plaques with both heavy plaque load and lipid-rich characteristics were at particularly high risk of MACE. The results of the ULTIMATE trial, published in the same year, confirm that IVUS-guided stenting improves the clinical prognosis of patients by significantly reducing TVF, with a more pronounced benefit especially in patients who meet the ideal criteria for IVUS evaluation, and that it plays an important role in guiding clinical treatment decisions [15].

The role of coronary physiological and functional tests in PCI therapy is also becoming more prominent, and the flow reserve fraction (FFR) and its numerous derivatives can better guide the development of treatment plans for critical lesions and provide a hemodynamic basis for interventional therapy. With the improved understanding of functional evaluation of coronary lesions, a series of FFR-based derivative techniques have been rapidly developed considering the limitations of FFR such as its invasive nature and the need for maximal vasodilation. Noninvasive flow reserve fraction (FFRCT) based on CT angiography allows noninvasive evaluation of coronary lesions based on CT angiography with both anatomical and functional advantages, significantly reducing the proportion of coronary angiograms and avoiding unnecessary invasive functional assessment of coronary arteries; it helps to decide the treatment strategy for complex lesions before invasive operations

[16]. Functional evaluation of coronary arteries also improves clinical prognosis of PCI patients, predicts the occurrence of MACE, and guides clinical decision-making. Quantitative flow fraction (QFR) based on coronary angiography, optical coherence tomography (OCT)-based flow reserve fraction (OFR) technology, and IVUS-FFR technology based on IVUS predict physiological functional indices by combining real-time morphology from angiography or intravital imaging with machine algorithms. At the same time, the above techniques do not require the use of vasodilating drugs such as adenosine, which simplifies the functional science detection of stenotic lesions, shortens the operation time, reduces the risk and cost of the procedure, and further expands the scope of the applicable population for physiological evaluation testing.

The combination of technology along with clinical application scenarios of endoluminal imaging and functional medicine allows for better prognosis prediction of coronary lesions and optimization of strategies. In the COMBINE FFR-OCT trial on lesions with FFR > 0.80 using OCT to identify vulnerable thin-cap fibroatheroma (TCFA) lesions, the results showed that the incidence of MACE was significantly higher in FFR-negative patients with TCFA than in the group without TCFA, with no significant difference from the FFR-positive hemodynamic reconstruction group. This study suggests that more than 25% of lesions with FFR > 0.80 in diabetic patients may actually be high-risk lesions [17]. The combination or even integration of intracavitary imaging techniques and physiological assessment, each with its own strengths in different lesions, is irreplaceable and will be of great importance in optimizing PCI and accomplishing precise treatment.

The stent platform of DES has been further upgraded in recent years, with a further reduction in the thickness of the stent beam and a more rational stent design, allowing for better throughput and support, as well as a greater drug/strut ratio. According to the latest results from the BIFLOW V trial, at 3 years of follow-up, ultrathin sirolimus-coated bioresorbable polymer stents continue to outperform everolimus-coated durable polymer stents in preventing target lesion failure [18]. A pooled analysis showed that for patients with small vessel lesions, the ultrathin stent beam OrsiroTM stent was more advantageous in reducing target lesion failure at 1 year. There has also been much exploration in the direction of developing durable polymer, biodegradable polymer, and polymer-free coated drug-eluting stents [19]. The ISAR-TEST-5 trial found no significant difference in outcomes with polymer-free versus durable polymer DES, which had a higher rate of distant follow-up event [20]. The SORTOUT IX trial compared BioFreedom stent is noninferior to a modern ultrathin strut biodegradable polymer cobalt-chromium sirolimus-eluting Orsiro stent for major adverse cardiovascular events at 12 months [21]. The HOST Reduce-Polytech-ACS trial enrolled ACS patients at high thrombotic risk and showed that the durable polymer DES (DP-DES) was not inferior to the biodegradable polymer DES (BP-DES) in terms of the primary endpoint (defined as all-cause death, non-fatal infarction). However, the incidence of device-related clinical endpoints (defined as cardiovascular death, target vessel-related infarction, and target vessel revascularization) was significantly lower in the DP-DES group at 1 year [22]. In terms of biodegradable stents, the MAGSTEMI trial showed that magnesium-based bioresorbable scaffolds (MgBRS) had a higher rate of target lesion revascularization and worse angiographic outcomes compared with sirolimus drug-eluting stents, despite the better performance of MgBRS in terms of drug-induced vasodilatory response. Half of the MgBRS restenoses were due to stent collapse [23].

There have also been some advances in the field of interventional devices without implantation and in the field of calcified lesion treatment. In coronary patients with new lesions in stable small vessels, the drug-coated balloon was no worse than everolimus drug-eluting stents in reducing in-lesion lumen loss. The DISRUPT-CAD III trial showed good near-term safety and efficacy of endovascular lithotripsy (IVL) applied to PCI in patients with severe calcified lesions. 99.5% of lesions had residual stenosis $< 30\%$. This technique is expected to be an effective solution for severely calcified lesions [24].

In parallel with the advances in devices, there have been changes in perioperative and long-term drug therapy, particularly in antiplatelet therapy regimens. In the era of "first-generation" drug-coated stents, there was a trend toward longer duration of dual antiplatelet therapy (DAPT) and more potent DAPT regimens based on new adenosine diphosphate (ADP) receptor antagonists to reduce the occurrence of late in-stent thrombosis. Then, with the widespread use of next-generation drug-coated stents and new ADP receptor antagonists, short-course DAPT followed by single-agent antiplatelet regimens became a new trend, and antiplatelet treatment strategies moved from aggressive to balanced. The results of several studies on antiplatelet treatment strategies after PCI (GLOBAL LEADERS trial, TICO trial) and another study on antiplatelet regimens (TWILIGHT trial) have shown that after a short course of DAPT (1–3 months), tegretol monotherapy substantially reduces the risk of bleeding compared with standard DAPT, without compromising ischemic benefit, providing a new treatment strategy for clinical high-risk PCI patients, especially those at high bleeding risk [25–27]. It is suggested that tegretol monotherapy antiplatelet regimen may be an optimal treatment strategy to balance ischemic and bleeding risks and may be applicable to ACS patients receiving next-generation DES therapy.

The development of antiplatelet therapy strategies based on genetic guidance has become a hot topic in recent years. The TAILOR PCI trial explored the prognostic impact of genotype-guided oral P2Y12 receptor antagonists on patients after PCI. Results showed a 34% reduction in the rate of the primary composite endpoint event in the genote-guided treatment group at 1 year after PCI, but did not reach a statistical significance [28]. However, given the higher rate of CYP2C19 mutations in the East Asian population, despite the failure of the TAILOR PCI trial to achieve statistically significant, genotype-guided individualized antiplatelet strategies may still be a good option for future research into precision therapy. Genotype-guided individualized antiplatelet strategies may be the way forward for further research into precision therapy.

In slowing down atherosclerosis, as statins have limited cholesterol reduction, the discovery of proprotein convertase bacillus subtilis protease/kexin type 9 (PCSK9) inhibitors at the beginning of this century has provided a new target for cholesterol-lowering therapy and has been hailed as an important advancement in lipid-lowering after statins. Two clinical trials on PCSK9 inhibitors (FOURIER trial, ODYSSEY OUTCOMES Trial) have confirmed the role of PCSK9 inhibitors in secondary prevention of ASCVD [29, 30]. Inclisiran is an siRNA that binds to the mRNA encoding the PCSK9 protein to reduce mRNA levels and prevent the liver from producing the PCSK9 protein through RNA interference, thereby enhancing the liver's ability to remove LDL-C from the blood and achieving the goal of lowering LDL-C levels. The ORION series trials showed that in adult patients with heterozygous familial hypercholesterolemia (HeFH), ASCVD, and/or ASCVD at high risk, inclisiran provided durable and effective reduction in LDL-C levels with a significant reduction in the incidence of MACE, greatly increasing patient compliance with lipid-lowering therapy [31]. In recent years, with the emergence of innovative cholesterol-lowering targets, lipid-modifying drugs have flourished and blossomed, and more and more innovative drugs can more effectively lower cholesterol and improve the prognosis of ASCVD patients.

1.4 Pacing and Electrophysiology

With the continuous development of biomedical engineering and electronic technology, the rapid development and wide application of cardiac pacing and electrophysiology technologies, which are closely related to these technologies, have had a profound impact on the treatment of cardiovascular diseases.

Since 1958, when the world's first fully embedded artificial pacemaker was implanted at Karolinska Hospital in Sweden, cardiac pacing technology has developed rapidly. Pacemakers are becoming smaller, more functional, with longer battery life, more physiological functions, and more intelligent and automated. From single-chamber pacemakers to dual-chamber pacemakers to triple-chamber pacemakers, from non-physiological to semi-physiological to physiological pacing, from treatment of bradycardia to synchronization of tachycardia to heart failure, from magnetic fields to magnetic resonance compatibility, from single-site pacing to multi-site pacing, the technology of pacing is changing rapidly.

As life expectancy increases worldwide, the number of pacemakers implanted continues to rise, with approximately one million implanted annually worldwide. Initially, pacemakers were mainly used for the treatment of slow arrhythmias, but with the rapid development of technology, the indications have been expanding; CRT has become an important non-pharmacological treatment for heart failure; buried automatic cardioverter-defibrillator (ICD) can significantly reduce the mortality rate of patients with malignant ventricular arrhythmias compared with antiarrhythmic drugs; in recent years, "physiological pacing" has been explored and improved. The exploration and improvement of "physiological pacing," the emergence of new means of structural heart disease treatment and the in-depth understanding of systemic diseases have led to a new era of cardiac pacing. The indications for pacing therapy are constantly being broadened, expanding from the treatment of simple disorders of cardiac electrical activity (such as pathological sinus node syndrome and other slow arrhythmias) to the treatment of disorders of cardiac electrical activity (such as atrial fibrillation), from the treatment of cardiac electrical disorders to the treatment of non-cardiac electrical disorders (such as hypertrophic obstructive cardiomyopathy, heart failure, etc.).

The pacing electrode leads have also progressed significantly, with increasingly thin leads, increasing resistance to compression and abrasion, increasing pacing polarity (from single-stage to bipolar to quadruple-stage), and increasing pacing vector combinations (dozens of vector combinations). However, there are still some unavoidable complications with electrode leads: lead wear and fracture, pacemaker capsular bag rupture and infection, thrombosis and blockage of the punctured vessels, and local pain from scarring. In some cases, the electrode leads cause massive tricuspid regurgitation, resulting in a dramatic deterioration of the patient's right heart function and even death. In addition, if the patient has an anatomical defect in the venous system or a complete blockage of the vessel at the time of pacemaker replacement, it is not possible to implant the pacemaker transvenously. These patients are either treated with more invasive open-heart lead placement or lost to pacemaker therapy. The problem of pacing leads caused great problems for clinicians and patients and inspired scientists to explore and innovate, leading to a new leap in the field of pacemakers in the twenty-first century—the birth of leadless pacemakers and subcutaneous ICDs. The great progress in the field of pacemakers has brought great life to the patients and saved millions of lives worldwide.

1.5 Heart Valve Disease

Heart valve disease is one of the most vibrant fields, and recent years have seen significant changes and advances in treatment concepts, indications, and approaches. Interventional therapies are becoming more and more available, and the accompanying pharmacologic options are being explored. Transcatheter aortic valve replacement remains the most rapidly developing area of valve disease treatment, and interventional approaches to other valves are gradually moving from clinical research to clinical dissemination. Below we present the following major research advances in heart valve disease.

1.5.1 The New Era of TAVR Treatment for All-Risk Populations

PARTNER 3 trial enrolled a total of 1000 patients with severe aortic stenosis at low surgical risk, and its 2-year follow-up data showed that the TAVR group had a lower rate of primary endpoint events than the SAVR group (11.5% vs. 17.4%, $P < 0.001$), with the advantage in the primary endpoint coming primarily from a lower readmission rate. From the 2-year follow-up results, TAVR remains superior to SAVR in terms of the primary endpoint, but the impact of valve thrombosis and new LBBB on the long-term prognosis of patients, as well as the long-term durability of TAVR valves, remains to be studied at further follow-up [32]. Long-term outcomes are particularly important because younger, low-risk patients have longer life expectancies, and we need to await the results of the PARTNER 3 study over a 10-year follow-up period.

The accumulation of clinical evidence has also driven guideline updates, and in December 2020, the ACC and the American Heart Association (AHA) jointly released the 2020 edition of the 2020 ACC/AHA Guideline for the Management of Patients With Valvular Heart Disease [33], in which treatment options for patients with aortic stenosis are undoubtedly one of the most talked-about updates in the new guidelines. The new guidelines abandon the previous practice of selecting interventions primarily based on the surgical risk of patients with aortic stenosis, with age becoming the primary reference factor for treatment decisions. According to the 2020 edition of the guidelines, the question is whether to choose TAVR or conservative treatment for patients who have a contraindication to surgery or are at high risk and have a life expectancy of >1 year. Conversely, for patients at moderate or low risk for surgery, the choice of SAVR or TAVR depends primarily on age. Patients under 65 years of age are preferred to SAVR; patients over 80 years of age are preferred to TAVR; and patients 65–80 years of age are decided by the heart team, taking into account life expectancy, anatomic conditions, comorbid cardiac and non-cardiac disease, frailty, and surgical risk, combined with the patient's wishes Choice of SAVR or TAVR (or TAVR if life expectancy is short, aortic porcelainization, comorbid severe lung disease, liver and kidney disease, anatomical conditions suitable for TAVR or low risk of TAVR). At this point, both European and US guidelines for the management of heart valve disease have effectively included low-risk patients in the TAVR indication population, fully opening a new era of TAVR treatment for all-risk populations.

1.5.2 Maturing of Transcatheter Mitral Valve Repair Devices

Mitral regurgitation (MR) is the most common heart valve disease with an extremely large patient population. However, because patients with mitral regurgitation are often combined with cardiac insufficiency or other comorbidities, only a minority of patients have undergone mitral valve surgery in clinical practice, suggesting that there is still a great need for treatment of patients with MR. Transcatheter mitral valve intervention offers a new option, but statistically, only about 10% of MR patients are judged to be suitable for transcatheter intervention after screening under current technical conditions, indicating that this technology still has great potential for development and upside.

Transcatheter mitral valve repair is mainly including edge-to-edge repair (e.g., MitraClip, PASCAL), direct annuloplasty (e.g., Cardioband), and artificial tendon implantation (e.g., NeoChord system). Of these, edge-to-edge repair is the most established and the only transcatheter mitral valve repair technique recommended by current guidelines for the treatment of primary and secondary mitral regurgitation. The COAPT trial showed that combining MitraClip with pharmacologic therapy significantly reduced the risk of heart failure readmission and all-cause mortality in patients with heart failure combined with severe secondary mitral regurgitation [34]. The COAPT trial was undoubtedly a landmark study in the field of structural cardiology and heart failure treatment, inspiring innovation in transcatheter mitral valve repair devices. Transcatheter mitral valve repair has matured. The MITRA-FR trial included the same patients as the COAPT study (severe secondary MR) and used the same device (MitraClip), but yielded very different results—transcatheter mitral edge-to-edge repair (MitraClip) did not reduce the risk of all-cause mortality and unplanned heart failure readmission. Although there were some differences between the two studies in terms of operator experience, degree of drug-optimized therapy, and sample size, none were sufficient to explain the differences in outcomes [35]. The prevailing view is that differences in left ventricular size and mitral regurgitation flow among the included patients may be the key to explaining the very different outcomes. How to interpret the difference in outcomes between the COAPT and MITRA-FR studies is a hotly debated topic. Whether the degree of MR is proportional to the change in LV end-diastolic volume, meaning the proportion of patients enrolled with secondary MR, may be the key point. A meta-analysis using patient-level data from the COAPT and MITRA-FR studies and ongoing studies such as RESHAPE-HF 2 is expected to help us clarify the optimal indication population for mitral valve intervention.

1.5.3 Transcatheter Tricuspid Valve Interventions Are on the Rise

Tricuspid regurgitation (TR) is another common heart valve disease, especially in the elderly population.TR is mostly secondary, often in patients with comorbid left heart system disease or atrial fibrillation, and its presence indicates a poor prognosis. However, treatment options for TR are limited, and studies have shown that surgical repair or replacement does not provide significant benefit and that drug therapy is less effective. With the development of transcatheter interventions for the aortic and mitral valves, transcatheter tricuspid valve interventions have also begun to emerge and have progressed rapidly in recent years. Tricuspid valve interventional techniques that have been used in clinical practice or are under clinical investigation include edge-to-edge repair (e.g., TriClip, PASCAL), valvuloplasty (e.g., Cardioband, Trialign), ectopic vena cava valve placement (e.g., TricValve), and transcatheter tricuspid valve replacement (e.g., NaviGate). The most commonly used technique is edge-to-edge repair. Of these, TriClip and PASCAL have received the CE mark for transcatheter tricuspid valve repair treatment. There are no completed randomized controlled trials in the field of transcatheter tricuspid valve interventions.

1.6 Heart Failure

Heart failure is a complex clinical syndrome with high prevalence, mortality, and rehospitalization rates. In recent years, novel drugs have been developed, new clinical studies have been published, and there has been rapid development in device therapy, resulting in a significant change in the treatment strategy for heart failure. The following is a review of the current status of heart failure treatment.

1.6.1 Pharmacological Treatment of Heart Failure

The major advances in the pharmacological treatment of heart failure have come from angiotensin receptor enkephalinase inhibitors (ARNI) and sodium-glucose co-transport protein 2 (SGLT2) inhibitors. The results of the PARADIGM-HF trial showed that sakubatril/valsartan further reduced the risk of the primary endpoint, composite endpoint, and post hoc analysis of the coronary composite endpoint compared with enalapril [36]. In the past 5 years, sakubatril valsartan has become the basis of a comprehensive pharmacological regimen for chronic HFrEF (class I) [37]. SGLT2 inhibitors are a new class of hypoglycemic agents that lower blood glucose by inhibiting glucose reabsorption by SGLT2 in the proximal tubule of the kidney. The EMPA-REG OUTCOME trial was the first to demonstrate that engramine reduced the incidence of major cardiovascular events in patients with type 2 diabetes [38]. The DAPA-HF trial evaluating the prognostic impact of SGLT2 inhibitors on heart failure in patients with HFrEF with a 3-year follow-up showed that dapagliflozin significantly reduced the risk of worsening or death in patients with HFrEF, independent of the presence of diabetes [39]. In the 2021 ESC Guidelines for the diagnosis and treatment of acute and chronic heart failure, SGLT2 inhibitors (canagliflozin, dapagliflozin, empagliflozin, ertugliflozin, sotagliflozin) are recommended in patients with diabetes at high risk of CV disease or with CV disease in order to prevent HF hospitalizations [37].

In addition, vericiguat, an oral soluble guanylate cyclase activator, may further increase intracardial cyclic guanosine (cGMP) concentrations by increasing soluble guanylate cyclase (sGC) concentrations, which may act as a vasodilator and inhibit myocardial remodeling. Vericiguat may be considered in patients in NYHA class IIIV who have worsening HF despite treatment with an ACE-I (or ARNI), a beta-blocker, and an MRA to reduce the risk of cardiovascular mortality or HF hospitalization, which is described in the 2021 ESC heart failure guidelines as a Class IIb recommendation [37, 40].

1.6.2 Device Therapy for Heart Failure

Current guidelines recommend device therapy for heart failure, including implantable cardioverter-defibrillators (ICDs) to prevent sudden cardiac death (SCD) and cardiac resynchronization therapy (CRT) to improve cardiac remodeling and clinical prognosis in some patients. Mechanical circulatory assist (MCS) therapy is still mainly used in transitional scenarios and needs further discussion as an indication for long-term use.

SCD is one of the leading causes of death in patients with heart failure. ICD is effective in reducing morbidity and mortality in people at high risk of SCD and is currently the most effective preventive measure for SCD in patients with heart failure and is recommended for patients with symptomatic HF (NYHA class II, III) of non-ischemic etiology and patients with LVEF $\leqq 35\%$ after $\geqq 3$ months of OMT [37, 41]. The fully subcutaneous implantable cardiac defibrillator (S-ICD) and wearable automated external cardiac defibrillation (WCD) devices are the major advances in the field of cardiac defibrillation devices in recent years, but further clinical data are needed to support them. More than 1/3 of patients with CHF develop cardiac asystole, which can seriously affect the prognosis of patients. CRT not only improves clinical symptoms and reduces hospitalization rates but also improves the long-term prognosis of patients and reduces the morbidity and mortality rate [37, 42]. Conventional CRT mainly consists of biventricular synchronous pacing (BiVP). In recent years, the new terms of pacing have emerged as new alternative to cardiac resynchronization therapy, including left ventricular multipoint pacing (MPP), left ventricular endocardial pacing (LVEP), his bundle pacing (HBP), and left bundle branch pacing (LBBP). All of these methods require long-term clinical data.

1.7 Conclusion

With the development of various basic disciplines such as biology, computer science, bioengineering, and pharmacology, the treatment strategies for cardiovascular diseases have undergone radical changes in the past 20 years, with many previous ideas being realized and misconceptions being corrected. The development of techniques for the treatment of cardiovascular disease has entered the fast lane. We are excited to see the challenges of cardiovascular disease overcome one by one and feel privileged to be a witness and a participant in this revolution!

References

1. Sevim Bayrak C, Zhang P, Tristani-Firouzi M, et al. De novo variants in exomes of congenital heart disease patients identify risk genes and pathways. Genome Med. 2020;12(1):9.
2. Lahm H, Jia M, Dreßen M, et al. Congenital heart disease risk loci identified by genome-wide association study in European patients. J Clin Invest. 2020;141837.
3. Izarzugaza JMG, Ellesøe SG, Doganli C. Systems genetics analysis identifies calcium-signaling defects as novel cause of congenital heart disease. Genome Med. 2020;12(1):76.
4. Richter F, Morton SU, Kim SW. Genomic analyses implicate noncoding de novo variants in congenital heart disease. Nat Genet. 2020;52(8):769–77.
5. Hsieh A, Morton SU, Willcox JAL. EM-mosaic detects mosaic point mutations that contribute to congenital heart disease. Genome Med. 2020;12(1):42.
6. Morton SU, Shimamura A, Newburger PE. Association of damaging variants in genes with increased cancer risk among patients with congenital heart disease. JAMA Cardiol. 2020.
7. Miao Y, Tian L, Martin M. Intrinsic endocardial defects contribute to hypoplastic left heart syndrome. Cell Stem Cell. 2020;27(4):574–89.
8. Sachdeva R, Valente AM, Armstrong AK. ACC/AHA/ASE/HRS/ISACHD/SCAI/SCCT/SCMR/SOPE 2020 appropriate use criteria for multimodality imaging during the follow-up care of patients with congenital heart disease. J Am Coll Cardiol. 2020;75(6):657–703.
9. Lopez KN, Morris SA, Sexson Tejtel SK. US Mortality attributable to congenital heart disease across the lifespan from 1999 through 2017 exposes persistent racial/ethnic disparities. Circulation. 2020;142(12):1132–47.
10. Baumgartner H, De Backer J, Babu-Narayan SV. 2020.
11. Waldmann V, Bouzeman A, Duthoit G. Long-term follow-up of patients with tetralogy of fallot and implantable cardioverter defibrillator: the DAI-T4F nationwide registry. Circulation. 2020;142(17):1612–22.
12. Koutroulou I, Tsivgoulis G, Tsalikakis D, Karacostas D, Grigoriadis N, Karapanayiotides T. Epidemiology of patent foramen ovale in general population and in stroke patients: a narrative review. Front Neurol. 2020;11:281.
13. Liu K, Song B, Palacios IF. Patent foramen ovale attributable cryptogenic embolism with thrombophilia has higher risk for recurrence and responds to closure. JACC Cardiovasc Interv. 2020;13(23):2745–52.
14. Erlinge D, Maehara A, Ben-Yehuda O. Identification of vulnerable plaques and patients by intracoronary near-infrared spectroscopy and ultrasound (PROSPECT II): a prospective natural history study. Lancet. 2021;397(10278):985–95.
15. Gao X-F, Ge Z, Kong X-Q. 3-year outcomes of the ULTIMATE trial comparing intravascular ultrasound versus angiography-guided drug-eluting stent implantation. JACC Cardiovasc Interv. 2021;14(3):247–57.
16. Zhou T, Wang X, Ting W. Clinical application of computed tomography angiography and fractional flow reserve computed tomography in patients with coronary artery disease: a meta-analysis based on pre- and post-test probability. Eur J Radiol. 2021;139
17. Kedhi E, Berta B, Roleder T. Thin-cap fibroatheroma predicts clinical events in diabetic patients with normal fractional flow reserve: the COMBINE OCT-FFR trial. Eur Heart J. 2021;
18. Kandzari DE, Koolen JJ, Doros G. Ultrathin Bioresorbable-polymer Sirolimus-eluting stents versus thin durable-polymer Everolimus-eluting stents for coronary revascularization: 3-year outcomes from the randomized BIOFLOW V trial. JACC Cardiovasc Interv. 2020;13(11): 1343–53.
19. Lefèvre T, Haude M, Neumann F-J. Comparison of a novel biodegradable polymer sirolimus-eluting stent with a durable polymer everolimus-eluting stent: 5-year outcomes of the randomized BIOFLOW-II trial. JACC Cardiovasc Interv. 2018;11(10):995–1002.
20. Kufner S, Ernst M, Cassese S. 10-year outcomes from a randomized trial of polymer-free versus durable polymer drug-eluting coronary stents. J Am Coll Cardiol. 2020;76(2):146–58.

21. Jensen LO, Maeng M, Raungaard B. Randomized comparison of the polymer-free biolimus-coated BioFreedom stent with the ultrathin strut biodegradable polymer sirolimus-eluting Orsiro stent in an all-comers population treated with percutaneous coronary intervention: the SORT OUT IX trial. Circulation. 2020;141(25):2052–63.
22. Kim H-S, Kang J, Hwang D. Durable polymer versus biodegradable polymer drug-eluting stents after percutaneous coronary intervention in patients with acute coronary syndrome: the HOST-REDUCE-POLYTECH-ACS trial. Circulation. 2021;143(11):1081–91.
23. Gomez-Lara J, Ortega-Paz L, Brugaletta S. Bioresorbable scaffolds versus permanent sirolimus-eluting stents in patients with ST-segment elevation myocardial infarction: vascular healing outcomes from the MAGSTEMI trial. EuroIntervention. 2020;16(11):e913–21.
24. Liang B, Ning G. Evaluation of the safety and efficacy of coronary intravascular lithotripsy for treatment of severely calcified coronary Stenoses: evidence from the serial disrupt CAD trials. Front Cardiovasc Med. 2021;19(8)
25. Takahashi K, Wang R, Kawashima H. Efficacy and safety of one-month DAPT followed by 23-month ticagrelor monotherapy in patients undergoing proximal LAD stenting: insights from the GLOBAL LEADERS trial. Int J Cardiol. 2020;1(320):27–34.
26. Kim B-K, Hong S-J, Cho Y-H. Effect of Ticagrelor monotherapy vs Ticagrelor with aspirin on major bleeding and cardiovascular events in patients with acute coronary syndrome: the TICO randomized clinical trial. JAMA. 2020;323(23):2407–16.
27. Mehran R, Baber U, Sharma SK. Ticagrelor with or without aspirin in high-risk patients after PCI. N Engl J Med. 2019;381(21):2032–42.
28. Pereira NL, Farkouh ME, So D. Effect of genotype-guided Oral P2Y12 inhibitor selection vs conventional Clopidogrel therapy on ischemic outcomes after percutaneous coronary intervention: the TAILOR-PCI randomized clinical trial. JAMA. 2020;324(8):761–71.
29. Giugliano RP, Pedersen TR, Park J-G. Clinical efficacy and safety of achieving very low LDL-cholesterol concentrations with the PCSK9 inhibitor evolocumab: a prespecified secondary analysis of the FOURIER trial. Lancet. 2017;390(10106):1962–71.
30. Szarek M, White HD, Schwartz GG. Alirocumab reduces total nonfatal cardiovascular and fatal events: the ODYSSEY OUTCOMES trial. J Am Coll Cardiol. 2019;73(4):387–96.
31. Kausik K. N Engl J Med. 2020;382(16):1507–19.
32. Mack MJ, Leon MB, Thourani VH. Transcatheter aortic-valve replacement with a balloon-expandable valve in low-risk patients. N Engl J Med. 2019;380(18):1695–705.
33. Otto CM, Nishimura RA, Bonow RO, et al. 2020 ACC/AHA guideline for the management of patients with valvular heart disease: executive summary: a report of the American College of Cardiology/American Heart Association joint committee on clinical practice guidelines. J Am Coll Cardiol. 2021;77(4):450–500.
34. Asch FM, Grayburn PA, Siegel RJ. Echocardiographic outcomes after Transcatheter leaflet approximation in patients with secondary mitral regurgitation: the COAPT trial. J Am Coll Cardiol. 2019;74(24):2969–79.
35. Messika-Zeitoun D, Iung B, Armoiry X. Impact of mitral regurgitation severity and left ventricular remodeling on outcome after MitraClip implantation: results from the Mitra-FR trial. JACC Cardiovasc Imaging. 2021;14(4):742–52.
36. McMurray JJV, Packer M, Desai AS. Angiotensin-neprilysin inhibition versus enalapril in heart failure. N Engl J Med. 2014;371(11):993–1004.
37. McDonagh TA, Metra M, Adamo M. 2021 ESC guidelines for the diagnosis and treatment of acute and chronic heart failure. Eur Heart J. 2021;42(36):3599–726.
38. Zinman B, Wanner C, Lachin JM. Empagliflozin, cardiovascular outcomes, and mortality in type 2 diabetes. N Engl J Med. 2015;373:2117–28.

39. McMurray JJV, Solomon SD, Inzucchi SE. Dapagliflozin in patients with heart failure and reduced ejection fraction. N Engl J Med. 2019;381:1995–2008.
40. Armstrong PW, Pieske B, Anstrom KJ. Vericiguat in patients with heart failure and reduced ejection fraction. N Engl J Med. 2020;382:1883–93.
41. Connolly SJ, Hallstrom AP, Cappato R. Meta-analysis of the implantable cardioverter defibrillator secondary prevention trials. Eur Heart J. 2000;21:2071–8.
42. Moss AJ, Hall WJ, Cannom DS, et al. Cardiac-resynchronization therapy for the prevention of heart-failure events. N Engl J Med. 2009;361:1329–38.

Chapter 2
Anatomy of Heart and Blood Vessel

Yong Zeng and Yaodong Din

2.1 Introduction

Currently, cardiovascular disease remains a serious global problem. With the development of interventional techniques, it gets more important to strengthen the impression of cardiac anatomy, mainly to avoid or reduce complications during interventional procedures. The purpose of this chapter is to give an anatomical introduction of the normal structure of the heart and blood vessels.

2.2 Overview of the Heart

The heart is covered by the pericardium and is located in the middle mediastinum. It is behind the sternum and rib cartilage, in front of the vertebrae, and on the upper surface of the diaphragm. The pericardium has an outer and an inner layer, with a lubricating fluid [1]. The internal anatomy of the heart consists of four myocardial chambers: the upper part is the left and right atria, and the lower part is the left and right ventricles. The right atrium and ventricle collect blood from the body and pump it to the lungs, while the left atrium and ventricle collect blood from lungs and pump it to the whole body.

The majority of anterior (upper) surface of the heart is formed by the right ventricle. The right atrium and left ventricle, respectively, make up the right and left sides of the heart [1]. The posterior surface is also located in the center of the midline surface, which consists of the left ventricle and left atrium [2].

There is a fibrous framework made of dense connective tissue passing transversely through the base of the heart [3]. The importance of the skeleton is that it

Y. Zeng (✉) · Y. Din
Beijing Anzhen Hospital, Capital Medical University, Beijing, China

Y.-D. Tang et al. (eds.), *Application of Biomaterials in the Treatment of Cardiovascular Diseases*, https://doi.org/10.1007/978-981-19-7712-1_2

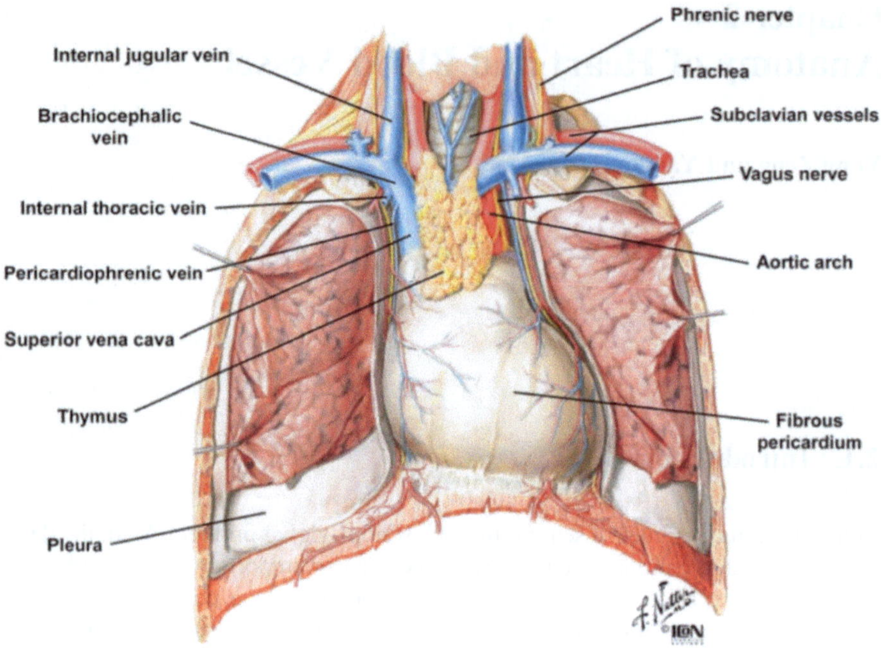

Fig. 2.1 The heart and structures in its neighborhood. Figure adapted from Handbook of Cardiac Anatomy, Physiology, and Devices [1], 2nd, and reused with permissions from Springer

provides a connection point for the myocardium of the atria and ventricles, anchors the four valves of the heart, and also electrically insulates the myocardium of the ventricles from the atria. The one-way flow of blood through the heart is dependent on the four valves. The mitral and tricuspid valves allow blood to flow only from the atria to the ventricles, while the pulmonary and aortic valves allow blood to flow only from the ventricles out of the heart and through the great arteries.

The myocardium is supplied by the coronary artery, which arises from the base of the aorta (coronary sinus) as the right and left coronary arteries [1] (Fig. 2.1).

2.3 Chambers of the Heart

2.3.1 Right Atrium

Located in right front of the left atrium, the right atrium extends downward from the left atrium. The outer wall of the right atrium constitutes by the right upper part of the thoracic costal surface, the right side of the heart bulges, and a small part of the right side of bottom of the heart [1, 4]. The superior vena cava opens at the top of the right atrium, and the inferior vena cava opens in the lower posterior part of the right atrium. The right auricle articulates with the right atrium, and the neck of its junction is slightly narrow and protrudes forward, obscuring the right side of the ascending aorta [1]. The coronary sulcus runs around the base of the right atrium separating it from the right ventricle. In the left posterior aspect of the right atrium are the right superior and inferior pulmonary veins and medially the ascending aorta, the beginning of the pulmonary artery and its bifurcation. On the lateral side of the right atrium, just below the lateral epicardium of the superior vena cava orifice, there is the sinus node, the pacing point of the cardiac conduction system [3].

The internal structure of the right atrium can be divided into three parts: the venous sinus part with a smooth inner wall, and the anterior part, the vestibular part in the area around the tricuspid orifice, and the right auricular part. The vestibular part has a smooth inner wall and a myocardial crest at the connection with the right auricle. There are myocardial trabeculae in the right auricle. The internal wall of the venous sinus section has no myocardial crest and is smooth, with openings for the superior and inferior vena cava and coronary sinus [1]. It is part of the venous sinus during the embryonic phase of heart development and later fuses and articulates with the atrial headquarters. The superior and inferior vena cava openings are located superiorly and inferiorly in the sinus portion of the right atrium, surrounded by thickened muscle bundles and without valves. The superior vena cava collects venous blood from the head and neck, upper branches, and thoracic reflux into the right atrium. The inferior vena cava collects venous blood from the lower part of the body, i.e., the abdomen, pelvis, and inferior branches, into the lowermost part of the right atrium. Anterior to the inferior vena cava orifice is a drape-like valve containing a small number of muscle fibers formed by the endocardial folds, the inferior vena cava valve, also known as the Eustachian valve.

There is a shallow oval fossa on the right side of the atrial septum, called the fossa ovalis, located to the upper left of the inferior vena cava orifice, and the base of the fossa is the primary septum of the primitive atrium [5]. The edge of the fossa is elevated and is formed by the secondary septum of the primitive atrium, also known as the Vieussens ring. The foramen ovale of the atrial septum is largely closed at the age of 3 years in the normal population, and if the foramen ovale is not closed, it becomes a common type of congenital secondary foramen type atrial septal defect (Fig. 2.2).

2.3.2 Right Ventricle

The anterior region of the rounded right ventricle forms the majority of the thoracic rib surface of the heart, which is separated from the pericardium and abuts the chest wall [3]. The lower part of the right ventricle is flat. The septal pericardium is adjacent to the central tendon of the diaphragm and a small portion of the diaphragmatic surface. The right ventricular myocardial funnel, the root of the pulmonary trunk, is connected to the root of the left posterior aorta by a band of collagenous tissue, which is called the funnel tendon or CONUS ligament. The right ventricle is a

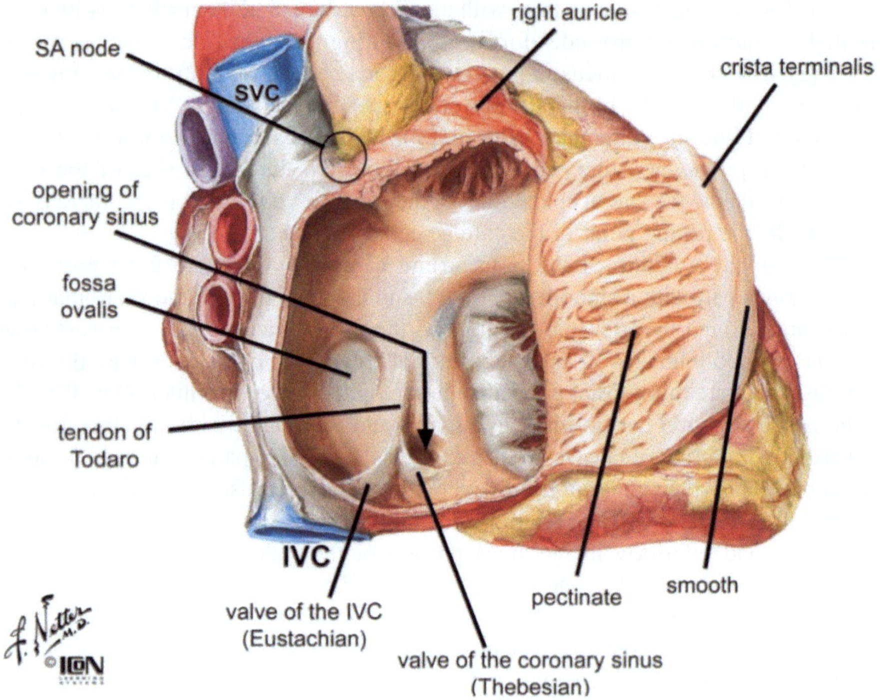

Fig. 2.2 The internal anatomy the right atrium has three regions: (1) the posterior part (venous sinus) with smooth walls; (2) the anterior part with walls consisting of horizontal, parallel muscular ridges known as pectus excavatum; and (3) the atrial septum. IVC = inferior vena cava; SA¼, sinoatrial; SVC¼, superior vena cava. Figure adapted from Handbook of Cardiac Anatomy, Physiology, and Devices [1], 2nd ed., and reused with permissions from Springer

"volume pump," and its anterior wall is significantly thinner than that of the left ventricle, with an average thickness of about 3–5 mm, which is only 1/3 of the wall thickness of the left ventricle [6].

The right ventricle extends from the right atrioventricular orifice to the left as a funnel, reaching the pulmonary artery orifice, with three pulmonary valves attached to the root of the pulmonary artery [3]. The right ventricle is divided into three regions: the inflow tract region, the apical region, and the outflow tract region. Some authors divide the right ventricle into three parts: sinus, meatus, and funnel (CONUS). The funicular septum is the muscular tissue of the funicular part of the posterior wall of the right ventricular funnel separating the pulmonary trunk from the beginning of the ascending aorta [3]. It is located in the subvalvular triangle of the combined left and right annular regions of the pulmonary valve and the muscular septum between the upper and lower branches of the septal meatus [3]. The thickness of the septum fungoides is normally 2–3 mm.

The swept marginal trabecula is located on the right ventricular surface of the septum in a "Y"-shaped structure and is the main component of the septal trabecula,

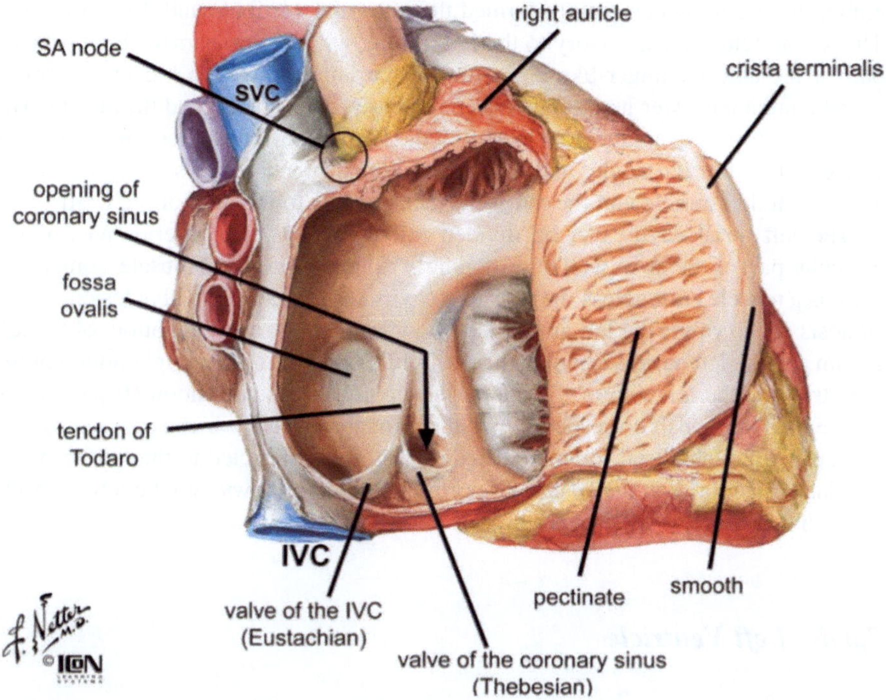

Fig. 2.3 The internal anatomy of the right ventricle. Thick trabeculae are characteristic of the right ventricular wall. The arterial cone forms the majority of the outflow tract. The right atrium or tricuspid valve is composed of papillary muscles, tendons, and three sets of cusps. Figure adapted from Handbook of Cardiac Anatomy, Physiology, and Devices [1], 2nd, and reused with permissions from Springer

divided into upper and lower branches and the body [4]. The fibers of its upper branch reach below the left semilunar ring of the pulmonary artery, and the fibers of the lower branch are located in the lowest part of the funnel septum toward the posterior inferior part of the septal membrane and are connected to the sinus wall muscle bundle in the right ventricular sinus. The septal sarcomeres are fixed to the septum at both ends, and the middle part is often free (Fig. 2.3).

2.3.3 *Left Atrium*

The body of the left atrium is roughly rectangular, it is located behind the left side of the right atrium, and the two atria are separated by the septum of the atrium in an inclined position [4, 5]. Its volume is smaller than that of the right atrium, but the wall is thicker than that of the right atrium, with an average thickness of 3 mm. The left atrium evolved from the proximal segment of the pulmonary vein during the embryonic stage, which mainly formed the left atrial vestibule and the left auricle. The left auricle is an accessory to the left atrium, with a long, narrow, hook-shaped part of the anterior finger-like projection of the left atrium with a deeply serrated margin and a narrower junction between the base of the auricle and the atrium. The left auricle forms an angle with the wall of the left atrium. The left coronary artery is concealed within the angle, and the left rotational branch travels along the lower edge of the left auricle and the atrioventricular sulcus to the base of the heart.

The left atrial cavity can be divided into two parts, the anterior part of the auricular part of the inner wall has irregularly running comb-like muscle, sometimes there is a muscle crest across the auricular cavity when the mitral valve lesion causes stenosis of the valve, the left atrial blood flow is stagnant, and the bottom of the left atrium and the left auricle can often have thrombus formation. The right region of the left atrium is the left side of the interatrial septum, and a half-moon-shaped crease can be seen in the part corresponding to the right atrial fossa, which is the remnant of the embryonic interatrial foramen. The left atrium is connected to the left atrioventricular orifice through the vestibular portion, and blood flows into the left ventricle through this orifice.

2.3.4 Left Ventricle

The left ventricle is located behind the upper left side of the right ventricle; it starts from the base of the heart in the plane of the coronary sulcus and extends to the apical part [1, 4]. The ventricular cavity is ovoid or nearly circular, with a wall three times as thick as that of the right ventricle, approximately 8–12 mm. The structural features are adapted to its strong contraction and its function of maintaining a high-pressure pulsation of the arterial system. On the surface of the heart, there is an atrioventricular sulcus (coronary sulcus) separating the base of the left ventricular cone from the left atrium [4]. The coronary sinus is located posterior to the left atrioventricular sulcus and goes straight to the right atrium through the area of the atrioventricular junction. The ventricular septum, which travels opposite to the anterior and posterior interventricular sulci, is the dividing line between the left and right ventricles [4].

The left ventricle has a mitral valve device attached to the inlet and an aortic valve device attached to the outlet. The upper border of the left ventricle is the mitral valve, the left lateral wall is the obtuse margin, the medial wall is the left ventricular surface of the septum, and the anterior and posterior walls are the thoracic rib and diaphragmatic surfaces of the left ventricular free wall, respectively [4]. The lower end is the left ventricular apex. The left ventricle is also divided into three parts: sinus, meatus, and funnel. The sinus is the projection area of the mitral valve and tendon structures with a smooth inner wall. Compared with the right ventricular wall, the fleshy column of the left ventricular wall is relatively small; the arrangement structure tends to be parallel, less cross braiding; the fleshy column near the apex is particularly developed; and the fleshy column is the boundary between the left ventricular

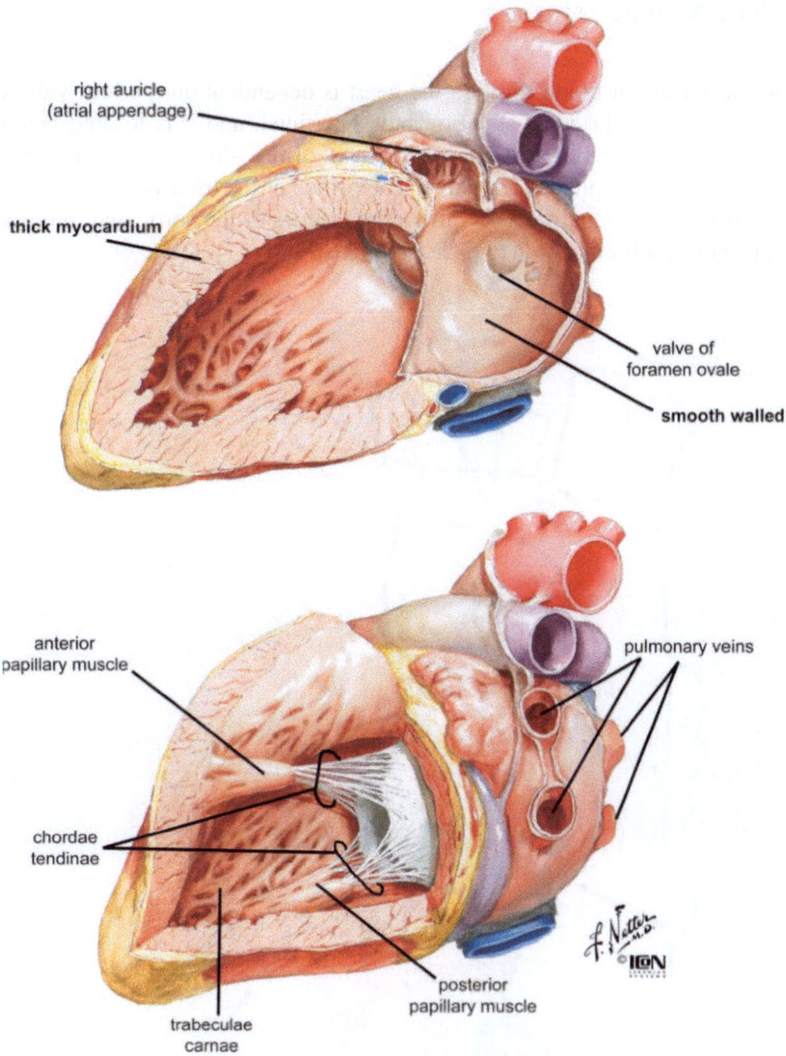

Fig. 2.4 Internal anatomy of the left atrium and ventricle. The majority of the left surface of the heart is formed by the left ventricle. The wall is lined with trabeculae, and the myocardium is much thicker than that of the left ventricle. The septum bulges toward the right ventricle, forming a barrel-shaped left ventricle. Figure adapted from Handbook of Cardiac Anatomy, Physiology, and Devices [1], 2nd, and reused with permissions from Springer

inflow tract and outflow tract. During embryonic development, most of the muscle bundles of the left ventricular funnel disappear, leaving only a nearly triangular-shaped muscular tissue. The inner wall is smooth, called the anterior smooth portion, and is located below the right and left semilunar valves of the aorta [7] (Fig. 2.4).

2.4 The Valves

The one-way flow of blood through the heart is dependent on the four valves. The pulmonary valve is the most superior and superficial, and it is located behind the second and third rib cartilages. The aortic valve orifice slopes downward, posterior and to the right of the pulmonary valve. The orifice of the tricuspid valve is separate from the orifice of the pulmonary valve. The orifices of the aortic and mitral valves are adjacent to each other [6] (Figs. 2.5 and 2.6).

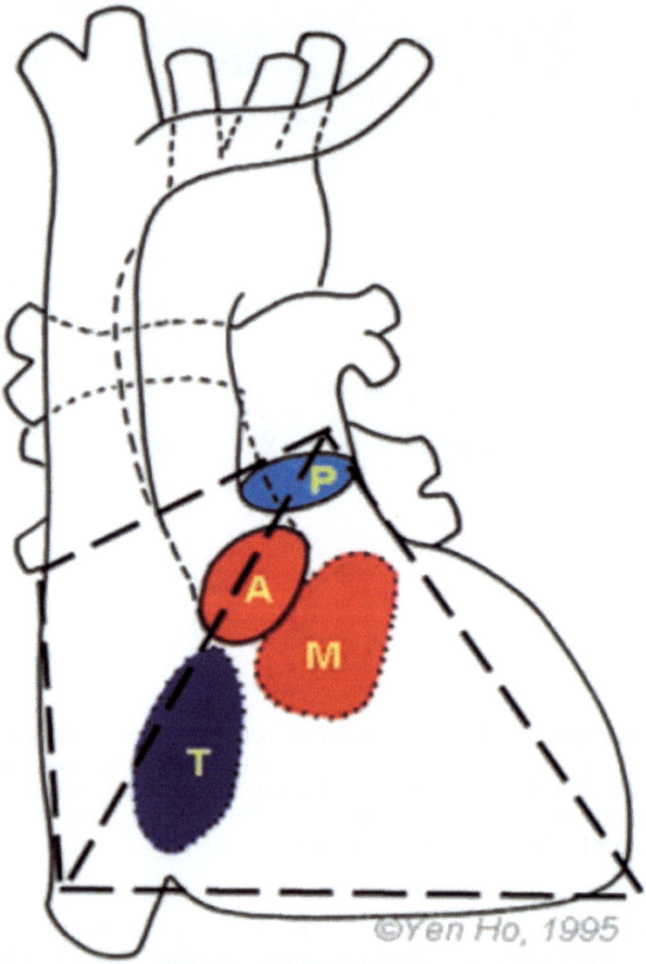

Fig. 2.5 The positions and planes of the four heart valves are visible in the anterior-posterior projection. A = aortic, M = mitral, P = pulmonary, and T = tricuspid valves. Figure adapted from Anatomy for Cardiac Electrophysiologists A Practical Handbook by S. Yen Ho, Sabine Ernst [8]

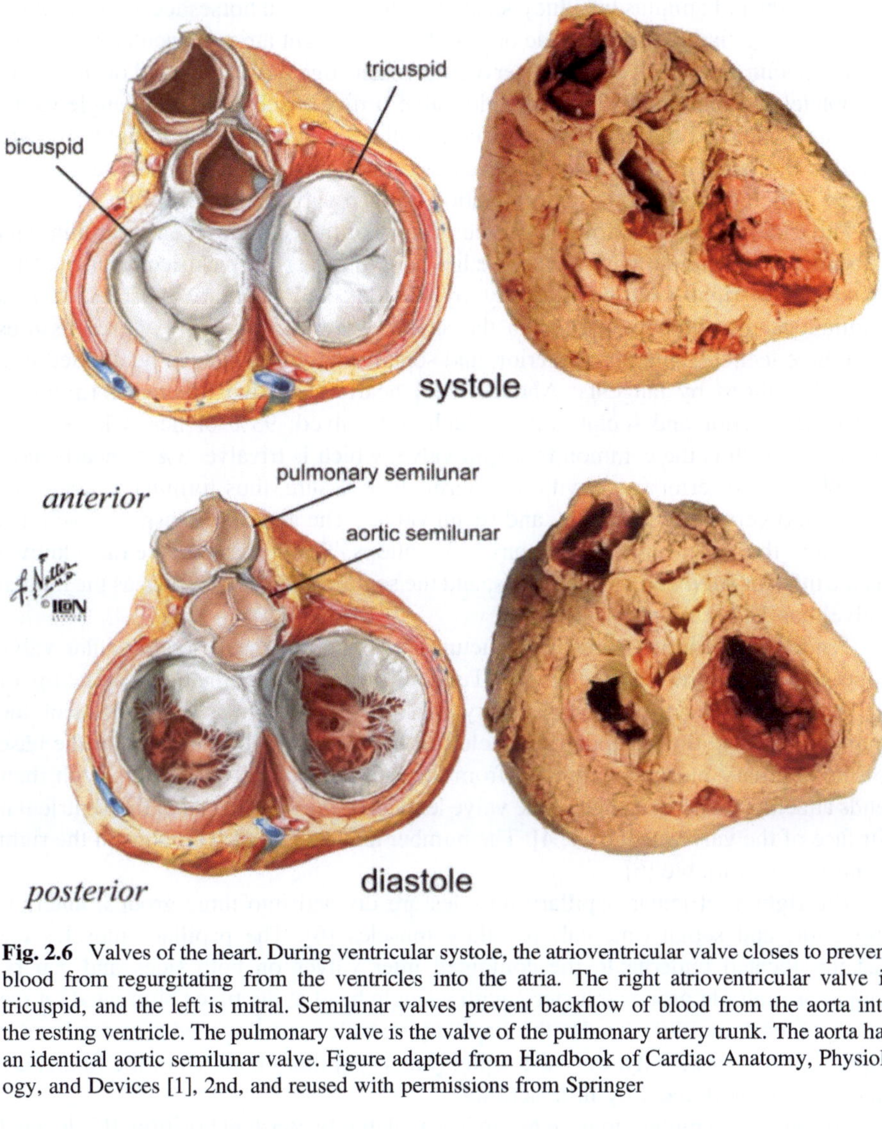

Fig. 2.6 Valves of the heart. During ventricular systole, the atrioventricular valve closes to prevent blood from regurgitating from the ventricles into the atria. The right atrioventricular valve is tricuspid, and the left is mitral. Semilunar valves prevent backflow of blood from the aorta into the resting ventricle. The pulmonary valve is the valve of the pulmonary artery trunk. The aorta has an identical aortic semilunar valve. Figure adapted from Handbook of Cardiac Anatomy, Physiology, and Devices [1], 2nd, and reused with permissions from Springer

2.4.1 Tricuspid Valve

The tricuspid valve has four parts, annulus, valve, tendon, and papillary muscle, and is called the right atrioventricular valvar complex. The tricuspid annulus is attached to the right atrioventricular foramen, and the right atrioventricular foramen is oval from the lateral view of the right atrium, with an average long diameter of 2.7 cm in men and 2.5 cm in women and a short diameter of 2.2 cm in men and 2.1 cm in women.

The tricuspid annulus is kidney-shaped on the front and horseshoe-shaped on the side and is the largest intracardiac orifice. From the right atrioventricular ring to the valve attachment line, it can be observed that each edge of the tricuspid orifice is not completely in one plane. The tricuspid valve orifice plane is at a 45° angle to the sagittal plane, and the ventricular surface of the valve orifice faces slightly downward to the left. The orifice has three edges called the anterior, posterior, and septal edges, which correspond to the orientation of each valve attachment line.

The tricuspid valve is formed by the folding of the right atrial endocardium and subendocardial tissue, with the free edge hanging over the ventricular cavity and fixed in the right ventricular cavity by tendons and papillary muscles. It is a continuous membranous curtain in the shape of a cuff, divided by three fissures into three leaflets, anterior, posterior, and septal, each of which can be divided into several scallops by tangents. About 4% of hearts have two fissures, forming an enlarged anterior and septal valve, which is bivalved; 93% of hearts have three fissures, which is the common tricuspid valve, which is trivalve; 3% of hearts have an enlarged posterior valve with an intermediate fissure, thus forming a quadruple valve; and very rarely, five, six, and seven valves. The anterior tricuspid valve (also known as the large, funnel valve) has 1–3 scallops. The posterior valve (also known as the inferior valve) has 1–4 scallops, and the septal valve (also known as the medial valve) has 1–3 scallops.

Tendons are fibrous cord-like structures that support the atrioventricular valve lobes attached to the ventricular wall. Tendons originate from the tip, near the tip, or at the edge of the tip of the papillary muscle, either from a single head of the papillary muscle tip (a papillary muscle can have multiple heads) or from the base of the papillary muscle or directly from the ventricular wall or septum, with their ends attached to the free edge of the valve leaflets or the rough zone of the ventricular surface of the valve leaflets [1, 4]. The number and size of such tendons in the right ventricle are variable [6].

The right ventricular papillary muscles are divided into three groups: anterior, posterior, and septal (internal) papillary muscles [6]. The papillary muscles are usually located underneath the conjoined inter valves on both sides and can be single or multiple groups (clusters) with a single head at the tip or bifurcated on the head to form multiple heads; the papillary muscles are finger-like; most have a certain degree of activity; only the septal (inner) papillary muscles of the funnel are excessively short and very little activity.

The anterior papillary muscle group has a relatively constant position. It is located near the sharp edge, in the anterior wall of the right ventricle, distal to the septal sarcomere nodal band (septal sarcomere regulating bundle). It is the largest one among the three groups of papillary muscles of the tricuspid valve [9]. The posterior papillary muscle is located below the right ventricular diaphragm close to the sharp border of the posterior septal septum, with two to four papillary muscles, some of which are bifurcated at the head. The papillary muscles have a forked head. The CONUS papillary muscle is a group of muscles also known as the septal papillary muscle, it has a short body, and some of them only have a cephalic end, or cephalic end is absent. The tendon cord starts directly from the upper edge of the lower branch of the septal meat column and the tip, it sends out tendon cords to the anterior septal conjunction and its adjacent scallops, and many segments of the septal papillary muscle start from the septum. The group of funnel papillary muscles (1–4) originating from the upper edge of the lower branch of the septal meat column and the apical part of the septal meat column and the group of parapapillary muscles (0–5) originating from its lower edge are collectively called Luschka or Lancisi papillary muscles.

The average area of the tricuspid orifice was 683(506~829) mm^2, and the average area of the total valve was 1635(1277~2773) mm^2, which was 1:2.4. The mean circumference of the tricuspid annulus was 11.4 cm in men and 10.8 cm in women, 3.5~6.1 cm in infants before 1 year of age, 6.2~6.9 cm in infants and children aged 1~4 years, 7.5~8.9 cm in children aged 8~12 years, and 9~10.3 cm in children aged 15~17 years.

2.4.2 *Mitral Valve*

There is a mitral valve complex on the left ventricular inflow tract that includes the mitral annulus at the left atrioventricular orifice, mitral leaflets, tendons, and papillary muscles together. The mitral annulus is densest on the medial aspect of the left and right fibrous triangle, from which emanates the coronary bundle, a progressively thinner fibrous subendocardial tendon that surrounds the mitral orifice at the atrioventricular orifice and separates the atrium from the ventricular muscle. In the absence of connective tissue support, the posterior mitral annulus enlarges causing a reduction in leaflet coaptation, resulting in mitral valve insufficiency [4].

The mitral valve is fixed to the inner wall of the left ventricle by tendon cords and papillary muscles in the shape of a cylinder sleeve and is divided by two fissures into an anterior (large) and a posterior (small) valve, which in turn is divided into several scallops by a notch [3, 4]. Since the depth of the fissure and notch do not reach the annulus, the valve tissue between the top of the fissure and the annulus is the connecting zone, namely, anterolateral commissure (ALC) and posteromedial commissure (PMC), and adhesions and incomplete closure of the valve leaflets occur mostly in the connecting zone.

The anterior valve, also known as the major valve (main valve), consists of a scallop with a valve starting from the ventricular surface of the left and right fibrous triangle and the posterior intervalvular triangle between the left and posterior semilunar annulus of the aorta and part of the left and posterior annulus, forming a fibrous link with the aorta, called the subaortic curtain. The anterior valve has only rough bands and hyaline bands and lacks basal bands. Both atrial surfaces of the valve leaflets are smooth, and the ratio of the rough band to the hyaline band is 0.6 (rough band: hyaline band = 0.6:1), with a closed edge on the atrial surface between the two bands. The anterior flap attachment margin accounts for 1/3 of the annulus, but the height of the anterior flap is about double that of the posterior flap.

The posterior valve is shaped like an apron, also known as a small valve, and its free edge is usually divided into three small scallops with the middle scallop being larger and the anterior outer and posterior inner scallops being smaller. Each leaflet has a crescent-shaped roughened band with tendon attachment on the ventricular surface, and from the roughened band to the attachment of the valve on the fibrous ring is the hyaline band of the leaflet with less tendon attachment. The ratio of the rough zone to the hyaline zone in the median area of the posterior valve is 1.4 (rough zone: hyaline zone = 1.4:1), suggesting that when the mitral valve is closed, the posterior leaflet has a larger counter surface than the anterior leaflet.

The anterior-posterior commissure is a segment of valve tissue from the tip of the cleft to the annulus, mostly arch-shaped, membranous, and rarely a valve fan leaflet. The head of the tip of the anterior papillary muscle group in the cardiac cavity points toward the anterior external commissure, which is located on the surface of the heart rather than at the origin of the spinous branch of the left coronary artery [1].

The mitral tendon cords are structurally and functionally similar to the tricuspid valve, but there are two types of tendon cords: true and false. False tendon cords are often found in the left ventricle and have no certain location or morphological pattern. True tendon cords include two major categories: detrusor tendon cords and basal tendon cords. The damage to the papillary muscle that maintains the left ventricle results in decreased ventricular elasticity and systolic function. Also, there are fan-shaped tendons and free edge tendons (class I), which have been shown to cause leaflet prolapse in areas without tendon attachment at the free edge of the valve. The basal tendon (class III) is only found in the posterior valve; the anterior valve has no basal tendon, which is an isolated myocardial tendon structure that emanates from the posterior wall of the left ventricle to the base of the posterior wall and supports the posterior valve.

In the adult normal heart, the average mitral orifice area was 547(440~676) mm^2, and the total valve area was 1394(1107~1578) mm^2, with an average ratio of 1:2.6. Mitral annulus area compared to mitral orifice area (AMO/AMV) = 1. 5/1~2. 2/1.

2.4.3 Pulmonary Valve

The pulmonary valve is located at the root of the pulmonary artery, at the apex of the right ventricular outflow tract, at a certain distance from the other three valve complexes of the heart, and is the most superficial of the heart valves. During embryonic life, the proximal bulbar septum divides the heart bulb into the pulmonary artery and the aortic channel. Due to rotation, the heart bulb and artery are transposed at the junction of the aorta and the pulmonary artery, with the pulmonary trunk sloping back to the left and the aorta sloping forward to the right, resulting in the aortic valve and pulmonary valve not being in one plane and forming an angle with each other.

The plane of the pulmonary valve orifice is oriented toward the upper left posterior, with three semilunar leaflets, called semilunar valves, and a mildly dilated pocket-shaped gap between its corresponding pulmonary artery walls called the pulmonary sinus. The three semilunar valves are attached in the pulmonary sinus, enclosing three crescent-shaped pockets, and the upper edge of the sinus is attached to the wall of the pulmonary artery at a point called the sinus crest, the average circumference of the pulmonary sinus crest level is 6.3 cm in men and 6.1 cm in women, and the height of the pulmonary sinus is the same as the height of the semilunar valves, averaging 1.5 cm in men and 1.3 cm in women [4]. The three semilunar valves are dense collagen-elastomeric membrane-like structures that form a triangular portion of the pulmonary artery wall between their lines of attachment, which becomes part of the right ventricular outflow tract, called the arterial as part of the ventricle, and is the intervalvular triangle of the three semilunar valves [4]. The fibrous annulus of the pulmonary orifice is three wave-shaped, with the pulmonary valve attached, and consists of three curved annuli, anterior, left, and right, which are connected to the fibrous annulus of the aortic orifice (aortic annulus) by the funnel tendon.

The anterior semilunar valve is located in the conical prominence of the pulmonary artery and is the most superficial and obvious; the left semilunar valve is located mostly facing the upper segment of the anterior interventricular sulcus, adjacent to the main stem of the left coronary artery and the beginning of the anterior descending branch; the right semilunar valve is located in the right posterior wall of the conical part of the pulmonary artery, near the right coronary sulcus and the beginning of the right coronary artery [3]. Because the pulmonary valve annulus is higher than the aortic valve annulus, both the left and right pulmonary sinuses are higher than the left and right aortic sinuses.

Each semilunar valve is formed by folds of endocardium with a fibrous tissue core (plate) sandwiched between the two endothelial and subendothelial layers. Each semilunar valve is divided into two parts, with the upper part being the contact zone and the lower part being the free zone. When the diastolic semilunar valve is closed, the three sinuses of the pulmonary artery are filled with blood and enclose three crescentic pockets. The contact zone is the lateral wall of the valve edge, which is close to each other, and the Arantius tubercle enhances the tightness of the intervalvular closure. The bottom of the blood-filled pocket becomes the main weight-bearing part.

The area of the pulmonary artery orifice is 389(311~600) mm^2, the sum of the mean areas of the pulmonary valve leaflets is 861(625~1064) mm^2, and the ratio of the leaflet to orifice area is 2.3:1. Simple enlargement of the pulmonary valve annulus does not necessarily cause valve closure insufficiency if its area does not exceed the mean area of the valve leaflets [7].

2.4.4 Aortic Valve

The aortic valve is located at the center of the heart's four groups of orifices, at the apex of the left ventricular outflow tract (the vestibular portion of the aorta) [4]. The structure is similar to that of the pulmonary valve but is stronger than the pulmonary valve [3]. The three semilunar valves of the aorta are attached to the aortic wall and partially to the ventricular wall. In addition to the right semilunar valve, the left and posterior semilunar valves are continuous with the anterior mitral leaflet, forming the subaortic curtain, and the corresponding continuous ends are thickened, forming the left and right fibrous triangle. The sac-like cavity formed between each semilunar valve and its attached aortic wall is called the aortic sinus (also called Valsalva sinus).

Depending on the presence or absence of coronary artery openings in the arterial wall, each semilunar valve of the aorta is called the left coronary semilunar valve, the right coronary semilunar valve, and the posterior coronary semilunar valve, and its corresponding sinus is called the left coronary semilunar sinus, the right coronary semilunar sinus, and the absent coronary semilunar sinus [6]. Between the semilunar attachments (annulus) of adjacent leaflets is a square triangular area called the "intervalvular triangle." The three intervalvular triangles in the aortic wall are anatomically part of the aortic wall rather than part of the left ventricle but are part of the ventricular wall when the semilunar valve is closed, also known as the intraventricular arterial wall (artery as part of the ventricle).

The semilunar valve consists of an endocardial fold sandwiched by an intervening fibrous tissue core. The basal edge of each semilunar valve is thickened, the aortic surface is deeply depressed, and its surface is rougher than the ventricular surface. When the valve leaflets close during diastole, the aortic sinuses begin to fill, and each protrudes into the aortic vestibule of the left ventricle. The three Arantius tubercles and the six equal and corresponding valve margins are close together, enhancing the tightness and firmness of the valve closure.

The upper border of the aortic sinuses exceeds the level of the free edge of the semilunar valve and forms an annular sinus ridge. The left and right coronary arteries open on the sinus wall below the sinus ridge, and the sinus wall at the attachment edge of the semilunar valve contains a large number of collagen fibers [1]. At the level of the aortic sinus, the lumen is approximately twice the diameter of the ascending aorta [4].

The mean aortic orifice area was 370 mm^2 (272–521), the sum of the mean areas of the three semilunar valves of the aorta was 849 mm^2 (537–1073), and the ratio of valve leaflet to orifice area was 2.3:1.

2.5 Vasculature of the Heart

The myocardium is supplied by a separate vascular itself. The arterial supply to the heart arises from the base of the aorta and runs in the coronary sulcus. The left coronary artery (LM) originates from the left coronary sinus of the aorta and subsequently divides into the left anterior descending branch (LAD) and the left circumflex artery (LCX). Right coronary artery (RCA) emanating from the right

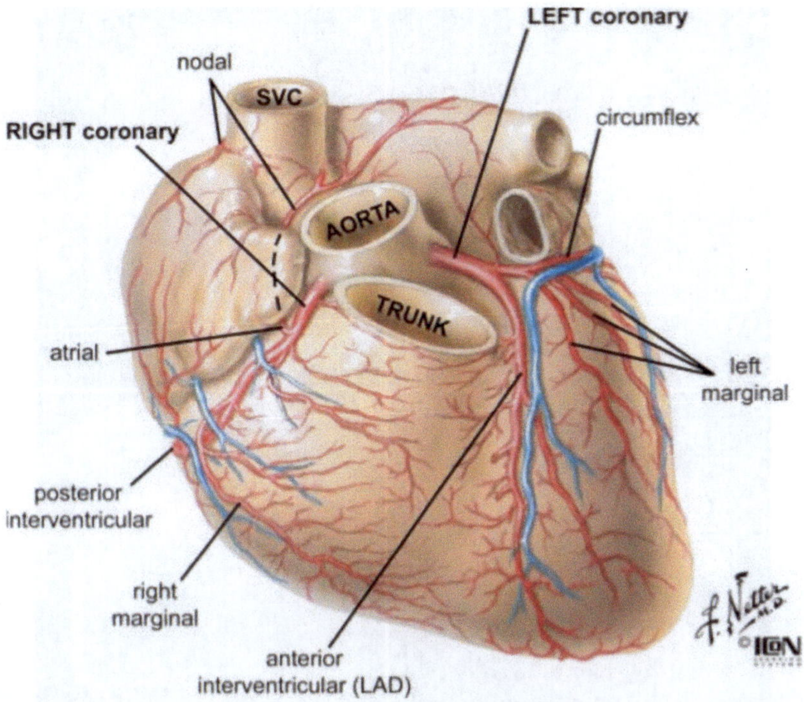

Fig. 2.7 Vasculature of the heart. The heart is supplied arterially through the right and left coronary arteries and their branches. Venous drainage occurs through cardiac veins. Figure adapted from Handbook of Cardiac Anatomy, Physiology, and Devices [1], 2nd, and reused with permissions from Springer

coronary sinus. The venous drainage is via cardiac veins that return deoxygenated blood to the right atrium [1] (Fig. 2.7).

2.5.1 Right Coronary Artery

RCA emanating from the right coronary sinus [6]. Passing though the adipose tissue, the space between the outflow tract and the right atrial appendage is occupied by the initial portion of the RCA. The RCA produces the conus arteriosus that supplies the base of the right ventricle [5]. The first atrial branch is primarily the sinus node artery [6]. The RCA enters the right atrial sulcus and is very close to the tricuspid annulus (TA).

The front wall of the RV is supplied by the right marginal branch [6]. RCA is divided distally into a posterior descending coronary artery (PDA) and a posterior left ventricular branch. In 70–80% of cases, the PDA arises from the right coronary artery and moves posteriorly along the interventricular sulcus [6]. The PDA supplies

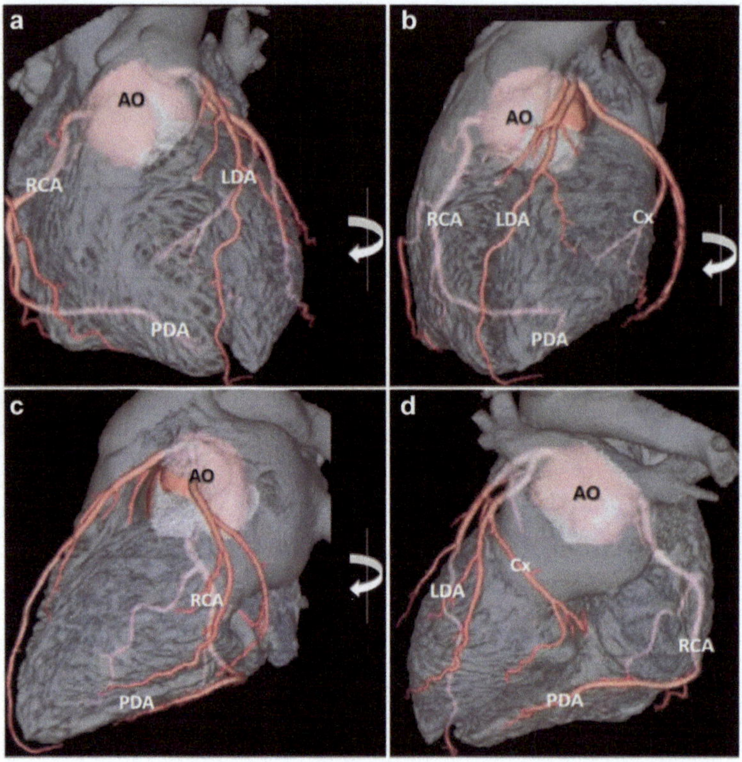

Fig. 2.8 CT volume-rendering modality. The four images (**a–d**) have been rotated around the Y-axis from left to right to show different angles of the coronary tree. Figure adapted from Atlas of Non-invasive Imaging in Cardiac Anatomy [6], (2020), and reused with permissions from Springer

the inferior half of the septum. At the critical site, the dominant RCA makes atrioventricular node artery. Atrioventricular node artery penetrates the fat-filled posterior conus space to reach the node [6]. Finally, before ending, RCA gives off an unequal number of posterolateral artery that supply the posterior medial mastoid muscle of mitral valve and posterior medial wall of the left ventricle [6] (Fig. 2.8).

2.5.2 *Left Coronary Artery*

Left coronary sinus emanating from the LAD. The left main coronary artery (LM) has an initial diameter of 4–5 mm and extends 5–15 mm to the left [6]. Then it is divided into two main vessels: the left descending coronary artery (LAD) and the circumflex artery (LCx). The intermediate ramus arises occasionally between the LAD and LCx, forming a tridentate of LM [6].

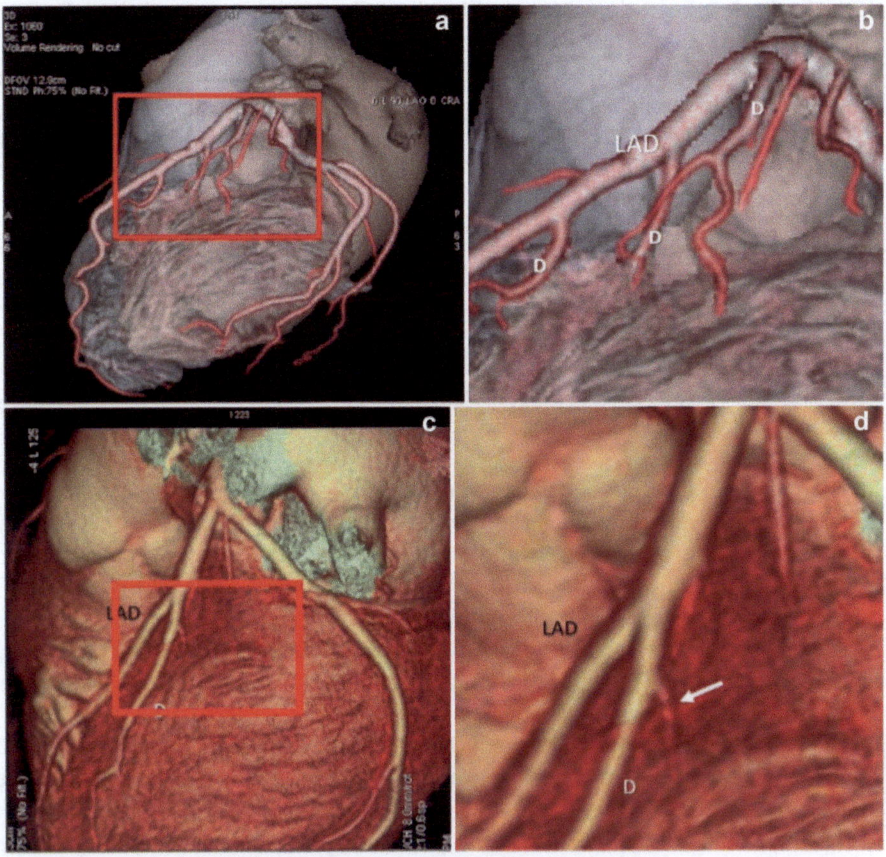

Fig. 2.9 (**a**) CT volume-rendering with transparency in lateral perspective. (**b**) The image shows three diagonal branches. (**c**) Volume-rendering format showing large diagonal lines. The area in the red box is zoomed in panel. (**d**) In the magnified image, a very thin septal branch starts from the large diagonal. Figure adapted from Atlas of Non-invasive Imaging in Cardiac Anatomy [6], reused with permissions from Springer (2020)

The LAD supplies a large amount of myocardium, which runs in the interventricular sulcus and extends to the apex of the heart [6]. The LAD produces three types of branches [6]. Diagonal branches, which go to the left, supply the anterolateral wall of the left ventricle [7]. The septal branches run perpendicular to the LAD and supply the ventricular septum [7]. The right ventricular branch is a small artery that runs from the LAD to the right and supplies the anterolateral wall of the right ventricle (Fig. 2.9).

In normal form, LCx comes from LM and is at right angles to LAD, forming a bifurcation [6]. After its origin, the LCx enters the atrioventricular sulcus and then descends as a marginal branch to the obtuse edge of the left ventricle. In right coronary-dominant hearts, it is possible that the atrioventricular section of LCX is short. In the left heart-dominant system, the LCx continues around the left atrial sulcus to the crux and gives rise to the posterior descending branch artery [6]. Atrioventricular node is supplied by its branches. In the case of balanced innervation, the main two or three branches of the LCx are marginal branches supplying the lateral wall of the left ventricle [7]. In nearly 30% of hearts, the atrioventricular node is fed by a small vessel that arises from the LCx and travels upward into the left atrium [6].

2.5.3 Cardiac Veins

Cardiac veins converge to form coronary sinuses, precordial veins, and minimal cardiac veins. Up to 49% of cardiac drainage is through coronary sinuses and cardiac veins, 25% through the front cardiac vein, and 17% through the smallest cardiac vein [3].

The great cardiac vein becomes a coronary sinus at the posterior aspect of the heart in the atrioventricular sulcus and then drains into the right atrium [5]. Tributary veins draining the posterior left and right ventricles and the ventricular septum converge to form the middle cardiac vein. The coronary sinus receives a small cardiac vein which arises anteriorly/laterally/inferiorly from the right ventricle.

The precordial vein, formed by two to four veins, drains to the anterior wall of the right ventricle, travels upward, crosses the right atrioventricular sulcus, and enters directly into the right atrium.

The smallest cardiac venous system consists of numerous small intra-myocardial veins, also known as the Thebesian veins [3]. These tiny vessels start from the capillary bed of the myocardium, which lead directly to the ventricles. Thebesian vein is most common in the atria and ventricular septum and more common on the right.

2.5.4 Myocardial Bridges

Coronary arteries usually run under the epicardium. Often, a portion of the artery runs along the myocardium (within the wall), either for a considerable length of time within the myocardium or in a muscle slip arrangement ("myocardial bridge"). Myocardial bridges are most common in the midsection of the anterior interventricular artery [1]. The myocardial fibers covering the anterior interventricular artery cross the artery in a vertical direction [1]. The incidence of myocardial bridges has been reported to be 0.5–16%, as measured by angiography in the catheterization laboratory. Stenoses appear in systole and disappear in diastole. The presence of straight or systolic stenosis appears to be an important diagnostic technique to detect coronary artery segments in angiography [1]. Myocardial bridging is usually a benign condition.

2.6 The Great Vessels

"Great vessels" refer to the large arteries and veins directly entering and exiting the heart mostly in the superior mediastinum, including the superior vena cava and inferior vena cava, the aorta, pulmonary trunk, and pulmonary veins [1] (Fig. 2.10).

The inferior vena cava crosses the diaphragm into the right atrium [1]. The pulmonary vein enters the left atrium. The jugular vein of the right arm and the jugular vein of the left arm merge to form the superior vena cava [7]. The superior vena cava also receives the jugular vein from behind and drains into the upper part of the right atrium [1].

The pulmonary artery trunk rises from right ventricle and passes anterior to the base of the aorta [1]. The pulmonary artery trunk is divided into the left and right pulmonary arteries as it emerges from the pericardium. The aortic arch is formed by the aorta rising from the left ventricle to the right and then curving to the left and posterior [1]. The aorta passes down the left side of the trachea and esophagus to become the descending aorta and enters the posterior mediastinum. The semilunar valve prevents backflow of blood from the aorta and pulmonary artery trunk [1]. The semilunar valves are located at the base of these large vessels, and each has three lobules. The "aorta and sinuses" are the areas of arterial dilatation, which are adjacent to the valves.

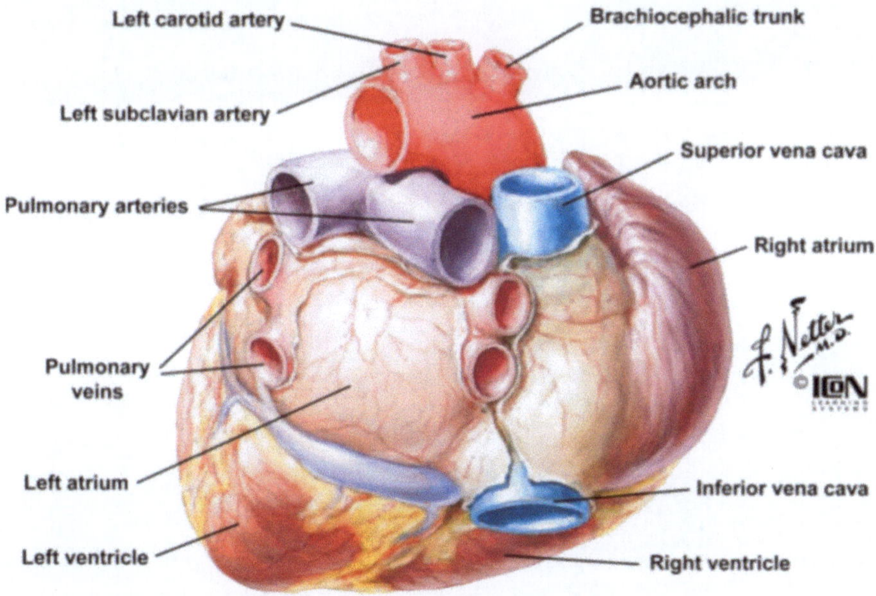

Fig. 2.10 The great blood vessels viewed from the posterior side of the heart. Figure adapted from Atlas of Human Anatomy [1], 2nd, and reused with permissions from Springer

References

1. Iaizzo PA. Handbook of cardiac anatomy, physiology, and devices. Totowa: Humana Press; 2005.
2. Clinical Cardiac MRI. 2005.
3. Iaizzo, Paul A. Handbook of cardiac anatomy, physiology, and devices; 2009.
4. Ziemer G, Haverich A. Cardiac surgery. Cardiac Surgery; 2017.
5. Filipoiu M. Atlas of heart anatomy and development. Atlas of Heart Anatomy and Development, 2014.
6. Faletra FF, Narula J, Ho SY. Atlas of non-invasive imaging in cardiac anatomy. Atlas of Non-Invasive Imaging in Cardiac Anatomy, 2020.
7. Widlansky ME, Malik MMA. PanVascular Medicine; 2015.
8. Francis J. Anatomy for cardiac electrophysiologists - a practical handbook. Authors: S. Yen Ho, Sabine Ernst. Cardiotext Publishing, LLC. Indian Pacing Electrophysiol J. 2013;13:239.
9. Daubert JP. Cardiac mapping. In: Shenasa M, Hindricks G, Borggrefe M, Breithardt G, Josephson ME, editors. Pacing and clinical electrophysiology, vol. 38. 4th ed. Hoboken. (ISBN: 97870046 966 pages: Wiley Blackwell; 2015.

Chapter 3
Physicochemical Properties of the Ideal Materials for Cardiac Devices

Yufeng Zheng

3.1 Physicochemical Properties of the Ideal Cardiac Materials for Non-implantable Interventional Devices

Non-implantable interventional devices, like guidewires and catheters, assist or directly complete the interventional treatment of heart disease in a minimally invasive way. Generally, these devices are designed to be slender and streamlined in order to facilitate delivery in the blood vessels, whose outer layer or main body is made of flexible polymer material, the core or the middle layer is made of rigid metal wire or metal mesh, and the head end will be made of precious metal development mark. According to different expected functions, balloon, electrode, and other components are added to realize the transport scaffold, physiological signal measurement, ablation, etc.

3.1.1 Guidewires

Guidewires play an important role in PCI, which act as the railway assisting in the insertion and location of the catheter into coronary artery.

The commercially available guidewires usually consist of four parts: core, tip, body, and coating (Fig. 3.1). With diverse designs, guidewires can achieve expected interventions in different lesions (Fig. 3.2). The components of guidewires include:

- *The core*: As a basic platform of guidewire, the type of the core is dependent on material, diameter, and taper, which determines flexibility, supportability,

---

Y. Zheng (✉)
School of Materials Science and Engineering, Peking University, Beijing, China
e-mail: yfzheng@pku.edu.cn

Y.-D. Tang et al. (eds.), *Application of Biomaterials in the Treatment of Cardiovascular Diseases*, https://doi.org/10.1007/978-981-19-7712-1_3

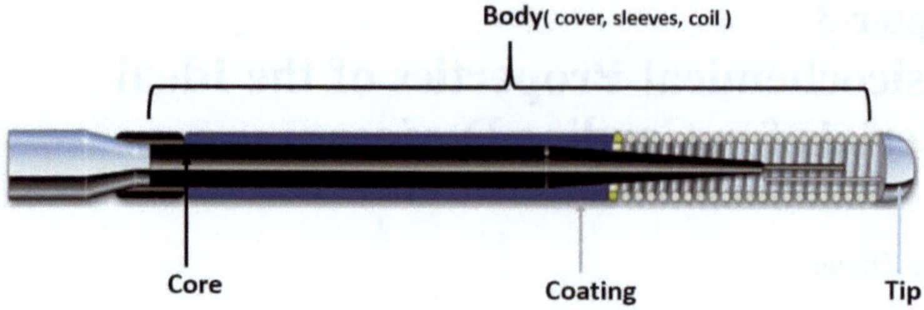

Fig. 3.1 The structure of a typic guidewire

Cougar LS (made by Medtronic)	
Core: Nitinol	Design: Shaping ribbon
Tip: Spring coil	Coating: Hydrophilic

Choice floppy (made by Boston Scientific)	
Core: Stainless steel	Design: Core to tip
Tip: Spring coil	Coating: Hydrophilic

Asahi fielder (made by Asahi)	
Core: Stainless steel	Design: Core to tip
Tip: Coil	Coating: Hydrophilic

HT all star (made by Abbott)	
Core: Stainless steel	Design: Core to tip
Tip: Spring coil	Coating: Hydrophobic

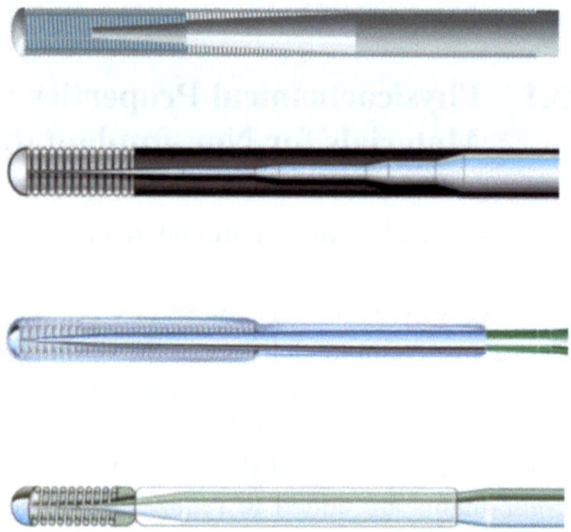

Fig. 3.2 Some examples of commercial guidewires

torqueability, as well as the visibility of wires. In-market guidewires are usually made of nitinol or steel. In comparison, the former is more flexible while less supportiveness. The latter offers strong support but is prone to bending or kinking [1, 2].

- *The tip*: As the most distal part of a guidewire, it is responsible for maneuvering and contact force transmission, commonly made of polymers or alloys.
- The body: It usually consists of coil, elastic polymer cover, or their hybrid, which contributes to wire's trackability. Hydrophilic cover is extremely lubricious and could provide smooth tracking through arteries, while it is at the coast of less tactile feedback.
- *The coating*: Hydrophilic or hydrophobic film coated on wire's body. Generally, hydrophilic coatings made of gel enhance lubricity, while hydrophobic coatings made of PTFE improve the trackability.

An optimal guidewire not only ensures minimal disruption to vessel wall during its smooth passage but also offers excellent steerability and clear visibility, facilitating easy navigation to the intended region. According to these requirements of guidewires, the key metrics for them can be described as follows:

- *Supportability*: the capability of a guidewire to resist bend and help deliver devices through vessel. Supportability is dominated by the stiffness of core's material and diameter [2]. The stiffer and lager the core is, the better supportiveness could be obtained at the sacrifice of its flexibility. For convenience, it is usually classified as light, medium, and extra support.
- *Torquability*: the ability to effectively transmit torque from proximal end to distal tip. The desirable transmission ratio is 1:1, which means that the distal end could perform the same action immediately when the proximal end is operated. Similar to supportability, the overall rigidity significantly influence wire's torquability [3].
- *Crossability*: the ability of crossing the lesion without penetration and obvious resistance, even in stenosis and occlusion. This performance is affected by complex parameters, such as the core's characteristics (material, diameter, and taper structure), tip stiffness, and coatings.
- *Tactile feedback*: the feedback of tactile information from tip to operator's hands. High sensitivity and real feedback help operator make correct decision to avoid penetrating vessel wall or atheromatous plaque. The more flexible the distal end is, the more distortional tactile feedback operators receive. Moreover, the hydrophobic coating enhances the real tactile sense, while the hydrophilic coating weakens it because of slippery gel-like surface.
- *Flexibility*: the characteristic of safe and successful navigation through tortuous vessels without perforation and dissection, which relies on the mechanical properties of the core and tip. Flexibility is in a manner opposite to the overall stiffness, that is, the stiffer a wire is, the less flexible it becomes [4].
- *Trackability*: the wire's ability to navigate a curved path, to follow the tip around curves and bends, as well as to navigate smoothly along the vessel. It can be measured by determining the forces required to advance the wire through a predetermined path [5].
- *Radiopacity*: the ability of clear vision for locating the guidewire. Au, Pt, and W generally as the radiopaque marks are placed on the proximal end, which is significant to improve safety and guarantee appropriate usage of device.

ISO 11070 is the main test standard used to ensure the guidewire's safety in use. There are several general requirements for sterile single-use intravascular devices, for instance, sterilization (testing in the reference of ISO 17665, ISO 11135 and ISO 11137), biocompatibility (testing under ISO 10993), corrosion resistance (testing in the reference of ISO 11070 Annex B), and radio-detectability (testing in the reference of ASTM F640-12). Besides, fracture test, flexing test, and peak tensile force test are three additional mechanical tests for guidewires. These test methods for biocompatibility, corrosion resistance, and mechanical properties of catheters will be described as follows:

- *Biocompatibility test*: As a short-time direct contact with blood equipment, hemocompatibility is very important, which is tested according to the ISO 10993-4. Evaluation of guidewires' hemocompatibility includes thrombosis tests, coagulation tests, and platelet and hematologic tests. Thrombosis capacity can be demonstrated by measuring the weight of the thrombus or by observing the aggregation of cells and fibrinogen on the surface of the guidewires using light microscopy and scanning electron microscopy. As for coagulation tests, whole blood or anticoagulant whole blood is used to contact with guidewires to detect the coagulation through the part of the blood coagulation activation time, prothrombin time, and thrombin time. Platelet testing is required to assess platelet count, aggregation, activation, and function in the blood after contacting with guidewires. Hematological test is to assess white blood cell activation, hemolysis, and reticulocyte count. The hemolysis test can detect the percentage of free hemoglobin in the total hemoglobin at the beginning of the test by the contact between the guidewire with the highly diluted red blood cell suspension.
- *Radio-detectability test*: According to ASTM F640, samples and user-defined standard samples are placed in the body mimic environment and exposed to X-rays. The optical density of the sample is compared with that of user-defined standard samples, and the pixel densities are compared after digital processing.
- *Corrosion resistance test*: Corrosion resistance requirements for single-use intravascular introducers according to the standard ISO 11070. First, the guidewire is immersed in 0.15 mol/l NaCl solution at 22 °C for 5 h. Then, the sample is transferred to boiling distilled or deionized water for 30 min. Allow to cool down and dry up at room temperature. The sample is visually inspected for signs of corrosion.
- *Fracture test*: The test instruments refer to ISO 11070. The distal end of the guidewire is fixed into the hole of a cylinder, and then the wire is tightly wrapped around the cylinder for at least 8 times. The wire is uncoiled and inspected for breakage except for the fixed and first turn. No separation or surface rupture is supposed to exist during the test.
- *Flexing test*: First, the distal around one cylinder of the experimental apparatus is bended, and the second cylinder is wrapped in the opposite direction. Then, the guidewire from cylinders is removed and straightened. Repeat the bending-straightening cycle for 20 times to observe whether the guidewire shows any defects, coating peeling, fracture, and other failures.
- *Peak tensile force test*: Selecting each key joint part as the test section. Fixing one end of the guidewire to a movable wedge clamp, the midpoint of guidewire is fixed to the immovable clamp, and applying tensile force at a speed of 10 mm/min until failure occurs, recording peak tensile force and failure position of guidewire.

3.1.2 Catheter

The advent of catheters has pushed the treatment and diagnosis of heart disease toward minimally invasive treatment, especially which provides a less expensive and safer alternative to coronary bypass grafting, known as percutaneous coronary angioplasty (PTCA) [6]. Depending on the intended design, the catheter may perform multiple complex functions. For example, the guiding catheter can transport and support interventional devices. Angiographic catheters are supposed to inject contrast agents and drugs. Ablation catheters are used to eliminate fibrillation [7–10]. Although the design of catheters varies greatly due to their intended functions, there are common requirements for catheters. The visible and soft tip ensures accurate positioning and keeps the vessel wall atraumatic. The remaining parts of the catheter are expected that not only are flexible enough to remain coaxial with the vessel but also have good torsion resistance to transmit torque accurately and provide stable support. A radiopaque ring made of gold, platinum, or tungsten is usually loaded on the tip. Table 3.1 summarizes the characteristics of common catheters. Accordingly, the wall of guiding and angiographic catheters are generally non-degradable polymer materials, such as nylon, polyvinyl chloride (PVC), polyethylene (PE), polyurethane (PU), and polytetrafluoroethylene (PTFE). To adjust the stiffness and torqueability of catheters, braided metal mesh is added to catheters, which exhibit a greater stiffness with a lower braiding angle between the vertical axes. However, the braided mesh with a higher angle could be more resistant to kinking [11]. The characteristics of ablation catheters will be introduced in Sect. 3.1.3.

ISO 10555-1 specifies the technical requirements and test methods for catheters. Similar to the requirements for guidewires in terms of biocompatibility, radiometric detectability, and corrosion resistance, catheters are required to be free of biological hazards, radiopaque, and corrosion attack. Besides, special requirements for mechanical and physical property also need to be evaluated, including peak tensile force, liquid leakage, air leakage, flow rate, burst pressure, and peak pressure. These devices, conditions, and parameters for test are recommended by ISO 10555-1.

- *Peak tensile force test*: All parts of the catheter need to be tested, including each tubular part and joints. The components are stored in an appropriate aqueous medium (at 37 ± 2 °C) for the appropriate duration of clinical use. Then, the test sample is fixed on the tensile test device. The distance between the two fixtures is measured. The samples are tested at a constant strain rate of 20 mm/min until stretched to the fracture [12]. The peak tensile force is recorded. The peak tensile force of catheters shall conform to the requirements in Table 3.2.
- *Leakage test*: Leakage of the catheter is not desired either under liquid pressure or during the suction process. When testing for fluid leakage under pressure, the catheter is connected to a syringe or pressure equipment. Then, the syringe or pressure equipment is filled with deionized water (22 ± 5°C), the air is drained, and the distal end of the catheter is sealed. At least 300 kPa to the catheter for 30 s are applied to check leakage. During aspiration, air leakage into hub assembly is

Table 3.1 The characteristics of commercial catheter [13]

Company	Product	Intended use	Material	Maximum guidewire size/ID (inches)	Maximum flow rate/ pressure limit
Angiodynamics	Accu-Vu sizing pigtail	Angiographic catheter	Co-extruded nylon	0.038	26cc/1050 psi
	Mariner Headhunter	Angiographic catheter	Duration coating/ nonbraided/co-extruded nylon	0.038	8cc/272 psi
	Mariner Cobra 2	Angiographic catheter	Duration coating/braided/ co-extruded nylon	0.038	10cc/254 psi
	Mariner straight art	Wire exchange angiographic catheter	Duration coating/braided/ co-extruded nylon	0.038	8cc/238 psi
Boston Scientific	Imager II TR Flush	Angiographic catheter	Braided/pebax	0.038	26cc/1050 psi
	Impulse IM	Angiographic catheter	Braided/pebax	0.038	8cc/1200 psi
	Radiofocus straight and angled Glidecath	Wire exchange/ shaped angio-graphic catheter	Hydrophilic coated nylon/ polyurethane	0.038	19cc/1000 psi
Biocardia	Morph/Uni-versal Deflectable Guide Catheter	Guiding catheter	Braided pebax shaft/PTFE liner	0.068	50 psi
Cordis	Vista Brite-Tip Hockey Stick Guide Catheter	Guiding catheter	Stainless steel braided/blended nylon shaft/ PTFE liner	0.088	N/A
	Vista Brite-Tip IM guide catheter	Guiding catheter	Stainless steel braided/blended nylon shaft/ PTFE liner	0.088	N/A

Table 3.2 Peak tensile force of catheter test section [12]

Smallest outer diameter of test section (mm)	Minimum peak tensile force (*N*)
≥0.55, <0.75	3
≥0.75, <1.15	5
≥1.15, <1.85	10
≥1.85	15

measured. The catheter is connected to a partially filled syringe, and whether there are any bubbles entering the syringe is checked.

- *Flow rate testing*: The constant-level tank is supplied with distilled water or deionized water and connected to the catheter by a male 6% (Luer) taper fitting. When the solution flows through the catheter, the outflow is collected for at least 30 s. The flow rate of the catheter is expressed as ml/min.
- *Burst pressure test*: The catheter is connected to the pressure generating device, and the fluid temperature in the tank is maintained at 37 ± 2 °C. The fluid flows into the catheter at a constant rate of 1 ml/s, and the pressure in the catheter is monitored until leak or explosion occurs. The maximum pressure value during explosion is recorded.
- *Power injection test*: This test is only applicable to indicate the product used for power injection. The catheter is connected to a real or simulated connection line that is fixed to a constant pressure device filled with the injection or the simulated injection. Flow rate and peak pressure are measured by mass balance and pressure sensor, respectively.

3.1.3 Electrophysiological Electrode

Electrophysiological electrodes are the basis of accurate and effective treatment of arrhythmia. Electrophysiological electrodes could determine the intracardiac electrocardiogram and analyze the origin of arrhythmia to guide the ablation operation [14]. Electrophysiological electrodes are often presented as electrode catheters (Fig. 3.3) for electrocardiogram mapping [15]. Electrode catheters are consisted of electrodes and insulated wires. At the head of the catheter, each wire is connected to an electrode exposed to the inner surface of the heart. At the proximal end of the

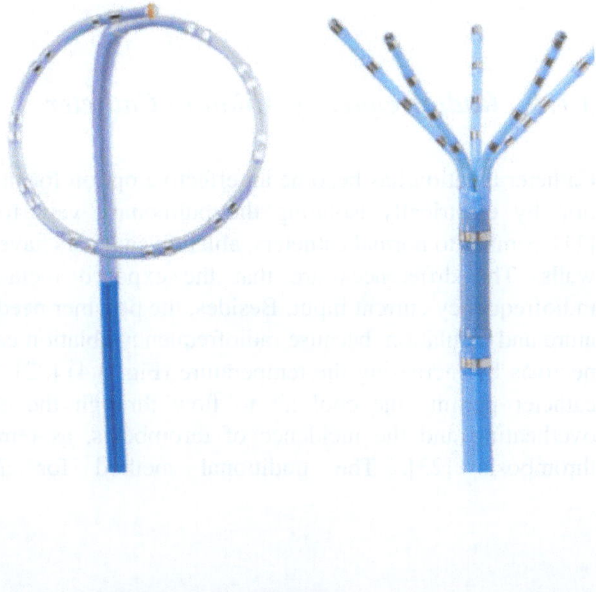

Fig. 3.3 Multipolar ring and star-shaped mapping catheters [15]

catheter, each wire is connected to an external recording device [15, 16]. When the wire is connected to the radiofrequency device, it becomes the radiofrequency ablation catheter, which will be described later. Electrodes have developed from unipolar or bipolar to multipolar, in order to achieve different signal recordings [16]. In addition, the multipole design reduces the size of electrode, providing several advantages for labeling scar-related arrhythmias (e.g., higher labeling resolution, stronger electrocardiosignal, and pace signal capture at lower output) [17]. It is noteworthy that the conductive nanomaterials (e.g., carbon nanotubes, gold nanoparticles) could lower the impedance between the electrode and tissue interface. Thus, in the process of recording electrophysiological signals, these nanomaterials are also loaded to reduce the background noise and obtain a lower signal threshold [18]. In order to interact with cardiac tissue better and make the electrode contact the tissue in a more flexible manner, the balloon have become a candidate platform for electrode loading. Balloon catheters with semiconductor sensors are not only capable of measuring temperature, flow, tactile, optical, and electrophysiological data, but can also control local tissue ablation [19]. It is reported that balloon could accurately feed back the electrophysiological status of the heart after exposure, which had a low signal-to-noise ratios of 60 dB and an impedance of 26 kΩ ± 8% at 1 kHz in normal saline (0.9%) solution [19]. However, epicardial monitoring method is not the mainstream clinical method.

Since cardiac electrophysiological electrodes are mostly used in the form of catheters, the electrode catheters need to meet the requirements of ISO 10555-1, which is generalized as follows: (1) no corrosion attack on the metal part of the conduit after corrosion test; (2) sterile and no harm to the biological tissue and cells in contact with it; and (3) the maintenance of structural integrity during use. Considering the function of intracardiac electrogram, it is necessary to utilize the electrode catheter to map the electrocardiogram in vivo, which usually carries out on myocardial infarction (MI) animal model [17, 20]. If appropriate, the impedance and signal-to-noise ratio of the electrode catheter in the service environment may be evaluated [19].

3.1.4 Radiofrequency Ablation Catheter

Catheter ablation has become an effective option for the treatment of atrial fibrillation by electrically isolating the pulmonary vein from the atrial myocardium [21]. Similar to normal catheters, ablation catheters have metallic tips and polymeric walls. The differences are that the exposed metallic parts are required for radiofrequency current input. Besides, the polymer needs to withstand high temperature and insulation, because radiofrequency ablation causes tissue coagulation and necrosis by increasing the temperature (Fig. 3.4) [22]. The open-irrigated ablation catheter permits the cool air to flow through the electrode surface to reduce overheating and the incidence of thrombosis, as temperature rise will promote thrombosis [23]. The traditional method for ablation is point-by-point

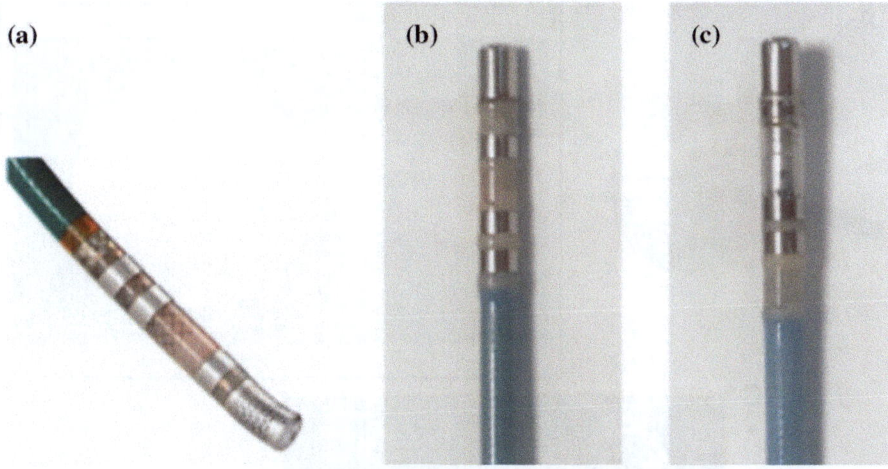

Fig. 3.4 The tips of the irrigated ablation catheters: (**a**) FlexAbility (St. Jude Medical, St. Paul, MN, USA). (**b**) ThermoCool Surround Flow (Biosense Webster, Diamond Bar, CA, USA). (**c**) ThermoCool SmartTouch Surround Flow (Biosense Webster) [22]

radiofrequency ablation guided by electroanatomical mapping. However, this method requires a long period of experience, which limits its wide application. To eliminate these problems, pulmonary vein isolation is required with simplified ablation tool. On the condition, the balloon device is expected to reduce the operational difficulty by one-time contact with the target pulmonary vein at once. The Heliostar (Biosense Webster, Diamond Bar, California) is a balloon-based irrigated catheter that uses radiofrequency energy. It arranges ten flexible gold-plated electrodes on the surface of balloon. Each electrode can independently deliver RF energy to the targeted tissue using a multielectrode RF generator and sense the temperature during ablation (Fig. 3.5). The clinical trial results of Heliostar showed that all targeted pulmonary veins (PVs) had been successfully isolated and 75.7% patients had remained in sinus rhythm without antiarrhythmic drug treatment during 12-month follow-up [24, 25]. In addition to balloon, expandable nitinol mesh has also been applied to spheroid-shaped lattice catheter, which allowed higher energy delivery at a lower current density and achieved continuous ablation in short time compared to a standard ablation catheter [26]. In first-in-human trial, the spheroid-shaped lattice catheter, which required a mean radiofrequency ablation of 2.7 ± 0.70 min, showed a pleasant result that pulmonary veins isolation was achieved in 64 of 65 (98.5%) [27].

General requirements for catheter equipment have been described in the previous section. Ablation catheters also need to be evaluated for corrosion resistance, biocompatibility, suspected detectability, and mechanical properties. In addition, since the ablation catheter is an active device, it is necessary to evaluate the electrical performance, such as direct current resistance, inter-electrode insulation resistance, insulation resistance between the electrode and the catheter's wall, as well as high-

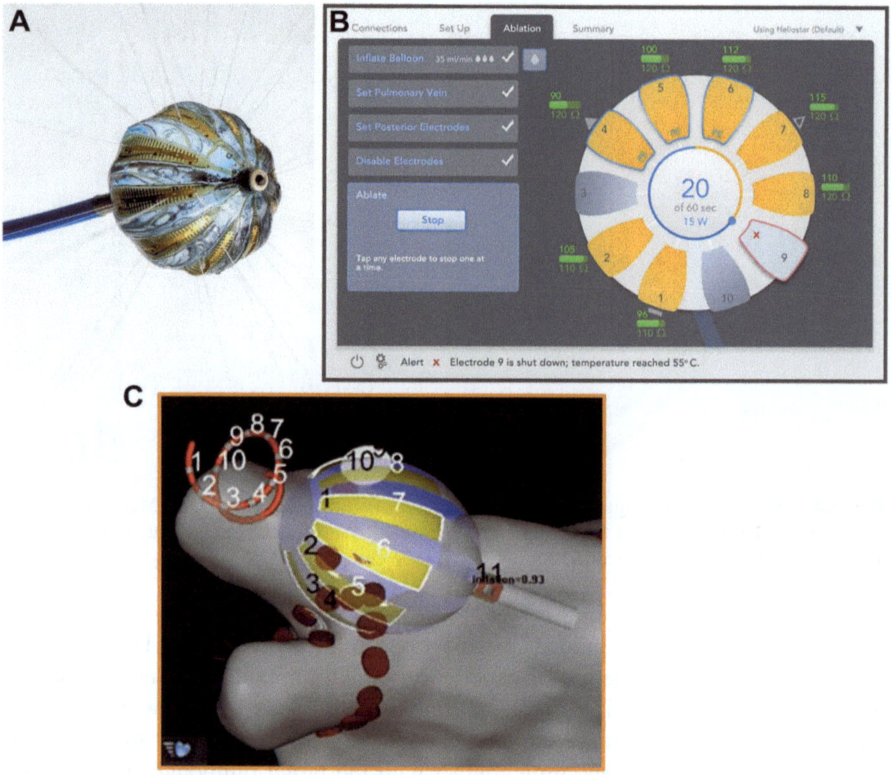

Fig. 3.5 The Heliostar radiofrequency balloon ablation catheter (Biosense Webster, Diamond Bar, California) [29]: (**a**) The design of Heliostar. (**b**) Local, segmental, or circumferential ablation can be performed with independent electrodes. (**c**) Integrated corresponding mapping system

frequency conductivity. Currently, internationally recognized standard for radiofrequency ablation devices is still in development. Thus, the test methods for safety and efficacy of ablation catheter could only refer to published articles at present.

- *In vitro setup*: During the construction of the in vitro ablation procedure, cardiac tissue is adopted [23]. Fresh cardiac tissue is immersed in the heparinized blood with pulsating circulation at 37 °C. The ablation catheter is placed vertically in contact with the myocardial tissue, and the ablation is performed after reaching contact stability. And then, it is necessary to access dimension of lesion after delivery of radiofrequency energy, including surface maximal and minimal dimensions, depth, and maximal cross-sectional width of the ablation area [23]. Also, it is essential to observe whether there is any thrombus or penetration after completion.
- *Ablation protocol*: Similar to the previously mentioned in vitro test mode, the power delivery, impedance, and temperature at the tip of electrode are recorded continuously at different ablation times, contact forces, and power [28]. After

each radiofrequency application, the catheter is left in place and the electrode-tissue interface is examined for thrombosis, charring, and disruption. The electrode catheter is then removed, cleaned, and placed in a new location for another RF application.

3.2 Physicochemical Properties of the Ideal Cardiac Materials for Implantable Interventional Devices

Common heart implant devices include vascular stents, artificial heart valves, vascular prostheses, heart patches, etc. In the process of developing implant devices, metallic, inorganic, and polymeric materials will be applied to different implant devices according to their different physicochemical properties. Unlike non-implantable devices, it is necessary to take long-term performance changes of the materials used to make implantable devices because they serve in the body for a long time.

3.2.1 Metallic Interventional Devices

As the most widely used implantable cardiac materials at present, metals are indispensable for the manufacturing of many instruments, such as coronary artery stents, aortic stents, mechanical heart valves, occluders, the shell of pacemaker, and defibrillator. According to their degradability, they can be divided into non-degradable alloys (stainless steel, CoCr alloys, TiNi alloys, and PtCr alloys) and degradable alloys (magnesium alloys, zinc alloys, and ferroalloys). With the development of metallic cardiovascular stents and the withdrawal of polymer stents in recent years [30–32], metallic materials have become the representative materials for the commercial stents. Here, the physical and chemical properties of metallic cardiovascular stents, as well as the technical standards and test methods of stents, will be introduced.

3.2.1.1 Non-degradable Metallic Stents

Most first-generation drug-eluting stents (DES) are made of 316L stainless steel (316L ss), while various alloys have been applied to stent manufacture, such as CoCr alloys, PtCr alloys, and high nitrogen steels. The mechanical properties, physical properties, chemical properties, and biocompatibility of these materials are shown in Table 3.3. The first-generation DES are almost made of 316L ss with an average strut thickness of about 120 μm, while the average strut thickness of the second-generation DES drops to 80 μm, which have great significance to reduce stent

Table 3.3 the physicochemical properties of nonbiodegradable alloys [42–46]

Material	Mechanical properties			Physical properties		Chemical properties	Biocompatibility
	YS (MPa)	UTS (MPa)	TE (%)	Density (g/cm^3)	Radiopacity		
316L ss	331	586	55–65	7.9	Radiolucent	$I_{corr} = 2.706*10^{-7}$ μA/cm^2 in Hank's solution at 37 °C	Initial clotting time: 40 min
HNS	546	941	52	N/A	Radiolucent	$I_{corr} = 2.334*10^{-7}$ μA/cm^2 in Hank's solution at 37 °C	Initial clotting time: 63 min
Co-28Cr-6Mo	492	1013	19	N/A	Radiopaque	$I_{corr} = 2.334*10^{-7}$ μA/cm^2 in Hank's solution at 37 °C	Initial clotting time: 45 min
L605 (Co–Cr–W–Ni)	629	1147	46	9.2	Radiopaque	$I_{corr} = 444.18$ /cm^2 in 0.9% NaCl solution at 37 °C	Cell lively growth rate > 80%
MP35N (Co–Cr–W–Mo)	414	930	45	8.43	Radiopaque		N/A
TiNi	560	1100–1200	14	6.45	Radiolucent	N/A	Hemolysis rate < 0.5%
PtCr	480	834	45	9.9	Radiopaque	$I_{corr} < 50$ nA	

thrombosis and enhance re-endothelialization after stent implantation [33, 34]. The reasons for the reduced strut thickness are listed as follows. On the one hand, the metallurgical and manufacturing processes are improved. On the other hand, the newly developed alloys (e.g., CoCr alloys), with higher yield strength and stronger corrosion resistance, are used as the backbone. For permanent stents, biocorrosion not only makes stents release heavy metal irons that may cause harmful reaction (accumulation in organs, allergy, and carcinoma) but also leads to strut fracture due to the biomechanical forces. Therefore, researchers around the world have focused on enhancing the corrosion resistance of inert biomedical alloys with alloying and surface modification. High-nitrogen nickel-free stainless steel (HNS) shows better corrosion and wear resistances and superior biocompatibility compared to the currently used 316L ss [35]. Different from Co–Cr–W–Ni alloy (L605), a novel nickel-free CoCr alloy, where N, Mn, and Fe are added to substitute for all of Ni and part of Co, shows higher pitting potential and tensile strength [36]. Surface engineering tools, such as iron implantation [37], atomic layer deposition [38], magnetron sputtering [39], electrochemical pretreatment [40], and nanosecond laser [41], which could construct a uniform and dense coating on the substrate, are helpful to improve the corrosion resistance of alloys. Besides, co-addition of the right amount of Co, Cr, and Pt into alloys could improve the radiopacity of stents.

3.2.1.2 Biodegradable Metallic Stents

Nowadays, the second-generation DES has completely replaced BMS and the first-generation DES, becoming the first choice of PCI worldwide. Despite major technical improvements, there are still concerns about the long-term safety of permanent stents, such as restenosis due to fatigue fracture of stent, the risk of hemorrhage during long-term DAPT treatment, and chronic inflammation due to long-term retention of foreign object in the body. Therefore, there is interest fully in the development of bioabsorbable stents (BRS), which has sparked the fourth revolution in interventional cardiology. Currently, there are only a few biodegradable metallic stents (e.g., Magmaris) in the market. Most of them are still under clinical research or in development. Table 3.4 shows the mechanical properties, biodegradability, and biocompatibility of various biodegradable metallic scaffolds. Mg alloys have good cytocompatibility in vitro and in vivo. But their development in thin-walled scaffolds is limited by their low mechanical properties and fast degradation rate compared with Fe alloys and Zn alloys [47, 48]. Pure iron and its alloys have excellent mechanical properties so that the wall thickness of nitrided Fe stent is as low as 53 μm [49]. However, the degradation rate of iron is too slow, and the long-term retention of degradation products may bring about potential safety problems. The ferromagnetism of Fe alloys can be transformed into paramagnetism, and the degradation rate can be improved by alloying with Mn. But the biocompatibility, especially cytocompatibility, will be significantly reduced with the increasing amount of Mn [50–52]. The standard electrode potential of pure zinc is between pure magnesium and pure iron, suggesting that the degradation rate may be between

Table 3.4 The physicochemical properties of biodegradable metallic stents

Materials	Mechanical properties			Biodegradable properties		Biocompatibility	References
	YS (MPa)	UTS (MPa)	TE (%)	In vitro	In vivo		
Pure Mg	12-58	48–162	5.9–15.9	0.1 mm/y SBF ≋	0.39 mm/year in rats	++	[62–64]
WE43	161–210	234–300	9–17.2	1.58–2.16 mm/year 0.9% NaCl ≋	Totally absorbed after 4 months in human	+	[65, 66]
AZ31	138–178	253–267	15–19	0.035 mm/year Hank's ϟ	0.02 mm/year in rats	–	[67]
Mg–Zn	169	279	18.8	0.07 mm/year SBF ≋	2.32 mm/year in rabbits	+	[63]
Mg–Li	75–150	107–235	15.6–46.1	0.10–0.34 mm/year Hank's ≋	N/A	+	[68]
Mg–Li–Zn	112–120	138–182	44–50	0.042–0.158 mm/year Hank's ϟ	0.04–0.4 mm/year in rats	+	[67]
Mg–Nd–Zn–Zr	104–315	205–324	12–31	0.327 mm/year Plasma ≋	Fully degrade in 4 months in rabbits	+	[69, 70]
Mg–Zn–Y–Nd	112–306	218–325	9.2–18	0.25–0.50 mm/year Hank's ≋	N/A	–	[71, 72]
Pure iron	130–700	169–900	8–37	0.16–0.85 mm/year Hank's ϟ	No major degradation within 28 days in pigs	+	[73, 74]
Fe–X (X=Mn, Co, Al, W, B, C and S)	350–450	420–850	5–10	0.087–0.187 mm/year Hank's ϟ	N/A	+	[75]
Nitride Fe	561	614	/	0.225 mm/year Hank's ϟ	Fully degrade in 4~5 years in pigs	+	[76, 77]
Fe-(20-35)Mn	234–421	428–701	7.5–32	0.4–1.3 mm/year Hank's ϟ	N/A	+	[78]

Fe–Mn–C–Cu	431–536	956–1078	35–68	0.3–0.4 mm/year Artificial urine ≋	N/A	+	[79]
Fe–Mn–Si	~180	~450	~16	N/A	N/A	–	[52]
Pure Zn		35–45	7.57–9.37	0.028–0.037 mm/year Hank's ≋	41.75 ± 29.72% of the stent was degraded after 12 months implantation in rabbits	+	[53, 80]
Zn–Mg	114–291	250–399	1–22	0.125–0.175 mm/year Hank's ≋	0.012–0.029 mm/year in rabbits	+	[81, 82]
Zn–Li	240–480	360–560	2–14	0.05–0.06 mm/year SBF ≋	0.008–0.045 mm/year in rats	Inflammatory response was slightly higher than pure zinc	[55, 83]
Zn–Cu	148–227	186–270	21–50	0.021–0.033 mm/year SBF ≋	28 ± 13% of the stent remained after 24 months implantation in pigs	The 100% extract was significantly cytotoxic	[56, 84]

≋, immersion test; ⚡, electrochemical test; ++, excellent biocompatibility; +, good biocompatibility; –, poor biocompatibility

Mg alloys and Fe alloys. Due to the poor mechanical strength of pure zinc as vascular stent, Mg, Cu, Li, and other elements are used to improve the mechanical properties of zinc alloy [53–56]. Among them, Zn-Li alloy has become the potential development direction of degradable stents because of its excellent mechanical properties and acceptable biocompatibility [57–60]. However, the work softening behavior of Zn alloys still needed to be solved, which may adversely affect the application of zinc alloy as stents [61].

3.2.1.3 Test Methods and Standards for Stents

The function of vascular stents is to support the vessels in the stenosis and occlusion segment, reduce the elastic retraction of the vessels, and keep the lumen blood flow without obstruction. The supporting effect is reflected in two aspects: the supporting force and the effective service time. The supporting force is generally obtained by measuring the radial strength and stiffness at room temperature and in air. Effective service time is generally measured in simulated body fluid with applied load. As for permanent stents, durability of stents needs to be considered since they are subjected to permanent pulsation and stress from surrounding tissues. Although the long-term durability of biodegradable stents is not required to be considered, the fatigue performance during the service period cannot be ignored. In addition, the corrosion resistance of permanent and biodegradable stents directly affects their effectiveness. What is more, since the service environment of stents gradually transits from blood to vascular tissue as endothelialization proceeds, hemocompatibility and cytocompatibility are essential to be evaluated. Testing methods for radial stiffness and strength, durability, corrosion, degradation, hemocompatibility, biocompatibility, visibility, and MRI compatibility of metallic stents are described as follows, which are in conjunction with ISO 25539-2:

- *Radial stiffness and strength tests*: Radial stiffness is external radial force when the diameter of stent uniformly changes, whereas radial strength is defined as the pressure that the stent was permanently deformed [85]. Two methods are often used to measure the radial force—to apply uniform pressure in the radial direction and to measure the force and displacement during compression by placing the stent between two parallel plates [86]. When the uniform radial pressure is applied to expanded stent at a compression rate of 0.1 mm/s, the strength at 10% compression of the nominal diameter is defined as radial strength [49].
- *Recoil test*: Due to the elastic phase of the material, excessive recoil will lead to disabled fixation of stent and operation failure [87]. For prevention, it is necessary to carry out the recoil testing and measure the percentage of acute stent recoil, which is calculated as $(L_1–L_2)/ L_1 \times 100\%$. L_1 refers to the average diameter of the stent expanded to nominal pressure, and L_2 refers to the average diameter of the stent when the balloon is withdrawn [85].
- *Fatigue and durability tests*: ISO 25539-2 points out that the minimal design life of permanent stents is 10 years, meaning that a minimum of 380 million cycles is

generally required under pulsating load. If the intended design life is less than 10 years, fatigue testing with shorter duration may be appropriate and shall be justified. In order to shorten test time, accelerated fatigue test is commonly used. As for fatigue testing, stent is placed in the mock vessel, and the lumen is filled with PBS. By simulating the diastole and systole of mock vessels, the pulsatile fatigue testing machine could apply circumferential force to stents. The pressure variations of the mock vessel need to be controlled within the range of 100 mmHg in order to simulate pulsation conditions. A referable accelerated circumferential pulsatile fatigue is recommended to carry out at 65 Hz in 37 ± 2 °C [88]. However, this accelerated test condition cannot be applied to biodegradable stents.

- *Electrochemical corrosion test*: The purpose of this assessment is to evaluate the susceptibility of metallic stents to corrosion. According to ASTM F2129, in phosphate buffer saline (PBS) (37 C, pH = 7.4), three-electrode cell system are recommended for use and the stent as working electrode. Before testing, the stent is expanded to its nominal size, and the area of stent exposed to electrolyte solution is recorded to calculate the corrosion current density [45, 77].
- *Immersion test*: Immersion tests, both dynamic and static, are commonly used in degradation assessment in vitro [89–92]. According to ASTM G31-72, the experimental specimens are soaked in simulated body fluid at 37 °C with an exposure ratio of 20 ml/cm^2. In a dynamic immersion test, the wall shear stress (0.68 Pa), temperature (36.9–37.1 °C), pH value (7.35–7.45), and dissolved oxygen (2.8–3.2 mg/L) were recommended to simulating the common values in the human coronary artery [75]. Finally, the corrosion products need to be removed to calculate the degradation rate.
- *Hemolysis test*: The hemolysis test is performed according to ASTM F756. Firstly, samples are incubated in PBS at 37 °C, followed by adding diluted blood to the sample at least for 3 h at 37 °C. After centrifugation, the absorbance (OD value) of the supernatant is measured by a microplate reader at the wavelength of 540 nm. The hemolysis ratio (HR) is calculated by the following formula: $HR = [(OD_t - OD_n)/(OD_p - OD_n)] \times 100\%$. The OD_t, OD_n, and OD_p are the OD values of tested group, negative, and positive controls, respectively [93]. If HR is lower than 5%, it is indicated the stent is non-hemolytic.
- *Thrombosis test*: Thrombosis test generally includes thrombus formation, plasma coagulation, platelet activation, and aggregation studies [94]. The samples are incubated in whole blood containing anticoagulant. $CaCl_2$ at a concentration of 0.1M is added to activate clotting [94]. After thrombosis, the quantity and morphology of thrombosis are measured and compared with the control group for evaluation. For plasma coagulation test, similar to thrombosis, the stents are incubated in the corresponding solution for different time duration, and the optical density (OD) value of free hemoglobin is measured at 540 nm. Generally, the initial clotting time and the whole clotting time are respectively defined as the time for OD < 0.1 and OD > 0.01 [44]. Platelet adhesion test is performed by exposing the specimen to platelet-rich plasma. After fixation and dehydration of platelets, the quantity of platelets and their activation level are evaluated [95].

- *Cytocompatibility test*: Cytocompatibility is the most basic and common in vitro biocompatibility test method which is divided into indirect contact cell assay and direct contact cell assay. The indirect contact cell assay involves the culture of cells in sample extract prepared by soaking the material in culture medium based on ISO 10993–5. As for the direct contact cell assay, cells are directly seeded on the surface of specimen. Common cell types include smooth muscle cells, endothelial cells, and fibroblasts. According to ISO10993–5, cytotoxicity can be evaluated by the results of MTT or CCK-8 assay [96, 97].
- *Visibility test*: According to ISO 25539-2, the stent system is placed in the phantom tissue model, and the imaging mode is as close to clinical conditions as possible. The degree of visibility may be assessed by locating the exact ends, orientation of critical points, and/or parts of the stent system. Alternatively, the degree of visibility may be compared to a specified material or device with known visibility. Additional information about visibility testing can be found in ASTM F640.

3.2.2 Inorganic Interventional Devices

Inorganic materials, such as low temperature isotropic pyrolytic carbon (LTIC), graphite, diamond-like carbon (DLC), Al_2O_3, SiC, TiO_2, and TiN, always exhibit excellent wear resistance and low density (Table 3.5). In addition, the biological stability of inorganic materials is also higher than that of common medical metal materials. S. Forti et al. [98] compared the protein adsorption and platelet adhesion on the surface of pyrolytic carbon and Ti–6Al–4V, respectively. In Fig. 3.6, although pyrolytic carbon absorbed more fibrinogen and von Willebrand factor (vWF), the total adsorbed plasma proteins, platelet adhesion, and activation were significantly less than that of Ti–6Al–4V. TiO_2 was regarded as the most biocompatible coating on TiNi alloys, which acted as a barrier preventing the dissolution of Ni^{3+} (Fig. 3.7), and the hemolysis rate decreased from 0.52% to 0.11% [99]. The combination of pyrolytic carbon and TiO_2 coating with anti-platelet adhesion property could prolong clotting time and only led to 1/8 thrombus comparing with that of LTIC, which was possibly due to the existence of $Ti^{2+}$and Ti^{3+} [100].

It is well-known that the aortic valve regulates blood supply to all major vessels of the body. Once the aortic valve tightens or narrows, the blood would be unable to flow easily, which is called aortic stenosis. In that case, the aortic valve allows blood to flow back, leading to heart failure. The most common heart valve surgery is aortic valve replacement due to aortic stenosis or aortic insufficiency, where the mechanical valves and biological valves are the dominant substitutes. However, mechanical valves made of metal and/or inorganic materials are far more durable than biological valves made of biological tissues. From 1969, pyrolytic carbon was first made into hollow spheres as part of a mechanical valve [101]. During the next three decades, pyrolytic carbon became the main biomaterials for almost new mechanical valves [102]. Bileaflet valves are predominantly used worldwide as cardiac valve

Table 3.5 Mechanical properties of inorganic materials [107–111]

Material	Density (g/cm^3)	Young' s modulus (GPa)	Hardness (DPH)	Flexural strength (MPa)	Fracture toughness ($MPa\cdot m^{1/2}$)	Strain-to-failure (%)	Wear resistance
LTIC	1.7–2.2	20–30	150–200	270–550	1.68	1.6–2.1	Excellent
Polycrystalline graphite	1.5–1.8	4–12	50–120	65–300	1.5	0.1–0.7	Poor
Glass carbon	1.4–1.6	24–28	150–200	70–205	N/A	2.0–5.0	Poor to excellent
Si-alloyed LTIC	1.7–2.2	230–270	230–370	350–530	0.9–1.1	1.5–2.0	Excellent
Al_2O_3	>3.90	380	>2000 (HV)	400	5	0.15	Excellent
SiC	3.0	410–422	30–32 (GPa)	N/A	1.48–1.58	N/A	Excellent
TiN	N/	550	31 (GPa)	N/A	N/A	N/A	Excellent

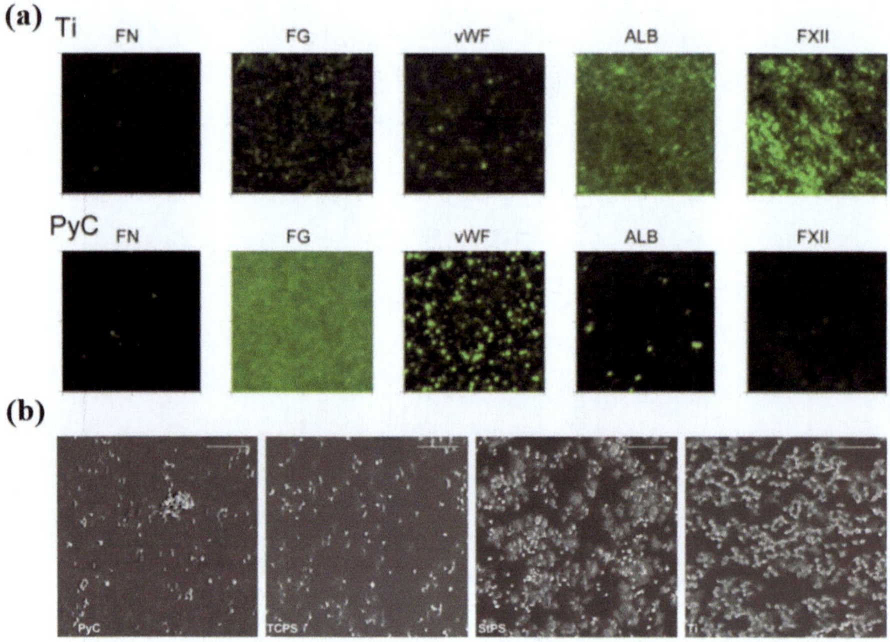

Fig. 3.6 (**a**) Protein adhesion on Ti–6Al–4V (Ti) and pyrolytic carbon (PyC) [98]. (fibronectin (FN), fibrinogen (FG), von Willebrand factor (vWF), albumin (ALB), and Hageman factor (FXII)). (**b**) Platelets adherent on PyC, tissue culture polystyrene (TCPS), polystyrene (StPS), and Ti [98]. (bar scale 20 μm)

substitutes, which significantly improve hemodynamics because its opening and closing principle is similar to that of natural valves compared to other types of mechanical valves. Some typical Bileaflet mechanical valves made of inorganic materials are shown in Fig. 3.8. The St. Jude Medical prosthesis (St. Jude Medical, Inc, St Paul, MN) had a durable record up to 30 years with housing and leaflets made of pyrolytic carbon [103]. The whole carbon-coated Sorin Bicarbon valve (Sorin Biomedica, Saluggia, Italy) showed no cases of intrinsic structural valve failure and durable effectiveness up to 17 years after aortic valve replacement [104]. The On-X aortic valve (CryoLife, Inc, Kennesaw, GA) made of pure pyrolytic (no-silicone) carbon reduced anticoagulation levels compared with other mechanical heart valves [105, 106].

The ideal artificial heart valves are expected that have the same functions as the natural one, including long-term durability, high reliability, safety, and biocompatibility. Since the heart valve prostheses need to provide on-off service in a dynamic high-pressure environment within blood for long periods of time, the hydrodynamic performance, durability assessment, as well as compatibility evaluation (including hemocompatibility, cytotoxicity, and sensitization, as previously mentioned in Sect. 3.2.1) are vital for inorganic heart valve. In order to simulate the hemodynamic conditions, the operational environment of heart valve substitute for adults shown in

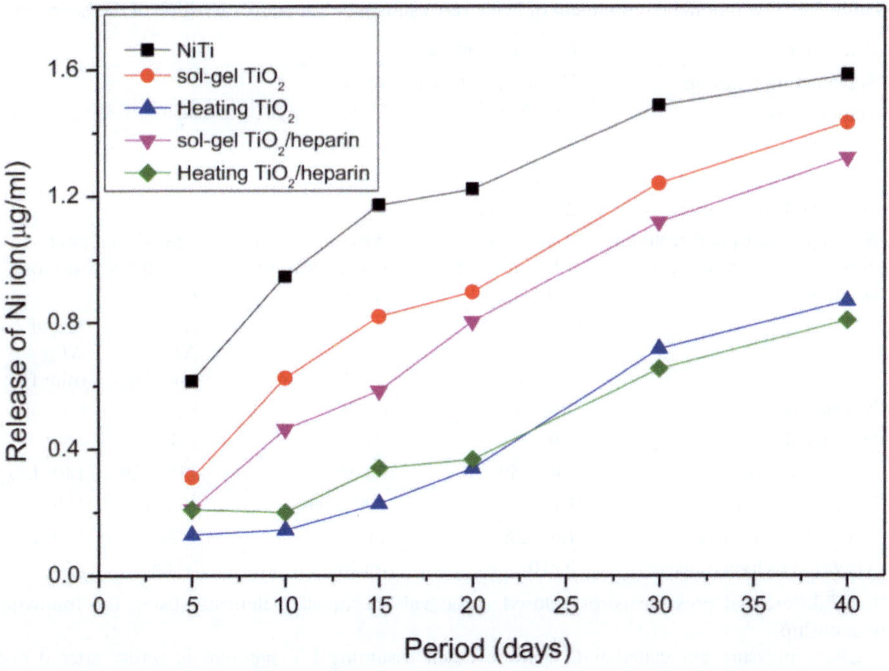

Fig. 3.7 The release of Ni iron from different TiO_2 coating and NiTi substrate [99]

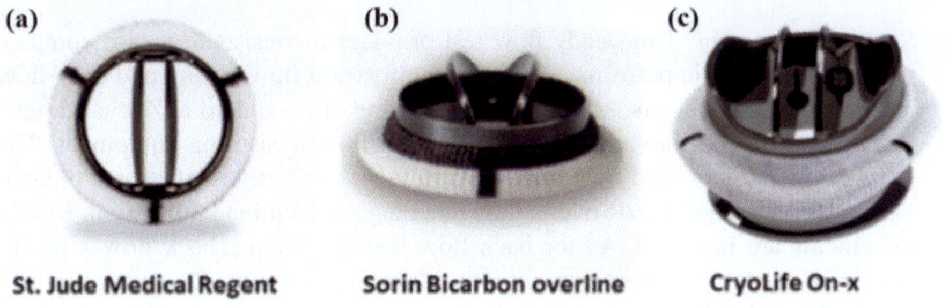

Fig. 3.8 Some typical Bileaflet mechanical valves [103]: (**a**) St. Jude Medical prosthesis with pyrolytic carbon (St. Jude Medical, Inc, St Paul, MN); (**b**) Sorin Bicarbon valve with whole carbon film (Sorin Biomedica, Saluggia, Italy); (**c**) On-X valve made of pyrolytic carbon coated graphite

Tables 3.6 and 3.7 needs to be considered and the recommended pressure values in Tables 3.8 and 3.9. Isotonic saline, blood, or blood equivalent fluids are adopted to develop the physical properties (e.g., specific gravity, viscosity at operating temperature) of the test conditions that is similar to the internal environment. ISO 5840 recommends several aspects of the evaluation of mechanical valve prostheses:

Table 3.6 Operational environment of heart valve substitute for adult's left side of the heart [112]

Parameter	General condition			
Surrounding medium	Human heart/human blood			
Temperature	34–42 °C			
Heart rate	30–200 bpm			
Cardiac output	3 l/min–15 l/min			
Forward flow volume	25–100 ml			
Blood pressures and resultant pressure loads by patient condition	Arterial peak systolic pressure (mm Hg)	Arterial end diastolic pressure (mm Hg)	Peak differential pressure across closed valve[a]	
			Aortic ΔP_A (mm Hg)	Mitral ΔP_M (mm Hg)
Normotensive	90–140	60–920	80–115	90–140
Hypotensive	<90	<60	<80	<90
Mild hypertensive	140–159	90–99	115–129	140–159
Moderate hypertensive	160–179	100–109	130–144	160–179
Severe hypertensive	180–209	110–119	145–164	180–209
Very severe hypertensive	≥210	≥120	≥165	≥210

[a]Peak differential pressure across closed aortic valve estimated clinically using the following relationship:
– $\Delta P_A \approx$ pressure associated with dicrotic notch assuming LV pressure is zero≈ arterial end diastolic pressure + 1/2 (arterial peak systolic pressure—arterial end diastolic pressure)
– Peak differential pressure across closed mitral valve estimated to be equivalent to arterial peak systolic pressure

- *Steady flow testing*: The steady flow test provides a consistent way to compare valve hydrodynamic performance, including forward flow testing and back flow leakage testing [112]. As for forward flow testing, the standard nozzle is adopted to characterize the forward flow pressure and flow measuring equipment. The steady flow rate, forward flow pressure differences, effective orifice area of heart valves, and standard nozzle over a flow rate range of 5 l/min to 30 l/min in 5 l/min increments are recorded. As for back flow leakage testing, back flow standard nozzle is used to measure the amount of leakage through the test valve and standard nozzles for five isometric back pressure measurements. Every sample need to be tested under each stage of back pressure that is appropriate for the intended device application in accordance with Tables 3.8 and 3.9.
- *Implant durability assessment*: In order to assess the minimum in vitro durability life and determine the expected durability-related failure mode of the heart valve replacement, as well as to provide information on the potential failure consequences, it is essential to conduct durability testing. Because the predictive effect of single durability test is limited, the complementary combination of accelerated wear testing (AWT), dynamic failure mode testing (DFM), and real-time wear testing (RWT) assessment provides a more comprehensive method to achieve the expected in vitro durability performance under the recommended pressure [112]. AWT of valves are carried out for 200 million cycles (biological valves)

Table 3.7 Heart valve substitute operational environment for adult's right side of the heart [112]

Parameter	General condition			
Surrounding medium	Human heart/human blood			
Temperature	34–42 °C			
Heart rate	30–200 bpm			
Cardiac output	3 l/min–15 l/min			
Forward flow volume	25–100 ml			
Blood pressures and resultant pressure loads by patient condition	Right ventricle peak systolic pressure (mm Hg)	Pulmonary artery end diastolic pressure (mm Hg)	Peak differential pressure across closed valve[a]	
			Pulmonary ΔP_P (mm Hg)	Tricuspid ΔP_T (mm Hg)
Normotensive	18–35	8–15	13–28	18–35
Hypotensive	<18	<8	<13	<18
Mild hypertensive	35–49	15–19	28–34	35–49
Moderate hypertensive	50–59	20–24	35–42	50–59
Severe hypertensive	60–84	25–34	43–59	60–84
Very severe hypertensive	≥85	≥35	≥60	≥85

[a]Peak differential pressure across closed pulmonic valve estimated clinically using the following relationship:
– ΔP_P approximately pressure associated with dicrotic notch assuming RV pressure is zero approximately pulmonary artery end diastolic pressure + 1/2 (right ventricle peak systolic pressure—pulmonary artery end diastolic pressure)
– Peak differential pressure across closed tricuspid valve estimated to be equivalent to right ventricle peak systolic pressure

Table 3.8 Recommended testing pressure values for left side of adult heart in vitro [112]

	Arterial peak systolic pressure (mm Hg)	Arterial end diastolic pressure (mm Hg)	Peak differential pressure across closed valve[a]	
			Aortic ΔP_A (mm Hg)	Mitral ΔP_M (mm Hg)
Normotensive	120	80	100	120
Hypotensive	60	40	50	60
Hypertensive				
Mild	150	95	125	150
Moderate	170	105	140	170
Severe	195	115	155	195
Very severe	210	120	165	210

or 400 million cycles (mechanical valves). Test frequency is in the range of 450–800 bpm, which varies according to valve response. Pressure transducers for the measurement of the transvalvular pressure difference have a natural frequency of at least 1000 Hz and a minimum measurement accuracy of ±0.65 kPa (±5 mmHg). Durability of heart valve replacement is characterized by the measurement of pressure difference/regurgitant volume and visual inspection under 5× magnification. DFM testing is designed to identify potential failure

Table 3.9 Recommended testing pressure values for right side of adult heart in vitro [112]

	Right ventricle peak systolic pressure (mm Hg)	Pulmonary artery end diastolic pressure (mm Hg)	Peak differential pressure across closed valve[a]	
			Pulmonary ΔPp (mm Hg)	Tricuspid ΔP_T (mm Hg)
Normotensive	25	10	20	25
Hypotensive	15	5	10	15
Hypertensive				
Mild	45	17	30	45
Moderate	55	22	40	55
Severe	75	30	50	75
Very severe	85	35	60	85

modes of heart valves under load conditions more severe than AWT. DFM testing is intended to provide a more thorough evaluation of implant durability after AWT. Therefore, samples of DFM can be AWT-survival or non-AWT circulating valves with appropriate pre-circulation. The closed differential pressure during the test is gradually increased, and the number of cycles is at least 50 million times for each pressure values. At the same time, valves are inspected and functionally evaluated by opening and closing behaviors, hydrodynamic performance inspections, and visual and SEM observation. If failure does not occur after 50 million cycles at maximum pressure, the test can be terminated at 200 million cycles. RWT could better simulate the biological conditions in vivo, whose results could be used to analyze the failure modes related to frequency and supplement the results of AWT [113, 114]. In addition, the load conditions need to be in line with the physiological waveform as far as possible. RWT is performed for at least 50 million cycles with test frequency below 200 bpm.

- *Fatigue assessment*: ISO 5840 required valve substitutes to maintain structural and functional integrity for at least 400 million cycles. First, it is necessary to determine the boundary conditions for the device to be used in vivo. Then, the fatigue strength of the materials is measured. Ultimately, fatigue test of the device is conducted to determine the fatigue safety factor and failure probability and analyze the stress/strain status of the structural components under the simulated physical conditions.
- *Corrosion assessment*: The potential for fretting (wear) and fretting corrosion post durability testing. As for mechanical heart valve made of metals or alloys, corrosion assessments are non-negligible. It would be desirable to choose the finished valve as test samples as far as possible because processing technology and structure design will cause unpredictable impact. Typically, a combination of qualitative observation, quantitative measurement, and statistical analysis is essential to provide an overall assessment of corrosion. The relevant methods have been described in Sect. 3.2.1.

3.2.3 Organic Interventional Devices

Organic interventional devices can be classified into nonbiodegradable and biodegradable devices depending on whether it can be absorbed by the body. Nonbiodegradable polymers, like PTFE, polyester, polyurethane (PU), polypropylene (PP), and polyvinyl alcohol (PVA), are often used to make artificial blood vessels and heart valves due to good biostability, mechanical properties, resistance to hydrolysis, and oxidation [115]. Because biodegradable polymer materials can be metabolized by the human body, they are often used in the field of vascular stents, occluders, tissue engineering stents, and drug delivery, such as polylactic acid (PLA), polyglycolic acid (PGA), polycaprolactone (PCL), polyhydroxyalkanoate (PHA), polydioxanone (PDS), and polydioxanone (PDO) [116].

3.2.3.1 Physicochemical Properties of Nonbiodegradable Organic Devices

As a permanent substitute for blood vessels in vivo, non-degradable artificial blood vessels are expected to maintain lasting strength, reliable degradation resistance, corrosion resistance, and good mechanical fatigue resistance, which is the key to the ability of artificial blood vessels to withstand long-term blood flow impact without deformation and rupture after transplantation. PTFE is a fluorocarbon polymer, and the expanded PTFE (ePTFE) with sufficient porosity is available for cellular ingrowth. The tensile strength and stiffness of ePTFE are 0.5 GPa and 14.0 MPa. A notable feature of ePTFE is excellent chemical stability due to little interaction between fluorocarbons and other molecules [117]. The ePTFE graft (diameter: 7–9 mm) can service effectively for 5 years [118]. Polyester with higher strength (tensile strength of 170–180 MPa, tensile modulus of 14 GPa) is primarily used to make woven graft that functioned well for a mean of 41 months [119]. Because the surfaces of ePTFE and polyester are electronegative, this minimizes their interaction with blood components. There was no significant difference in effectiveness between ePTFE and polyester according to meta-analysis and systematic review [120]. However, the clinical efficacy of ePTEF with a small-diameter (<6 mm) and polyester vascular prosthesis is unsatisfactory, since much lower blood flow is more likely to promote thrombosis and intimal hyperplasia for their surface condition [121, 122]. Therefore, on the basis of good biological stability, surface modifications of ePTFE and polyester are generally carried out, such as plasma treatments, heparinization, hydrophilic surface modification, and load angiogenic factors [123–125]. PUs are a kind of copolymers which consist of crystalline hard segments and amorphous soft segments and possesses urethane groups (–NHCOO–) in the chemical structure [126]. By adjusting the distribution and length of soft and hard segments, the tensile strength of PUs can be adjusted in a wide range from 29 to 257 MPa [127, 128]. Compared with PTFE, PUs have a biomimetic microphase separation structure of the soft and hard segments, which improves the state of serum protein adsorption and reduces platelet adhesion [129]. PUs could be made for small blood vessels which are generally attached to saphenous vein [130, 131]. The tensile strength of them is generally higher than 5 MPa, the compliance is about 4–7%, and the burst pressure is from 1710 to 2200 mmHg. Nevertheless, due to the hydrolytic soft segment of polyols, the first-generation PU grafts disintegrated in vivo, and thus the relative clinical trials were halted. In order to improve the biostability of PUs, siloxanes and carbonates have been selected as a soft segment component. When PU (polycarbonate soft segment) and ePTFE grafts were implanted in the abdominal aorta of rats, PU (polycarbonate soft segment) grafts achieved complete endothelialization within 4 weeks, while ePTFE grafts needed 6 months for endothelialization with increased intimal hyperplasia [132]. Further, siloxane-based PU and ePTFE grafts were implanted in sheep arteries. It came out that siloxane-based PU grafts could maintain unobstructed for 24 months, but half of ePTFE grafts occluded after implantation for 6 months [133]. Although the clinical results of PUs are superior to that of PTFE in animal experiments, further improvements in stability and biocompatibility of PUs grafts are needed to confirm clinical viability.

Nonbiodegradable organic heart valves combine the durability of mechanical valves with the flexibility of biological valves. They gradually become candidates for artificial heart valves especially with the development of transcatheter aortic valve replacement (TAVR) [115, 134]. However, most of non-degradable materials for artificial valves without modification brought unsatisfactory results. The failure of valve was mainly due to stiffening and calcification of ePTFE leaflets [135, 136]. Patients only survived for several months after the implantation of the first-generation PU valve owing to premature degradation [137]. Considering the instability of the first-generation PU, more stable polyurethanes were developed, like poly(carbonate-urea)urethane (PCU), polycarbonate-based materials that contain polyhedral oligomeric silsesquioxane nanoparticles (POSS), or polysiloxane soft segments (PSUs) [138]. The durability of PCU aortic trileaflet has reached 600 million cycles, and variable degree of calcification, degeneration, and a small amount of thrombus were observed in calves [139]. According to the result of in vitro testing, POSS-PCU valves showed comparable hydrodynamic performance to other clinically used tissue valves for TAVR [140].

3.2.3.2 Physicochemical Properties of Biodegradable Organic Devices

Degradable organic devices could assist temporary tissue recovery. After accomplishing their missions, they could gradually degrade, and the degradation products could be excreted or metabolized. According to their sources, polymers for the preparation of biodegradable organic devices can be divided into natural biodegradable polymers and synthetic biodegradable polymers. Natural biodegradable polymers derived from plants or the human body have excellent biomechanical and degradable properties, such as chitosan, collagen (and gelatin), carboxymethyl chitin, fibrin, and alginic acid [141]. However, since the quality of natural materials

Table 3.10 Physicochemical properties of biodegradable polymer [48, 149]

Material	Tg (°C)	Tm (°C)	Young's modulus (GPa)	UTS (MPa)	TE (%)	Degradation period (months)
PLA	60	180–190	2–4	65	2–6	18–30
PDLLA	55–60	Amorphous	1–3.5	40	1–2	3–4
PLLA	60–65	175	2–4	60–70	2–6	>24
PGA	35–40	255–230	6–7	90–110	1–2	4–6
PDLGA (50/50)	45	Amorphous	1–4.3	45	1–4	1–2
PLGA (82/12)	50	135–145	3.3–3.5	65	2–6	12–18
PCL	154	55–60	0.34–0.36	23	>4000	24–36
PDO	–10–0	N/A	1.5	N/A	N/A	6–12
PLA/PCL (70/30)	20	100–125	0.02–0.04	2–4.5	>100	12–24

is difficult to control, they need to be modified properly before biomedical application. Synthesized biodegradable polymers could be manufactured reproductively on a large scale by controlling the processing conditions. Easy to be modified, they are widely used in biomedical devices. In fact, some of them have been applied to coronary stents, occluders, artificial blood vessels, heart valves [142–146], including polylactic acid (PLA), polyglycolic acid (PGA), polycaprolactone (PCL), and polyethylene glycol (PEG). Table 3.10 summarizes the physicochemical properties of biodegradable polymer. In comparison with WE43 alloy, the degradable polymeric devices generally have higher compliance because the modulus of the polymers is more similar to soft tissue. The mechanical properties and degradation rate of the biodegradable polymer can be adjusted by controlling the degree of crystallinity. Taken polylactic acid as an example, poly-L-lactic acid (PLLA) is a semicrystalline polymer with higher strength and longer degradation period than amorphous poly-DL-lactic acid (PDLLA) [147, 148]. Their degraded monomer is lactic acid, which is a kind of sugar metabolite. One reason for the good biocompatibility of biodegradable polymers is that they could break down into the body's metabolites, which are further broken down into water and carbon dioxide and excreted from the body. For instance, the monomer glycolic acid of PGA and the monomer lactic acid of PLA are both products of respiration.

Table 3.11 sums up the biodegradable polymeric stents in preclinical and clinical testing. At present, most of the degradable polymer scaffolds which are in clinical research or have received CE approval generally adopt PLLA as the backbone. Their complete absorption time ranges from 1 to 4 years. As the most representative biodegradable stents, Absorb BVS (Abbott Vascular, Santa Clara, CA, USA) had obtained CE mark and FDA approval, which were withdrawn worldwide in 2017. Early on, Absorb BVS has shown lower incidence of target lesion failure and stent thrombosis compared to DES [150, 151]. However, The ABSORB III trial (5 years

Table 3.11 Biodegradable polymeric stents in preclinical and clinical testing [163–168]

Manufacturer	Device	Backbone	Strut thickness	Absorption time	Availability
Abbott (US)	Absorb	PLLA	157 μm	>3 year	Delisted
	Falcon	PLLA	99 μm	2.25 year	Clinical study
Elixir (US)	Desolve	PLLA	150 μm	2 year	CE mark approved
	DesolveCx Plus	PLLA	120 μm	2 year	CE mark approved
Reva (US)	Fantom	Tyrosine polycarbonate	125 μm	4 year	CE mark approved
	Fantom Encore	Tyrosine polycarbonate	95/105/ 125 μm	4 year	CE mark approved
Arterial Remodeling Technologies (France)	ART-BRS	PDLLA	170 μm	6 months	CE mark approved
Meril Life (India)	MeRes 100	PLLA	100 μm	1 year	CE mark approved
Amaranth (US)	Magnitude	PLLA	98 μm	2-3 year	Clinical study
Arterius (UK)	Arteriosorb	PLLA	95 μm	2 year	Clinical study
Manli Cardiology (Singapore)	Mirage	PLLA	125–150 μm	14 months	Clinical study
Lepu Medical Technology (Beijing)	Neovas	PLLA	170 μm	3 year	CFDA Mark approved
HuaAn Biotechnology	Xinsorb	PLLA	160 μm	5 year	CFDA Mark approved
MicroPort	Firesorb	PLLA	100/125 μm	5 year	Clinical study

of follow-up) showed that the thrombosis in BVS scaffold was significantly higher than that in DES (2.3% vs 0.7%, $P = 0.01$), indicating that the risk of very late thrombosis was increased [152]. The strut thickness of Absorb BVS is 150~160 μm, almost 3 times that of DES. It is noted that the degradable stents need to provide adequate support before the blood vessel recover. Due to the low strength of polymers, the most effective way to provide sufficient support is to increase the wall thickness. However, it will increase the risk of thrombosis and delayed endothelialization due to greater blood flow disturbance compared with thin-walled stents [153]. During the degradation of BVS, the greater thickness may also be the main reasons for the uneven degradation and partial structural prolapse, which further induce very late stent thrombosis. Considering the disadvantage of thick strut, the development of the second-generation BRS is mainly focused on reducing the strut thickness under the premise of maintaining sufficient supporting force. Firesorb (MicroPort, Shanghai, China), with thinner strut (100~125 μm), whose target lesion failure rate was 0% with no stent thrombosis according to 1-year clinical follow-up [154]. FANTOM (REVA Medical, Inc., San Diego, CA, USA) is a radiopaque stent made of tyrosine polycarbonate with a thickness of 125 μm. Due to the addition of iodine to the polymer backbone, the stent could be visualized throughout the degradation process, which is the main feature that could distinguish this stent from others. The reabsorption time of FANTOM was similar to that of most PLLA scaffolds, but the weight loss in the first year was more than 80% [155]. The MeRes100 (Meril Life Sciences Pvt. Ltd., India) is also made of PLLA with the strut thickness of 100 μm. It has a novel hybrid design (closed cell at the edges and open cell within the stent). The endothelial coverage of the MeRes100 at 6 months was 99.3% by optical coherence tomography (OCT) in First-in-Man clinical trials, and no stent thrombosis was reported according to 2-year follow-up [155, 156].

Atrial septal defect (ASD) is one of the most common congenital heart defect and accounts for about 30% of the congenital heart disease population. Double umbrella occluders made of superelastic NiTi wires have become the first choice for ADS treatment, but the dissolution of Ni ions under physiological conditions can lead to serious complications, such as allergy and perforation [157–159]. The ideal occluder might be a temporary device that can be gradually absorbed in the body and replaced by natural tissue, minimizing the possibility of long-term complications associated with the metallic occluder. The complete degradation period of self-expanding occluder is generally about 24 weeks in dog, which is made of polydioxanone (PDO), poly-L-lactic acid (PLA), and polyglycolic acid (PGA) [160]. In spite of no excrescence and thrombus formation on the occluders, severe inflammation around them occurred after implantation for 8 weeks. Then it decreased after 12 weeks' implantation, but did not completely disappear until 24 weeks [161]. Although PLLA is without shape memory effect, the PLLA occluder with self-expandable design can also be used in interventional treatment (Fig. 3.9), which showed no adverse events during the 2-year follow-up period of PLLA-made occluder in a 4-year-old boy [146]. With the development of 4D printing, PLA and nano Fe_3O_4 nanoparticles composite were made into the magnetically corresponding occluder by electrospinning [162]. This composite with shape memory effect has a maximum strength of 60 MPa and the shape recovery rate of approximately 95% [144]. However, there is much room to improve PLA-Fe_3O_4 occluder's biocompatibility, which was covered with a large number of thrombus after subcutaneous implantation of mice for 12 weeks [162].

3.2.3.3 Test Methods and Evaluation Criteria for Artificial Vascular Prosthesis

According to ISO 25539-1, the design properties of vascular prosthesis include at least the following factors:

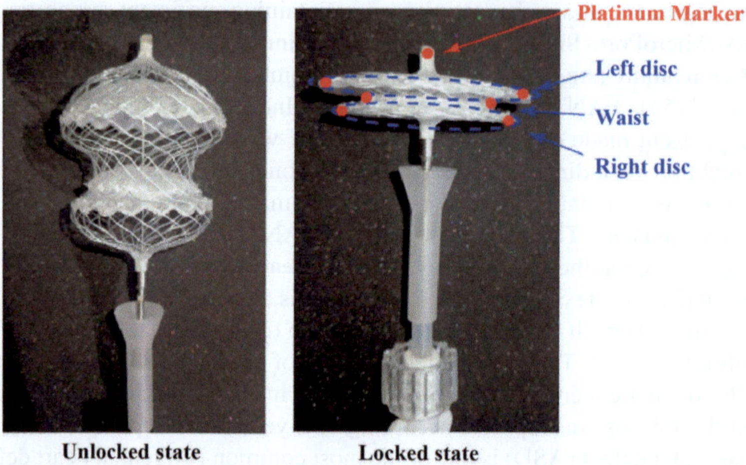

Fig. 3.9 Fully biodegradable PLLA occluder [146]

1. Enough strength to maintain structural integrity (e.g., burst strength, longitudinal tensile strength, fatigue, and durability) and effective fixation to the implant site.
2. Ability to achieve the expected function and isolate lesions. For example, it is essential to provide a barrier between the implant and the aneurysm in the blood vessel to prevent blood flow through the implant wall.
3. Proper permeability: porosity, water permeability, integral water leakage, and water entry pressure.
4. Ability to maintain unobstructed blood flow in the lumen.
5. Ability to safely use magnetic resonance imaging (MRI) in patients with intravascular prostheses.

Since biocompatibility and NMR compatibility have been described previously, in this section the testing methods for mechanical properties and physical properties are focused, including burst pressure, longitudinal tensile strength, suture retention, porosity, water osmotic pressure, and amount testing.

- *Burst pressure test*: The purpose of this test is to determine the pressurized burst strength of vascular prosthesis. The artificial vascular prosthesis needs to be attached to the pressurization apparatus and be filled with simulated fluids like PBS in lumen, continuingly applying a steady rise in pressure at a controlled rate 10 kPa/s and 70 kPa/s until the vascular prosthesis ruptures [169]. The rate of pressure rise and the pressure when prosthesis bursts are recorded by the pressure-measuring device.
- *Longitudinal tensile strength test*: The vascular prosthesis can be clamped on a mechanical tester with suitable furniture along the longitudinal direction. The tensile force at fracture is measured. The distance between fixtures is 50–150 mm, and the stretching rate is 50–200 mm/min [169].

- *Suture retention test*: The purpose of this test is to determine the force necessary to pull a suture from the prosthesis or cause the wall of the prosthesis to fail while pulling a suture inserted through the wall. The anastomotic suture technique determines the short-term and long-term patency rate after implantation to a certain extent. The wetted vascular prosthesis is first cut open and then sutured. The force required to pull the suture through the prosthesis, or cause the wall of the prosthesis to fail, and the suture size are recorded [170].
- *Porosity test*: Microscope is used to measure the surface node spacing and the number and diameter of voids per square millimeter on the surface of the vascular prosthesis after stretching.
- *Water osmotic pressure and amount test*: One end of the artificial blood vessel is stuffed, and the other end is filled with water and is pumped into a pipe that can transmit pressure. The pressure gradually increases from zero. When bubbles appear on the surface, the corresponding pressure is recorded as water osmotic pressure. When the pressure is raised to 16.0 kPa, the pressure is maintained for 60 s, and the water leakage is measured within 60 s. The inner surface area of the artificial vascular prosthesis that water flows through is measured, and then the water osmotic amount in the unit of mL/(min cm^2) is obtained [170].

3.3 Physicochemical Properties of the Ideal Cardiac Materials as Tissue Engineering Scaffold

Tissue engineering scaffolds can not only bind or transport bioactive molecules but also promote the repair and regeneration of tissue and cells [171]. As one of the most important organs of the human body, the heart has a complex structure, which is composed of myocardial tissue, valve tissue, and vessel tissue. Myocardial tissue is consisted of various cells (contractile cardiomyocytes, cardiac fibroblasts, nerve cells, smooth muscle cells, endothelial cells, and connective tissue), and fibroblasts account for about 75% of the volume in a network of collagen [172]. The main cell types of valve tissue are valve mesenchymal cells (VIC) in the leaflets and valve endothelium cells on the blood-contacting surfaces. Their extracellular matrix (ECM) is composed of collagen types I and III [173]. Endothelial cells, smooth muscle cells, and fibroblasts are the dominating cells in the intima, media, and adventitia of small-diameter arteries, respectively. The ECM of endothelial cells is composed of fibronectin and mature collagen type IV and laminin [174]. Considering various physiological environment of different cells and tissues, the raw materials for tissue engineering scaffolds are selected according to the implantation site. According to the source, tissue engineering scaffolds can be divided into natural scaffolds and synthetic polymer scaffolds. Natural polymer (collagen, algin, hyaluronic acid (HA), elastin, fibrin) and decellularized extracellular matrix (dECM) both belong to natural scaffolds [175]. Decellularized porcine pulmonary valve and bovine pericardium materials are often used to make biological valves

[176, 177]. Synthetic tissue engineering scaffolds have been investigated in myocardial patch, heart valve, artificial blood vessel and other fields because of their stable quality and adjustable performance. The synthetic biodegradable polymer (PLLA, PCL, PGA) and synthetic biodegradable hydrogel are commonly adopted for the fabrication of synthetic tissue engineering scaffolds [178–180].

3.3.1 Cardiac Tissue Engineering Scaffold

Heart failure occurs when cardiomyocytes are replaced by scar tissue after ischemia and the heart is unable to beat properly [181, 182]. Tissue engineering method aims to reestablish the structure and function of injured myocardium, where one of the major challenges is to select a suitable tissue engineering scaffold [183]. dECM can be obtained from a wide range of decellularized native organs of different species, promoting cells homing and lowering immunogenicity. dECM has developed into various forms of cardiac tissue engineering scaffold, which are either solid scaffolds that maintain native vasculature structure or soluble materials that can form injectable hydrogels as heart patches [184]. For regeneration of the whole heart, dECM is preferred because it preserves the full complexed three-dimensional structure of the heart (Fig. 3.10a) [185]. When the dECM scaffold was refilled with neonatal rat cardiomyocytes, spontaneous pulsating excitatory transmission was observed, but disordered excitatory transmission occurred due to a random mix of cardiomyocytes, endothelial cells, and smooth muscle cells [186]. In order to evaluate the mechanical properties of dECM scaffold, sinusoidal oscillation load was carried out on the left ventricle of original porcine heart and the dECM scaffold. It was found that the dECM scaffold could maintain the elastic modulus of the original heart tissue (271.8 ± 55.6 KPa) after proper decellularization treatment (272.4 ± 26.4 KPa) [187]. The human-derived dECM scaffold showed excellent cytocompatibility (Fig. 3.10b) and promoted stem cell differentiation [185]. The differentiated cardiomyocytes formed into new muscle bundles and exhibited mature calcium dynamics with electrical coupling [185]. Due to the limited source of human heart, it is necessary to find more engineering scaffolds to repair myocardiums. The heart patch is a device that provide a suitable cellular microenvironment and the most sufficient support for myocardial regeneration [188]. Based on good biocompatibility and mechanical matching of dECM, collagen, as the main extracellular matrix protein, has become an ideal natural biomaterial for cardiac engineering [189]. The cell-free collagen patch integrates with cardiomyocytes to reduce fibrosis [190]. The human bone marrow mesenchymal cells (HMSC-loaded) collagen patch was able to maintain more than 50% of the cell mass for 24 h [191]. Norahan MH et al. fabricated the heart patch with electroactive collagen and graphene oxide composite [192]. The electrical conductivity and tensile strength of the patch were, respectively, in a range of $1.5–29*10^{-4}$ S/m and 45–162 kPa, which regulated the expression of the cardiac genes compared with nonconductive patch [192]. Although cardiac patches have shown promising results, they require more invasive surgical

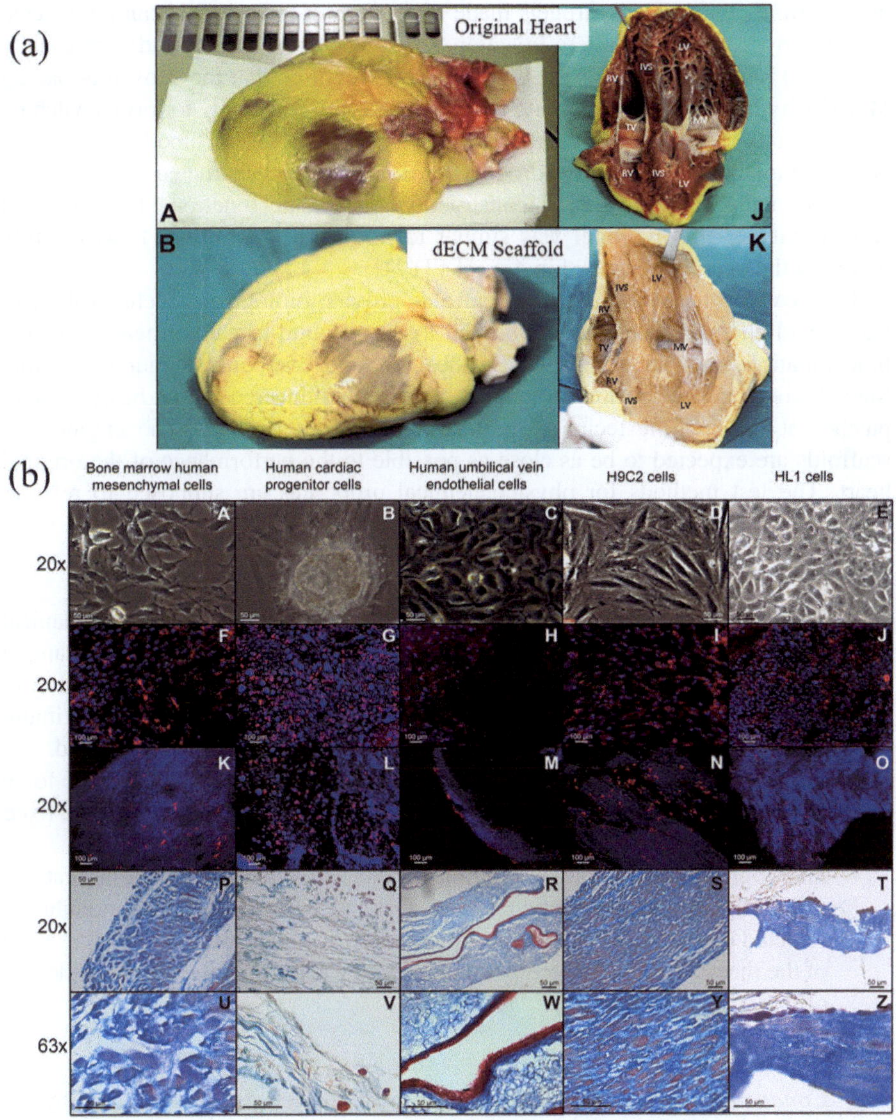

Fig. 3.10 (**a**) Comparison of integral structure between original human heart and dECM scaffold [185]. (**b**) Cytocompatibility of dECM scaffolds [185]. (The ability of dECM to drive cell commitment and differentiation were assessed using human cardiac-progenitor cells (hCPC) and bone marrow mesenchymal cells (HMSCs); the response of mature cells to dECM was evaluated using HUVECs and cardiac cell lines H9C2 and HL1)

interventions. Injectable hydrogels made of polymeric biomaterials can be injected into the myocardium by micro-injection. Cells and active compounds are usually embedded in the solution for gel transition. It is reported that the exosome-loading MA-HA hydrogel was injected into the porcine pericardial cavity to form a patch by catheter, and a few inflammatory cells were observed in the blood after 3 days [193]. A paintable conductive hydrogel (HPAE–Py/Geln) constructed by a variety of polymer chains enhanced the reconstruction of cardiac function. The electrical conductivity of the hydrogel was similar to normal heart muscle. It was tightly bound to the beating heart within 4 weeks [194].

Ideal cardiac tissue engineering scaffolds provide biological, mechanical, electronic, and chemical cues that are similar to those found in natural heart, which is biocompatible and biodegradable. Besides, it could overcome immune rejection. Since there is no international standard for tissue-engineered whole heart or heart patches at present, the technical requirements of myocardial tissue engineering scaffolds are expected to be as close as possible to the performance of the original heart. The test methods for physicochemical properties are supposed to refer to related published articles and mimic the human environment.

- *Mechanical properties*:
  - *Uniaxial compression test*: In order to evaluate the strength and biomechanical fitness of scaffolds, tensile strength and modulus tests are required. The sample needs to be wetted and then stretched at a rate of 10 mm/min according to ASTM D638M. According to the initial stress-strain slope and the maximum stress, the Young's modulus and the tensile strength could be calculated.
  - *Adhesion test*: In order to measure the fixation of the patch, an adhesive force test is needed. The patch overlaps the heart tissue, and the samples are fixed between the two film clamps at the tensile rate of 2 mm/min [195].
  - *Rheological test*: Rheological testing evaluates dynamic viscoelastic features of hydrogels, such as elastic (storage, G') and viscous (loss, G'') modulus of selected hydrogel. During the testing process, samples are placed on the plate of the rheometer, and the temperature was elevated from 20 to 60 °C at the rate of 0.5 °C/min [195].
- *Physical properties*:
  - *Conductivity test*: The conductivity of the hydrogels can be measured by the standard van der Pauw technique and four-point probe method [194]. Other electroactivity measurement can be tested by cyclic voltammetry (CV) measurements in the potential range of -1.0–1.0 V at the scan rate of 100 mV/s. The electrochemical impedance spectroscopy (EIS) measurements are carried out at the frequency range of 1 MHz~0.01 Hz [194].
  - *Swelling test*: The swelling behavior of hydrogels is evaluated by gravimetric method. After drying, the mass is denoted as W_d. Then the hydrogels are immersed in PBS at 37 °C until completely wetting, at which time the mass is denoted as W_s. The swelling ratio is calculated as follows: swelling ratio $= (W_s - W_d)/W_d * 100\%$ [195].

- *Biodegradation*: Before immersion, the weight of scaffold is recoded and denoted as W_i. Then the scaffold is immersed in simulated body fluid and is removed at various points. After lyophilizing for 12 h, the weight is recorded and denoted as W_d. The weight loss is calculated as follows: $(W_i - W_d)/W_i * 100\%$ [195].
- *Biocompatibility*:
  - *Cytocompatibility*: Cytocompatibility includes viability, attachment, spreading, and proliferation assessment of cells. The first step is to inoculate suitable cells on the scaffold and cultured for a period of time. CCK-8 assay, MTT method, and Live/Dead assay are commonly used to evaluate cell viability. Cell attachment and spreading can be observed by microscopy, such as confocal fluorescence microscopy and scanning electron microscopy [196]. To evaluate cell proliferation in scaffolds, total DNA content can be assayed and used to determine the number of cells in the constructs by a standard DNA digestion procedure [197].
  - *Biocompatibility in vivo*: Models of myocardial infarction (e.g., ligation of the left anterior descending branch) followed by tissue engineering stent implantation are required. For example, it is reported that the scaffolds were implanted continuously for 24–42 days in rat models. Afterward, the functional and histological analyses were carried out [198].

Currently, there are no available drugs to treat valvular heart disease. The mechanical valves and bioprosthetic valves dominate the market of valve replacement [199]. Tissue engineering is a promising approach to create valve prostheses that imitate valve tissue as much as possible and overcome fatal problems, including biomechanical mismatch of mechanical valve and calcification of biological valves. For instance, the decellularized human pulmonary valves seeded with autologous endothelial cells showed excellent hemodynamic performance during 10-year follow-up [200]. With the development of 3D printing technology, artificial heart valves can be designed in different scales. For example, biodegradable polymeric tissue engineering scaffolds exhibit improved mechanical properties and cell infiltration by synthesis of various materials as well as regulation of pore size and density [201]. It has been reported that the tensile strength and Young's modulus of an electrospun PEGdma-PLA scaffold as valve leaflets were, respectively, 2.1 MPa and 141 MPa [202]. Valvular interstitial cells (VICs) and valvular endothelial cells (VECs) were seeded on the scaffold. Besides, by mimicking the curved structure of collagen, PCL valve scaffolds exhibited good hydrodynamic performance. The mean transvalvular pressure drop was 2.45 ± 0.36 mmHg, and an effective orifice area was 3.3 ± 0.26 cm^2. SMCs could fill the pores of the scaffold after seeding for 2 weeks [203]. PGS-PCL reinforced HA hydrogel also showed improved mechanical properties and promoted cell proliferation [196]. There is no standard for tissue engineering valves, but the functions of valvular scaffold can be tested according to ISO 5840. The testing methods for mechanical properties, biocompatibility, and degradation properties have been mentioned in Sect. 3.2.2. Besides, the mean pressure gradient, effective orifice area, and durability could be assessed at physiological aortic conditions (flow rate, 5.0 L/min; frequency, 70 bpm; mean aortic pressure, 100 mmHg, 120–80 mmHg) [203]. Detailed testing methods for mechanical properties can be referred to the existing literature. Since tissue remodeling occurs primarily in vivo, evaluation of valvular function in large animal models is critical [204]. Besides, ISO 5840-2 requires the adoption of compliant chambers for the evaluation of tissue engineering valve devices. Besides, specific test is set according to the material and structural characteristics of tissue engineering valves, such as sewing ring integrity, environmental degradation, and calcification.

3.3.2 Artificial Blood Vessel

Vascular tissue engineering is an effective way for the source of vascular grafts and has become a research hotspot in recent years. Its basic principle is to combine the normal vascular endothelial cells with biodegradable scaffold materials to form vascular tissue with specific shape and function for repair of the diseased and injured vessels. The reconstruction of vascular function is one of the problems to be solved before the clinical application of vascular tissue engineering. The ideal tissue engineering blood vessels are expected to have good biocompatibility (including anti-coagulation, anti-inflammatory, non-immunogenicity, allowing cell adhesion and growth as well as other biological properties). Besides, the tissue engineering blood vessel is able to simulate the elasticity and compliance of natural blood vessel, be easy to process and shape, and be available for suture and sterilization. Degradable tissue engineering scaffolds for small-diameter vessels have an advantage over the permanent one, as the vascular prostheses are gradually replaced by autogenous cells, and the adverse immune responses as well as mismatches could be avoided. In terms of mechanical properties, vascular tissue engineering scaffold needs to reach high burst pressure of 1700 mmHg (equal to that of saphenous vein) [205]. Single-component tissue engineering scaffolds are difficult to meet this requirement, so tissue engineering vessels with multiple composite materials have been developed [206]. By increasing the fiber's diameter, the burst pressure of collagen/chitosan/poly(l-lactic acid-co-ε-caprolactone) scaffolds (diameter = 3 mm) could reach as high as 3365 mmHg [207]. In addition, collagen and chitosan enhanced the adhesion and proliferation of endothelial cells on the scaffold [207]. In a rat model, although the cell infiltration continued to increase on PCL scaffold with the burst pressure of 3280 ± 280 mmHg during 6 months' implantation, the cellular regression was observed after implantation for 12 months [208]. It indicated the necessity for further improving the scaffold's biocompatibility. The arginine-glycine-aspartic acid (RGD), a tripeptide derived from fibronectin, could effectively promote cell adhesion [125]. PCL-RGD scaffold had three times more endothelial coverage than that of PCL scaffold and still kept patent after implantation in the rabbit carotid artery for 4 weeks [209]. In order to simulate the multilayer structure of natural blood vessels, three-layered polyε-caprolactone (PCL) and chitosan (CS) scaffolds were fabricated, which loaded more heparin in the internal CS layer by an ionic bond. Heparin loading PCL/CS scaffolds promoted the secretion of VEGF, moderately inhibited the proliferation of SMC cells, and thus reduced intimal hyperplasia [210]. Moreover, SMCs and ECs were successively seeded on the scaffold made of PCL and type I collagen to realize recellularization that improved the structural integrity in preclinical trial [211].

Although tissue engineered blood vessels have been proven to be promising in animal studies, no product has yet been used in clinical trials. Methods for testing the mechanical and physical properties of non-degradable vascular prosthesis have been described in Sect. 3.2.3, while there are no specific evaluation methods for tissue-engineered blood vessels. According to existing reports, vascular tissue engineering scaffolds, especially scaffolds with cell seeding, the preconditioning by pulsatile bioreactor is indispensable. In in vitro studies, the scaffold needs to be precultured under pulsatile flow to simulate arterial hemodynamic conditions (shear stress of 9.9 dyne/cm^2 in diastole, 13.2 dyne/cm^2 in systole, and pulsatile flow waveform of 60 cycles/min) at 37 °C and under the atmosphere of 5% CO_2 [211]. After the trial in vivo, it is necessary to perform histological examination and microscopic observation, evaluating the safety and effectiveness of tissue-engineered blood vessel.

3.4 Physicochemical Properties of the Ideal Cardiac Materials as Drug Carrier

Drug carriers can improve drug availability and safety by delivering it to the target lesion and controlling the release rate of drug [212, 213]. Degradable polymers are currently the most widely used cardiac drug carrier material. They could deliver drugs in a variety of forms, such as coatings, patches, injectable hydrogels, nanoparticles, and microspheres [214–218]. The purpose of drug therapy itself is to avoid major surgical trauma, so the best delivery method of drug carrier is minimally invasive or even noninvasive surgery to promote clinical transformation [219]. Here, the drug carriers will be introduced, which are in the form of stent coatings, injectable hydrogels, and nanomaterials.

3.4.1 The Coating of Drug-Eluting Stents (DES)

Among the mainstream of commercial drug-eluting stents (DES), polymer coating is utilized to carry anti-intimal hyperplasia drugs that is local-eluting and releasing in the blood vessels, which effectively inhibit intimal hyperplasia and in-stent restenosis compared to bare metal stents (BMS) [220–222]. The first-generation DES adopted nonbiodegradable polymers as the drug carrier, such as parylene C (PC), poly(n-butyl methacrylate) (PBMA), poly(ethylene-co-vinyl acetate) (PEVA), poly (styrene-b-isobu-tylene-b-styrene), and polyvinylidenefluoride (PVDF) [223].

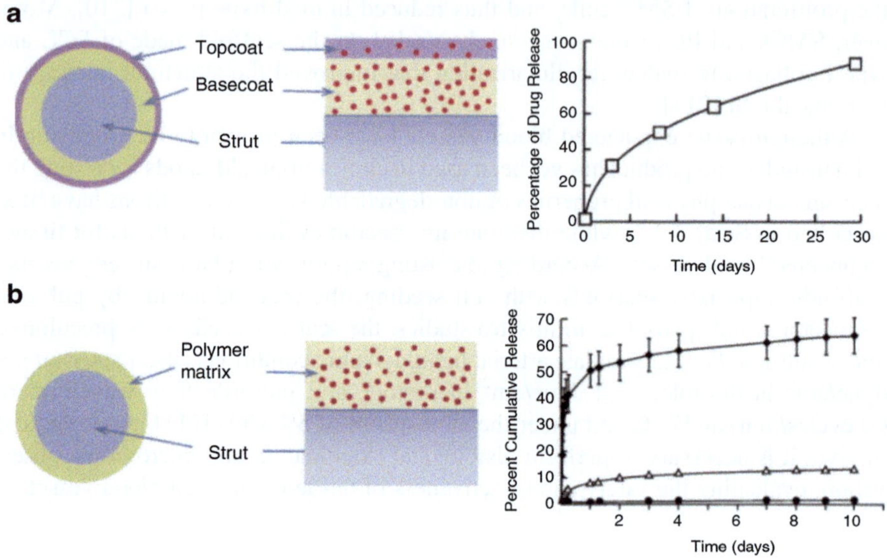

Fig. 3.11 Design of nonbiodegradable coating and drug release profile in vitro: (**a**) Cypher stent, (**b**) Taxus stent [225]

The controlled release mechanism of non-degradable polymer coating includes diffusion and swelling, and diffusion is the main mechanism [224]. In the case of non-degradable polymer coatings, the release rate is usually reduced as the drug gradually travels longer distances [221]. Diffusion-based controlled-release systems are often combined with a topcoat [225], resulting in a constant drug release rate in Fig. 3.11. As for swelling-controlled release, the polymer absorbs water and expands, loosening the drug-carrying coating network and releasing the drug into the environment [226].

However, non-degradable polymers have been linked to an increased risk of late thrombus and heart attacks [227]. Therefore, biodegradable polymers have been used in the second-generation DES to solve the problem of polymer retention over time. Poly-l-lactic acid (PLLA) and poly(lactic-co-glycolic) acid (PLGA) are the common choice for biodegradable coating because they have got the approval of FDA. By adjusting the molecular weight and thickness of the coating material, the degradation periods of the coating vary from 3 to 12 months, and the drug release time varies from 1 to 12 months (Table 3.12). PLLA and PLGA are both lipophilic with similar mechanical properties and degradation mechanism [228, 229]. Nevertheless, PLGA is more hydrophilic; the drug release time and degradation time are shorter than PLLA. In short, nonbiodegradable polymer coatings are based on these two mechanisms for controlled drug release, while as for biodegradable coatings, drug could be gradually released as the process of degradation proceeds [230]. The key to drug release control is that the concentration maintains in a proper range between the lowest effective concentration and the highest tolerated concentration

Table 3.12 Nonbiodegradable and biodegradable coating of commercial DES [221, 223]

Stent	Manufacture	Drug	Polymer	Coating thickness (μm)	Polymer degradation (months)	Drug release (months)
Cypher	Cordis	Sirolimus	PC, PEVA, PBMA	12.6	N/A	N/A
Xience-V	Abbott	Everolimus	PBMA, PVEF	7.8	N/A	N/A
Taxus	Boston Scientific	Paclitaxel	SIBS[a]	16	N/A	N/A
Biomatrix	Biosensors	Biolimus A9	PLLA	10	6–9	6
Biomime	Meril	Sirolimus	PLLA and PLGA	2	2	1
DESyne BD	Elixir	Novolimus	PLLA	N/A	6–9	3
MiStent	Micell Tech.	Sirolimus	PLGA	N/A	9	3
Nobori	Terumo	Biolimus A9	PLLA	10	6–9	6
Orsiro	Biotronik	Sirolimus	PLLA	11	14	12
Supraflex	Sahajanand Medical Tech.	Sirolimus	PLLA/ PLGA	N/A	9–12	1.5
Synergy	Boston Scientific	Everolimus	PLGA	4	4	3

[a]SIBS, styrene-b-isobutylene-b-styrene

for at least 3 weeks [231]. The release of sirolimus from PLGA can be divided into initial diffusion control stage with low release rate and later degradation control stage, which can be accelerated or delayed by adding PEG or topcoat, respectively (Fig. 3.12) [232]. According to the clinical data, compared with the non-degradable coating, the biodegradable polymer coating as a drug carrier could result in thinner new intima and would not cause in-stent thrombosis, which means that the biosecurity is improved [227, 233]. The functional realization of drug-carrying coating not only depends on the drug carrier itself but also requires comprehensive improvement of stent design, coating structure and composition, drug properties, and coating's preparation technology.

3.4.2 *Injectable Hydrogels and Nanomaterials*

Injectable hydrogels are formed in situ under physiological conditions after physical and/or chemical cross-linking, which facilitates minimally invasive treatment and avoids the disruption of biomolecular function by cross-linking catalysts. Hydrogels

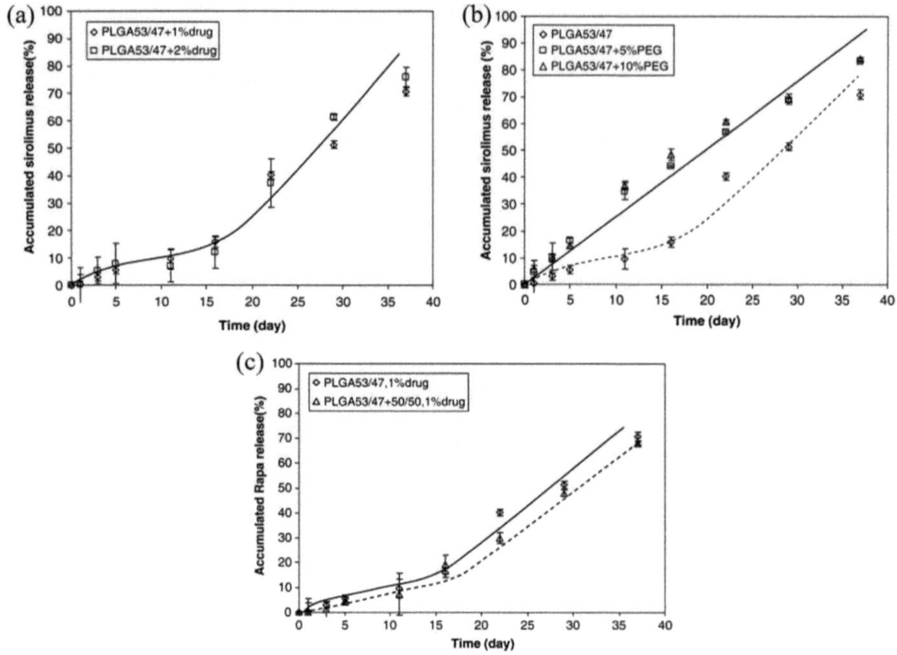

Fig. 3.12 Drug release characteristics of PLGA coating: (**a**) accumulated sirolimus released from PLGA; (**b**) PEG accelerating drug release, solid line, with PEG, and dotted line, without PEG; (**c**) topcoat slowing drug release, solid line, with topcoat, and dotted line, without topcoat [232]

can form into almost any size and shape. Their macroscopic design determines the diffusion way into the body. The mesh size of network determines the drug's diffusion way at nanoscale. The molecular and atomic scales affect the interactions between drugs and polymer [234]. During drug release, the mesh size determines how the drug diffuses through the hydrogel [234]. When the grid size is larger than the drug, the drug release process is dominated by diffusion, which is not able to realize the slow release. When the mesh size approaches the drug size, steric hindrance has a prominent effect on drug diffusion, achieving slow and prolonged release. When drug molecules are much larger than the grid size, the drug remains physically stuck in the network unless the network degrades (network biodegradation) or the mesh size increases (swelling or deformation by mechanical force). Similar to the design of drug-carrying coating, the injectable hydrogels as a drug carrier are also required to maintain the continuous drug release at effective concentration and reduce the dosage and harmful effects. In order to regulate cardiomyocyte proliferation, the hyaluronic acid hydrogel was used for the local continuous delivery of miR-302 mimics, which released miR-302 mimics over 2 weeks in vitro and reduced cardiac end-diastolic (39%) and end-systolic (50%) volumes in 4-week rat model [235]. A controllable NO-releasing redox injectable hydrogel (NO-RIG) formed by the electrostatic cross-linking between the polyion complex flower-type micelles can transform into hydrogel at body temperature (Fig. 3.13), which

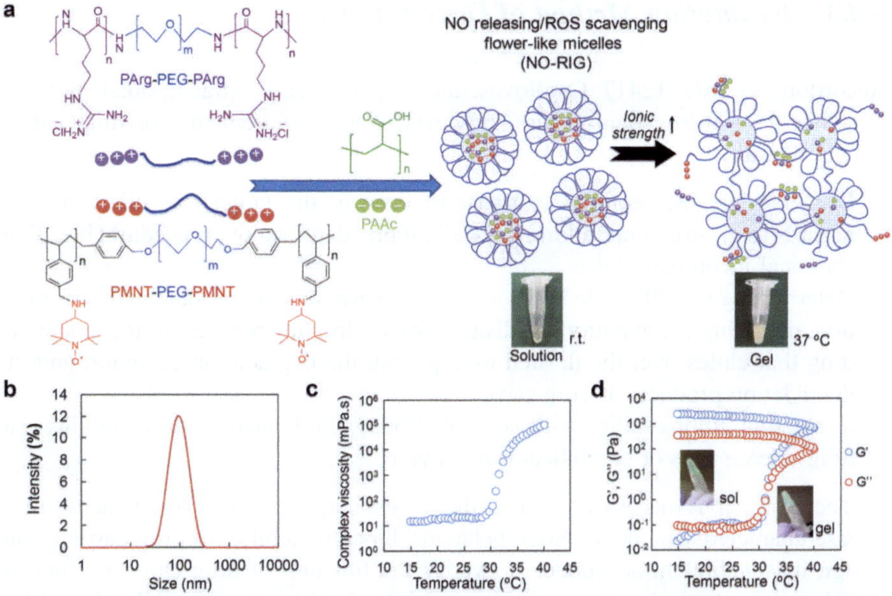

Fig. 3.13 Schematic illustration of flower-type polyion complex micelle [236]

significantly decreased infarction in serious infarction model [236]. Moreover, it also needs to consider the design of gel time for injectable hydrogels. The solution has good fluidity at the time of injection and will not be in the form of gel prematurely. Compared with bulk materials, nanomaterials are easily ingested by cells due to their small size and high specific surface area.

Nanomaterials are easy to realize multi-function by combining with cell-specific ligands, magnetic localization, and size-based selectivity for active targeting [237]. Drugs are usually encapsulated in nanomaterials or bonded to the surface of nanomaterials. However, nanomaterials are difficult to be fixed to the lesion site and can be used in combination with injectable hydrogels to improve the slow-release performance. Gelatin-treated PLGA nanoparticles (PLGA NPs) were able to release sirolimus continuously for up to 50 days after delivering to the site of injury via catheter device [238]. Injectable microporous alginate-based hydrogel was used to attenuate adverse cardiac remodeling. It could carry 5-azacytidine (5-Aza) in zein protein nanoparticle with MSCs. It came out that 35% of drug is released in 7 days, which reduced fibrosis and improved cardiac output in rat. Salvianolic acid B (SaB) is an ingredient of traditional Chinese medicine well recognized for its ability to protect cardiomyocytes. Chen et al. [239] fabricated a self-healing elastin-mimic peptide hydrogel (EMH) containing SaB-loaded polydopamine nanoparticles (SaB-PDA), which kept great mechanical strength the sustained release of 17% in 96 h in vitro.

3.4.3 Evaluation Method of Drug Carrier

According to ISO 12417 Cardiovascular implants and extracorporeal systems-Vascular device-drug combination products, some requirements for drug carrier are put forward:

1. Drug carriers are required to transport drug to the target site safely without hazards (e.g., structural failure, targeted failure, degradation, procedural bleeding, chemical incompatibility).
2. Determining the full cycle behavior of drug carriers from production to degradation, including the quantity and distribution of drug in drug carrier, the amount of drug that elutes over the desired time period, the degradation behavior, and the degradation products of drug carrier.
3. Evaluating compatibility with simulated body fluid, such as drug stability and drug release tests or preclinical tests in vivo.

Therefore, it is necessary to test the access capacity, fixation capacity, drug release characteristics, degradation behavior, biocompatibility of drug carriers, and design reasonable animal models to verify the function of drug carriers. Uniaxial stretching, rheological testing, cytocompatibility, and blood compatibility have been described previously, following test methods focusing on drug release testing and degradation behavior analysis. Among them, drug release test is an essential method to evaluate drug carrier.

- *Access capacity test*: Dislodgement force represents the force required to remove the pre-mounted DES from the non-dilating balloon, which can be tested by separating the stent from the delivery system at a constant speed (e.g., 200 mm/min) according to ISO 25539-2. The force values are recorded when the stent moves beyond a predetermined critical distance (e.g., the edge of the balloon) or reaches a clinically acceptable load.
- *Fixation capacity test*: The solidified sample is sandwiched between the lamellar target tissues and peeled using a tensile testing machine. The maximum peeling stress between the sample and the matrix is recorded as the bonding strength [240].
- *Drug release test*: In the in vitro static drug release experiment, the scaffold was immersed in PBS (pH = 7.4) containing other organic solvents at a certain concentration, which depends on the type of drug. After immersion at 37 °C for a period of time, the PBS is updated in a preset cycle to simulate human metabolism [241]. The quantitative and qualitative analyses of the drug concentration are performed using liquid chromatography analysis. The choice of release medium is based on the appropriate solubility while trying to use as much buffer as possible [242]. US Food and Drug Administration recommends that the elution profile is complete and cover at least 80% of the labeled amount of the drug release or up to the point when a plateau is reached. As the in vivo drug release experiments, wherever possible, the selection of animal models including the clinical site, vascular anatomy, physiology (e.g., the clotting system), and drug

metabolism is expected to be as close to the human body. The follow-up time depends on how long the drug is expected to stay in the body (e.g., limited 24 h, extended 24 h to 30 days, or permanent 30 days). Long-time or absorbable drug carriers require additional follow-up time. According to ISO 12417, long-term studies in each animal are necessary at least 26 weeks. Besides, the release, retention, distribution, and absorption of the drug need to be described including drug plasma levels, drug tissue levels, and the amount of drug in the drug carrier over time. As for the safety studies, it is essential to assess dose-dependent effects, including the effects of overdosage (no drug, notional drug dose, and overdose), regional (downstream), and systemic toxicity.

- *Degradation test*: As for degradation studies in vitro, physiologically relevant aqueous solutions are usually adopted, and appropriate physiological loading conditions (e.g., axial, pulsating, torsion, bending, pulse) are applied if necessary. ISO 10993-13 introduces two methods, real-time degradation test and accelerated degradation test. Real-time degradation test in a simulated environment is carried out at 37 ± 1 °C. For devices whose intended use is longer than 30 days, test periods include 1, 3, 6, and 12 months in real-time degradation test. For devices whose intended use is less than 30 days, four alternative test periods shall be used, including 30 days. The temperature of accelerated degradation test is about 70 °C. For equipment with an expected service time of over 30 days, test periods of 2 s and 60 days are selected in accelerated degradation test. In devices with an expected service time of less than 30 days, the test cycles are 2 and 7 days. Additional test periods may be optional, depending on the intended use of the polymer.

References

1. Burke MN. Wire strategy as a first option: properties of the tools. Interventional Cardiol Clin. 2012;1(3):309–14. https://doi.org/10.1016/j.iccl.2012.03.004.
2. Toth GG, Yamane M, Heyndrickx GR. How to select a guidewire: technical features and key characteristics. Heart. 2015;101(8):645–52. https://doi.org/10.1136/heartjnl-2013-304243.
3. Sarkissian C, Korman E, Hendlin K, Monga M. Systematic evaluation of hybrid guidewires: shaft stiffness, lubricity, and tip configuration. Urology. 2012;79(3):513–7.
4. Silva MB, Cheng CC. Guidewires, catheters, and sheaths. In: Moore WS, Ahn SS, editors. Endovascular surgery. 4th ed. Philadelphia: W.B. Saunders; 2011. p. 59–69. https://doi.org/10.1016/B978-1-4160-6208-0.10007-2.
5. Silva Jr MB, Cheng CC. Guidewires, catheters, and sheaths. In: Endovascular surgery, 59. 2010.
6. Goodman G. 94 - Cardiovascular techniques and technology. In: Dyro JF, editor. Clinical engineering handbook. Burlington: Academic; 2004. p. 417–20. https://doi.org/10.1016/B978-012226570-9/50103-4.
7. From AM, Bell MR, Rihal CS, Gulati R. Minimally invasive transradial intervention using sheathless standard guiding catheters. Catheter Cardio Inte. 2011;78(6):866–71.
8. Collet C, Sonck J, Leipsic J, Monizzi G, Buytaert D, Kitslaar P, Andreini D, De Bruyne B. Implementing coronary computed tomography angiography in the catheterization laboratory. JACC: Cardiovascular Imaging. 2020.

9. Kuck K-H, Reddy VY, Schmidt B, Natale A, Neuzil P, Saoudi N, Kautzner J, Herrera C, Hindricks G, Jaïs P. A novel radiofrequency ablation catheter using contact force sensing: Toccata study. Heart rhythm. 2012;9(1):18–23.
10. Sankar D, Hemalatha K. A non-Newtonian fluid flow model for blood flow through a catheterized artery—steady flow. Appl Math Model. 2007;31(9):1847–64.
11. Kocaturk O, Saikus CE, Guttman MA, Faranesh AZ, Ratnayaka K, Ozturk C, McVeigh ER, Lederman RJ. Whole shaft visibility and mechanical performance for active MR catheters using copper-nitinol braided polymer tubes. J Cardiovasc Magnetic Resonance. 2009;11(1): 1–10.
12. Standardization IOf. ISO 10555-1 Intravascular catheters - Sterile and single-use catheters- - Part 1: General requirements. Part 1: General requirements. Geneva: International Organization for Standardization; 2013.
13. Wheatley GH III, McNutt R, Diethrich EB. Introduction to thoracic endografting: imaging, guidewires, guiding catheters, and delivery sheaths. Ann Thorac Surg. 2007;83(1):272–8.
14. Hammill SC, Sugrue DD, Gersh BJ, Porter CBJ, Osborn MJ, Wood DL, Holmes DR. Clinical intracardiac electrophysiologic testing - technique, diagnostic indications, and therapeutic uses. Mayo Clin Proc. 1986;61(6):478–503.
15. Issa ZF, Miller JM, Zipes DP. 4 - electrophysiological testing: tools and techniques. In: Issa ZF, Miller JM, Zipes DP, editors. Clinical arrhythmology and electrophysiology. 3rd ed. Philadelphia: Elsevier; 2019. p. 81–124. https://doi.org/10.1016/B978-0-323-52356-1.00004-9.
16. Friedman PA. Novel mapping techniques for cardiac electrophysiology. Heart. 2002;87(6): 575–82.
17. Tschabrunn CM, Roujol S, Dorman NC, Nezafat R, Josephson ME, Anter E. High-resolution mapping of ventricular scar comparison between single and multielectrode catheters. Circ-Arrhythmia Elec. 2016;9(6)
18. Nick C, Joshi R, Schneider JJ, Thielemann C. Three-dimensional carbon nanotube electrodes for extracellular recording of cardiac myocytes. Biointerphases. 2012;7(1-4):58.
19. Kim DH, Lu NS, Ghaffari R, Kim YS, Lee SP, Xu LZ, Wu JA, Kim RH, Song JZ, Liu ZJ, Viventi J, de Graff B, Elolampi B, Mansour M, Slepian MJ, Hwang S, Moss JD, Won SM, Huang YG, Litt B, Rogers JA. Materials for multifunctional balloon catheters with capabilities in cardiac electrophysiological mapping and ablation therapy. Nat Mater. 2011;10(4):316–23.
20. Berte B, Relan J, Sacher F, Pillois X, Appetiti A, Yamashita S, Mahida S, Casassus F, Hooks D, Sellal JM, Amraoui S, Denis A, Derval N, Cochet H, Hocini M, Haissaguerre M, Weerasooriya R, Jais P. Impact of electrode type on mapping of scar-related VT. J Cardiovasc Electr. 2015;26(11):1213–23.
21. Kalla M, Sanders P, Kalman JM, Lee G. Radiofrequency catheter ablation for atrial fibrillation: approaches and outcomes. Heart Lung Circ. 2017;26(9):941–9.
22. Hamaya R, Miyazaki S, Kajiyama T, Watanabe T, Kusa S, Nakamura H, Hachiya H, Iesaka Y. Efficacy and safety comparison between different types of novel design enhanced open-irrigated ablation catheters in creating cavo-tricuspid isthmus block. J Cardiol. 2018;71(5–6): 513–6.
23. Guerra JM, Jorge E, Raga S, Galvez-Monton C, Alonso-Martin C, Rodriguez-Font E, Cinca J, Vinolas X. Effects of open-irrigated radiofrequency ablation catheter design on lesion formation and complications: in vitro comparison of 6 different devices. J Cardiovasc Electr. 2013;24(10):1157–62.
24. Reddy VY, Schilling R, Grimaldi M, Horton R, Natale A, Riva S, Tondo C, Kuck KH, Neuzil P, McInnis K, Bishara M, Zhang BH, Govari A, Abdelaal A, Mansour M. Pulmonary vein isolation with a novel multielectrode radiofrequency balloon catheter that allows directionally tailored energy delivery short-term outcomes from a multicenter first-in-human study (RADIANCE). Circ-Arrhythmia Elec. 2019;12(12)
25. Dhillon GS, Honarbakhsh S, Di Monaco A, Coling AE, Lenka K, Pizzamiglio F, Hunter RJ, Horton R, Mansour M, Natale A, Reddy V, Grimaldi M, Neuzil P, Tondo C, Schilling RJ. Use

of a multi-electrode radiofrequency balloon catheter to achieve pulmonary vein isolation in patients with paroxysmal atrial fibrillation: 12-month outcomes of the RADIANCE study. J Cardiovasc Electr. 2020;31(6):1259–69.
26. Barkagan M, Leshem E, Rottmann M, Sroubek J, Shapira-Daniels A, Anter E. Expandable lattice electrode ablation catheter A novel radiofrequency platform allowing high current at low density for rapid, titratable, and durable lesions. Circ-Arrhythmia Elec. 2019;12 (4).
27. Anter E, Neuzil P, Rackauskas G, Peichl P, Aidietis A, Kautzner J, Nakagawa H, Jackman WM, Natale A, Reddy VY. A lattice-tip temperature-controlled radiofrequency ablation catheter for wide thermal lesions first-in-human experience with atrial fibrillation. Jacc-Clin Electrophy. 2020;6(5):507–19.
28. Yokoyama K, Nakagawa H, Shah DC, Lambert H, Leo G, Aeby N, Ikeda A, Pitha JV, Sharma T, Lazzara R, Jackman WM. Novel contact force sensor incorporated in irrigated radiofrequency ablation catheter predicts lesion size and incidence of steam pop and thrombus. Circ-Arrhythmia Elec. 2008;1(5):354–62.
29. Maurer T, Schluter M, Kuck KH. Keeping it simple balloon devices for atrial fibrillation ablation therapy. Jacc-Clin Electrophy. 2020;6(12):1577–96.
30. Prasad K, Bazaka O, Chua M, Rochford M, Fedrick L, Spoor J, Symes R, Tieppo M, Collins C, Cao A. Metallic biomaterials: current challenges and opportunities. Materials. 2017;10(8):884.
31. Fu J, Su Y, Qin Y-X, Zheng Y, Wang Y, Zhu D. Evolution of metallic cardiovascular stent materials: a comparative study among stainless steel, magnesium and zinc. Biomaterials. 2020;230:119641.
32. Zhang J, Gao X, Kan J, Ge Z, Han L, Lu S, Tian N, Lin S, Lu Q, Wu X. Intravascular ultrasound versus angiography-guided drug-eluting stent implantation: the ULTIMATE trial. J Am Coll Cardiol. 2018;72(24):3126–37.
33. Kolandaivelu K, Swaminathan R, Gibson WJ, Kolachalama VB, Nguyen-Ehrenreich K-L, Giddings VL, Coleman L, Wong GK, Edelman ER. Stent thrombogenicity early in high-risk interventional settings is driven by stent design and deployment and protected by polymer-drug coatings. Circulation. 2011;123(13):1400–9.
34. Jiménez JM, Davies PF. Hemodynamically driven stent strut design. Ann Biomed Eng. 2009;37(8):1483–94.
35. Chen S, Yao Z, Guan Y, Yang H, Shahzad MB, Wu Y, Zhang B, Shen L, Yang K. High nitrogen stainless steel drug-eluting stent-assessment of pharmacokinetics and preclinical safety in vivo. Bioact Mater. 2020;5(4):779–86.
36. Wang Q, Ren Y, Shahzad MB, Zhang W, Pan X, Zhang S, Zhang D. Design and characterization of a novel nickel-free cobalt-base alloy for intravascular stents. Mater Sci Eng: C. 2017;77:565–71.
37. Chang L, Zhifeng Z, Kwok-Yan L. Improved corrosion resistance of CoCrMo alloy with self-passivation ability facilitated by carbon ion implantation. Electrochim Acta. 2017;241:331–40.
38. Abbass MK, Ajeel SA, Wadullah HM. Biocompatibility, bioactivity and corrosion resistance of stainless steel 316L nanocoated with TiO2 and Al2O3 by atomic layer deposition method. In: Journal of physics: conference series; 2018. vol 1. IOP Publishing, p. 012017.
39. Wang L, Zhao X, Ding M, Zheng H, Zhang H, Zhang B, Li X, Wu G. Surface modification of biomedical AISI 316L stainless steel with zirconium carbonitride coatings. Appl Surf Sci. 2015;340:113–9.
40. Parsapour A, Khorasani SN, Fathi MH. Effect of surface treatment and metallic coating on corrosion behavior and biocompatibility of surgical 316L stainless steel implant. J Mater Sci Technol. 2012;28(2):125–31.
41. Lu Y, Guan Y, Li Y, Yang L, Wang M, Wang Y. Nanosecond laser fabrication of superhydrophobic surface on 316L stainless steel and corrosion protection application. Colloids Surf A. 2020;604:125259.

42. Mani G, Feldman MD, Patel D, Agrawal CM. Coronary stents: a materials perspective. Biomaterials. 2007;28(9):1689–710.
43. Wang RX, Wang RX, Chen DF, Qin GW, Zhang EL. Novel CoCrWNi alloys with Cu addition: microstructure, mechanical properties, corrosion properties and biocompatibility. J Alloy Compd. 2020;824.
44. Ren Y, Wan P, Liu F, Zhang B, Yang K. In vitro study on a new high nitrogen nickel-free austenitic stainless steel for coronary stents. J Mater Sci Technol. 2011;27(4):325–31.
45. O'Brien BJ, Stinson JS, Larsen SR, Eppihimer MJ, Carroll WM. A platinum-chromium steel for cardiovascular stents. Biomaterials. 2010;31(14):3755–61.
46. Armitage DA, Parker TL, Grant DM. Biocompatibility and hemocompatibility of surface-modified NiTi alloys. J Biomed Mater Res Part A. 2003;66(1):129–37.
47. Ma J, Zhao N, Betts L, Zhu DH. Bio-Adaption between Magnesium alloy stent and the blood vessel: a review. J Mater Sci Technol. 2016;32(9):815–26.
48. Ang HY, Huang YY, Lim ST, Wong P, Joner M, Foin N. Mechanical behavior of polymer-based vs. metallic-based bioresorbable stents. J Thorac Dis. 2017;9(Suppl 9):S923.
49. Lin WJ, Zhang DY, Zhang G, Sun HT, Qi HP, Chen LP, Liu ZQ, Gao RL, Zheng W. Design and characterization of a novel biocorrodible iron-based drug-eluting coronary scaffold. Mater Design. 2016;91:72–9.
50. Hermawan H, Purnama A, Dube D, Couet J, Mantovani D. Fe-Mn alloys for metallic biodegradable stents: degradation and cell viability studies. Acta Biomater. 2010;6(5): 1852–60.
51. Hermawan H, Dubé D, Mantovani D. Degradable metallic biomaterials: design and development of Fe–Mn alloys for stents. J Biomed Mater Res Part A. 2010;93(1):1–11.
52. Liu B, Zheng YF, Ruan LQ. In vitro investigation of Fe30Mn6Si shape memory alloy as potential biodegradable metallic material. Mater Lett. 2011;65(3):540–3.
53. Yang HT, Wang C, Liu CQ, Chen HW, Wu YF, Han JT, Jia ZC, Lin WJ, Zhang DY, Li WT, Yuan W, Guo H, Li HF, Yang GX, Kong DL, Zhu DH, Takashima K, Ruan LQ, Nie JF, Li X, Zheng YF. Evolution of the degradation mechanism of pure zinc stent in the one-year study of rabbit abdominal aorta model. Biomaterials. 2017;145:92–105.
54. Jin HL, Zhao S, Guillory R, Bowen PK, Yin ZY, Griebel A, Schaffer J, Earley EJ, Goldman J, Drelich JW. Novel high-strength, low-alloys Zn-Mg (< 0.1 wt% Mg) and their arterial biodegradation. Mat Sci Eng C-Mater. 2018;84:67–79.
55. Zhao S, Seitz JM, Eifler R, Maier HJ, Guillory RJ, Earley EJ, Drelich A, Goldman J, Drelich JW. Zn-Li alloy after extrusion and drawing: structural, mechanical characterization, and biodegradation in abdominal aorta of rat. Mat Sci Eng C-Mater. 2017;76:301–12.
56. Tang ZB, Niu JL, Huang H, Zhang H, Pei J, Ou JM, Yuan GY. Potential biodegradable Zn-Cu binary alloys developed for cardiovascular implant applications. J Mech Behav Biomed. 2017;72:182–91.
57. Guo H, Hu J, Shen Z, Du D, Zheng Y, Peng J. In vitro and in vivo studies of biodegradable Zn-Li-Mn alloy staples designed for gastrointestinal anastomosis. Acta Biomater. 2021;121:713–23.
58. Li Z, Shi Z-Z, Hao Y, Li H-F, Liu X-F, Volinsky AA, Zhang H-J, Wang L-N. High-performance hot-warm rolled Zn-0.8 Li alloy with nano-sized metastable precipitates and sub-micron grains for biodegradable stents. J Mater Sci Technol. 2019;35(11):2618–24.
59. Dai Y, Zhang Y, Liu H, Fang H, Li D, Xu X, Yan Y, Chen L, Lu Y, Yu K. Mechanical strengthening mechanism of Zn-Li alloy and its mini tube as potential absorbable stent material. Mater Lett. 2019;235:220–3.
60. Zhao S, Seitz J-M, Eifler R, Maier HJ, Guillory RJ II, Earley EJ, Drelich A, Goldman J, Drelich JW. Zn-Li alloy after extrusion and drawing: structural, mechanical characterization, and biodegradation in abdominal aorta of rat. Mater Sci Eng: C. 2017;76:301–12.
61. Li GN, Yang HT, Zheng YF, Chen XH, Yang JA, Zhu DH, Ruan LQ, Takashima K. Challenges in the use of zinc and its alloys as biodegradable metals: perspective from biomechanical compatibility. Acta Biomater. 2019;97:23–45.

62. Cheng J, Liu B, Wu YH, Zheng YF. Comparative in vitro study on pure metals (Fe, Mn, Mg, Zn and W) as biodegradable metals. J Mater Sci Technol. 2013;29(7):619–27.
63. Zhang SX, Zhang XN, Zhao CL, Li JA, Song Y, Xie CY, Tao HR, Zhang Y, He YH, Jiang Y, Bian YJ. Research on an Mg-Zn alloy as a degradable biomaterial. Acta Biomater. 2010;6(2): 626–40.
64. Li Z, Zhou SJ, Huang N. Effects of ECAE processing temperature on the microstructure, mechanical properties, and corrosion behavior of pure Mg. Int J Min Met Mater. 2015;22(6): 639–47.
65. Dobatkin S, Martynenko N, Anisimova N, Kiselevskiy M, Prosvirnin D, Terentiev V, Yurchenko N, Salishchev G, Estrin Y. Mechanical properties, biodegradation, and biocompatibility of ultrafine grained magnesium alloy WE43. Materials. 2019;12(21):3627.
66. Waksman R, Erbel R, Di Mario C, Bartunek J, de Bruyne B, Eberli FR, Erne P, Haude M, Horrigan M, Ilsley C. Early-and long-term intravascular ultrasound and angiographic findings after bioabsorbable magnesium stent implantation in human coronary arteries. JACC. 2009;2 (4):312–20.
67. Wu JY, Zhao DL, Ohodnicki JM, Lee B, Roy A, Yao R, Chen SN, Dong ZY, Heineman WR, Kumta PN. In vitro and in vivo evaluation of multiphase ultrahigh ductility Mg-Li-Zn alloys for cardiovascular stent application. Acs Biomater Sci Eng. 2018;4(3):919–32.
68. Zhou WR, Zheng YF, Leeflang MA, Zhou J. Mechanical property, biocorrosion and in vitro biocompatibility evaluations of Mg-Li-(Al)-(RE) alloys for future cardiovascular stent application. Acta Biomater. 2013;9(10):8488–98.
69. Wang SJ, Han Z, Nie YJ, Dai JW, Li X, Cheng JL, Zhang XB. Modified mechanical properties of Mg-Nd-Zn-Ag-Zr alloy by solution treatment for cardiovascular stent application. Mater Res Express. 2019;6(8)
70. Zhang J, Li HY, Wang W, Huang H, Pei J, Qu HY, Yuan GY, Li YD. The degradation and transport mechanism of a Mg-Nd-Zn-Zr stent in rabbit common carotid artery: a 20-month study. Acta Biomater. 2018;69:372–84.
71. Du BN, Xiao ZP, Qiao YX, Zheng L, Yu BY, Xu DK, Sheng LY. Optimization of microstructure and mechanical property of a Mg-Zn-Y-Nd alloy by extrusion process. J Alloy Compd. 2019;775:990–1001.
72. Wang P, Liu J, Shen S, Li QY, Luo XJ, Xiong P, Gao S, Yan JL, Cheng Y, Xi TF. In vitro and in vivo studies on two-step alkali-fluoride-treated Mg-Zn-Y-Nd alloy for vascular stent application: enhancement in corrosion resistance and biocompatibility. Acs Biomater Sci Eng. 2019;5(7):3279–92.
73. Peuster M, Hesse C, Schloo T, Fink C, Beerbaum P, von Schnakenburg C. Long-term biocompatibility of a corrodible peripheral iron stent in the porcine descending aorta. Biomaterials. 2006;27(28):4955–62.
74. Schinhammer M, Hanzi AC, Loffler JF, Uggowitzer PJ. Design strategy for biodegradable Fe-based alloys for medical applications. Acta Biomater. 2010;6(5):1705–13.
75. Liu B, Zheng YF. Effects of alloying elements (Mn, Co, Al, W, Sn, B, C and S) on biodegradability and in vitro biocompatibility of pure iron. Acta Biomater. 2011;7(3): 1407–20.
76. Lin WJ, Qin L, Qi HP, Zhang DY, Zhang G, Gao RL, Qiu H, Xia Y, Cao P, Wang X, Zheng W. Long-term in vivo corrosion behavior, biocompatibility and bioresorption mechanism of a bioresorbable nitrided iron scaffold. Acta Biomater. 2017;54:454–68.
77. Feng QM, Zhang DY, Xin CH, Liu XD, Lin WJ, Zhang WQ, Chen S, Sun K. Characterization and in vivo evaluation of a bio-corrodible nitrided iron stent. J Mater Sci-Mater M. 2013;24(3): 713–24.
78. Hermawan H, Dube D, Mantovani D. Degradable metallic biomaterials: design and development of Fe-Mn alloys for stents. J Biomed Mater Res A. 2010;93a(1):1–11.
79. Ma Z, Gao M, Na D, Li YD, Tan LL, Yang K. Study on a biodegradable antibacterial Fe-Mn-C-Cu alloy as urinary implant material. Mat Sci Eng C-Mater. 2019;103.

80. Liu XW, Sun JK, Yang YH, Pu ZJ, Zheng YF. In vitro investigation of ultra-pure Zn and its mini-tube as potential bioabsorbable stent material. Mater Lett. 2015;161:53–6.
81. Mostaed E, Sikora-Jasinska M, Mostaed A, Loffredo S, Demir AG, Preuitali B, Mantouani D, Beanland R, Vedani M. Novel Zn-based alloys for biodegradable stent applications: design, development and in vitro degradation. J Mech Behav Biomed. 2016;60:581–602.
82. Jin H, Zhao S, Guillory R, Bowen PK, Yin Z, Griebel A, Schaffer J, Earley EJ, Goldman J, Drelich JW. Novel high-strength, low-alloys Zn-Mg (< 0.1 wt% Mg) and their arterial biodegradation. Mater Sci Eng: C. 2018;84:67–79.
83. Zhao S, McNamara CT, Bowen PK, Verhun N, Braykovich JP, Goldman J, Drelich JW. Structural characteristics and in vitro biodegradation of a novel Zn-Li alloy prepared by induction melting and hot rolling. Metall Mater Trans A. 2017;48a(3):1204–15.
84. Zhou C, Li HF, Yin YX, Shi ZZ, Li T, Feng XY, Zhang JW, Song CX, Cui XS, Xu KL, Zhao YW, Hou WB, Lu ST, Liu G, Li MQ, Ma JY, Toft E, Volinsky AA, Wan M, Yao XJ, Wang CB, Yao K, Xu SK, Lu H, Chang SF, Ge JB, Wang LN, Zhang HJ. Long-term in vivo study of biodegradable Zn-Cu stent: a 2-year implantation evaluation in porcine coronary artery. Acta Biomater. 2019;97:657–70.
85. Chichareon P, Katagiri Y, Asano T, Takahashi K, Kogame N, Modolo R, Tenekecioglu E, Chang C-C, Tomaniak M, Kukreja N. Mechanical properties and performances of contemporary drug-eluting stent: focus on the metallic backbone. Expert Rev Med Devic. 2019;16(3): 211–28.
86. Kim DB, Choi H, Joo SM, Kim HK, Shin JH, Hwang MH, Choi J, Kim DG, Lee KH, Lim CH. A comparative reliability and performance study of different stent designs in terms of mechanical properties: foreshortening, recoil, radial force, and flexibility. Artificial Organs. 2013;37(4):368–79.
87. Wiktor DM, Waldo SW, Armstrong EJ. Coronary stent failure: fracture, compression, recoil, and prolapse. Interventional Cardiol Clin. 2016;5(3):405–14.
88. Li J, Luo Q, Xie Z, Li Y, Zeng Y. Fatigue life analysis and experimental verification of coronary stent. Heart Vessels. 2010;25(4):333–7.
89. Huang T, Cheng J, Bian D, Zheng YF. Fe-Au and Fe-Ag composites as candidates for biodegradable stent materials. J Biomed Mater Res B. 2016;104(2):225–40.
90. Guo H, He Y, Zheng Y, Cui Y. In vitro studies of biodegradable Zn-0.1 Li alloy for potential esophageal stent application. Mater Lett. 2020;275:128190.
91. Lin S, Ran X, Yan X, Wang Q, Zhou JG, Hu T, Wang G. Systematical evolution on a Zn–Mg alloy potentially developed for biodegradable cardiovascular stents. J Mater Sci: Mater Med. 2019;30(11):1–12.
92. Dong H, Lin F, Boccaccini AR, Virtanen S. Corrosion behavior of biodegradable metals in two different simulated physiological solutions: comparison of Mg, Zn and Fe. Corros Sci. 2021;182:109278.
93. Wang Q, Tan L, Yang K. Cytocompatibility and hemolysis of AZ31B magnesium alloy with Si-containing coating. J Mater Sci Technol. 2015;31(8):845–51.
94. Mohan C, Chennazhi K, Menon D. In vitro hemocompatibility and vascular endothelial cell functionality on titania nanostructures under static and dynamic conditions for improved coronary stenting applications. Acta Biomater. 2013;9(12):9568–77.
95. Moradi S, Hadjesfandiari N, Toosi SF, Kizhakkedathu JN, Hatzikiriakos SG. Effect of extreme wettability on platelet adhesion on metallic implants: from superhydrophilicity to superhydrophobicity. Acs Appl Mater Inter. 2016;8(27):17631–41.
96. Lin WJ, Zhang G, Cao P, Zhang DY, Zheng YF, Wu RX, Qin L, Wang GQ, Wen TY. Cytotoxicity and its test methodology for a bioabsorbable nitrided iron stent. J Biomed Mater Res B. 2015;103(4):764–76.
97. Zhang Z, Jia B, Yang H, Han Y, Wu Q, Dai K, Zheng Y. Biodegradable ZnLiCa ternary alloys for critical-sized bone defect regeneration at load-bearing sites: In vitro and in vivo studies. Bioact Mater. 2021;6(11):3999–4013.

98. Forti S, Lunelli L, Della Volpe C, Siboni S, Pasquardini L, Lui A, Canteri R, Vanzetti L, Potrich C, Vinante M. Hemocompatibility of pyrolytic carbon in comparison with other biomaterials. Diam Relat Mater. 2011;20(5-6):762–9.
99. Huang J, Dong P, Hao W, Wang T, Xia Y, Da G, Fan Y. Biocompatibility of TiO2 and TiO2/heparin coatings on NiTi alloy. Appl Surf Sci. 2014;313:172–82.
100. Kwok SC, Yang P, Wang J, Liu X, Chu PK. Hemocompatibility of nitrogen-doped, hydrogen-free diamond-like carbon prepared by nitrogen plasma immersion ion implantation–deposition. J Biomed Mater Res Part A. 2004;70(1):107–14.
101. De Bakey M. Discussion of Braunwald NS, Tatooles C, Turina M, et al. New development in the design of fabric covered prosthetic heart valves. J Thorac Cardiovasc Surg. 1971;62:673–82.
102. Gott VL, Alejo DE, Cameron DE. Mechanical heart valves: 50 years of evolution. Ann Thorac Surg. 2003;76(6):S2230–9.
103. De Martino A, Milano AD, Thiene G, Bortolotti U. Diamond anniversary of mechanical cardiac valve prostheses: a tale of cages, balls, and discs. Ann Thorac Surg. 2020;110(4): 1427–33.
104. Celiento M, Filaferro L, Milano AD, Anastasio G, Ferrari G, Bortolotti U. Single center experience with the Sorin Bicarbon prosthesis: a 17-year clinical follow-up. J Thorac Cardiovasc Surg. 2014;148(5):2039–44.
105. Puskas JD, Gerdisch M, Nichols D, Fermin L, Rhenman B, Kapoor D, Copeland J, Quinn R, Hughes GC, Azar H. Anticoagulation and antiplatelet strategies after On-X mechanical aortic valve replacement. J Am Coll Cardiol. 2018;71(24):2717–26.
106. Chaudhary R, Garg J, Krishnamoorthy P, Shah N, Feldman BA, Martinez MW, Freudenberger R. On-X valve: the next generation aortic valve. Cardiol Rev. 2017;25(2):77–83.
107. Ritchie RO. Fatigue and fracture of pyrolytic carbon: a damage-tolerant approach to structural integrity and. J Heart Valve Dis. 1996(5 Suppl I):S9–31.
108. Bokros J. Carbon biomedical devices. Carbon. 1977;15(6):353–71.
109. Marti A. Inert bioceramics (Al2O3, ZrO2) for medical application. Injury. 2000;31:D33–6.
110. Reddy JD. Mechanical properties of Silicon Carbide (SiC) thin films. University of South Florida; 2008.
111. Stone D, Yoder K, Sproul W. Hardness and elastic modulus of TiN based on continuous indentation technique and new correlation. J Vacuum Sci Technol A: Vacuum Surf Films. 1991;9(4):2543–7.
112. Standardization IOf. ISO 5840 Cardiovascular implants—cardiac valve prostheses. Geneva: International Organization for Standardization; 2021.
113. Butterfield M, Fisher J. Fatigue analysis of clinical bioprosthetic heart valves manufactured using photooxidized bovine pericardium. J Heart Valve Dis.2000;9(1):161–6; discussion 167.
114. Iosif M, Gabbay S. Comparative in-vitro fluid dynamics characterization of heart valve bioprostheses under accelerated fatigue conditions and under physiologic conditions. Biomed Instrument Technol. 1993;27(6):495–505.
115. Ghanbari H, Viatge H, Kidane AG, Burriesci G, Tavakoli M, Seifalian AM. Polymeric heart valves: new materials, emerging hopes. Trends Biotechnol. 2009;27(6):359–67.
116. Banoriya D, Purohit R, Dwivedi R. Advanced application of polymer based biomaterials. Mater Today: Proceedings. 2017;4(2):3534–41.
117. Dhanumalayan E, Joshi GM. Performance properties and applications of polytetrafluoroethylene (PTFE)-a review. Adv Compos Hybrid Ma. 2018;1(2):247–68.
118. Prager M, Polterauer P, Böhmig H-J, Wagner O, Fügl A, Kretschmer G, Plohner M, Nanobashvili J, Huk I. Collagen versus gelatin-coated Dacron versus stretch polytetrafluoroethylene in abdominal aortic bifurcation graft surgery: results of a seven-year prospective, randomized multicenter trial. Surgery. 2001;130(3):408–14.
119. Damme HV, Deprez M, Creemers E, Limet R. Intrinsic structural failure of polyester (Dacron) vascular grafts. A general review. Acta chirurgica belgica. 2005;105(3):249–55.

120. Roll S, Müller-Nordhorn J, Keil T, Scholz H, Eidt D, Greiner W, Willich SN. Dacron® vs. PTFE as bypass materials in peripheral vascular surgery–systematic review and meta-analysis. BMC Surg. 2008;8(1):1–8.
121. Kieffer E, Gomes D, Chiche L, Fléron M-H, Koskas F, Bahnini A. Allograft replacement for infrarenal aortic graft infection: early and late results in 179 patients. J Vasc Surg. 2004;39(5): 1009–17.
122. Kowalewski R, Zimnoch L, Wojtukiewicz MZ, Glowinski S, Glowinski J. Coagulation activators and inhibitors in the neointima of polyester vascular grafts. Blood Coagulation Fibrinolysis. 2003;14(5):433–9.
123. Chandy T, Das GS, Wilson RF, Rao GHR. Use of plasma glow for surface-engineering biomolecules to enhance bloodcompatibility of Dacron and PTFE vascular prosthesis. Biomaterials. 2000;21(7):699–712.
124. Lindholt JS, Gottschalksen B, Johannesen N, Dueholm D, Ravn H, Christensen ED, Viddal B, Florenes T, Pedersen G, Rasmussen M, Carstensen M, Grondal N, Fasting H. The Scandinavian Propaten (R) trial-1-year patency of PTFE vascular prostheses with heparin-bonded luminal surfaces compared to ordinary pure PTFE vascular prostheses - a randomised clinical controlled multi-centre trial. Eur J Vasc Endovasc. 2011;41(5):668–73.
125. Ren X, Feng Y, Guo J, Wang H, Li Q, Yang J, Hao X, Lv J, Ma N, Li W. Surface modification and endothelialization of biomaterials as potential scaffolds for vascular tissue engineering applications. Chem Soc Rev. 2015;44(15):5680–742.
126. Li JG, Chen ZB, Yang XN. State of the art of small-diameter vessel-polyurethane substitutes. Macromol Biosci. 2019;19(5).
127. Wu Q, Henriksson M, Liu X, Berglund LA. A high strength nanocomposite based on microcrystalline cellulose and polyurethane. Biomacromolecules. 2007;8(12):3687–92.
128. Spaans C, De Groot J, Dekens F, Pennings A. High molecular weight polyurethanes and a polyurethane urea based on 1, 4-butanediisocyanate. Polymer Bull. 1998;41(2):131–8.
129. Anderson JM, Hiltner A, Wiggins MJ, Schubert MA, Collier TO, Kao WJ, Mathur AB. Recent advances in biomedical polyurethane biostability and biodegradation. Polymer Int. 1998;46 (3):163–71.
130. Hasan A, Memic A, Annabi N, Hossain M, Paul A, Dokmeci MR, Dehghani F, Khademhosseini A. Electrospun scaffolds for tissue engineering of vascular grafts. Acta Biomater. 2014;10(1):11–25.
131. Hiob MA, She S, Muiznieks LD, Weiss AS. Biomaterials and modifications in the development of small-diameter vascular grafts. ACS Biomater Sci Eng. 2017;3(5):712–23.
132. Jeschke MG, Hermanutz V, Wolf SE, Köveker GB. Polyurethane vascular prostheses decreases neointimal formation compared with expanded polytetrafluoroethylene. J Vasc Surg. 1999;29(1):168–76.
133. Soldani G, Losi P, Bernabei M, Burchielli S, Chiappino D, Kull S, Briganti E, Spiller D. Long term performance of small-diameter vascular grafts made of a poly (ether) urethane–polydimethylsiloxane semi-interpenetrating polymeric network. Biomaterials. 2010;31(9): 2592–605.
134. Butany J, Ahluwalia MS, Munroe C, Fayet C, Ahn C, Blit P, Kepron C, Cusimano RJ, Leask RL. Mechanical heart valve prostheses: identification and evaluation (vol 12, pg 322, 2003). Cardiovasc Pathol. 2003;12(6):322–44.
135. Braunwald NS, Morrow AG. A late evaluation of flexible teflon prostheses utilized for total aortic valve replacement - postoperative clinical hemodynamic and pathological assessments. J Thorac Cardiov Sur. 1965;49(3):485.
136. Imamura E, Kaye MP, Davis GD. Radiographic assessment of leaflet motion of gore-tex laminate trileaflet valves and Hancock xenograft in tricuspid position of dogs. Circulation. 1977;56(6):1053–8.
137. Braunwald E. Nina Starr Braunwald: some reflections on the first woman heart surgeon. Ann Thorac Surg. 2001;71(2):S6–7.

138. Bezuidenhout D, Williams DF, Zilla P. Polymeric heart valves for surgical implantation, catheter-based technologies and heart assist devices. Biomaterials. 2015;36:6–25.
139. Daebritz SH, Fausten B, Hermanns B, Schroeder J, Groetzner J, Autschbach R, Messmer BJ, Sachweh JS. Introduction of a flexible polymeric heart valve prosthesis with special design for aortic position. Eur J Cardio-Thorac. 2004;25(6):946–52.
140. Rahmani B, Tzamtzis S, Sheridan R, Mullen MJ, Yap J, Seifalian AM, Burriesci G. In vitro hydrodynamic assessment of a new transcatheter heart valve concept (the TRISKELE). J Cardiovasc Transl. 2017;10(2):104–15.
141. Gyles DA, Castro LD, Silva JOC Jr, Ribeiro-Costa RM. A review of the designs and prominent biomedical advances of natural and synthetic hydrogel formulations. Eur Polymer J. 2017;88:373–92.
142. Im SH, Jung Y, Kim SH. Current status and future direction of biodegradable metallic and polymeric vascular scaffolds for next-generation stents. Acta Biomater. 2017;60:3–22.
143. Pfeiffer D, Stefanitsch C, Wankhammer K, Müller M, Dreyer L, Krolitzki B, Zernetsch H, Glasmacher B, Lindner C, Lass A. Endothelialization of electrospun polycaprolactone (PCL) small caliber vascular grafts spun from different polymer blends. J Biomed Mater Res A. 2014;102(12):4500–9.
144. Lin C, Liu L, Liu Y, Leng J. 4D printing of bioinspired absorbable left atrial appendage occluders: a proof-of-concept study. Acs Appl Mater Inter. 2021;13(11):12668–78.
145. Zhai W, Lj Q, Xm M, Wang S, Yf X, Peng B, Liu M, Huang J, Gc W, Zheng J. Coaxial electrospinning of P (LLA-CL)/heparin biodegradable polymer nanofibers: potential vascular graft for substitution of femoral artery. J Biomed Mater Res Part B: Appl Biomater. 2019;107 (3):471–8.
146. Li B, Xie Z, Wang Q, Chen X, Liu Q, Wang W, Shen Y, Liu X, Li A, Li Y. Biodegradable polymeric occluder for closure of atrial septal defect with interventional treatment of cardiovascular disease. Biomaterials. 2021;120851.
147. Renouf-Glauser AC, Rose J, Farrar DF, Cameron RE. The effect of crystallinity on the deformation mechanism and bulk mechanical properties of PLLA. Biomaterials. 2005;26 (29):5771–82.
148. Stoppato M, Stevens HY, Carletti E, Migliaresi C, Motta A, Guldberg RE. Effects of silk fibroin fiber incorporation on mechanical properties, endothelial cell colonization and vascularization of PDLLA scaffolds. Biomaterials. 2013;34(19):4573–81.
149. Middleton JC, Tipton AJ. Synthetic biodegradable polymers as orthopedic devices. Biomaterials. 2000;21(23):2335–46.
150. Ormiston JA, Serruys PW, Regar E, Dudek D, Thuesen L, Webster MWI, Onuma Y, Garcia-Garcia HM, McGreevy R, Veldhof S. A bioabsorbable everolimus-eluting coronary stent system for patients with single de-novo coronary artery lesions (ABSORB): a prospective open-label trial. Lancet. 2008;371(9616):899–907.
151. Farag M, Spinthakis N, Gorog DA, Prasad A, Sullivan K, Akhtar Z, Kukreja N, Srinivasan M. Use of bioresorbable vascular scaffold: a meta-analysis of patients with coronary artery disease. Open heart. 2016;3(2):e000462.
152. Kereiakes DJ, Ellis SG, Metzger DC, Caputo RP, Rizik DG, Teirstein PS, Litt MR, Kini A, Kabour A, Marx SO, Popma JJ, Tan SH, Ediebah DE, Simonton C, Stone GW, Investigators AI. Clinical outcomes before and after complete everolimus-eluting bioresorbable scaffold resorption five-year follow-up from the ABSORB III trial. Circulation. 2019;140(23): 1895–903.
153. Sakamoto A, Jinnouchi H, Torii S, Virmani R, Finn AV. Understanding the impact of stent and scaffold material and strut design on coronary artery thrombosis from the basic and clinical points of view. Bioengineering. 2018;5(3):71.
154. Xu B, Guan C, Gao R. TCT-13 a first-in-man study of the firesorb sirolimus target eluting bioresorbable vascular scaffold in patients with coronary artery disease (FUTURE-I): one-year clinical and imaging outcomes. J Am Coll Cardiol. 2017;70(18S):B6.

155. Regazzoli D, Leone PP, Colombo A, Latib A. New generation bioresorbable scaffold technologies: an update on novel devices and clinical results. J Thorac Dis. 2017;9(Suppl 9):S979.
156. Seth A, Onuma Y, Costa R, Chandra P, Bahl VK, Manjunath CN, Mahajan AU, Kumar V, Goel PK, Wander GS, Kalarickal MS, Kaul U, Kumar VKA, Rath PC, Trehan V, Sengottuvelu G, Mishra S, Abizaid A, Serruys PW. First-in-human evaluation of a novel poly-L-lactide based sirolimus-eluting bioresorbable vascular scaffold for the treatment of de novo native coronary artery lesions: MeRes-1 trial. Eurointervention. 2017;13(4):415–23.
157. Raghu A, Kawalsky D, Feldman M. Embolic stroke due to a left atrial thrombus two years after placement of an atrial septal defect closure device. Am J Cardiol. 2006;98(9):1294–6.
158. Crawford GB, Brindis RG, Krucoff MW, Mansalis BP, Carroll JD. Percutaneous atrial septal occluder devices and cardiac erosion: a review of the literature. Catheter Cardio Inte. 2012;80 (2):157–67.
159. Happel CM, Laser KT, Sigler M, Kececioglu D, Sandica E, Haas NA. Single center experience: implantation failures, early, and late complications after implantation of a partially biodegradable ASD/PFO-device (BioStar (R)). Catheter Cardio Inte. 2015;85(6):990–7.
160. Zamiri P, Kuang Y, Sharma U, Ng T, Busold R, Rago A, Core L, Palasis M. The biocompatibility of rapidly degrading polymeric stents in porcine carotid arteries. Biomaterials. 2010;31 (31):7847–55.
161. Yu-feng Z, Xin-miao H, Cao J, Hu J-Q, Bai Y, Jiang H-B, Zhao-Feng L, Chen Y, Wang W, Yong-Wen Q. Animal experimental study of the fully biodegradable atrial septal defect (ASD) occluder. BioMed Res Int. 2012.
162. Lin C, Lv J, Li Y, Zhang F, Li J, Liu Y, Liu L, Leng J. 4D-printed biodegradable and remotely controllable shape memory occlusion devices. Adv Funct Mater. 2019;29(51):1906569.
163. Omar WA, Kumbhani DJ. The current literature on bioabsorbable stents: a review. Curr Atheroscler Rep. 2019;21(12).
164. Wu YZ, Shen L, Yin JS, Chen JH, Qian JY, Ge L, Ge JB, Trial XRC. Twelve-month angiographic and clinical outcomes of the XINSORB bioresorbable sirolimus-eluting scaffold and a metallic stent in patients with coronary artery disease. Int J Cardiol. 2019;293:61–6.
165. Song L, Sun ZW, Guan CD, Yan HB, Yu MY, Cui JY, Mu CW, Qiu H, Tang YD, Zhao YY, Qiao SB, Suryapranata H, Gao RL, Xu B, Investigators F-I. First-in-man study of a thinner-strut sirolimus-eluting bioresorbable scaffold (FUTURE-I): Three-year clinical and imaging outcomes. Catheter Cardio Inte. 2020;95:648–57.
166. Xu K, Fu GS, Xu B, Zhou YJ, Su X, Liu HL, Zhang Z, Yu B, Wang XZ, Han YL, Investigators NO. Safety and efficacy of the novel sirolimus-eluting bioresorbable scaffold for the treatment of de novo coronary artery disease: one-year results from a prospective patient-level pooled analysis of NeoVas trials. Catheter Cardio Inte. 2019;93:832–8.
167. Tenekecioglu E, Torii R, Katagiri Y, Chichareon P, Asano T, Miyazaki Y, Takahashi K, Modolo R, Al-Lamee R, Al-Lamee K, Colet C, Reiber JHC, Pekkan K, van Geuns R, Bourantas CV, Onuma Y, Serruys PW. Post-implantation shear stress assessment: an emerging tool for differentiation of bioresorbable scaffolds. Int J Cardiovas Imag. 2019;35(3): 409–18.
168. Jinnouchi H, Torii S, Sakamoto A, Kolodgie FD, Virmani R, Finn AV. Fully bioresorbable vascular scaffolds: lessons learned and future directions. Nat Rev Cardiol. 2019;16(5): 286–304.
169. Standardization IOf. ISO 25539-1 Cardiovascular implants—Endovascular devices Part 1: Endovascular prostheses. Geneva: International Organization for Standardization; 2017.
170. Kong XY, Han BQ, Wang HX, Li H, Xu WH, Liu WS. Mechanical properties of biodegradable small-diameter chitosan artificial vascular prosthesis. J Biomed Mater Res A. 2012;100a (8):1938–45.
171. Simon CG Jr, Yaszemski MJ, Ratcliffe A, Tomlins P, Luginbuehl R, Tesk JA. ASTM international workshop on standards and measurements for tissue engineering scaffolds. J Biomed Mater Res Part B: Appl Biomater. 2015;103(5):949–59.

172. Manabe I, Shindo T, Nagai R. Gene expression in fibroblasts and fibrosis - involvement in cardiac hypertrophy. Circ Res. 2002;91(12):1103–13.
173. Courtney T, Sacks MS, Stankus J, Guan J, Wagner WR. Design and analysis of tissue engineering scaffolds that mimic soft tissue mechanical anisotropy. Biomaterials. 2006;27 (19):3631–8.
174. Stegemann JP, Kaszuba SN, Rowe SL. Review: advances in vascular tissue engineering using protein-based biomaterials. Tissue Eng. 2007;13(11):2601–13.
175. Hirt MN, Hansen A, Eschenhagen T. Cardiac tissue engineering: state of the art. Circ Res. 2014;114(2):354–67.
176. Voges I, Brasen JH, Entenmann A, Scheid M, Scheewe J, Fischer G, Hart C, Andrade A, Pham HM, Kramer HH, Rickers C. Adverse results of a decellularized tissue-engineered pulmonary valve in humans assessed with magnetic resonance imaging. Eur J Cardio-Thorac. 2013;44(4): E272–9.
177. Lei Y, Guo G, Jin W, Liu M, Wang Y. Riboflavin photo-cross-linking method for improving elastin stability and reducing calcification in bioprosthetic heart valves. Xenotransplantation. 2019;26(2):e12481.
178. Chen QZ, Harding SE, Ali NN, Lyon AR, Boccaccini AR. Biomaterials in cardiac tissue engineering: ten years of research survey. Mat Sci Eng R. 2008;59(1–6):1–37.
179. Crapo PM, Wang Y. Small intestinal submucosa gel as a potential scaffolding material for cardiac tissue engineering. Acta Biomater. 2010;6(6):2091–6.
180. Li Z, Guan J. Hydrogels for cardiac tissue engineering. Polymers. 2011;3(2):740–61.
181. Robison P, Caporizzo MA, Ahmadzadeh H, Bogush AI, Chen CYX, Margulies KB, Shenoy VB, Prosser BL. Detyrosinated microtubules buckle and bear load in contracting cardiomyocytes. Science. 2016;352(6284).
182. Mattera R, Benvenuto M, Giganti MG, Tresoldi I, Pluchinotta FR, Bergante S, Tettamanti G, Masuelli L, Manzari V, Modesti A, Bei R. Effects of polyphenols on oxidative stress-mediated injury in cardiomyocytes. Nutrients. 2017;9(5)
183. Vunjak-Novakovic G, Tandon N, Godier A, Maidhof R, Marsano A, Martens TP, Radisic M. Challenges in cardiac tissue engineering. Tissue Eng Part B: Rev. 2010;16(2):169–87.
184. Bejleri D, Davis ME. Decellularized extracellular matrix materials for cardiac repair and regeneration. Adv Healthc Mater. 2019;8(5):1801217.
185. Sanchez PL, Fernandez-Santos ME, Costanza S, Climent AM, Moscoso I, Gonzalez-Nicolas MA, Sanz-Ruiz R, Rodriguez H, Kren SM, Garrido G, Escalante JL, Bermejo J, Elizaga J, Menarguez J, Yotti R, del Villar CP, Espinosa MA, Guillem MS, Willerson JT, Bernad A, Matesanz R, Taylor DA, Fernandez-Aviles F. Acellular human heart matrix: a critical step toward whole heart grafts. Biomaterials. 2015;61:279–89.
186. Yasui H, Lee J-K, Yoshida A, Yokoyama T, Nakanishi H, Miwa K, Naito AT, Oka T, Akazawa H, Nakai J. Excitation propagation in three-dimensional engineered hearts using decellularized extracellular matrix. Biomaterials. 2014;35(27):7839–50.
187. Ferng AS, Connell AM, Marsh KM, Qu N, Medina AO, Bajaj N, Palomares D, Iwanski J, Tran PL, Lotun K. Acellular porcine heart matrices: whole organ decellularization with 3D-bioscaffold & vascular preservation. J Clin Transl Res. 2017;3(2):260.
188. Izadifar M, Chapman D, Babyn P, Chen X, Kelly ME. UV-assisted 3D bioprinting of nanoreinforced hybrid cardiac patch for myocardial tissue engineering. Tissue Eng Part C: Methods. 2018;24(2):74–88.
189. Wu WQ, Peng S, Song ZY, Lin S. Collagen biomaterial for the treatment of myocardial infarction: an update on cardiac tissue engineering and myocardial regeneration. Drug Deliv Transl Re. 2019;9(5):920–34.
190. Serpooshan V, Zhao M, Metzler SA, Wei K, Shah PB, Wang A, Mahmoudi M, Malkovskiy AV, Rajadas J, Butte MJ. The effect of bioengineered acellular collagen patch on cardiac remodeling and ventricular function post myocardial infarction. Biomaterials. 2013;34(36): 9048–55.

191. Roche ET, Hastings CL, Lewin SA, Shvartsman DE, Brudno Y, Vasilyev NV, O'Brien FJ, Walsh CJ, Duffy GP, Mooney DJ. Comparison of biomaterial delivery vehicles for improving acute retention of stem cells in the infarcted heart. Biomaterials. 2014;35(25):6850–8.
192. Norahan MH, Amroon M, Ghahremanzadeh R, Mahmoodi M, Baheiraei N. Electroactive graphene oxide-incorporated collagen assisting vascularization for cardiac tissue engineering. J Biomed Mater Res A. 2019;107(1):204–19.
193. Zhu D, Li Z, Huang K, Caranasos TG, Rossi JS, Cheng K. Minimally invasive delivery of therapeutic agents by hydrogel injection into the pericardial cavity for cardiac repair. Nat Commun. 2021;12(1):1–10.
194. Liang S, Zhang YY, Wang HB, Xu ZY, Chen JR, Bao R, Tan BY, Cui YL, Fan GW, Wang WX, Wang W, Liu WG. Paintable and rapidly bondable conductive hydrogels as therapeutic cardiac patches. Adv Mater. 2018;30(23).
195. Baei P, Jalili-Firoozinezhad S, Rajabi-Zeleti S, Tafazzoli-Shadpour M, Baharvand H, Aghdami N. Electrically conductive gold nanoparticle-chitosan thermosensitive hydrogels for cardiac tissue engineering. Mater Sci Eng: C. 2016;63:131–41.
196. Eslami M, Vrana NE, Zorlutuna P, Sant S, Jung S, Masoumi N, Khavari-Nejad RA, Javadi G, Khademhosseini A. Fiber-reinforced hydrogel scaffolds for heart valve tissue engineering. J Biomater Appl. 2014;29(3):399–410.
197. Baei P, Jalili-Firoozinezhad S, Rajabi-Zeleti S, Tafazzoli-Shadpour M, Baharvand H, Aghdami N. Electrically conductive gold nanoparticle-chitosan thermosensitive hydrogels for cardiac tissue engineering. Mat Sci Eng C-Mater. 2016;63:131–41.
198. Jackman CP, Ganapathi AM, Asfour H, Qian Y, Allen BW, Li YZ, Bursac N. Engineered cardiac tissue patch maintains structural and electrical properties after epicardial implantation. Biomaterials. 2018;159:48–58.
199. Kheradvar A, Groves EM, Dasi LP, Alavi SH, Tranquillo R, Grande-Allen KJ, Simmons CA, Griffith B, Falahatpisheh A, Goergen CJ, Mofrad MRK, Baaijens F, Little SH, Canic S. Emerging trends in heart valve engineering: Part I. Solutions for future. Ann Biomed Eng. 2015;43(4):833–43.
200. Dohmen PM, Lembcke A, Holinski S, Pruss A, Konertz W. Ten years of clinical results with a tissue-engineered pulmonary valve. Ann Thorac Surg. 2011;92(4):1308–14.
201. Ahn J-H, Kim J, Han G, Kim D, Cheon K-H, Lee H, Kim H-E, Kim Y-J, Jang T-S, Jung H-D. 3D-printed biodegradable composite scaffolds with significantly enhanced mechanical properties via the combination of binder jetting and capillary rise infiltration process. Addit Manuf. 2021;41:101988.
202. Hinderer S, Seifert J, Votteler M, Shen N, Rheinlaender J, Schaffer TE, Schenke-Layland K. Engineering of a bio-functionalized hybrid off-the-shelf heart valve. Biomaterials. 2014;35 (7):2130–9.
203. Saidy NT, Wolf F, Bas O, Keijdener H, Hutmacher DW, Mela P, De-Juan-Pardo EM. Biologically inspired scaffolds for heart valve tissue engineering via melt electrowriting. Small. 2019;15(24):1900873.
204. Steinhoff G, Stock U, Karim N, Mertsching H, Timke A, Meliss RR, Pethig K, Haverich A, Bader A. Tissue engineering of pulmonary heart valves on allogenic acellular matrix conduits: in vivo restoration of valve tissue. Circulation. 2000;102(Suppl_3):Iii-50-Iii-55.
205. Seifu DG, Purnama A, Mequanint K, Mantovani D. Small-diameter vascular tissue engineering. Nat Rev Cardiol. 2013;10(7):410.
206. Awad NK, Niu HT, Ali U, Morsi YS, Lin T. Electrospun fibrous scaffolds for small-diameter blood vessels: a review. Membranes-Basel. 2018;8 (1).
207. Yin A, Zhang K, McClure MJ, Huang C, Wu J, Fang J, Mo X, Bowlin GL, Al-Deyab SS, El-Newehy M. Electrospinning collagen/chitosan/poly (L-lactic acid-co-ϵ-caprolactone) to form a vascular graft: mechanical and biological characterization. J Biomed Mater Res A. 2013;101(5):1292–301.

208. de Valence S, Tille J-C, Mugnai D, Mrowczynski W, Gurny R, Möller M, Walpoth BH. Long term performance of polycaprolactone vascular grafts in a rat abdominal aorta replacement model. Biomaterials. 2012;33(1):38–47.
209. Zheng W, Wang Z, Song L, Zhao Q, Zhang J, Li D, Wang S, Han J, Zheng X-L, Yang Z. Endothelialization and patency of RGD-functionalized vascular grafts in a rabbit carotid artery model. Biomaterials. 2012;33(10):2880–91.
210. Yao Y, Wang J, Cui Y, Xu R, Wang Z, Zhang J, Wang K, Li Y, Zhao Q, Kong D. Effect of sustained heparin release from PCL/chitosan hybrid small-diameter vascular grafts on anti-thrombogenic property and endothelialization. Acta Biomater. 2014;10(6):2739–49.
211. Ju YM, Ahn H, Arenas-Herrera J, Kim C, Abolbashari M, Atala A, Yoo JJ, Lee SJ. Electrospun vascular scaffold for cellularized small diameter blood vessels: a preclinical large animal study. Acta Biomater. 2017;59:58–67.
212. Ulbrich K, Hola K, Subr V, Bakandritsos A, Tucek J, Zboril R. Targeted drug delivery with polymers and magnetic nanoparticles: covalent and noncovalent approaches, release control, and clinical studies. Chem Rev. 2016;116(9):5338–431.
213. Kikuchi A, Okano T. Pulsatile drug release control using hydrogels. Adv Drug Delivery Rev. 2002;54(1):53–77.
214. Liu Y, Liu Q, Liu Y, Ju F, Ma Q, He Q. In vivo evaluation of enhanced drug carrier efficiency and cardiac anti-hypertrophy therapeutic potential of nano-curcumin encapsulated photo-plasmonic nanoparticles combined polymerized nano-vesicles: a novel strategy. J Photochem Photobiol B: Biology. 2019;199:111619.
215. Singh B, Garg T, Goyal AK, Rath G. Recent advancements in the cardiovascular drug carriers. Artif Cells Nanomed Biotechnol. 2016;44(1):216–25.
216. Gupta P, Garcia E, Sarkar A, Kapoor S, Rafiq K, Chand HS, Jayant RD. Nanoparticle based treatment for cardiovascular diseases. Cardiovasc Haematol Disorders-Drug Targets. 2019;19(1):33–44.
217. Ferreira MP, Talman V, Torrieri G, Liu D, Marques G, Moslova K, Liu Z, Pinto JF, Hirvonen J, Ruskoaho H. Dual-drug delivery using dextran-functionalized nanoparticles targeting cardiac fibroblasts for cellular reprogramming. Adv Funct Mater. 2018;28(15):1705134.
218. Such GK, Yan Y, Johnston AP, Gunawan ST, Caruso F. Interfacing materials science and biology for drug carrier design. Adv Mater. 2015;27(14):2278–97.
219. Sahoo S, Kariya T, Ishikawa K. Targeted delivery of therapeutic agents to the heart. Nat Rev Cardiol. 2021.
220. Hoffmann R, Mintz GS, Dussaillant GR, Popma JJ, Pichard AD, Satler LF, Kent KM, Griffin J, Leon MB. Patterns and mechanisms of in-stent restenosis - a serial intravascular ultrasound study. Circulation. 1996;94(6):1247–54.
221. Nogic J, McCormick LM, Francis R, Nerlekar N, Jaworski C, West NE, Brown AJ. Novel bioabsorbable polymer and polymer-free metallic drug-eluting stents. J Cardiol. 2018;71(5):435–43.
222. Windecker S, Haude M, Neumann F-J, Stangl K, Witzenbichler B, Slagboom T, Sabaté M, Goicolea J, Barragan P, Cook S. Comparison of a novel biodegradable polymer sirolimus-eluting stent with a durable polymer everolimus-eluting stent: results of the randomized BIOFLOW-II trial. Circulation. 2015;8(2):e001441.
223. Hu T, Yang J, Cui K, Rao Q, Yin T, Tan L, Zhang Y, Li Z, Wang G. Controlled slow-release drug-eluting stents for the prevention of coronary restenosis: recent progress and future prospects. Acs Appl Mater Inter. 2015;7(22):11695–712.
224. Commandeur S, van Beusekom HM, van der Giessen WJ. Polymers, drug release, and drug-eluting stents. J Intervention Cardiol. 2006;19(6):500–6.
225. Acharya G, Park K. Mechanisms of controlled drug release from drug-eluting stents. Adv Drug Delivery Rev. 2006;58(3):387–401.

226. Zhao HQ, Jayasinghe D, Hossainy S, Schwartz LB. A theoretical model to characterize the drug release behavior of drug-eluting stents with durable polymer matrix coating. J Biomed Mater Res A. 2012;100a(1):120–4.
227. Garg S, Bourantas C, Serruys PW. New concepts in the design of drug-eluting coronary stents. Nat Rev Cardiol. 2013;10(5):248–60.
228. Parker T, Dave V, Falotico R. Polymers for drug eluting stents. Curr Pharmaceut Des. 2010;16 (36):3978–88.
229. Loo SCJ, Tan ZYS, Chow YJ, Lin SLI. Drug release from irradiated PLGA and PLLA multi-layered films. J Pharmaceut Sci. 2010;99(7):3060–71.
230. Pan CJ, Tang JJ, Weng YJ, Wang J, Huang N. Preparation and in vitro release profiles of drug-eluting controlled biodegradable polymer coating stents. Colloid Surface B. 2009;73(2): 199–206.
231. Babapulle MN, Eisenberg MJ. Coated stents for the prevention of restenosis: Part II. Circulation. 2002;106(22):2859–66.
232. Wang XT, Venkatraman SS, Boey FYC, Loo JSC, Tan LP. Controlled release of sirolimus from a multilayered PLGA stent matrix. Biomaterials. 2006;27(32):5588–95.
233. Raungaard B, Jensen LO, Tilsted H-H, Christiansen EH, Maeng M, Terkelsen CJ, Krusell LR, Kaltoft A, Kristensen SD, Bøtker HE. Zotarolimus-eluting durable-polymer-coated stent versus a biolimus-eluting biodegradable-polymer-coated stent in unselected patients undergoing percutaneous coronary intervention (SORT OUT VI): a randomised non-inferiority trial. The Lancet. 2015;385(9977):1527–35.
234. Li JY, Mooney DJ. Designing hydrogels for controlled drug delivery. Nat Rev Mater. 2016;1 (12).
235. Wang LL, Liu Y, Chung JJ, Wang T, Gaffey AC, Lu MM, Cavanaugh CA, Zhou S, Kanade R, Atluri P, Morrisey EE, Burdick JA. Sustained miRNA delivery from an injectable hydrogel promotes cardiomyocyte proliferation and functional regeneration after ischaemic injury. Nat Biomed Eng. 2017;1(12):983.
236. Vong LB, Bui TQ, Tomita T, Sakamoto H, Hiramatsu Y, Nagasaki Y. Novel angiogenesis therapeutics by redox injectable hydrogel - regulation of local nitric oxide generation for effective cardiovascular therapy. Biomaterials. 2018;167:143–52.
237. Panyam J, Labhasetwar V. Biodegradable nanoparticles for drug and gene delivery to cells and tissue. Adv Drug Delivery Rev. 2003;55(3):329–47.
238. Zweers MLT, Engbers GHM, Grijpma DW, Feijen J. Release of anti-restenosis drugs from poly(ethylene oxide)-poly (DL-lactic-co-glycolic acid) nanoparticles. J Control Release. 2006;114(3):317–24.
239. Chen R, Zhu C, Xu L, Gu Y, Ren S, Bai H, Zhou Q, Liu X, Lu S, Bi X. An injectable peptide hydrogel with excellent self-healing ability to continuously release salvianolic acid B for myocardial infarction. Biomaterials. 2021;120855.
240. Wang W, Xiang L, Gong L, Hu W, Huang W, Chen Y, Asha AB, Srinivas S, Chen L, Narain R. Injectable, self-healing hydrogel with tunable optical, mechanical, and antimicrobial properties. Chem Mater. 2019;31(7):2366–76.
241. Zhao HQ, Jayasinghe D, Hossainy S, Schwartz LB. A theoretical model to characterize the drug release behavior of drug-eluting stents with durable polymer matrix coating. J Biomed Mater Res A. 2012;100(1):120–4.
242. Alexis F, Venkatraman SS, Rath SK, Boey F. In vitro study of release mechanisms of paclitaxel and rapamycin from drug-incorporated biodegradable stent matrices. J Control Release. 2004;98(1):67–74.

Chapter 4
Drug Delivery Systems and Cardiovascular Disease

Xiaoli Wang, Xiaoyu Liang, Jianwei Duan, Youlu Chen, Huiyang Li, Kaijing Liu, and Jing Yang

4.1 Heart Disease

4.1.1 Atherosclerosis and Coronary Artery Disease

4.1.1.1 Atherosclerosis

The most common form of cardiovascular disease (CVD) is atherosclerosis (AS), which is a chronic inflammatory process of blood vessel walls caused by changes in cholesterol metabolism, coupled with endothelial cell dysfunction, resulting in the retention of low-density lipoprotein (LDL) under the endothelium [1]. These captured lipoproteins are modified through various mechanisms, such as reactive oxygen oxidation (ROS) and enzymatic cleavage, to convert them into pro-inflammatory particles [2]. The modified LDL further stimulates endothelial cells to produce adhesion molecules, leading to the adhesion and recruitment of monocytes on the arterial wall and differentiation into macrophages. Modified LDL was taken in by macrophages to form foam cells [3]. The accumulation of foam cells, cholesterol, and other substances forms the atherosclerotic plaque. Gradually, the plaque, or atheroma, becomes a more complex, advanced plaque [4]. Unstable plaque rupture can cause clinical events in the form of thrombotic occlusions [5].

The drugs used in the treatment of atherosclerosis are divided into five categories: vasodilators, such as calcium channel blockers or calcium antagonists; adjust lipide, antiplatelet drugs, thrombolytic and drugs and anticoagulants, and another. Unfortunately, the general oral medication is limited to the treatment of early atherosclerosis due to its non-specific distribution, adverse reactions, and low bioavailability

X. Wang · X. Liang · J. Duan · Y. Chen · H. Li · K. Liu · J. Yang (✉)
Tianjin Key Laboratory of Biomaterial Research, Institute of Biomedical Engineering, Chinese Academy of Medical Science and Peking Union Medical College, Tianjin, China
e-mail: wangxl@bme.pumc.edu.cn; yangjing@bme.cams.cn

Y.-D. Tang et al. (eds.), *Application of Biomaterials in the Treatment of Cardiovascular Diseases*, https://doi.org/10.1007/978-981-19-7712-1_4

[6, 7]. By using different kinds of drug delivery system, all these problems will be solved effectively.

Liposomes

Liposome is a kind of nanovesicles composed of bilayer phospholipid membrane. Liposomes have become drug delivery vehicles due to their biocompatibility, ability to trap hydrophilic or hydrophobic pharmaceutical preparations, and easy functionalization [8]. They may transport a variety of drugs, such as low molecular weight drugs, imaging agents, peptides, proteins, and nucleic acids. Liposomes can release the drug slowly, thus continuously exposing to the target area and improving the therapeutic effect.

Based on the fact that atherosclerosis is a chronic inflammatory disease, Chono et al. [9] investigated the efficacy of dexamethasone (DXM) incorporated into liposomes (DXM-liposomes) on atherosclerosis; DXM-liposomes were prepared to three different particle sizes to clarify the influence of particle size on the drug delivery to atherosclerotic lesions and the effect on atherosclerosis. DXM-liposomes with the size of 200 nm (55 μg DXM/kg) were significantly more potent on antiatherosclerosis than that of free-DXM (55 μg DXM/kg), suggesting that efficient delivery of DXM to the atherosclerotic lesions by DXM-liposomes induces an excellent antiatherosclerosis effect at a lower dose. In addition, targeted molecules are used to modify liposomes for selective delivery of therapeutic drugs.

Poh et al. [10] reported the potential of folic acid-modified liposomes as a carrier to deliver imaging and therapeutic drugs. As shown in Fig. 4.1, in order to determine whether folic acid-coupled nanocarriers can be used to target the treatment of inflammatory diseases, a PEG-coated liposome was prepared. The liposome contains folic acid-coupled PEG and can also transport imaging or therapeutic drugs. Experimental results prove that these folate liposomes specifically bind to folate receptor-positive cells and accumulate in the inflammation site of the atherosclerosis mouse model.

Early, noninvasive detection and diagnosis of atherosclerotic lesions are necessary to prevent and treat clinical endpoints. The combined use of specific anti-inflammatory proteins, high-dose contrast agents, and liposome particles can enable existing imaging technologies to better identify and visualize plaques, providing a basis for research and prospective treatment strategies [11]. Exteriorized phosphatidylserine (PS) residues present in apoptotic cells promote macrophage recognition by macrophage scavenger receptor/CD36 followed by rapid phagocytosis. Maiseyeu et al. investigated the utility of paramagnetic gadolinium liposomes enriched with PS (Gd-PS) in imaging atherosclerotic plaque [12]. Gd-PS-enriched liposomes enhance atherosclerotic plaque and colocalized with macrophages in experimental atherosclerosis. Studies of other liposomes used in atherosclerosis are shown in Table 4.1.

Micelles

Amphiphilic polymers (AMPs) are composed of branched hydrophobic regions and hydrophilic polyethylene glycol (PEG), which showed a good anti-atherosclerotic

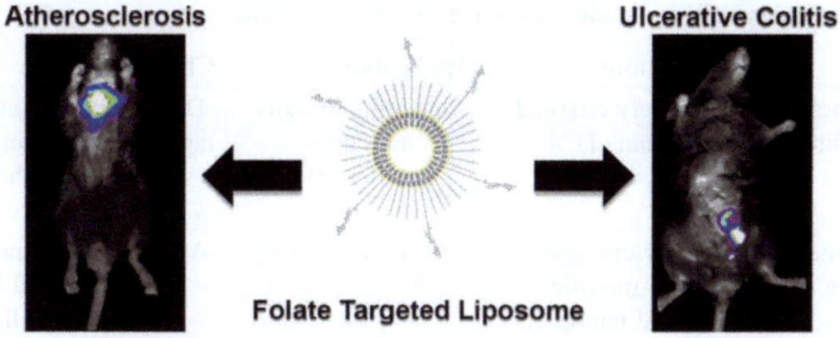

Fig. 4.1 Schematic illustration of folate-targeted liposomes for diagnostic and therapeutic applications. Reproduced with permission from [10]

effect through directly inhibiting the uptake of oxLDL by macrophages. AMPs with ether bonds between the hydrophobic arm and the sugar backbone show enhanced degradation stability and storage stability and enhanced anti-atherosclerotic biological activity [18]. Inflammation and oxidative stress are the two main factors involved in the pathogenesis of atherosclerosis. Some studies have constructed an intelligent drug delivery system that responds to the oxidative microenvironment of atherosclerotic plaque. As shown in Fig. 4.2, using polyethylene glycol and polythiopropylene block copolymer (PEG-PPS) as ROS-responsive materials, *Andrographis*-loaded micelles were prepared to simultaneously reduce inflammation and ROS levels to treat atherosclerosis [19]. Hou et al. [20] designed a novel dual ROS-sensitive and CD44 receptors targeting amphiphilic carrier material, oligomeric hyaluronic acid-2′-[propane-2,2-diyllbls (thio)] diacetic acl-hydroxymethylferrocene (oHA-TKL-Fc); oligomeric hyaluronic acid (oHA) was the linear polysaccharide that had high affinity for macrophage-specific. Thioketal linkages and ferrocene implemented dual ROS-sensitive strategy. Then, curcumin was loaded with oHA-TKL-Fc into nano-micelles by self-assembling method, becoming a promising strategy for atherosclerosis.

Metal Nanoparticles

Owing to the unique physical and chemical properties of metal nanoparticles, scientists have developed optical, electrical, and magnetic methods to regulate cell signaling pathways. Transient receptor potential vanilloid subfamily 1 (TRPV1) is a thermally sensitive cationic channel whose activation may reduce lipid storage and the formation of atherosclerotic lesions. Using the near-infrared light absorption characteristics of copper sulfide (CuS), scientists developed copper sulfide (CuS) nanoparticles coupled with antibodies targeting TRPV1 that can act as a photothermal switch for TRPV1 signaling in vascular smooth muscle cells (VSMCs). As shown in Fig. 4.3, after irradiation, local temperature rises to open the thermosensitive TRPV1 channel to prevent the progression of atherosclerosis [21].

Magnetic drug targeting is an effective method to deliver drugs to lesions in the body. Zhang et al. [22] developed a DNA-coated superparamagnetic iron oxide nanoparticles (DNA-SPIONs), which can be effectively accumulated in the

Table 4.1 Summary of drug-loaded liposome in atherosclerosis

Drug	Formulations	Application	Characteristic
Cyclopentenone prostaglandins	Negatively charged liposome [13]	Anti-inflammatory and anti-proliferative	Deliver cyclopentenone prostaglandin to the injured arterial wall cells of atherosclerotic mice
Prednisolone phosphate	Polyethylene glycol (PEG)-modified liposomal nanoparticle [14]	Anti-inflammatory	A single intravenous administration resulted in a strong vessel wall inflammation reduction in atherosclerotic rabbits
Prednisolone	PEG-coated liposome [15]	Anti-inflammatory	Half-life of prednisolone in vivo is prolonged by 63 h. In a randomized, placebo-controlled trial, intravenous LN-PLP appeared in 75% of the macrophages in the iliofemoral plaques of vascular surgery patients
Telmisartan	Folic acid modified liposomes [16]	Stabilize the advanced atherosclerotic lesions	Deliver angiotensin receptor blocker telmisartan (Tel) to macrophages
Atorvastatin calcium and curcumin	E-selectin binding peptide modified liposomes [17]	Hypolipidemic, anti-inflammatory, and antioxidative effects	Deliver liposomes to overexpress e-selectin of endothelial cells, leading to the synergistic inhibition of adhesion molecules (e-selectin and ICAM-1) and blood lipid levels

macrophages of atherosclerotic plaques and promote nanosystemic delivery of particles to atherosclerotic plaque after intravenous injection into ApoE(-/-) mice. Studies of other metal nanoparticles used in atherosclerosis are shown in Table 4.2. However, magnetic nanoparticles are potentially toxic to the cardiovascular system, especially atherosclerosis. Fe(3)O(4) nanoparticle surface coated with aminoguanidine (Fe(3)O(4)-AG NPs), which are absorbed by HUVECs and induce HUVEC apoptosis, remarkably promoted plaque vulnerability at low dose (0.5 mg/kg) [23]. The researchers selected iron oxide nanoparticles (Fe_2O_3 and Fe_3O_4) as two widely used nanomaterials to evaluate their potential risks to the human endothelial system. The results summarized that intravascular iron oxide nanoparticles can induce inflammation and dysfunction of the endothelial system through the following three ways: (1) nanoparticles can escape the phagocytosis that directly interacts with the endothelial monolayer; (2) nanoparticles are covered by monocyte phagocytosis and dissolution, impacting endothelial cells in the form of free iron ions; or (3) nanoparticles are swallowed by monocytes, triggering oxidative stress [24]. Studies have also shown that nano-ZnO particles may promote the occurrence of atherosclerosis and accelerate the formation of foam cells [25]. Based on these shortcomings, researchers continue to improve the function of materials, optimizing the delivery system.

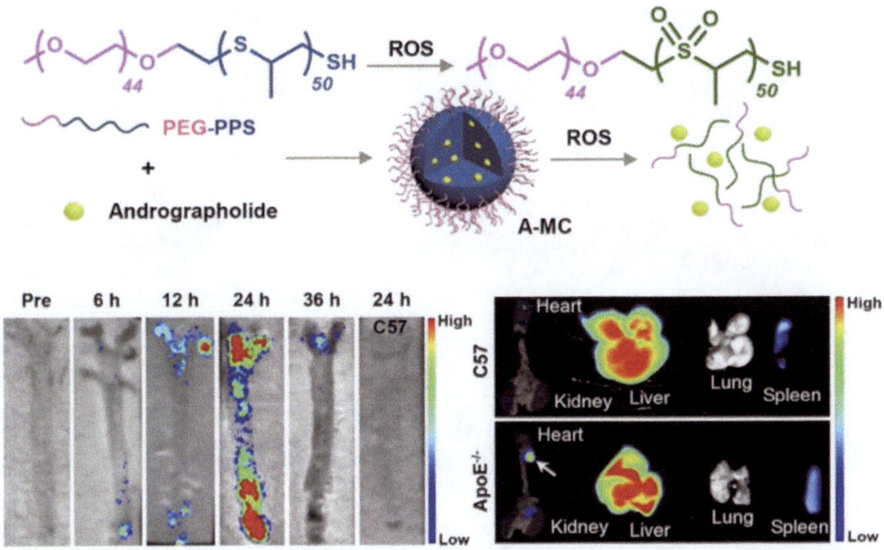

Fig. 4.2 Schematic illustration of the in vivo therapeutic mechanism for reducing ROS and pro-inflammatory cytokines in atherosclerotic plaques. Reproduced with permission from [19]

Polymers Nanoparticles

Scientists have also been interested in using synthetic polymers as nanocarriers for atherosclerotic treatment. The successful delivery of inflammation-resolving mediators to atherosclerotic plaques also requires an effective targeting strategy. The scientists chose to utilize biodegradable controlled-release polyesters in the form of poly (D-lactic acid) (PLA) and poly lactic-co-glycolic acid (PLGA) polymers to form protective NPs. The surface of the nanoparticles was modified with collagen IV (Col IV) for delivering anti-inflammatory cytokine interleukin 10 (IL-10) to atherosclerotic plaques in mice. The delivery process is shown in Fig. 4.4 [26].

Multifunctional nanoparticles are also used in atherosclerosis. Chmielowski et al. [27] used a kind of ROS-responsive polymer materials to prepare ferric acid poly (anhydride-ester) nanoparticles, which reduced the formation of foam cells by regulating the lipogenesis of macrophages and the production of reactive oxygen species. In addition, a redox-sensitive poly(oligo-l-arginine) (rsPOLA) was designed, which can be used as a targeted gene carrier for endothelium-2 adrenergic receptor (nitric oxide synthase, eNOS) and substrate [28].

In the atherosclerosis model of ApoE(-/-) mice, Alaarg et al. [29] compared three clinically relevant nanomedicines, including the high-density lipoprotein ([S]-HDL), polymer micelles ([S]-PM), and liposomes ([S]-LIP). These three formulas showed different biological behaviors in Apoe(-/-) mice. Although [S]-PM and [S]-LIP have longer circulatory half-lives, the cumulative levels of these three platforms in atherosclerotic plaques are similar. In addition, [S]-HDL and [S]-PM are taken up by plaque macrophages higher than [S]-LIP, while [S]-PM is taken up most by Ly6C

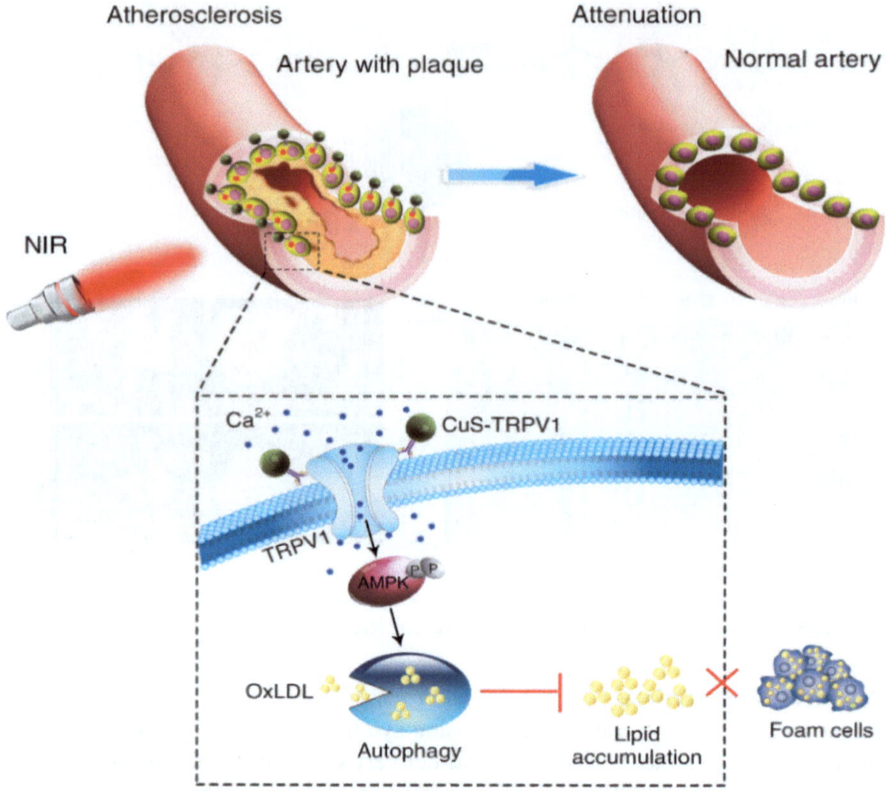

Fig. 4.3 Illustration of CuS-TRPV1 switch for photothermal activation of TRPV1 signaling to attenuate atherosclerosis. Reproduced with permission from [21]

(high) monocytes. Among the three dosage forms, [S]-PM has the best effect in reducing the load of macrophages in advanced atherosclerotic plaques. In summary, our data shows that [S]-PM is a promising targeted drug delivery system that can be used to treat atherosclerosis and other clinical inflammatory diseases. Studies of other polymers nanoparticles used in atherosclerosis are shown in Table 4.2.

Bionic Nanoparticles

Membrane-coated carriers are currently being widely used because of the function of surface protein of its membrane. Therefore, researchers use nanotechnology and biological structure to design bionic nanoparticles to better realize the possibility of reduced immunogenicity and targeted drug delivery.

The red blood cell membrane has good biocompatibility, long circulatory half-life ($\approx$120 days), and excellent performance [30], and it is also used in the research of encapsulating nanoparticles. As the Fig. 4.5 shows, the polylactic acid-glycolic acid copolymer (PLGA) nanoparticles loaded with rapamycin (RAP) and cloaked with the cell membrane of red blood cells (RBCs), to form a nanocomposite with highly complex functionalized biological activity. The biomimetic properties of the RBC

Table 4.2 Summary of drug-loaded nanoparticles in atherosclerosis

Class	Drug	Formulations	Characteristic
Metal nanoparticles	Dexamethasone phosphate	Superparamagnetic iron oxide nanoparticles (SPION) [34]	The study realized the magnetic targeting effect of SPION on the abdominal aorta. However, the use of dexamethasone-conjugated SPION to target vascular injury in a high-fat environment may lead to accelerated inflammation
	Rapamycin	Paramagnetic iron-oxide nanoparticles in a complex with low pH-sensitive cyclodextrin [35]	The system served not only as multifunctional imaging probes but also as nanocarriers for the treatment of atherosclerosis
Polymers nanoparticles	D-PDMP (a glycosphingolipid synthesis inhibitor)	Nanoparticles composed of polyethylene glycol (PEG) and sebacic acid [36]	The stay time of D-PDMP was extended from less than 1 h to at least 4 h (up to 48 h or longer), which increased the intervention effect on atherosclerosis by an order of magnitude
	(-)-Epigallocatechin gallate (EGCG)	Self-assembled nanoparticles of chitosan (CS) and aspartic acid (PAA) [37]	The average proportion of lipid deposition (16.9 ± 5.8%) after EGCG-CS-PAA nanoparticles treatment is significantly reduced, which is close to oral simvastatin (15.6 ± 4.1%); the curative effect on rabbit atherosclerosis is significantly improved
	Pioglitazone	Poly (lactic-co-glycolic-acid) (PLGA) nanoparticles [38]	The system inhibited the activation of macrophages in hyperlipidemia ApoE (-/-) mice and the rupture of atherosclerotic plaques
	Methotrexate	Lipid and polymer chains are nano-precipitated to form 100 nm spherical polymer nanostructures labeled with Cu [39]	Systemic delivery of methotrexate to macrophages via multifunctional, hybrid nano-constructs constitutes an effective strategy to inhibit atherosclerosis progression and induce, potentially, the resorption of vascular lesions

(continued)

Table 4.2 (continued)

Class	Drug	Formulations	Characteristic
Bionic nanoparticles	Rapamycin	Membrane proteins of J774 macrophages Bionic nanoparticles [40]	The number of proliferating macrophages is reduced, accompanied by a decrease in key pro-inflammatory cytokines and changes in plaque morphology
	Atorvastatin	Amphiphilic oxidation-sensitive chitosan oligosaccharide nanoparticles coated with macrophage membranes [41]	The macrophage membrane not only prevents NP from being cleared from the reticuloendothelial system but also guides NP to inflamed tissues, where the ROS responsiveness of NPs enables specific payload release

interface can reduce the phagocytosis mediated by macrophages in the blood and enhance the accumulation of nanoparticles in atherosclerotic plaques to achieve targeted drug release [31]. Dehaini et al. [32] successfully prepared a new type of cell membrane-encapsulated nanoparticles that combined the functions of two different cell types, which retained the characteristics of the platelet membrane and the red blood cell membrane, and the red blood cell membrane can significantly extend the cycle period.

In addition, the platelet membrane can also be researched in bionic nanoparticles. Here, a novel kind of drug-coated balloon is reported (Fig. 4.6). The drug-loaded porous nanomotors based on Janus aminated mesoporous silica (JAMS) that was obtained by asymmetric modification of platinum (Pt) nanoparticles are prepared. The platelet membrane is used to wrap the nanomotors to reduce the leakage of drugs before reaching the plaque. After the NIR irradiation, the nanomotors can not only penetrate into the plaque but also be effectively taken up by activated endothelial cells and inflammatory macrophages. The nanomotor can effectively decrease the vascular proliferation area through the combination treatment of photothermal and drug therapy in vivo, providing more clues for drug balloon coating with high efficiency in the treatment of atherosclerosis [33]. Studies of other bionic nanoparticles used in atherosclerosis are shown in Table 4.2.

Dendrimers

The application of dendrimers in the field of biomedicine with their unique dendritic structure and various surface properties has attracted widespread attention. Due to the large number of terminal groups, drug molecules can be attached to the surface of dendrimers through covalent bonds [42, 43], and the internal cavity can encapsulate small molecules [44].

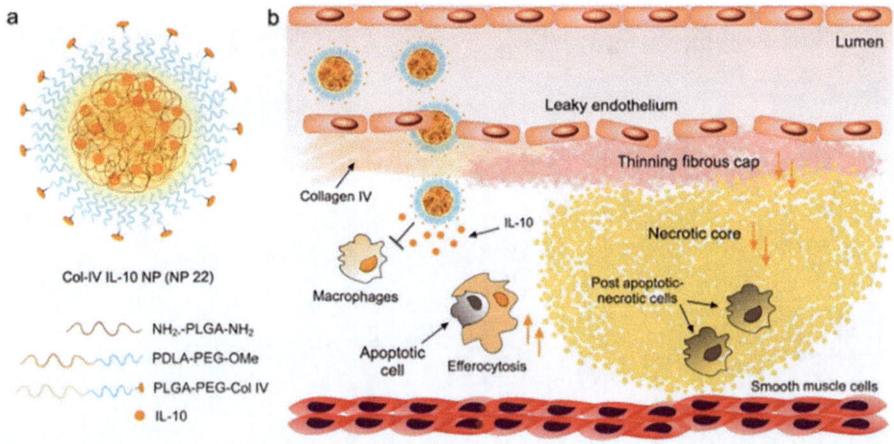

Fig. 4.4 Targeted anti-inflammatory NP design and application to resolution of inflammation in atherosclerotic plaques. Reproduced with permission from [26]

Targeting dendrimers show excellent delivery capabilities. Poly(ethylene glycol) coated, acetic anhydride-terminated, and folate-targeted poly(amidoamine) (PAMAM) dendrimers interact with macrophages expressing folate receptors in vitro cells bind and accumulate selectively in inflammatory areas in the body [45]. Due to the relatively low toxicity of dendritic macromolecules and the negligible activation of complementary reactions, they are considered to be one of the most effective and promising drug carriers. In this work, the aggregation of PAMAM zero generation dendrimers (G0) acting as drug delivery carriers, as well as conjugated G0 PAMAM dendrimers with a ZnPc photosensitizer, to symptomatic and asymptomatic human carotid tissues was investigated by using atomic force microscopy (AFM). [46]. Galactose functionalized PAMAM dendrimer generation 5 (Gal-G5)-specific delivery hepatic cholesteryl ester hydrolase (CEH) expression vector [47]. PAMAM dendrimers can solubilize lipophilic drugs and extend their circulation time; in addition, PAMAM dendrimers seem to have some accumulation on the inflammation side. The complex of indomethacin and 4-carboxymethoxy-pyrrolidone PAMAM dendrimer shows improved anti-inflammatory properties and temperature-dependent binding and release properties [48]. Mannose-functionalized dendritic nanoparticles selectively deliver LXR ligands to atherosclerotic plaque-associated macrophages [49].

Hydrogel

Yao et al. [50] developed a composite hydrogel drug delivery system. The hydrogel is a high molecular weight hyaluronic acid (HA) modified with glycidyl methacrylate as a hydrogel precursor; a cross-linkable block copolymer (CBC) is used as the drug coating material and a microscopic cross-linking agent. The composite hydrogel drug delivery system can not only release drugs mechanically and sensitively but also target inflammatory macrophages at thrombus.

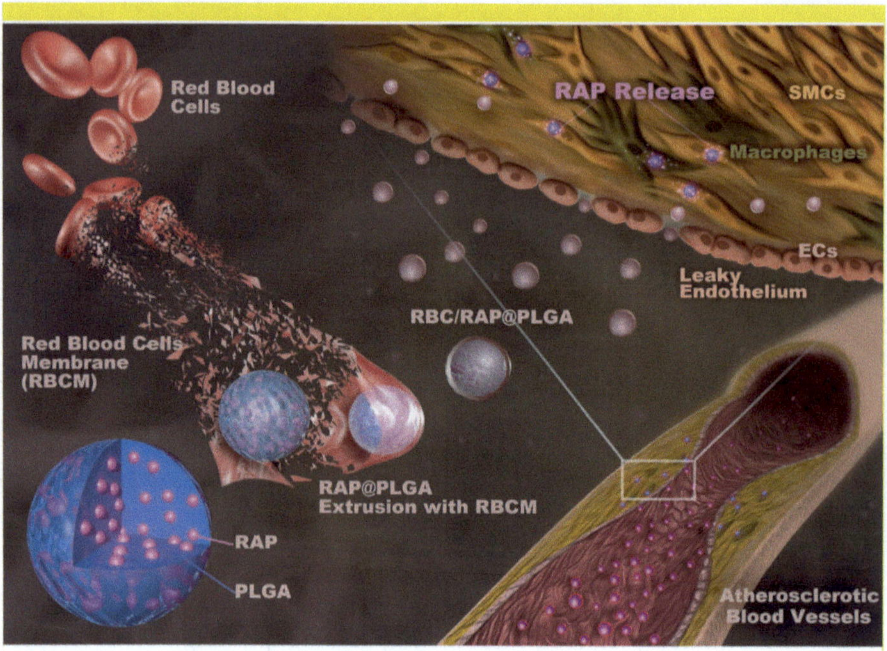

Fig. 4.5 Illustrations displaying the preparation of RBC/RAP@PLGA for the treatment of atherosclerosis. Reproduced with permission from [31]

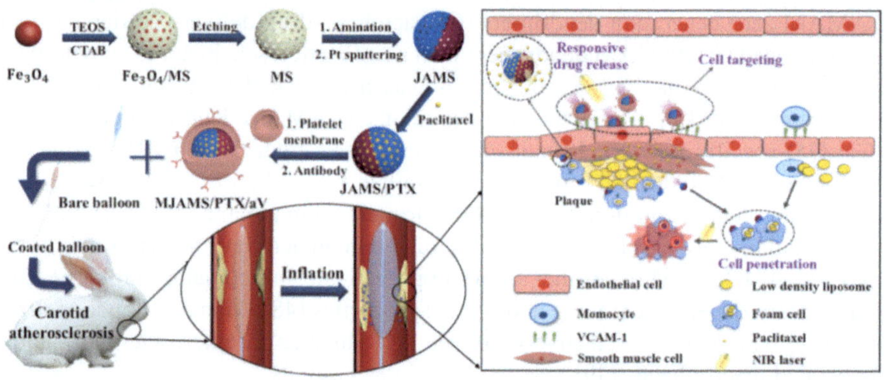

Fig. 4.6 Schematic illustration of the synthesis process of MJAMS/PTX/aV and the mechanism of treatment of atherosclerosis using the MJAMS/PTX/aV coated balloon. Reproduced with permission from [33]

Cell Carriers

Cell carriers are also used in the delivery of atherosclerotic drugs. There are multiple interactions between neutrophils (NEs) and atherosclerosis. One study used NEs as cell carriers loaded with cationic liposomes. Cell carriers based on NEs can effectively internalize cationic lipids. The plastids are sensitive to the chemotactic response of atherosclerotic inflammatory cells and finally achieve in vitro targeted transport into target cells. In addition, after injection of NE vector into AS animal model, the accumulation of NEs in atherosclerotic plaque was significantly enhanced in vivo. Therefore, NE-based cell carriers may become a new strategy to target the delivery of payloads to atherosclerotic plaques and promote the research and development of anti-atherosclerosis drugs to control diseases [51].

DDS has been widely used in atherosclerosis. Nanomaterials may serve as a powerful platform for diagnostic imaging and therapeutic control of atherosclerotic inflammatory cells, but their efficacy is limited by the non-specific uptake of mononuclear phagocytic system cells. The poly sulfonic mucopolysaccharide cells located in the liver, spleen, blood, lymph nodes, and kidneys, regardless of surface functionalization or targeting ligand coupling, can remove most of the intravenously injected nanomaterials from the blood circulation [52]. Nano-toxicity of nanoparticles to cardiovascular diseases has received special attention, and endothelial dysfunction caused by nanoparticles is considered to be the main mechanism of disease development. Researchers will continue to optimize function of materials and improve delivery systems. In the near future, clinical trials may prove the safety and effectiveness of DDS in patients with cardiovascular diseases.

4.1.1.2 Myocardial Infarction and Myocardial Ischemia Reperfusion Injury

Acute myocardial infarction (AMI) is myocardial necrosis caused by acute and persistent ischemia and hypoxia of the coronary artery. Because of some inducements, coronary atherosclerotic plaques rupture, platelets gather on the surface of ruptured plaques to form thrombus, and blocking the coronary artery lumen leads to myocardial ischemia and necrosis [53].

The main treatment strategies of MI include reducing myocardial oxygen consumption, reducing blood lipid, increasing coronary blood flow, improving left ventricular remodeling, and improving myocardial metabolic pathway. Drug treatments include the following: (1) Venous access should be established as soon as possible after admission; (2) small dose of morphine is the most effective analgesic.

Myocardial ischemia-reperfusion injury refers to the pathological process in which the ischemic myocardium can recover normal perfusion after partial or complete acute coronary artery occlusion, but the tissue damage is progressively aggravating. The mechanism is mainly related to the production of intracellular oxygen free radicals, calcium overload, leukocyte inflammation, and the lack of high-energy phosphate compounds [54].

The application of nano DDS in coronary heart disease is a feasible strategy, because the inflammatory environment enhances the permeability of target lesions and endows nano DDS with "targeted" characteristics.

Liposome

At present, the lesions of cardiovascular diseases usually show an increase in vascular permeability. Liposomes loaded with biomacromolecules are more likely to enter the circulation through the vascular wall directly and reach the lesions passively. Therefore, researchers can design and modify the carriers to increase the chemotaxis and the surface of cells and enhance the active targeting ability. Researchers designed the platelet-like protein liposomes (PLPS) to simulate the interaction between platelets and circulating monocytes. Compared with normal liposomes, PLPS did not accumulate on the uninjured endothelial cells, but accumulated at the injured site 72 h after injury. PLPS enhance the cardiac targeting of drugs [55]. Related studies are shown in Table 4.3. With the continuous combination of cardiovascular drug delivery technology and pathological mechanism, more and more new liposome drugs will be used in the diagnosis and treatment of cardiovascular diseases (Fig. 4.7).

Thrombosis is an important event in blood vessels, which is not only related to MI and stroke but also a large proportion in incidence rate and mortality rate. Thrombolytic drugs are usually used to treat AMI, acute cerebral infarction, and pulmonary embolism. However, due to the short half-life of plasma and systemic side effects, the clinical efficacy of thrombolytic drugs is limited. It is meaningful to improve the effectiveness of thrombolytic drugs clinically. Researchers prepared cyclic RGD (cRGD) functionalized liposomes to encapsulate urokinase and achieved good thrombolytic effect [56]. cRGD liposome can significantly reduce the dosage of urokinase by 75% in mice with thrombosis and achieve the same thrombolytic effect as free urokinase.

Reperfusion may eventually lead to irreversible damage of myocardial function. A variety of cells are involved in the multi-phase repair process, including inflammatory cells, vascular cells, and cardiac fibroblasts. For the treatment of these cell types in the infarct margin area, researchers developed liposomes with surface modification to contain peptide PARP-1 inhibitors that are affinity to myocardial cell type [57]. When all cell types were evaluated with liposomes, the inhibitory efficiency of PARP-1 was about 3 times higher than that of negative control peptide liposome.

The researchers tried to determine whether liposome nanoparticles could be displayed by different imaging methods. The results showed that compared with the control group, the T1-weighted signal intensity of ischemic and non-ischemic myocardium of mice receiving liposome was significantly increased; liposome group preferentially arrived at the site of myocardial injury, which improved the diagnostic rate of myocardial injury [58].

Nanoparticles

Nanoparticles are new strategies for the prevention, diagnosis, and treatment of cardiovascular diseases and are expected to become a new method for the treatment of cardiovascular diseases. Nanosystems can be used as tools for the treatment and diagnosis of coronary artery disease and MI. Nanoparticles are widely used in noninvasive imaging, targeted drug delivery, and photothermal therapy [65].

Table 4.3 Summary of AMI drug-loaded liposome

Drug	Formulation	Characteristic	Application
CCR2 silencing siRNA	C12-200, disteroylphosphatidyl choline, cholesterol, PEG-DMG liposome [59]	Decrease of atherosclerotic plaque and infarct area in mice with MI	Targeting cardiac inflammatory cells
Cobalt protoporphyrin	Platelet-like proteoliposomes [55]	Accumulated at the injured site 72 h after injury	Targeting: biomimicking platelet interactions with circulating monocytes
VEGFR2 inhibitor (PTK787)	Cardiomyocyte (I-1) and endothelial targeted (B-40) liposomes [60]	Significantly reduced angiogenesis; reduce the minimum effective dose of systemic administration	Targeted intra-organ drug delivery through cell type-specific administration
Eplerenone; MR antagonists	Phospholipid liposome [61]	Improved the healing response; prevent cardiac remodeling and deterioration of function	Targeting: identify the macrophage MR (mineralocorticoid receptor) as a crucial pathogenic player in cardiac wound repair after MI
Plasminogen activators	Platelet microparticle (PMP)-inspired nanovesicle (PMIN) liposome [62]	Anchor actively onto thrombus via PMP-relevant molecular mechanisms and allow drug release via thrombus-relevant enzymatic trigger	Thrombus-targeted delivery and site-specific action of thrombolytic drugs
Sphingosine-1; myoblast	Sphingosine-1-phosphate (S1P) liposomes [63]	Apoptosis of transplanted cells and infarct size were significantly reduced	Myocardial ischemia reperfusion injury
pH (low) insertion peptide (pHLIP)	pH-sensitive pHLIPs [wild type (WT) and Var7] or WT-pHLIP-coated liposomes [64]	Enables targeting of cells in tissues with low extracellular pH	Delivery of pharmaceutical agents or imaging probes to the myocardial ischemia regions

In atherosclerosis, MI, and restenosis, many innate immune cells increase, resulting in a strong inflammatory response. Monocytes and macrophages play a key role in post-traumatic healing. Polymer nanoparticles were used for the treatment of MI and restenosis, to develop monocytes as a strategy for biological targeted therapy and diagnosis of nano drug delivery system [66]. Researches showed good therapeutic effects by targeting macrophages with nanoparticles shown in Table 4.4.

The instability and rupture of atherosclerotic plaque is the main cause of acute MI. Inflammation is a mediator of poor ventricular remodeling after MI, which may lead to reduced ejection fraction and subsequent heart failure. The application of nano DDS is a unique and promising method to prevent cardiovascular events including AMI by regulating cardiovascular system inflammation. For instance, nanoparticle-mediated pioglitazone targeting inflammatory monocytes [67]

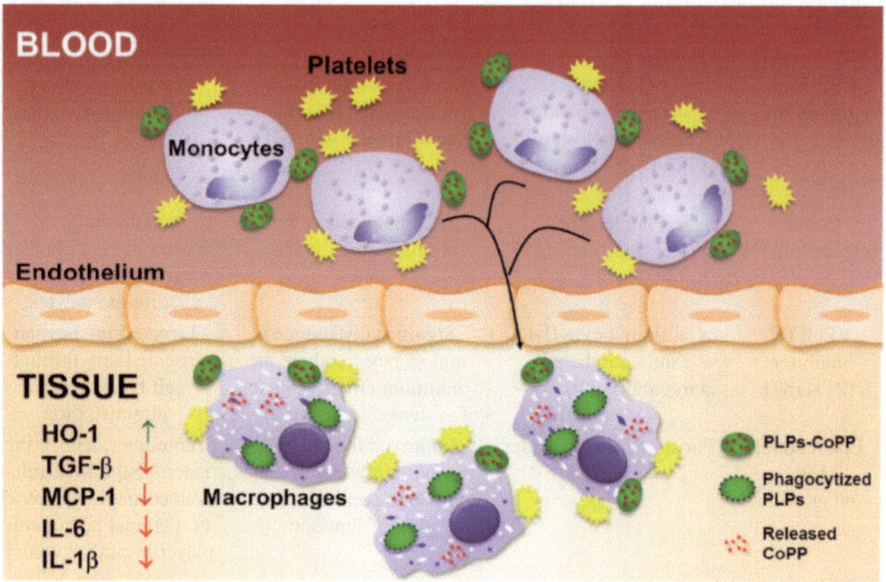

Fig. 4.7 Biomimicking platelet-monocyte interactions as a novel targeting strategy for heart healing. Reproduced with permission from [55]

antagonized monocyte-/macrophage-mediated acute inflammatory response; it can promote cardiac healing after MI and protect the heart from IR injury and myocardial remodeling. Other studies are listed in Table 4.4.

After AMI, due to the low proliferation of myocardial cells and limited self-healing ability, cardiac function will decline and cannot recover as before. Conventional myocardial blood supply cannot restore the apoptotic cardiomyocytes. Therefore, stem cell therapy has become a new treatment. Iron oxide superparamagnetic nanoparticles, due to their unique magnetism and good biocompatibility, can be used to guide and monitor the therapeutic effect of stem cells on AMI. It has been recognized as one of the most promising stem cell markers [68]. Blocki et al. designed MSCs pre-labeled with iron oxide nanoparticles which can be traced by magnetic resonance imaging (MRI) within weeks after transplantation into infarcted myocardium [69]. Histological analysis confirmed that the transplanted cells were integrated in the infarct site. Some researchers have shown that AuNP/GMT/PEI nanocomposites can effectively transform into cardiomyocytes after MI, thus effectively restoring cardiac function and scar area, which can be used to develop effective therapy for cardiac regeneration in patients with heart disease (see Fig. 4.8) [70].

Controllable induction of angiogenesis is an important therapeutic target for ischemic diseases, which is beneficial to various normal physiological processes. The researchers found that nanoparticles of celecoxib effectively induced therapeutic angiogenesis in normal and ischemic organs in animal models [71]. VEGF's half-life is short. When encapsulated in nanoparticles, VEGF was released for 31 days, which improved the left ventricular function of infarcted mice heart and increased the therapeutic effect of VEGF [72].

Nanoparticles also have an application in the treatment of MI, that is, polymer-coated stent—slow release of drug—increases the long-term elution of stent. The development of coronary drug-eluting stents (DES) has introduced new structural changes of bioabsorbable polymers. These advances are expected to improve long-term effectiveness and safety profile of DES. In a long-term clinical trial, researchers compared patients with long-term use of eluting polymer stents (81 μ BP and everman). In a large randomized trial, significant differences in TLF and target vessel-related MI persisted for 2 years, and BP-SES was more suitable for treatment than DP-EES. The cumulative target lesion revascularization and late/very late stent thrombosis of BP-SES were also significantly reduced [73].

Heart tissue engineering is the key to the treatment of MI. The intelligent materials release will be helpful. A small molecule (3i-1000) was loaded into a new biodegradable conductive patch for application in infarcted myocardium [74]. The high-density myoblasts on the patch could survive for 1 month at least. 3i-1000 could induce cell proliferation. The conductive patch showed high blood wettability and drug slow release. The developed biomaterials meet the mechanical, electrical, and biological requirements of cardiac therapy.

Nanoparticles have also been used in the diagnosis and detection of MI. The photoelectrochemical (PEC) sensor based on carbon dots (CDs) has been successfully applied to the detection of GSH in serum and MI [75]. Cardiac magnetic resonance (CMR) is used to evaluate the structure, function, perfusion, and viability of the heart due to its high spatial resolution and no ionizing radiation. However, it has limitations in the assessment of tissue and myocardial pathophysiological changes after MI. Nanoparticle-based contrast agents provide tracking MI at the cellular and molecular levels [76]. Other studies are listed in Table 4.4.

Hydrogel

Hydrogel as a carrier system is particularly beneficial because they are structurally like the extracellular matrix (ECM) [87]; it can be processed under relatively mild conditions and can be delivered in a minimally invasive manner. In addition, hydrogels can be designed to degrade in the process of angiogenesis. For these reasons, hydrogel shows great potential as a substrate for angiogenesis.

Injectable biopolymer hydrogel as a scaffold has attracted widespread attention in promoting cardiac function and preventing left ventricular negative remodeling after MI. Fan et al. have prepared a bifunctional MI-responsive hydrogel, which promoted angiogenesis and inhibited myocardial remodeling by targeting upregulating matrix metalloproteinase -2/9 (MMP-2/9) as shown in Fig. 4.9 [88].

The myocardium is a relaxing tissue alternatively with electric pulses. Soft and conductive hydrogels may be conductive to the transmission of mechanical and electrical signals. Mihic et al. synthesized a biocompatible conductive biomaterial (PPy-chitosan), which can enhance the biological conductivity [89]. PPy-chitosan

Table 4.4 Summary of AMI drug-loaded nanoparticles

Drug	Formulation	Characteristic	Application
MiRNA-21	Hyaluronan sulfate (HAS) Ca^{2+}-miRNA NPs [77]	Control macrophage phenotype; reduce remodeling and heart failure	Target cardiac macrophages at the infarct zone
Toll-like receptor 4 inhibitor TAK-242	Polylactic acid, glycolic acid nanoparticles containing TAK-242 [78]	Enhanced the drug delivery to monocytes/ macrophages	Target monocyte/ macrophage-mediated inflammation
(18) F-emodin	Polyglucans nanoparticle [79]	Increasing the signal of infarct focus and atherosclerotic plaque	Enriched in macrophages of heart and plaque
Statins	Polylactic acid (PLA), polyglycolic acid (PGA), and poly lactic-co-glycolic acid (PLGA) [80]	Reduce plaque instability and rupture, myocardial ischemia-reperfusion injury, and ventricular remodeling	Anti-inflammatory effect
Pioglitazone	Pioglitazone—PLGA nanoparticles [81]	Reduce $Ly6C^{high}$ inflammatory monocytes and induced M2 macrophage-associated genes	Regulating inflammation in the cardiovascular system
siRNA targeting Icam1, Icam2, Vcam1, Sele, and Selp (siCAM5)	7C1/C14PEG2000 poly vascular endothelial affinity nanoparticles [82]	Effectively silence five key adhesion molecules	Reduce the recruitment of neutrophils and monocytes in atherosclerotic lesions
Deoxyribozyme (DNAzyme) which silence TNF-α	DNAzyme functionalized gold nanoparticles [83]	50% knockdown of TNF-α in the rat myocardium	Significant anti-inflammatory effects
Paramagnetic quantum dot tripeptide Asn-GlyArg (cNGR)	Calcium-binding compound gadolinium oxide [84]	Target of tripeptide CGR is CD13, strongly upregulated on the surface of vascular endothelial cells	Imaging of myocardial neovascularization
Iron oxide	DSPE-PEG coated magnetic field-responsive nanocubes (MFRFs) [85]	Selective targeting of macrophages in the infarcted tissue and MRI monitoring	Perform qualitative and quantitative MRI of MI
Anionic soybean peroxidase (SBP)	Nanoprobes (Ab2-SiO_2-SBP) [86]	Immunosensing biochip for simultaneous detection of three biomarkers	Chemiluminescent imaging

injection is better than non-conductive polymer in maintaining cardiac function after MI. Other conductive hydrogels are listed in Table 4.5.

Sometimes, hydrogels tested in preclinical studies are not suitable for minimally invasive catheter delivery because of the excessive viscosity of materials, the excessive gelation time, and the possibility of blood compatibility and embolization.

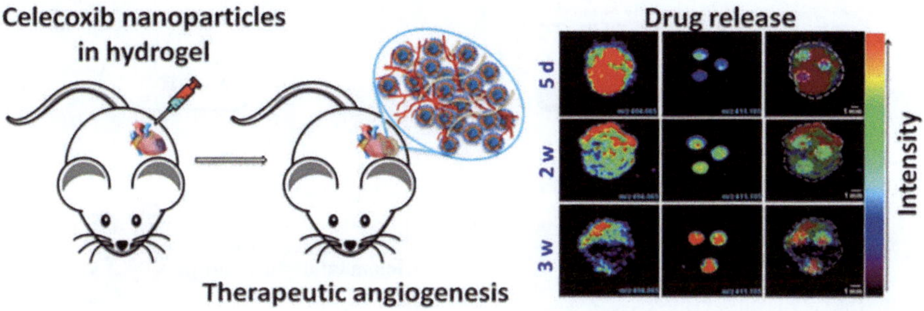

Fig. 4.8 Efficient in vivo direct conversion of fibroblasts into cardiomyocytes using a nanoparticle-based gene carrier. Reproduced with permission from [70]

Researchers have developed a technology for the preparation of space-restricted cyclic peptides, low resistance injections, and rapid assembly into hydrogels when the disease-related enzymes are linearized [90]. The mobility of hydrogels proved its usefulness in vivo. Functionalization allows the material to flow through the cardiac injection catheter without clogging without reducing blood compatibility. These studies laid the foundation for developing structural hydrogel therapeutic biomaterials.

For cardiac tissue engineering, encapsulation of stem cells from explants in protective hydrogels can promote long-term cell survival before intramyocardial injection [79]. By changing the physical properties of hydrogels, the paracrine signal of encapsulated cells can be enhanced, thus promoting the endogenous repair of damaged myocardium. To study the structure of core shell gel hydrogel composed of fibrin-specific poly (N-isopropylacrylamide) nanomaterials (FSNs), a dual drug delivery system was designed to meet the need of reconstructing blood flow and inhibiting cardiac fibrosis after I/R damage [91]. Other hydrogels for cardiac tissue engineering are listed in Table 4.5.

The miRNA nanoparticle-hydrogel system has great potential for further development to improve treatment. Injectable hydrogel/miNP complex can deliver miRNA safely and effectively to repair damaged myocardium [92]. This shear thinning elastin like hyaluronic acid hydrogels enables smooth and rapid injection to targets, effectively releasing miNPs and achieving local retention.

Viral Vector

The virus is also used in the treatment of myocardial infarction due to its high transfection efficiency. Viral drug delivery can reduce bleeding from antithrombotic therapy. Adeno-associated virus (AAV) has become one of the most promising tools for gene transfer because it has low pathogenicity and can proliferate effectively with high titer. As a new therapeutic strategy, regenerative therapy is highly valued but has obstacles, such as potential tumor formation risk and low survival rate. However, through viral vector with therapy gene, new cardiomyocytes can be produced, and scar tissue can be reduced to restore cardiac function. Related studies are listed in Table 4.6.

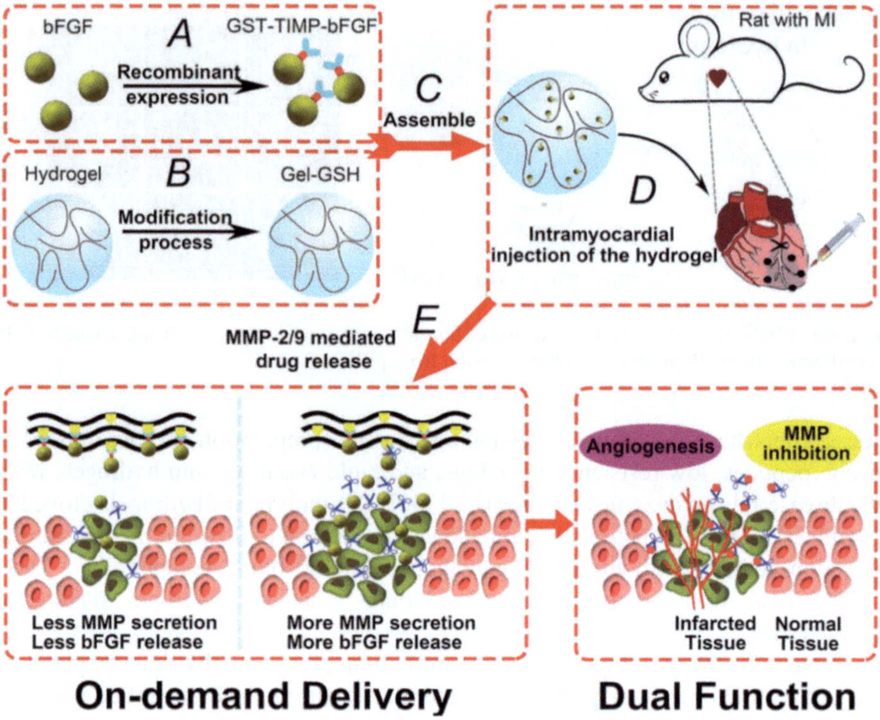

Fig. 4.9 Myocardial infarction-responsive smart hydrogels targeting matrix metalloproteinase for on-demand growth factor delivery. Reproduced with permission from [88]

4.1.1.3 Angina Pectoris

Angina pectoris is a clinical syndrome caused by coronary artery insufficiency, AMI, and hypoxia [105]. At present, the therapeutic drugs include lipid-regulating drugs, nitrates, beta-blockers, calcium channel blockers, antiplatelet agents, vasodilators, and so on [106]. The use of α-blockers in coronary heart disease usually combines with β-blockers [107]. However, the use is severely limited by hypotension, decreased coronary perfusion pressure, and risk of reflex oxygen-consuming tachycardia.

The development of drugs provides a variety of new DDS for improving drug efficacy. Angina pectoris is in urgent need of treatment and drug must be absorbed within a few minutes. Therefore, a rapid release system has been developed, such as orally disintegrating tablets and nasal spray. For drugs with short blood half-life, the concentration of drugs appears obvious peak valley phenomenon. A sustained-release system or pulse drug delivery system with delayed pulse release should be developed.

Table 4.5 Summary of AMI drug-loaded hydrogel

Drug	Formulation	Characteristic	Application
Late-gadolinium for MRI	Injectable hyaluronic acid-based hydrogel [93]	Reduced left ventricular volume, improved ejection fraction, and increased wall thickness	Improve cardiac structure and function
Graphene oxide (GO)	PEG-MEL/HA-SH/GO hydrogel [94]	Has fatigue resistance and electrical conductivity	Conductive hydrogel
Conductive polypyrrole	Fe^{3+} sprayed hydrogel covalently linked to pyrrole and dopamine [95]	Electrophysiological signal transmission was effectively promoted	Conductive and adhesive hydrogel
Tetraaniline (a conductive oligomer), adipose-derived stem cells (ADSCs), H_2S	ALG-CHO and gelatin—conductive H_2S releasing hydrogel [96]	Anti-inflammatory and stimulating angiogenesis	Conductive and adhesive hydrogel
Neonatal rat ventricular myocytes (NRVMs)	Injectable reverse-thermo gel (RTG)-AuNP hydrogel [97]	Targeted survival of NRVMs for up to 21 days, leading to an increase in connexin 43 (Cx43)	Biomimetic and conductive hydrogel for cardiac tissue engineering
Brown adipose-derived stem cells (BADSCs)	Clad-based zwitterionic starch hydrogel [98]	High efficiency of spontaneous myocardial differentiation	Cardiac tissue engineering
BADSCs	Fullerenol/alginate hydrogel [99]	Reduce ROS in lesion; improve the retention and survival rate of BADSCs, induce angiogenesis	Cardiac tissue engineering

Liposomes

The application of liposomes in angina pectoris may include contrast agents in imaging techniques, delivery of drugs to plaque sites. Intravenous liposomal prostaglandin E1 (Lipo-PGE1) reduced the incidence of perioperative myocardial injury in patients with unstable angina undergoing selective PCI [108]. Researchers developed a pre-liposome formulation to reduce protocatechuic aldehyde through the liver for the first time through metabolism and then coated with granules to change drug release in chronic treatment of angina pectoris, improving oral bioavailability and expected plasma concentration time course.

Nanoparticles

Many drugs for angina pectoris are sublingual. However, many drugs have low mucosal utilization. Nanoparticles can be used to improve the bioavailability of drugs. Lercanidipine is a selective dihydropyridine calcium antagonist, which is mainly used in the treatment of hypertension and angina pectoris. However, its bioavailability is less than 10% [109]. The drug loaded by nanoparticles permeability was increased about 4.5 times through the oral mucosa of pigs. Verapamil is a calcium channel blocker, which is effective in the treatment of angina pectoris. However, bioavailability is low due to first pass effect. Research and development

Table 4.6 Summary of virus vector for AMI

Encapsulated genes/drugs	Viral vector	Characteristic	Application
Tissue plasminogen activator (tPA)	Plant viruses (TMV) [100]	Less effect on hemostasis and shorter bleeding time	Target cardiovascular thrombosis
Cardiac-specific troponin T (TnT)	AAV9 [101]	High transgene expression levels were found in the heart	Increased the transcriptional activity
Insulin-like growth factor EA (IGF-1Ea)	AAV [102]	Helpful for left ventricular remodeling	Promote myocardial growth and regeneration
Tissue protective factor of array cDNA library from mouse secretion	AAV2 or 9 [103]	Sustain autophagy and remove dysfunctional mitochondria	Protects skeletal muscle and heart from ischemic damage
Cardiac-specific transcription factors (GATA4, MEF2C, TBX5, collectively known as GMT)	Sendai virus (SeV) [104]	Reprogram fibroblasts into nonintegrated cardiomyocytes	Regenerative therapy, restore cardiac function

of hybrid verapamil dextran nanostructured lipid carriers significantly increased the cell uptake of verapamil [110].

4.1.2 *Hypertension*

Hypertension means that the pressure in blood vessels continues to exceed the normal range, that is, the systolic blood pressure exceeds 140 mmHg, or the diastolic blood pressure exceeds 90 mmHg. It is a chronic cardiovascular disease. If not treated in time, serious organ damage and lesions may occur, such as atherosclerosis, cerebral stroke, myocardial infarction, kidney failure, etc. And complications such as cerebral thrombosis and cerebral hemorrhage may be produced.

At present, the pathogenesis of hypertension is not yet clear, which may be the result of the combined effect of genetics and environment. Some survey results show that 30–50% of hypertensive patients have a genetic background, which may be caused by polygenic inheritance. In addition, with the increase of age, the elasticity of arterial blood vessels weakens, and lipids remain in the walls of blood vessels, and there will be a rise in blood pressure, which is related to mood swings, insulin resistance, or obesity. The pathophysiological mechanism may be related to sympathetic nerve excitation and nerve mediator release.

In clinical terms, the principles of treating hypertension are controlling blood pressure, maintaining treatment, improving symptoms, and preventing complications. Existing treatment methods include basic treatment, drug treatment, and device treatment. Basic treatment is to improve the patients' life status from diet, exercise, etc., such as less salt when eating, more exercise, and reducing alcohol consumption. Drug treatment is to use β-blockers, diuretics, calcium antagonists, and so on to achieve the effect of reducing blood pressure. Device therapy includes radio-frequency therapy, baroreflex therapy, etc. The above methods have improved the state of hypertension in patients to a certain extent, but it is worth noting that hypertension is a lifelong disease and it is not easy to cure. Even if the blood pressure is controlled within the normal range after blood pressure reduction, many risk factors other than increased blood pressure still have an important impact on prognosis.

As we all know, drug delivery system delivers right amount of drug to right place at right time, thereby increasing the efficiency of drug use, improving the curative effect, reducing the cost, and reducing the toxic and side effects. Therefore, the combination of the drug delivery system and drug for treating hypertension can achieve slow and controlled release of the drug, thereby reducing the number of administrations and reducing the risk of toxic and side effects. In addition, through direct modification of water-soluble polymers (such as PEG), or the use of micelles, liposomes, and other carriers to encapsulate insoluble drugs, the solubility and dissolution rate of insoluble drugs can be improved. The application of drug delivery system in the treatment of hypertension diseases will be discussed in detail below.

Nanoparticles

Diuretics are first-line antihypertensive drugs. If patients use three antihypertensive drugs at the same time, one of them must be diuretics. In the process of lowering blood pressure, the dosage of diuretics is generally small, and patients rarely have side effects. The advantages of existing drug delivery are controlling release and extending action time. For example, the preparation of lecithin/chitosan nanoparticles loaded with hydrochlorothiazide (HCT) and the complex of hydrochlorothiazide and β-cyclodextrin (HCT-β-CD) was carried out in vitro, which showed sustained release [111]. In addition, in vivo activity of DOCA-induced hypertensive rats showed a 1.5% reduction in systolic blood pressure and prolonging action time.

Calcium antagonists are a class of drugs that selectively block Ca^{2+} from entering cell through calcium channel on the cell membrane at channel level, reducing intracellular Ca^{2+} concentration. Generally, calcium antagonists can be absorbed orally, but due to the strong first-pass effect, the bioavailability is low. Therefore, the use of drug delivery systems can improve this problem. Because nifedipine has poor water solubility, and cyclodextrin has high solubility and strong inclusion ability in water, preparation of cyclodextrin-based nanoparticles [112] can increase the oral solubility of drugs (such as nifedipine), which have poor water solubility, thereby increasing bioavailability.

Combining the above studies, we can draw the conclusion that changing the ratio of components of carrier or modifying or grafting the components can achieve sustained and controlled release of drugs and improve the drugs' bioavailability.

In addition, with the development of science and technology, new processes are emerging. The carrier prepared by new process can achieve local targeted delivery of drugs to a certain extent, improve bioavailability, and achieve a better therapeutic effect. Eun-Sol Ha et al. developed lercanidipine-hydroxypropyl methylcellulose (HPMC) nanoparticles to improve oral bioavailability, using supercritical antisolvent (SAS) process [113]. Furthermore, Taguchi design (TgD) can used to improve oral bioavailability of antihypertensive drugs. The research shows that solid lipid nano-bioparticles (ANbp) loaded with a new bio-enhancing agent (rutin) can reach the above effects [114].

Angiotensin-converting enzyme inhibitors control hypertension by inhibiting angiotensin II biosynthesis. Captopril is a commonly used medicine of its type. Aliskiren was encapsulated in magnetic poly(D, L-lactide) nanoparticles by a modified nanoprecipitation method, and encapsulated aliskiren can lower the blood pressure of male spontaneously hypertensive rats, which is more significant than ordinary administration.

Angiotensin receptor blockers bloke the binding to angiotensin receptors, which can directly cause vasodilation, reducing vasopressin secretion, aldosterone synthesis, secretion reduction, etc., which reduce blood pressure in combined effect. Satan drugs such as valsartan and olmesartan are its representatives. The research of Bapi Gorain et al. and Nilima T. Pandya et al. were based on olmesartan to prepare nanoemulsion and solid lipid nanoparticles [115, 116], respectively, to improve the oral bioavailability of relatively insoluble OM, thereby ameliorating clinical application.

Another study found that mRNA encoding angiotensin-converting enzyme I (ACE1) related to angiotensin II synthesis was selected as the target, and the oligonucleotide-containing polylysine was fixed to design nanocomposites on TiO_2 nanoparticles (TiO_2 central point PL-DNA), and further introducing the materials into hypertensive ISIAH rats can significantly reduce systolic blood pressure (reduced by 20–30 mmHg) [112]. Therefore, it can be seen that gene therapy for RNA may become a new and effective method for treating cardiovascular diseases.

In summary, we can see that the main challenges of antihypertensive drug therapy are its extremely hydrophobic and extremely low oral bioavailability, which will lead to higher doses/frequency and related drug side effects, while using drugs delivery systems can improve this problem, increase bioavailability, and ameliorate clinical application.

Liposome

Among the documents searched, the earliest was the preparation of liposomes by infiltration of propranolol hydrochloride and lecithin in methanol-chloroform into microporous sorbitol in 1995 [113]. The liposome has good fluidity and slow-release characteristics by controlling the ratio of drug/lecithin/sorbitol and the particle size of sorbitol and realizes continuous administration of propranolol.

Liposome drug delivery system has a broad application prospect. After liposome modification, liposomes can be given many functions, such as improving efficacy, reducing adverse reactions, penetrating biological barriers, etc. At present, the laboratory research of liposomes is becoming more and more mature, but the industrialization of liposome drug delivery system is still a key factor restricting its development, which is reflected in the high cost of membrane materials such as phospholipids used. However, we believe that with the development of technology and the continuous updating of new equipment, liposome drug delivery system will have a good development prospect. Other drug delivery systems using liposome preparation are as Table 4.7.

Virus Vector

Protein products derived from animals, plants, or applied biotechnology research and development that have certain biological activities and can be used to prevent and diagnose human, animal, and plant diseases are called protein drugs. Protein has more than 100 amino acids. Compared with small-molecule chemical drugs, protein drugs have stronger pharmacological activity and lower toxicity, so they have attracted widespread attention. However, protein drugs have poor stability in vivo; are easily denatured and inactivated; furthermore, have poor fat solubility; and are not easy to penetrate biofilms. And most of them need to be administered for a long time, resulting in poor compliance. Therefore, it is necessary to research and develop new dosage forms or combine protein drugs with new technologies to improve the efficacy of protein drugs in the body.

Using a drug delivery system to deliver proteases can improve the stability of protein drugs in gastrointestinal tract, which is very beneficial for improving the efficacy of the drug. In 2018, Irisa Trianti et al. use tetrameric glycoprotein neuraminidase (NA) from influenza A virus (H5N1) as a baculovirus surface protein to construct double carrier for delivery peptide antigens and DNA vaccines [122]. Finally, the recombinant NA-AngII pseudotype baculovirus stimulates the production of anti-AngII antibodies and lowers systolic blood pressure (SBP), and it is used for the treatment of hypertension by subcutaneous or intranasal immunotherapy. However, the current delivery technology is limited to preclinical research stage, and it is hoped that protein preparations will be available in the near future.

Compared with other viral vectors, adeno-associated virus (AAV) has many advantages. First, AAV is non-pathogenic, does not stimulate inflammation or immune response, does not enter dividing cells, and can provide stable long-term gene expression. Second, different serotypes of AAV have different tissue and organ targeting, which provides help for the safety and effectiveness of preparation. Based on the current experiments of AAV in animals, the effect can be maintained for several years after one injection, and it is the most long-acting gene therapy vector. Therefore, in the field of gene delivery, AAV has received widespread attention.

In 1999, Dagmara Mohuczy et al. developed an adeno-associated virus (AAV) vector with AT1-R antisense DNA [117], which can deliver AT1-R antisense to inhibit AT1-R in VSMC. The mechanism of this research is to inhibit the vasoconstriction of angiotensin II (Ang II) type 1 angiotensin receptor (AT1-R) and to promote the growth of peripheral target-vascular smooth muscle, thereby reducing the role of Ang II to achieve the effect of suppressing high blood pressure. In addition to the mechanism of reducing the role of Ang II, reducing hypertension

Table 4.7 Other liposome delivery systems.

Drug	Formulations	Characteristic
Carvedilol	Liposomes	Change oral bioavailability by changing the type and concentration of bile salts and carrier surface charge during the preparation process [114]
Amlodipine	Liposomes	Enhance the penetration of deep skin and improve the bioavailability of amlodipine [115]
Ladipine	Liposomes using thin-film hydration technology	Showing a significant decrease in blood pressure and had a better therapeutic effect [116]

can also be accomplished by regulating vascular activity. Human tissue kallikrein (HK) cleaves kininogen to produce effective vasoactive peptides, which play an important role in regulating cardiovascular and renal functions. T Wang et al. conducted research using HK [118]. They combined human kallikrein cDNA with a recombinant adeno-associated virus (rAAV) vector to explore the potential therapeutic effect of kallikrein on hypertension and related secondary complications [118]. The results showed that the systolic blood pressure was significantly reduced after single-vein injection of rAAV-HK vector, and during the 20-week experiment, weekly blood pressure monitoring showed a stable antihypertensive effect. As we all know, gene therapy for hypertension requires long-term expression of therapeutic genes to steadily lower blood pressure. Thus, ELISA and reverse transcription-polymerase chain reaction also confirmed the continuous expression of the transgenic product HK. Later, Jiang-tao et al. used a common carrier to obtain similar results [119]. They delivered rAAV (rAAV-HK) encoding HK to spontaneously hypertensive rats (SHRs) and cultured human embryonic kidney (HEK) 293 cells and found that it not only can lower blood pressure but also treat high blood pressure by directly inhibiting end organ damage, which shows that this research result is expected to contribute to the development of good gene therapy.

Studies have found that vasodilatory peptides, such as atrial natriuretic peptide (ANP), may be excessively produced during the treatment of hypertension, which limits gene therapy to a certain extent. Therefore, the gene therapy method for hypertension based on ANP needs to be completed by controlling the gene expression of ANP. Kurt J. Schillinger et al. prepared the helper-dependent adenoviral vector AV with mifepristone (Mfp)-inducible gene regulation system to control the expression of ANP in vivo [120]. It was observed that Mfp-induced ANP expression persisted for more than 120 days after vector administration. The physiological effects of ANP, including the reduction of systolic blood pressure, are controlled by Mfp. This research result shows that the adenovirus vector AV is a promising vector in the field of gene expression control.

Local adenovirus (Ad) can mediate gene transfer into the carotid artery of spontaneously hypertensive rats (SHRSP) to improve endothelial function. Endothelial nitric oxide synthase (eNOS) is the main enzyme involved in the synthesis of nitric oxide. Ad and eNOS both have important roles in adjusting the structure of blood vessel walls and participating in cardiovascular diseases. William H. Miller et al. combine the two and explore the potential of vector-targeted delivery of

Ad-coded endothelial nitric oxide synthase (AdeNOS) to prevent SHRSP from increasing blood pressure [121]. First they used bispecific antibodies (anti-ACE/anti-Ad capsid, Fab9B9) to rearrange Ad transduction, blocking the binding of Ad and adenovirus receptors, and at the same time retargeting AdeNOS to angiotensin-converting enzyme, making eNOS highly overexpress in pulmonary vasculature and continue to lower blood pressure. This research on eNOS provides new ideas for the treatment of cardiovascular diseases, and at the same time, it also reflects the possibility of using targeted vectors to achieve gene therapy.

In the early gene therapy research, retroviral vectors were also widely popular. In 1999, Hongwei Wang et al. constructed a retroviral vector (LNSV) containing ACE sense (LNSV-ACE-S) and ACE antisense (LNSV-ACE-AS) sequences, and they were used in rat pulmonary artery endothelial cells (RPAEC) [122]. After eight consecutive passages, the transcripts of ACE-S and ACE-AS were steadily expressed, but the expression was higher in ACE-AS. However, in spontaneously hypertensive rat model, blood pressure decreased by about 15 mmHg for a long time. These observations indicate that the delivery of ACE-AS by retroviral vectors leads to permanent suppression of ACE and long-term reduction of hypertension in spontaneously hypertensive rats, with a stable and sustained antihypertensive effect.

Later, M.I. Phillips proposed two strategies: a) antisense oligodeoxynucleotides (AS-ODN) and b) antisense DNA delivered in viral vectors, which can suppress the vasoconstriction-related properties gene [123]. In the experiment, AS-ODN was injected in naked form or liposome form. AS-ODN targets receptors and effectively reduces hypertension. And when delivered with liposomes, a single dose is valid for up to 1 month. No side effects or toxic effects were found. For the vector, the use of adenovirus (AAV) with the construct showed that single-dose administration can reduce and slow the development of hypertension. Comparing the two, and considering the long-term efficacy, it is concluded that ODN may be developed first. Because the delivery of viral vectors requires more engineering techniques to determine its safety.

In addition to using viral vectors for drug delivery systems, scholars have also studied small-molecule RNA. Anne-Sophie Arnold et al. developed small interfering RNA (siRNA) targeting β(1)-AR [124], reducing blood pressure in spontaneously hypertensive rats (SHR), and reducing cardiac hypertrophy after one injection, which shows the possibility of using siRNA to treat cardiovascular diseases. Later, vector was prepared to target GAX [125], remodeling blood vessels by inhibiting the expression of GAX. These results have opened up new directions for gene therapy of cardiovascular diseases. Therefore, it can be seen that gene therapy for RNA may become a new and effective method for treating cardiovascular diseases.

Others

It is worth mentioning that one of the five principles for the treatment of hypertension—the joint principle. If patients often use a drug that cannot effectively control blood pressure, and in order to increase the effect of reducing blood pressure without increasing adverse reactions, two or more different antihypertensive drugs can be used in combination therapy. For example, Sartan drugs or public drugs

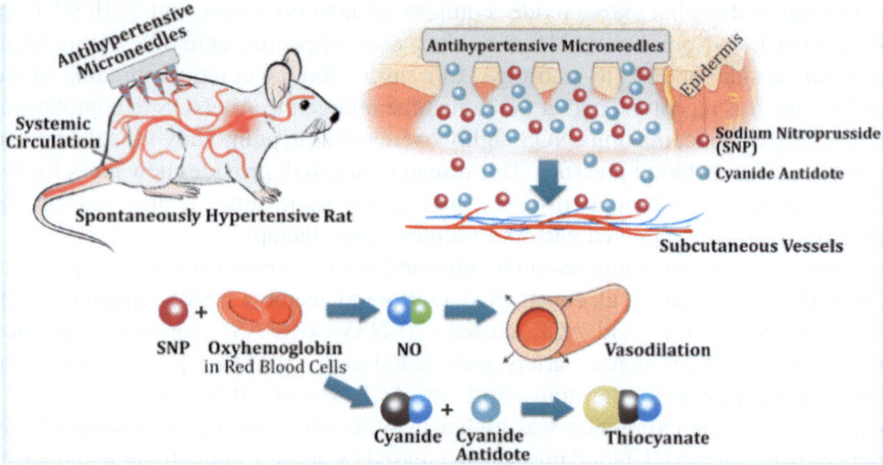

Fig. 4.10 Scheme of experiment. Reproduced with permission from [126]

associated with calcium ion antagonists have a better combined effect. Similarly, the combination of antihypertensive drugs and antidote can reduce toxic to a certain extent. Yan Li et al. verifies this assertion that they prepared soluble microneedles loaded with sodium nitroprusside (SNP) and sodium thiosulfate (ST) and demonstrated their antihypertensive effect in a spontaneously hypertensive rat model (see Fig. 4.10) [126]. The results showed that combined administration effectively inhibited the side effects caused by continuous intake of SNP, including skin irritation and target organ damage. This study lifted the restrictions on the use of SNP for first aid or self-rescue and introduced a controllable and self-managing method for the treatment of hypertension emergencies, which has good application prospects. Other drug delivery systems in hypertension were showed in Table 4.8.

Although current antihypertensive drugs are effective, they have poor compliance, expensive price, and a short duration (a few hours or a day). Drug delivery systems deliver drugs that can prolong drug efficacy and improve bioavailability. Gene therapy provides a way to produce lasting antihypertensive effect (weeks, months, or years). However, the success of gene transfection, efficiency of transfection and influence on other functions of the body are uncertain. Therefore, a large amount of research is still needed for drug delivery systems to deliver drugs or gene therapy before they can be applied to the clinic.

4.1.3 Heart Failure

Heart failure is a clinical syndrome from the occurrence of cardiovascular disease to final severe stage. When myocardium is damaged or heart is overloaded for a long

Table 4.8 Other drug delivery systems in hypertension

Drug	Formulations	Characteristic
Carvedilol	Carvedilol-coated SLN loaded with n-carboxymethyl chitosan (MCC) [127]	Protect the rapid release of carvedilol in an acidic environment, increase the bioavailability
	Osmotic pump capsule of carvedilol nanosuspension [128]	Controlling the release of carvedilol, the plasma concentration minimized fluctuations improve the oral bioavailability of poorly soluble drugs
	Nanoemulsions [129]	Enhance bioavailability by enhancing the transdermal permeability of carvedilol
	Supramolecular gels [130]	
Nimodipine	PCL-PEO-PCL micelles [131]	Control release of drugs is achieved by adjusting the PEO/PCL ratio
Nitrendipine	Chitosan-modified nitrendipine nanocrystals [132]	Improve the physical stability of nitrendipine, showing a certain degree of slow-release performance and improving the drug's bioavailability
	Microsphere-coated acrylamide-grafted chitosan nitrendipine [133]	Show lower drug release characteristics in stomach, while in intestine, the drug release is extended for 14 h
Felodipine	Self-microemulsifying drug delivery system (SMEDDS) [134]	Achieve rapid drug lag when the expected time is reached freed
	Nanosuspension [135]	Improve the dissolution rate, thereby improving its oral bioavailability
Candesartan cilexetil (CC)	Nanosuspension [136]	Improve the oral bioavailability
Irbesartan (IRB) and valsartan	Self-nano-emulsified drug delivery systems (SNEDDS) [137, 138]	Improving the poor water solubility dissolution rate and increase its oral bioavailability
Amlodipine besylate (AB)	Nanoemulsion (NE) using spontaneous emulsification [139]	Enhance the solubility and oral bioavailability of AB and achieve local delivery of drugs at the target site (total residence time was increased three times)
Captopril, amlodipine, and valsartan	Chitosan (CS) polymer using ion gel-assisted ultrasound [138]	Improving the currently oral administration of available hydrophobicity drugs and also providing extended release function
Captopril and aliskiren	SBA-15 powder matrix with MgO [137]	Safely and effectively processed the adsorption of captopril and aliskiren and vitro release

time, the structure of the heart may change, resulting in reduced blood pumping to the heart and reduced blood reaching some tissues or organs, or make venous blood unable to return, so as to deposit in the lungs, and a series of symptoms were produced; ultimately heart failure has formed.

Heart failure may often cause difficulty in breathing, especially when it is active. And symptoms such as edema, bloating, nausea, and vomiting may also occur, which will not only affect the function of circulatory system but also cause damage to breathing and liver and kidney functions and will seriously endanger life.

In clinic, the treatment of heart failure is based on medication, and surgery is performed when necessary. But the curability is low, and only a small part can be cured. Therefore, the core of current heart failure treatment is to control and delay the development of heart failure, thereby alleviating various symptoms and improving the quality of life.

With the development of technology, controlled release technology of drug delivery system can not only realize the sustained release of drugs but also more accurately and intelligently regulate the space and time of drug release. In addition, drug delivery system can target drugs to specific lesions, form a relatively high concentration locally, reduce damage to normal tissues and cells, and achieve a targeted effect. Based on this, the use of drug delivery systems for heart failure treatment will bring new hope to heart failure patients.

Nanoparticles

Left ventricular (LV) remodeling after myocardial infarction (MI) can lead to heart failure. By protecting the heart from ventricular remodeling after infarction, heart failure can be reduced. Statins can regulate blood lipids and can also be used in acute coronary syndrome to inhibit vascular endothelial inflammation and improve vascular endothelial function. Therefore, statins can play anti-inflammatory and anti-thrombotic effects in cardiovascular system diseases. Yajing Mao et al. had studied the role of statins in anti-heart failure [140]. They prepared pitavastatin nanoparticles that could target monocytes/macrophages and used pitavastatin's anti-inflammatory effects to protect the heart from left ventricular remodeling after infarction. About weakening myocardial infarction to reduce heart failure, we have discussed it in detail in Sect. 4.1.1.2.

Hydrogel

Neuregulin-1 beta (NRG) is a member of the epidermal growth factor family and plays a key role in the development and proliferation of cardiomyocytes. Systemic administration of NRG has shown efficacy in animal models of cardiomyopathy, but the use of NRG requires daily infusions, which hinders the implementation of this method. Therefore, if it is possible to achieve continuous NRG delivery, increasing the dosing interval and reducing the number of dosing are of great benefit to the implementation of this method. In 2014, Jeffrey E. Cohen MD et al. wrapped NRG in a hydrogel to enhance myocardial mitotic activity, reduce apoptosis, and enhance left ventricular function, thereby preventing and treating heart failure [141].

Others

The study found that apoptosis secondary to acute ischemia and chronic remodeling is associated with heart failure. Reducing cardiomyocyte apoptosis can reduce heart failure. Kyle I. Mentkowsk et al. designed heart cell-derived cells (CDC) to express Lamp2b (an exosome membrane protein), fused with cardiomyocyte-specific peptide (CMP) WLSEAGPVVTVRALRGTGSW as shown in Fig. 4.11 [142]. The results showed that uptake of cardiomyocytes increased and apoptosis decreased.

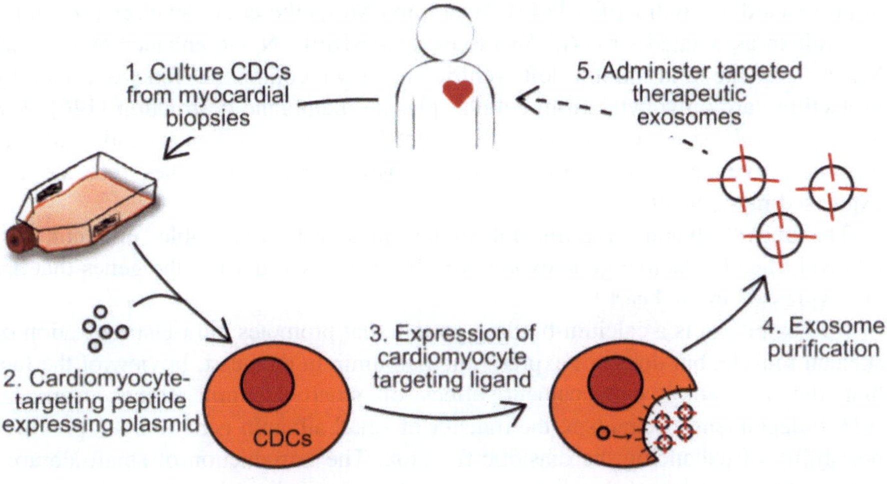

Fig. 4.11 Schematic diagram demonstrates the methods for development of CDC-derived cardiomyocyte targeted exosomes and their future therapeutic potential. Reproduced with permission from [142]

Based on the above, we can see that enhancing cardiomyocyte function and reducing cardiomyocyte apoptosis have potential clinical applications in the treatment of heart failure.

Gene Vector

In the above section, we mentioned that enhancing cardiomyocyte function can reduce heart failure to a certain extent. As early as 1999, John P. Maurice et al. used the intracoronary method to deliver adenovirus gene-mediated Abeta-β2AR to rabbit myocardium [143], producing diffuse multi-chamber myocardial expression. Myocardial expression reached a peak 1 week after gene transfer, and its contractile force measurement parameters were significantly enhanced and showed increased responsiveness to the β-agonist isoproterenol.

In addition to Abeta-β2AR affecting cardiomyocytes, the effects of other genes on cardiomyocytes have been gradually explored, such as SERCA gene. SERCA is a myocardial sarcoplasmic reticulum Ca^{2+}-ATPase, which plays an important role in the regulation of calcium homeostasis in cardiac muscle cells. If Ca^{2+}-ATPase (SERCA2a) activity is reduced, it will lead to abnormal calcium treatment. Disorders of calcium metabolism have been shown to significantly promote contractile dysfunction observed in heart failure. Therefore, SERCA can be expressed by delivering a vector carrying the SERCA gene, and the contractile function of heart can be restored to a normal level. In 2000, Michael I. Miyamoto et al. delivered an adenovirus with SERCA2a gene to an animal model of heart failure and found that SERCA2a overexpression in vivo restored the heart's systolic and diastolic function to normal levels [144]. Later, Federica del Monte et al. used the antisense lamban (asPL) adenovirus gene to reduce the inhibitory effect on SERCA2a and improve cardiac contractility [145]. Hoshijima Masahiko et al. develop a recombinant adeno-associated virus (rAAV) to express S16EPLN and enhance myocardial SRCa^{2+} uptake and inhibit left ventricular (LV) contractile function, thereby protecting cardiomyocytes from cellular plasma membrane destruction [146]. The above several experimental studies have achieved stable, efficient, and relatively heart-selective gene expression. Next, we will discuss the genes that are not expressed in the heart.

The above several experimental studies have achieved stable, efficient, and relatively heart-selective gene expression. Next, we will discuss the genes that are not expressed in the heart.

Microalbumin is a calcium-binding protein that promotes ultra-fast relaxation of skeletal muscle, but does not express microalbumin in the heart. In view of the fact that the relaxation enhancement effect of small albumin comes from an ATP-independent mechanism, the transfer of small albumin gene to heart provides possibility of enhancing the diastolic function. The introduction of small albumin gene in the heart found that it has caused physiologically relevant acceleration of cardiac diastolic performance and enhanced diastolic performance in animal models of slow myocardial diastole [147]. Therefore, microalbumin gene transfer may provide a new treatment to correct cellular disorders in calcium signaling pathway that leads to abnormal growth or damage to the heart or other organs.

The above experiments proved the possibility of controlling genes to adjust Ca^{2+} homeostasis to control heart failure, but it is unknown whether the gene dosage is beneficial to heart failure. In 2008, MJ Byrne et al. studied this problem. They administered adeno-associated virus 2 (AAV2)/1 SERCA2a to pacing-induced HF sheep and determined the dose-dependent improvement in cardiac performance through left ventricular pressure analysis, demonstrating the potential for rapid conversion of gene therapy to clinical application in HF patients [148].

It is known that at cellular level, the contractile function of myocardial cells can be improved by improving perfusion, adjusting Ca^{2+} homeostasis, inhibiting myocardial hypertrophy, remodeling fibrosis, and thereby treating heart failure. Cardiomyocyte apoptosis is one of the important mechanisms involved in the development of heart failure. Over the years, intervention on cardiomyocyte function to prevent and treat heart failure is a new way of experiment and clinic, but there are still many problems remain to be resolved, such as the selection of vectors in gene therapy, and the effectiveness of gene transfection is limited to in vitro experiments or animal experiments, and it is still too early to be used in clinical practice.

Regulation of cardiomyocyte apoptosis is one of the measures to treat heart failure, and on the other hand, we can control the contraction of cardiomyocytes to improve heart failure. Delivery of the adenovirus S100A1 gene in a post-infarct rat heart failure model showed normalization of S100A1 protein expression and reversal of systolic dysfunction in the failing myocardium both in vivo and in vitro [149]. Adenoviral vector (AV) was developed to drive (sh) RNA against PKCa (Ad-shPKC alpha) and increase the transfection efficiency to improve the contractility of silenced neonatal rat cardiomyocytes and engineered heart tissue (EHT) [150]. Adeno-associated virus (rAAV6-PLC beta 1b-CT) can also be used to make a

Table 4.9 Treatment of heart failure by controlling the contraction of myocardial cells.

Formulations	Characteristic
Deliver the adenovirus S100A1 gene	Normalized S100A1 protein expression and reversed the contractile dysfunction of the failed myocardium in vivo and in vitro [149]
Adenovirus vector (AV) to drive (sh)RNA targeting PKCa	Increase transfection efficiency and improve the silencing of neonatal rat cardiomyocytes and engineered heart tissue contraction [150]
Adeno-associated virus (rAAV6-PLC beta 1b-CT)	Preventing the loss of contractile function [151]

small gene express inhibitory peptide PLC beta 1b-CT in the heart, thus preventing loss of contractile function [151]. These results suggest that improving the contractile function of cardiomyocytes in the failing heart represents a therapeutic strategy for heart failure. Treatment of heart failure by controlling the contraction of myocardial cells was shown in Table 4.9.

Moreover, the ventricular remodeling after myocardial infarction causes heart failure, and improving the myocardial infarction and ventricular remodeling can attenuate the clinical manifestations of heart failure. Researchers used coronary arteries in congestive heart failure pigs as a model and found that adenovirus encoding adenylate cyclase was associated with enhanced left ventricular (LV) function and that gene transfer increased left ventricular function in congestive heart failure and attenuated deleterious left ventricular remodeling [152]. Other researchers have prevented the ventricular remodeling [153]. They used HF rats and mediated β-ARKct to intramyocardial delivery via AAV6, significantly improving cardiac contractility and preventing left ventricular remodeling. In recent years, researchers have developed a gene therapy tool using recombinant adeno-associated viral vector (rAAV) encoding chain acyl-coenzyme A dehydrogenase (MCAD) in the heart, attenuating pathological remodeling in cardiac disease models [154]. This follows that improving ventricular remodeling by gene delivery may represent a novel strategy to improve cardiac function. Studies aiming at improving myocardial infraction in Table 4.10.

Etiological treatment is also one of the strategies for treating heart failure. Olli Tenhunen et al. found that local adenovirus-mediated wild-type (WT) p38 kinase and MKK3b co-transfection can reduce infarct size and increase ejection fraction of the disease [155], suggesting that the use of p38 kinase is a potential new approach for the treatment of heart failure. In addition, Yuxin Wang et al. used gene transfection to express cTnl-targeted short peptides on the surface of mesenchymal stem cells to obtain cTnl-targeted exosomes, which were shown to be succesfully engulfed by cardiomyocytes, proliferation of cardiomyocyres around the infartion were promoted and heart function was ultimately restored (see Fig. 4.12) [156].

Table 4.10 Improving myocardial infarction and ventricular remodeling to reduce heart failure

Formulations	Characteristic
Adenovirus encoding type VI adenylate cyclase	Increase left ventricular function of congestive heart failure and reduce harmful left ventricular remodeling [152]
Deliver β-ARKct into the myocardium via AAV6	Improving cardiac contractility and reversing left ventricular remodeling [153]
Recombinant adeno-associated viral vector (rAAV) encoding cardiac mid-chain acyl-CoA dehydrogenase (MCAD)	Reduce pathological remodeling in heart disease models [154]

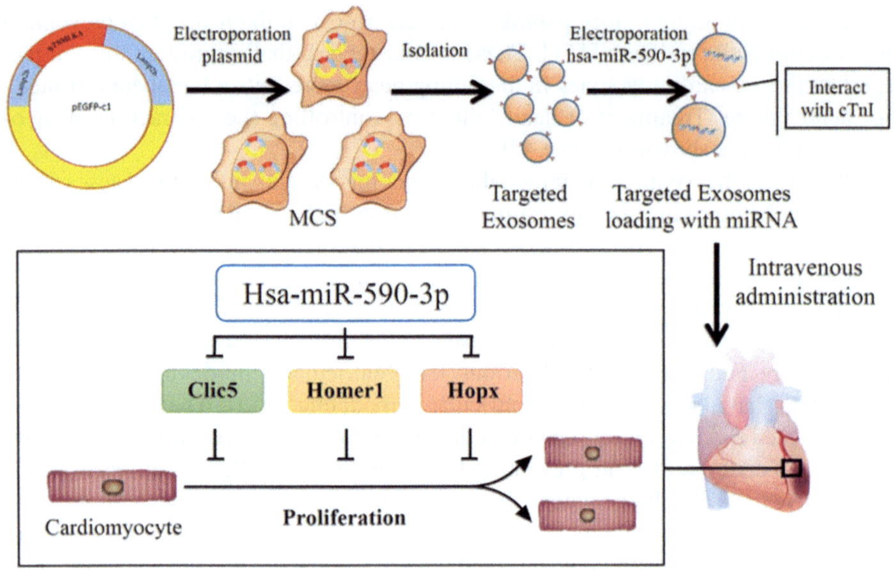

Fig. 4.12 The schematic diagram of plasmid construction, MSC electroporation, exosome isolation, miRNA loading, intravenous administration, and cardiomyocyte proliferation based on the function of hsa-miR-590-3p. Hsa-miR-590-3p inhibited the expression of genes inhibiting cell proliferation, i.e., Hopx. As a result of it, hsa-miR-590-3p promoted cardiomyocyte proliferation. Reproduced with permission from [156]

Heart failure is caused by a decline in heart function. Therefore, its treatment includes maintaining cardiac function, rapidly restoring myocardial function, and inducing myocardial regeneration [157]. Above we talked about improving myocardial cell contraction, interfering with myocardial cell apoptosis to improve heart failure, and below from improving myocardial failure by inducing myocardial regeneration (Table 4.11) and inhibiting fibrosis.

On the other hand, myocardial fibrosis is a pathological change that many heart diseases develop to a certain stage. If it continues to develop, it will aggravate heart

Table 4.11 Improve myocardial failure by inducing myocardial regeneration

Formulations	Characteristic
An improved lentiviral vector containing mRNA stabilizer and nuclear import sequences	Significantly improve gene transduction in growth arrest cells and induce stable gene expression in adult myocardium [158]
Expression of cell cycle regulator cyclin A2 (adenovirus vector)	Initiate myocardial mitosis, induce endogenous myocardial regeneration, and improve myocardial function [159]
Delivery of modified RNA to the myocardium	Improve myocardial function by making it express efficiently and quickly in the heart [160]

failure, cause arrhythmia and sudden death, and thus endanger life and health. Therefore, improving heart function and myocardial fibrosis can reduce heart failure. A dimer myocardium-associated adenovirus vector (rAAV9-shPLB) was developed [161], intravenously delivered RNAi to the heart, restored heart expansion, and reduced heart hypertrophy, myocardial cell diameter, and heart fibrosis. More importantly, in the course of RNAi treatment, no microRNA was out of control, or liver toxicity was found, and the treatment effect was better.

In 2011, S Sundararaman et al. conducted a rodent model test of heart failure using transient stromal cell-derived factor 1α (SDF-1) and luciferase plasmid driven by the cytomegalovirus (CMV) promoter [162]. They found that the non-viral gene transfer of hSDF-1 can promote the improvement of angiogenesis and cardiac function in rats with ischemic heart failure, accompanied by scar remodeling, reducing the trend of myocardial fibrosis. These data indicate that independent non-viral hSDF-1 gene transfer is a strategy to improve the cardiac function of ischemic cardiomyopathy.

In addition, the use of new technologies can transfer genes to specific anatomical and pathological targets, release targets, improve the myocardial microenvironment, and thus promote the recovery of myocardial function [163].

Through ultrasound-targeted microbubble destruction (UTMD) technology and ultrasound microbubble contrast-mediated gene transfection (TCUMGT) technology, combined with pcDNA (3.1) for P-selectin-human vascular endothelial growth factor 165 plasmid phospholipid micro MBs selectin, they transfer therapeutic genes to ischemic myocardium, increase myocardial blood vessel density, and improve heart function [163].

In summary, the current treatment for heart failure is only to relieve symptoms. Although there is no cure, it can extend the life expectancy of patients. Gene therapy involves the process of gene transfer. Once the gene is in place, it can restore heart function and provide new hope for the cure of heart failure patients.

Currently, heart failure cannot be completely cured. The existing medical treatment and other methods are only to relieve the uncomfortable symptoms. Although gene therapy can improve the patient's heart function to a certain extent, its use in humans still requires a lot of clinical research. In the future, scholars can develop efficient and uniform methods for gene transfer to the heart muscle and apply it to large mammals and eventually to humans. At the same time, in the future development, the treatment of heart failure should be more comprehensive, which is not only conducive to the recovery of patients, but also reduces the patient's complication and pain.

4.1.4 Arrhythmia

A normal heart rhythm is called a sinus rhythm, and a heart that beats too fast, too slow, or irregularly is considered arrhythmia, including ventricular tachycardia, ventricular fibrillation, and cardiac arrest. Antiarrhythmic drugs are pharmaceuticals prescribed to treat abnormal rhythms of the heart and can be categorized into five classes based on their general effect (Table 4.12) [164].

Due to the side effects and low bioavailability of many antiarrhythic drugs, nanotechnology has been widely used to increase their bioavailability and reduce side effects. For example, amiodarone is widely used to treat or prevent some types of arrhythmia including supraventricular and ventricular ones. However, the drug

Table 4.12 Classification of antiarrhythmic drugs

Drug	Class	Mechanism	Side effect
Lidocaine [165] Quinidine Mexiletine	I:Sodium-channel blockers	Depress phase 0 (rapid depolarization) and prolong, shorten, or do not affect repolarization	Heart weakness, digestive discomfort, autoimmune reactions, hypotension, anti-muscarinic effects, congestive heart failure, hyperthyroidism, and hypothyroidism
Esmolol Propranolol Sotalol Atenolol	II:Beta-blockers	Decrease the slope of phase 4 and reduce the heart rate and conduction	Bradycardia, ventricular atrial block, severe asthma attacks, fatigue and sleep disorders
Amiodarone Beryllium Suhalol	III:Potassium-channel blockers	Prolong phase 3 and delay repolarization, increase the duration of action potential and ERP	Severe hypotension, asthma, new arrhythmias, and sinus bradycardia
Verapamil Bipridil Diltiazem	VI:Calcium channel blockers	Reduce atrioventricular conduction, terminate paroxysmal supraventricular tachycardias, and decrease the ventricular rate in atrial fibrillation	Pulmonary fibrosis, microcrystalline deposition in the cornea and skin, tremor, paresthesia, and congestive heart failure
Digoxin Magnesium sulfate Adenosine Guan-fu base A	V:Other antiarrhythmic agents	–	

has considerable side effects in the lung, liver, and thyroid resulting from the prolonged circulation time and off-target tissue accumulation, such as amiodarone pulmonary toxicity (APT) and hypo- or hyperthyroidism [166].

Various carriers have been applied to deliver these antiarrhythic drugs specifically to the site of action and circumvent their side effects. Use of carriers in drug delivery provides various benefits including control of solubility, vascular-circulation time, and specific targeting of the drug and, consequently, reduction of the side effects of the drug [167]. In year 2013, Elgart's group reported the use of a self-nano-emulsifying drug delivery system (SNEDDS) for improving the oral bioavailability of amiodarone [168]. Currently, antiarrhythic drug carriers can be classified into liposome, polymeric particle, inorganic particle, and hydrogel.

Liposome

Liposomes were in the size range of 30 nm to several micrometers and can be categorized into uni-lamellar and multi-lamellar vesicles base on the number of the phospholipid bilayers surrounding the aqueous solution. Liposomes can usually be prepared by a thin film method, injection method, reverse evaporation method, and double emulsion method [169].

Low water solubility drug, for example, amiodarone [170], can be directly encapsulated into liposomes during their preparation, and the encapsulation efficiency is governed by drug-lipid interactions and the solubility of the drug in the liposome membrane. In general, the encapsulation efficiency of hydrophobic drugs can easily reach 100%. High water-soluble drugs, for example, lidocaine hydrochloride [171], adenosine [172], and propranolol hydrochloride [173], can be loaded into liposomes during their preparation and can also be actively entrapped by using interaction force after liposome formation. And their entrapment efficiency is determined by the ability of liposomes to trap aqueous buffer containing a dissolved drug during formation or the interaction strength [174].

It is worth noting that the gains in the therapeutic index are mostly related to the reduction of toxicity. This reduction in the toxicity of liposomal drug compared with the free drug is originated from the poor extravasation of liposomes into the tissues which have tight endothelial junctions [175]. The summary of antiarrhythic drug-loaded liposome can be found in Table 4.13.

Polymeric Particle

"Polymer-based particle" is a general term given to any polymeric particles including polymeric capsules, micelles, dendrimers, colloids, etc. Because of their high biocompatibility and biodegradability properties, various synthetic polymers such as polyurethane and poly(lactic-co-glycolic acid) and natural polymers, such as chitosan (CS), cyclodextrin, and silk, are extensively used in the fabrication of particles. And the summary of antiarrhythic drug-loaded polymeric particle can be found in Table 4.14.

Hydrogel

Hydrogel is a three-dimensional network of hydrophilic polymers, which is able to imbibe a large amount of water while maintaining the structure. Owing to their

Table 4.13 Summary of antiarrhythmic drug-loaded liposome

	Drug	Formulations	Characteristic
Low water solubility	Amiodarone	Nano-sized PEGylated liposomes [170]	Nano-sized liposomes improve the targeted delivery of liposomal amiodarone
	Amiodarone	Ascorbyl palmitate-doped dimyristoylphosphatidylcholine liposomes [176]	Ascorbyl palmitate as antioxidant can minimize or avoid the adverse effects of amiodarone
	Guanfu base A	Nanolipids composed of poloxamer 188, lecithin, and medium-chain fatty acid [177]	Nanolipids can help drug achieve longer circulation time in blood and better heart specificity
High water solubility	Lidocaine hydrochloride	TAT modified liposomes [171]	TAT facilitated the uptake of the liposomes across the cell membrane with no considerable toxicity
	Adenosine	PEGylated liposomes [172]	PEGylated liposomes exhibited a prolonged blood residence time
	Propranolol	Liposomes-in-PLGA microsphere [178]	The microspheres (~77.8 μm) achieved sustained drug release for 40 days

Note: TAT, a cell-penetrating peptide, transactivation transcriptional activator

Table 4.14 Summary of antiarrhythmic drug-loaded polymeric particle

Polymer	Drug	Formulations	Characteristic
CS	Adenosine	CS-TPP NPs [179]	Biocompatible and biodegradable, high mucoadhesiveness, immunostimulating properties
	Verapamil	CS microspheres [180]	High entrapment efficiency and small burst release were observed
Cyclodextrin	Amiodarone	Nanoparticles composed of L-lysine cross-linked succinyl-β-cyclodextrin [181]	A 250% enhancement in the selective delivery of the drug to cardiac tissue
Silk	Adenosine	Adenosine powder reservoirs coated with silk fibroin [182]	Adenosine release was sustained for 14 days with a linear release profile
Polyurethane	Sotalol	Sotalol-polyurethane matrix formulation [183]	The implantation of formulation optimized efficacy in the myocardium and minimized toxic systemic effects
PLGA	Atenolol	Atenolol-loaded PLGA nanoparticles using nanoprecipitation method [184]	The nanoparticle for oral administration had a size of 192.6 ± 3.5 nm and a drug entrapment efficiency of 71.65 ± 1.8%

tunable physical properties, controllable degradability, and capability to protect labile drugs from degradation, hydrogels serve as a platform in which various physiochemical interactions with the encapsulated drugs control their release. An amiodarone-containing polyethylene glycol-based hydrogel was formulated and placed on the atrial epicardium [185, 186]. Epicardial application of the hydrogel was well tolerated and allows antiarrhythmic efficacy at very low systemic drug levels, which reduced toxicity from lower off-target tissue distribution.

Except drugs, gene therapy is also one of the commonly used methods for treating arrhythmia, including to restore or increase gene expression and to reduce expression of defective alleles or specific pathways and in situ DNA correction or manipulation. Those carriers for drug delivery can also be used to deliver genes. However, the commonly used carriers for gene delivery are viruses, such as adenovirus.

The possibility of transferring a gene into cardiac cells has been investigated since the early 1990s. The initial strategy was direct myocardial injection of naked DNA plasmids. Using this method, the transgene was expressed for only 30–60 days, but these studies proved the feasibility of gene transfer into cardiomyocytes, indicating the possibility of treating heart diseases with gene therapy.

Viral Vectors

Given their high gene transfer efficiency and long-term transgene expression, viral vectors are the most suitable vehicles for gene therapy designed to treat inherited and acquired arrhythmias. Several viral vectors have been investigated for gene delivery to the heart, including lentivirus, adenovirus, and recombinant adeno-associated virus (Table 4.15) [187].

Plasmid DNA is widely used for in vitro transfection experiments as it is easy to produce and has a large packaging capacity. However, its main shortcomings include a very low transfection efficiency in vivo and only transient gene expression, limiting its application [188]. Adenovirus is the most widely used viral vector in cardiac gene therapy experiments, which is able to transduce the mammalian heart with a very high efficiency and provides a great way to undertake proof-of-principle studies. Lentiviruses are capable of long-term gene expression by virtue of their ability to integrate into the host cell genome. However, lentiviral vectors have a low capacity for cardiac transduction. The most promising vector for cardiac gene therapy is based on recombinant adeno-associated virus (rAAV), which is derived from a non-pathogenic parental virus.

Both the drug carriers (liposome, polymeric particle, and hydrogel) and viral vectors are facing the same problem; the elimination of both introduced carriers and vectors by immune system would not only affect the therapeutic effect but also induce production of pro-inflammatory cytokines and chemokines with harmful effects.

Exosomes

Exosomes were a minimally invasive tool for therapeutic drug/gene delivery. Exosomes are nanoscale membrane vesicles particles (30–100 nm in diameter) that are secreted from multivescicular endosomes by almost all types of cell. The

Table 4.15 Vectors for cardias gene therapy [187]

Vector for gene delivery	Genome	Packaging capacity	Advantages	Disadvantages
Plasmids	DNA	Not applicable	Low-cost production, simple design, mild immune response	Low transduction efficiency, short-term expression
DNA-lipid complexes	DNA	Not applicable	Low-cost production, systemic delivery, large packaging size, minimal biosafety concern	Limited myocardial transfection, limited organ specificity, short-term expression
Lentiviruses	Single-strand RNA	Up to 8kb	Efficient infection, long-term expression, low immunogenic response	Inflammatory response, risk of insertional mutagenesis
Adenoviruses	Double-strand DNA	Up to 35 kb	Efficient infection, high packaging capacity, quick transgene expression, no host genome integration	Inflammatory response, immunogenicity, short-term expression
Adeno-associated virus (AAVs)	Single-strand DNA	Up to 5 kb	Efficient infection, long-term expression, serotype modulation of organ specificity, weak immune response	Difficult high titer production, limited packaging capacity, endogenous human anti-AAV antibodies can reduce therapy efficacy

exploitation of exosomes for delivery of drug/gene therapy in cardiovascular disease is still in its infancy, and major questions remain unanswered [189, 190]. For example, which human cells are best for deriving exosomes, and whether the type of source exosomes affects their efficacy? Besides, whether a large-scale manufacturing process can be achieved to produce therapeutic exosomes.

4.1.5 Congenital Heart Disease

Congenital heart disease (CHD) is the most common type of congenital malformations, accounting for about 28% of congenital malformations. Congenital heart disease occurs in 0.4–1% of newborns, which is not to be underestimated [191]. The treatments include operation, interventional therapy, and drugs.

Valve replacement is the first choice for the treatment of CHD. The existing valve replacements are far from ideal, which need anticoagulant therapy or have limited durability. As displayed in Fig. 4.13, heart valve tissue engineering (HVTE) is a promising alternative, with the potential to synthesize mechanically robust tissues with growth, repair, and remodeling capabilities [192]. The clinical realization of bioengineering valves depends on the appropriate combination of cells, biomaterials, and bioreactors. Biomechanical regulation of the valve in vitro promotes the

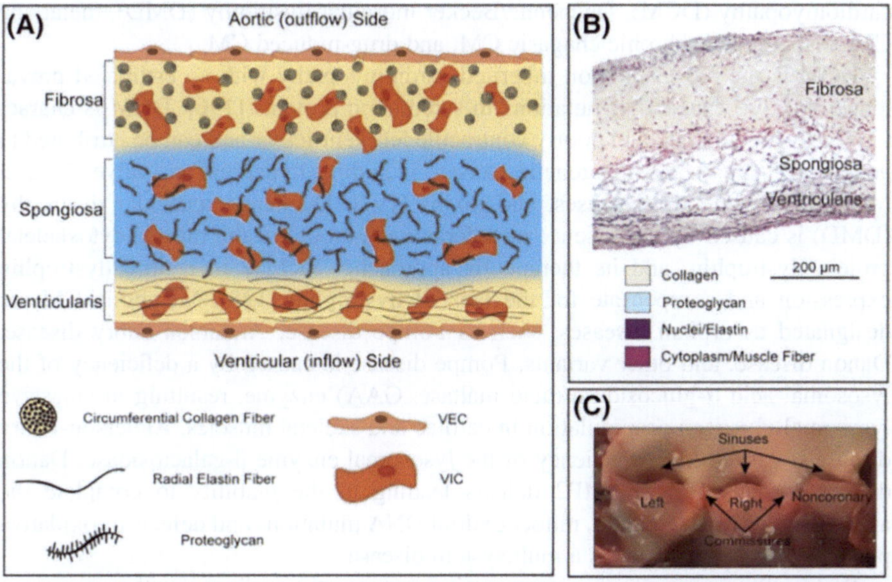

Fig. 4.13 Biomechanical conditioning of tissue-engineered heart valves. Reproduced with permission from [192]

differentiation of progenitor cells into tissue synthetic myofibroblasts and prepares structures that can withstand the complex hemodynamic environment of natural valves.

Impaired systolic function caused by acute injury or congenital defects can lead to cardiac complications and heart failure. Current treatments slow the progression of the disease, but cannot save heart function. Researchers injected a mixture of adeno-associated virus serotype 6 (rAAV6) into the orbit of wild-type mice [193]. Increasing the level of gene through rAAV-mediated cardiac specific expression provides a new method of gene therapy, which can potentially enhance cardiac systolic function. Similarly, researchers constructed an AAV2-carrying mouse Skl gene for expression in vivo [194] which can effectively inhibit inflammation and myofibroblast transformation and alleviate aortic valve fibrosis.

4.1.6 Other Heart Diseases

4.1.6.1 Other Subtypes of Cardiomyopathies

Cardiomyopathies are defined as diseases of the myocardium associated with cardiac dysfunction ranging from lifelong symptomless forms to major health problems such as progressive heart failure, arrhythmia, thromboembolism, and sudden cardiac death [195]. They are classified as hypertrophic cardiomyopathy (HCM), dilated cardiomyopathy (DCM), Duchenne/Becker muscular dystrophy (DMD), metabolic CM, diabetic CM, chronic chagasic CM, and drug-induced CM.

HCM is the most common inherited cardiomyopathy with an estimated prevalence of 1:200 caused by mutations in sarcomeric proteins [196]. DCM is characterized by an enlarged and poorly contractile left ventricular and can be attributed to genetic (25–35%) and nongenetic causes, including hypertension, valve disease, inflammatory/infectious causes, and toxins [197]. Duchenne muscular dystrophy (DMD) is caused by the absence or a decreased amount of the muscle cytoskeletal protein dystrophin, and its therapeutic approaches include to restore dystrophin expression and compensate for the lack of dystrophin [198]. Metabolic CM are designated as orphan diseases, such as Pompe disease, Anderson-Fabry disease, Danon disease, and other variants. Pompe disease is caused by a deficiency of the lysosomal acid α-glucosidase (acid maltase, GAA) enzyme, resulting in amassive lysosomal glycogen accumulation in cardiac and skeletal muscles. Anderson-Fabry disease is caused by a deficiency of the lysosomal enzyme β-galactosidase. Danon disease is caused by LAMP2 defects leading to the inability to complete the autophagic process. Besides, mitochondrial DNA mutations and defects of oxidative phosphorylation manifest as a multisystem disease.

So far, there is no curative treatment against genetic cardiomyopathy, as current therapies are focused on symptoms relief by pharmacological intervention and not on the cause of cardiomyopathy. The idea of gene therapy as a therapeutic technique achieving permanent correction of hereditary diseases has been introduced by Clyde Keeler in 1947 [199]. In general, the goal of gene therapy is to deliver the therapeutic gene to the place where it is missing. Probably the biggest hurdle for human gene therapy is sufficient and safe delivery of the gene therapy medicinal product into the patient's body. For gene therapy of cardiac diseases, the currently used viral vectors are adeno-associated virus (AAV), which possessed defective replication and low immunogenicity, with long-term transgene expression over many years in nondividing cells and no known pathogenicity [200].

Diabetic CM is characterized by early impairments in diastolic function, accompanied by the development of cardiac hypertrophy, myocardial fibrosis, and cardiomyocyte apoptosis [201]. However, the pathophysiology of diabetic cardiomyopathy is incompletely understood, and currently no specific therapy is available. Chagas disease is a leading cause of heart disease affecting approximately ten million people in Latin America and elsewhere worldwide. The two major methods available for the treatment of Chagas disease are chemotherapy and therapeutic vaccine to delay or prevent chagasic cardiomyopathy in patients [202]. Cardiomyopathy can also be induced by a wide variety of drugs including DOX, cyclophosphamide, chloroquine, and hydroxychloroquine. The mechanisms and treatment strategies of main subtypes of cardiomyopathies were listed in Table 4.16.

Table 4.16 Mechanisms and treatment strategies of main subtypes of cardiomyopathies

Type	Pathogenesis	Delivery system-based treatment
Hypertrophic CM	Mutations in sarcomeric proteins; mostly autosomal dominant	Delivery of SERCA2a by adeno-associated virus (AAV) [203]
		Allele-specific knockdown of a dominant-negative MYH6 variant with an AAV9 vector resulted in the amelioration of hypertrophic cardiomyopathy in a mouse model [204]
		Myocardial injection of a lentivirus-sencoding mouse Mybpc3 cDNA improved myofilament contractile functions in Mybpc3-targeted-deficient mice, which is a viable approach for treatment of hypertrophic cardiomyopathy because of mutations in cMyBPC [205]
Dilated CM	Viral myocarditis progressed to dilated CM [206]; mostly autosomal dominant: results from defects in multiple cellular components with a final common pathway of deranged cardiomyocytes contraction [197]	Receptor ligand-linked packaging RNA (pRNA) dimers as a coxsackievirus B3 (CVB3) protease 2A carrier can specifically deliver the gene to target cells via its ligand and specific receptor interaction and inhibit virus replication effectively [207, 208]
		Electrospun nanofibrous collagen scaffolds seeded with cardiomyocytes derived from human-induced pluripotent stem cells (hiPS-CM) were epicardially delivered in a mouse model of dilated CM [209]
		Qili qiangxin capsules, a Chinese traditional medicine for DCM therapy, resulted in a significant improvement in the 6 min walk test and EF
Duchenne/Becker muscular dystrophy (DMD)	X-linked inheritance: dystrophin gene mutations [210]	Recombinant adeno-associated virus type 9 (rAAV9)-mediated myocardial microdystrophin transduction for the prevention of cardiac pathology in dystrophic mice [211]
		A highly abbreviated microdystrophin gene was delivered to the heart using AAV9, aiming to ameliorating functional deficiency in a phenotypic model of Duchenne CM [212]
		Three clinical trials conducted in the USA, two were initiated in 2017 (NCT03368742, NCT03375164)

(continued)

Table 4.16 (continued)

Type	Pathogenesis	Delivery system-based treatment
		and one in 2018 (NCT03362502), which are estimated to end in 2021 and 2024, respectively. The serotypes of AAV are AAV-9 in the NCT03368742 and NCT03362502 studies and AAV-rh74 in the NCT03375164 study [213]
Metabolic CM	Autosomal recessive, X-linked or maternal (in mitochondrial diseases). Caused by enzyme deficiency or defects in oxidative phosphorylation	The lysosomal enzyme α-galactosidase-A (α-Gal-A) is chemically modified with a homobifunctional polyethylene glycol (PEG, 2000 Da) cross-linker attached to the two protein subunits of the α-Gal-A enzyme, resulting in a PEGylated and covalently bound homo-dimer, which is a novel enzyme for the therapy of Fabry disease named as PRX-102 (Protalix, Carmiel, Israel) [214]
Diabetic CM	Type 1 and type 2 diabetes mellitus caused cardiomyocyte apoptosis, autonomic neuropathy, and altered autophagic response	AAV9 combined with cardiac troponin T promoter were employed to treat diabetic CM by manipulating miR-21 expression [215]
		miR155-conjugated AuNP was used to improve cardiac function by therapeutically reducing miR155 in macrophages [216]
		Curcumin was encapsulated with poly n-butyl cyanoacrylate, poly (lactide-co-glycolide), chitosan, albumin, or acrylamide polymers to alleviate diabetic CM and diabetic complication [217, 218]
		The combination therapy of ultrasound-targeted microbubble destruction (UTMD) technique with aFGF-loaded nanoparticles was used to prevent diabetic CM in a diabetes animal [219, 220]
		The delivery of nerve growth factor (NGF) gene with AAVs can prevent diabetic cardiomyopathy in type 1 diabetic mice [221]
Chronic chagasic CM	A chronic parasitosis caused by the Kinetoplastid *Trypanosoma cruzi*	Antichagasic drugs were loaded in polymeric nanocapsules (PCL, PLA-PEG) and liposomes, or complexed with metal and cyclodextrins, or dissolved with PEG 400, increasing their solubility, bioavailability, and effectivity [202]

(continued)

Table 4.16 (continued)

Type	Pathogenesis	Delivery system-based treatment
		A therapeutic vaccine, which contained both the antigen and the CpG ODN adjuvant in the PLGA nanoparticle delivery system, could bolster the immune response, thereby slowing or stopping the progression of chagasic CM [222]
		A bivalent therapeutic vaccine comprised of two recombinant *T. cruzi* antigens, Tc24 and TSA-1, formulated on alum together with the TLR-4 agonist, E6020, was developed to treat chronic Chagas disease [223]
Drug-induced CM	Drug-induced cardiotoxicity with a dose-dependent manner	Polyoxalate containing vanillyl alcohol nanoparticle was designed to rapidly scavenge H_2O_2 and release vanillyl alcohol with antioxidant, anti-inflammatory, and antiapoptotic properties for treating DOX-induced cardiac toxicity [224]
		Liposome-encapsulated DOX and/or tumor-targeted delivery of DOX can decrease systemic drug concentration and improve the targeting of the treatment, therefore reducing drug cardiotoxicity [225]
		DOX-loaded biodegradable PLGA NPs with the average size of 185 nm demonstrated superior performance in vivo as evident by enhanced bioavailability and lower toxicity [226]

4.1.6.2 Valvular Heart Disease

Valvular heart disease is characterized by damage to or a defect in one of the four heart valves: the mitral, aortic, tricuspid, or pulmonary. Valvular heart disease is the third most common cause of cardiovascular disease, and its incidence is increasing with increasing longevity [227]. There are many different types of valve disease; some types can be present at birth, while others may be acquired later in life. The treatment of valvular heart disease includes medical treatment, surgical treatment, and interventional therapy, and most valve conditions can't be treated with medication alone.

Calcific aortic valve disease (CAVD) is the most common VHD in the elderly, and miRNAs is an attractive therapeutic alternative way to intervene early during the disease process and potentially delay or even reverse disease processes after initiation [228]. Prior to applying these oligonucleotides as therapeutic agents, direct chemical modification and/or the use of delivery vehicles can improve stability and efficacy of miR and antimiRs and reduce their degradation by serum nucleases. There are many approaches to aid in the delivery of (anti)miRs, including liposome-, polymeric particle-, inorganic particle-, hydrogel-, and viral vectors-based approaches.

Adeno-associated virus (AAV) is a clinically proven gene delivery vector capable of transducing many cell types and tissues. AAV serotypes 1–9 were tested for their ability to deliver a green fluorescent protein (GFP) reporter into primary valvular interstitial cells (VICs) in vitro [229]. AAV2 and AAV3 were capable of transducing VICs more efficiently than other serotypes, yielding up to 98% successfully transduced cells. Such effective delivery vectors for valve cells may be broadly useful for numerous applications, including the study of valvular cell biology, development of valve disease therapies, and regulation of genes for tissue engineering heart valves.

4.1.6.3 Pericardial Disease

Pericarditis

The cause of pericarditis may be infectious (such as viral and bacterial) or non-infectious (such as systemic inflammatory disease, cancer, and post-heart injury syndrome) [230]. The typical and most common manifestation of pericarditis is chest pain. The most common treatment for idiopathic and viral pericarditis is nonsteroidal anti-inflammatory drugs and colchicine [231]. At present, DDS has not been applied in this area, and further prospective studies are needed to clarify the etiology and epidemiology of pericarditis and the best treatment methods, especially for patients with recurrent pericarditis and for nonsteroidal anti-inflammatory drugs and colchicine ineffective.

Pericardial Effusion

Malignant pericardial effusion is a common manifestation that seriously affects the prognosis of patients with advanced malignant tumors. Effective control of pericardial effusion is of great significance for alleviating patients' pain and prolonging their life. Pericardiocentesis combined with catheter drainage and intracavitary chemotherapy are the main treatments for malignant pericardial effusion. China has independently developed implantable sustained-release fluorouracil, which has been used in the treatment of malignant pericardial effusion and has achieved certain positive effects [232].

4.2 Aortic and Peripheral Vascular Disease

4.2.1 Aortic Disease

4.2.1.1 Aortic Dissection

As the largest blood vessel in the human body, aorta can be divided into four anatomical regions: aortic arch, ascending aorta, thoracic aorta, and abdominal aorta. Aortic dissection (AD) is a tear in the intima of the aorta or bleeding in the aortic wall, which can lead to separation (dissection) of the walls of the aorta and can be life-threatening [233, 234]. AD is most common in people between the ages of 65 and 75 and occurs in 35 cases per 100,000 people per year. Despite surgical or interventional treatment, AD has a 30% mortality rate in the first year [235]. Therefore, it is essential to develop effective treatment strategies.

In recent years, drug delivery systems have been designed for specific targeted delivery of AD, sustained or triggered drug release, and synergistic drug administration to achieve better therapeutic outcomes. The application of drug delivery system can not only improve the selective delivery of drugs to aortic tissues but also improve the bioavailability of drugs and reduce drug side effects. Drug delivery systems are emerging for the treatment of AD, and some design strategies are described below.

Polymeric Particle

MicroRNAs (miRNAs) are a series of highly conserved small noncoding RNA molecules that have been explored for effective treatment of various diseases, including AD [236]. Studies have shown that the promotion of miR-145 enables SMCs to maintain a contractile phenotype and ultimately achieve the protection of AD [237–240]. However, due to its own instability and complex environment, miR-145 can't be delivered directly. Polymer is a widely studied non-viral vector that can condense negatively charged nucleic acids into nanocomplexes for further cellular transfer and intracellular release [241, 242]. Chen Xu et al. prepared a multifunctional delivery nanosystem (TP-GD/mirNA-ColIV) consisting of gadolinium chelating tannic acid (TA), low toxic cation PGEA(ethanolamine-amino poly (glyceryl methacrylate)), and type IV collagen-targeting peptide (ColIV), which can be used to deliver different nucleic acids effectively (see Fig. 4.14) [243]. Among these structural components, natural TA has been widely used to target heart tissues; low toxic cation PGEA can load negatively charged nucleic acid by surface charge adsorption; ColIV can specifically target the early target type IV collagen of intimal AD to enhance targeting. Ultimately, this multifunctional nucleic acid delivery nanosystem can deliver miR-145 effectively to stabilize vascular structure and prevent the progression of aortic dissection.

Inorganic Particle

More and more researches focus on metal-organic frameworks (MOFs). As an emerging multifunctional material, MOFs are constructed by metal ions/clusters

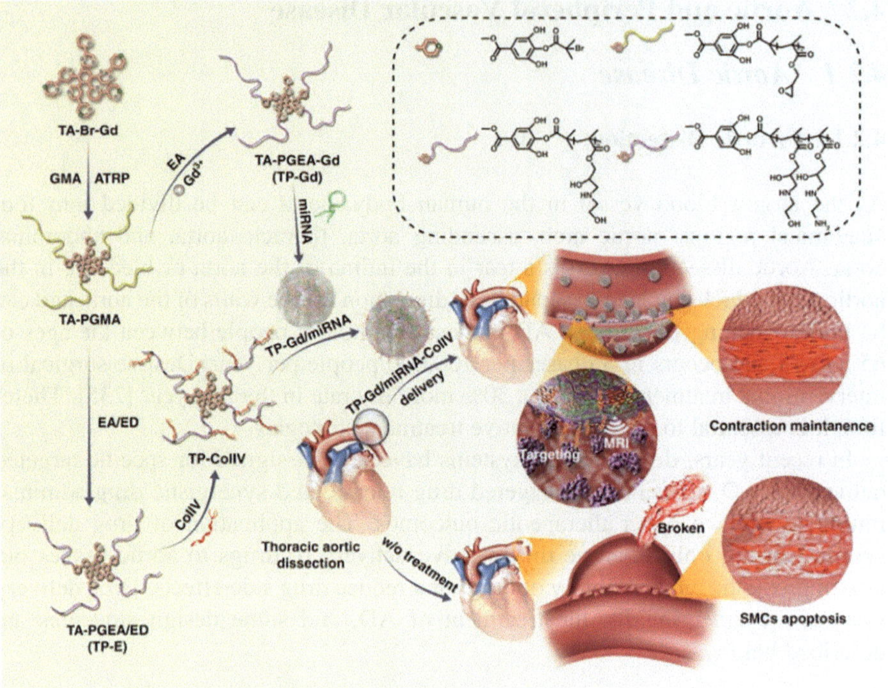

Fig. 4.14 Schematic diagram of tP-GD/mirNA-COLIV complex preparation and targeted gene therapy for thoracic aortic dissection. Reproduced with permission from [243]

and organic ligands through coordination bonds [244]. As a drug carrier, MOFs have many excellent characteristics, such as adjustable size, diverse functions, high loading rate, and good biocompatibility. Moreover, MOFs are an ideal material for effective hemostasis and targeting vascular damage in vivo [245]. Based on this, Yang Liu et al. prepared Tb(III)-MOF based on fluorescent metal organic matrix. This compound could reduce the relative expression of the β-receptor and MMP-9 genes in cardiomyocytes and play a protective role on AD [246]. In addition, Yaping He et al. synthesized a cuboidal and biocompatible γ-cyclodextrin metal-organic frameworks (CD-MOFs) [247]. The synthesized cubic nanoparticles have an effective hemostatic effect, reducing the bleeding time and blood loss by 90%, and have a strong targeting ability for vascular damage in vivo.

Hydrogels

Before the onset of AD, aortic root dilation was simple. Aortic root dilatation is the most common cardiovascular symptom. Its complications include aortic regurgitation, aortic dissection, and aortic rupture, which are the main causes of morbidity and mortality. Hydrogels are unique three-dimensional biomedical materials that can withstand large amounts of water while remaining intact, making them suitable for drug delivery applications. J Durlach et al. found that the hydrogel coated balloon angioplasty could serve as an aortic carrier, so they were locally transmitting the

Mg-dependent hammerhead ribozyme, specifically controlling the genetic alterations leading to fibrin 1, which played a role in relieving and treating aortic complications successfully [248].

Virus vector can transfer genetic material into cells effectively by using the natural ability of it, which is a promising drug delivery tool [249]. The levels of matrix metalloproteinase-2 (MMP2) and C3a (Cleavage pro-inflammatory substance of C3 invertase) are significantly increased in patients with AD [250–253]. C3aR antagonist can inhibit the development of AD before and after the development of AD mice [254]. C3aR knockout can reduce the expression of matrix metalloproteinase-2 (MMP2) in AD mice. Based on this, Weihong Ren et al. used recombinant adeno-associated virus (rAAV) to produce MMP2 knockout mice via MMP2 short hairpin RNA in vivo and found that MMP2 deletion significantly reduced the formation of TAD [255].

4.2.1.2 Aortic Aneurysm

Aortic aneurysm (AA) is a type of CVD with local or diffuse abnormal expansion of aortic wall, whose blood vessels exceed 50% of normal blood vessel diameter so that they compress surrounding organs to cause symptoms, usually developing in abdominal aorta, but also locating in thoracic aorta. AA causes weakness of aortic wall and increases the risk of aortic rupture.

If not treated in time, there will be a possibility of rupture and bleeding with serious consequences. At present, there are no clear drugs that can inhibit the occurrence and development of AA. Clinical drugs are mostly used to control hypertension and treat concomitant diseases such as diabetes, hyperlipidemia, coronary heart disease, and cardiac insufficiency. Therefore, developing new drugs, targeting the pathogenesis of AA, is of great significance to the prevention and treatment of AA. In terms of surgical treatment, endovascular stent treatment is often adopted, because its efficacy and risk are better than previous open chest and open surgery.

These measures have been successful in short term, but may fail in the later stages due to the formation of thrombosis and restenosis of blood vessels. Therefore, the development of a nanoparticle-based system for specifically delivering drug molecules to aneurysms is necessary.

In the treatment of AA, polymerized nanoparticles are used to deliver drugs ranging from exogenous NO, collagen oligomers to MMP inhibitors by inhibiting vascular inflammation, protecting elastin, or inducing elastin synthesis to inhibit AA growth.

At present, the pathogenesis of aortic aneurysms is not fully understood, but it is generally believed that the pathophysiological processes involved in the onset of disease mainly include oxidative stress, vascular inflammation, and extracellular matrix degradation.

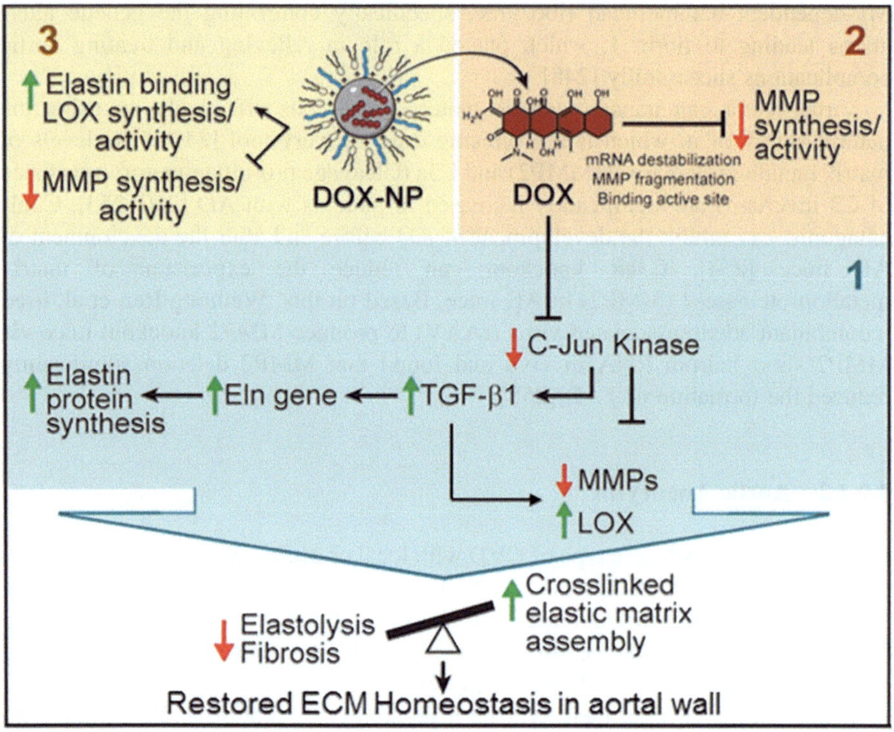

Fig. 4.15 Schematic detailing the mechanistic pathways exhibited by long-term, systemic release of DOX from biodegradable polymeric nanoparticles in regenerating the elastic matrix. The upregulation of TGF β-1 upon DOX inhibition of the JNK pathway enables a pro-regenerative metric for modulating the extracellular matrix. Reproduced with permission from [258]

Nanoparticles

Doxycycline is the most used drug in laboratory research for targeted treatment of aneurysms, which achieves therapeutic effects by inhibiting the MMP activity of aneurysms and inhibiting vascular inflammation [256].

Doxycycline was locally delivered through PLGA nanoparticles(NPs) [257], which were functionalized by cationic polyester method, increasing cell uptake and inhibiting MMP activity. Andrew Camardo et al. conducted a similar study, and the released DOX inhibited the expression and phosphorylation of JNK, thereby increasing elastogenesis and lysyl oxidase-mediated elastic matrix cross-linking as shown in Fig. 4.15 [258]. Lavanya Venkataraman et al. also used PLGA NPs, using DOX and TGF-1 for local control and continuous delivery to smooth muscle cells in human aortic collagen three-dimensional gel to promote the elasticity of collagen construct generated [259]. On the basis of the previous, superparamagnetic iron oxides (SPIONs) and DOX PLGA NPs are combined to enhance targeting [260]. Under the external magnetic field, DOX-SPION NPs can move while

Table 4.17 Summary of studies on other nanoparticles for aortic aneurysms

Polymer	Drug	Characteristic
Nanoparticles	Hyaluronic acid oligomer (HA-o)	Deliver HA-o over a longer period (>30 days), increasing elastic matrix synthesis and lysyl oxidase activity [265]
Albumin NP	PGG	Reverse the development of moderate aneurysms by administering a chelating agent EDTA and a polyphenol PGG [266]
		Deliver PGG to the aneurysm site and prevent aneurysm growth by protecting elastin, resulting in macrophage aggregation, reduced matrix metalloproteinase (MMP) activity, elastin degradation, calcification, and aortic aneurysm development [267]
	Exogenous NO	Inhibiting the release of inflammation markers and cell proliferation, promoting the synthesis of elastic matrix and improving the vascular tone(see Fig. 4.16) [268]

enhancing the elastic matrix deposition, significantly inhibiting the synthesis of MMP-2 and the activity of MMP-2 and MMP-9.

Another kind of drug used in laboratory research is rapamycin, which achieves therapeutic purposes by retaining aortic elastin and smooth muscle cell populations and reducing density of wall macrophages and promoting neovascularization [261].

Takuro Shirasu et al. designed a polyethylene glycol shell NPs containing rapamycin [262], which was colocalized with macrophages to inhibit the activity of matrix metalloproteinases and the expression of inflammatory cytokines, thereby inhibiting the formation of AAA. And the researchers prepared a rapamycin-loaded ROS-responsive NPs [263], which can release drug molecules under the trigger of high levels of ROS, significantly inhibit calcification, and weaken ROS-mediated oxidation stress and apoptosis. Later, the NP was modified with peptide ligand cRGDfK and macrophage cell membrane, further enhancing the targeting and therapeutic effect of aneurysm.

In addition to DOX and RAP, Nasim Nosoudi et al. target BB-94-loaded NPs coupled with elastin antibody to aneurysms [264], inhibiting MMP activity, elastin degradation, calcification, and aneurysm development in the aorta at low doses.

Studies of other nanoparticles used in aortic aneurysm are shown in Table 4.17.

Hydrogel

Brenton Jennewine et al. developed cationic amphiphile (DMAB) modified submicron particles (SMPs) to deliver DOX [269], enhancing elastin regeneration repair. And, in order to provide greater specificity for SMP targeting, the DOX-SMP surface is bound to an antibody against cathepsin K. Moreover, Fatemeh Zehtabi et al. and Maomao Yu et al. prepared hydrogels. Fatemeh Zehtabi et al. developed an injectable closed chitosan (CH) hydrogel containing DOX [270], which has two stages of DOX release: one is explosive and the other is slow continuous. The released DOX can inhibit the activity of MMP-2. In vivo tests in porcine renal arteries have shown that the success rates of immediate embolization and delayed

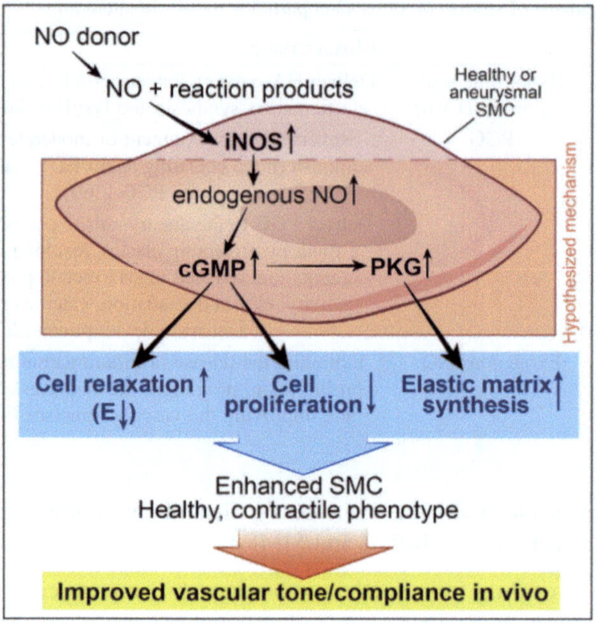

Fig. 4.16 Schematic of the hypothesized mechanism by which exogenously delivered nitric oxide (NO) donors such as GSNO ultimately influence smooth muscle cell (SMC) phenotype and genotype in both healthy and aneurysmal aorta. Sequentially, the NO released by donors upregulates iNOS gene expression and manifests as endogenous NO release, which in turn increases cyclic guanosine monophosphate (cGMP) and protein kinase G (PKG) expression, thereby enhancing cell relaxation, suppressing proliferation, and promoting elastic matrix synthesis. These together contribute to heightened contractile SMC phenotype and improved vasculature tone and compliance in vivo. Here E refers to elastic modulus, an estimate of the cell stiffness. Reproduced with permission from [268]

embolization are 96% and 86%, respectively. Besides, Maomao Yu et al. studied the DOX/PECT hydrogel to improve the bioavailability of DOX [271]. In general, DDS delivers DOX to enhance targeting, improve bioavailability, inhibit MMP activity, and increase elastic matrix formation. Furthermore, the results of above research have shown promise in vivo and vitro studies.

Others

Protein medicine is a research hot spot in the field of biomedicine. Compared with traditional small-molecule drugs, protein drugs have the advantages of good safety, strong targeting, and remarkable curative effect. However, due to the large molecular weight of protein and the small amount of surface charge, it cannot enter cell through the cell membrane. At the same time, because the protein structure is complicated and easy to degrade or inactivate, protein therapy is facing severe challenges. Therefore, how to efficiently and safely deliver protein molecules to target cells is crucial for protein therapy and basic biological research. At present, various delivery vehicles (such as polymers, liposomes, inorganic nanoparticles, vesicles, etc.) have been widely developed and applied to the field of protein delivery and therapy and have achieved remarkable results.

In the treatment of aneurysms, Alexa A. Yrineo et al. developed self-assembled type I collagen oligomers to be delivered to mice as injectable therapeutic delivery vehicles [272]. On the 14th day, the tissue was taken out for histological analysis, displaying that the degradation of self-assembled fibril collagen in the body was minimal and the inflammation and cell invasion of most implants were also minimal, which further confirms the potential of this material as a treatment. However, further preclinical and clinical studies are needed to determine whether the specific therapies incorporated in the self-assembled type I collagen matrix described herein can be delivered near the aorta and locally limit AAA expansion.

On the whole, the free drug may reach the whole body, which has an adverse effect on normal tissues. The DDS is used to target the drug to the treatment site, which increases the utilization rate of the drug, reduces toxic and side effects, and improves the curative effect.

Targeted delivery of drugs and MMP inhibitors using NPs may be an attractive strategy for inhibiting aneurysm progression. At present, many targeted vascular drug delivery systems have not yet reached the clinic. Before the clinicalization of these technologies, there is still a need to better optimize the release kinetics of bioactive molecules and identify synergistic biomolecule combinations. In the future, multifunctional nanotherapy, dual-agent-loaded NP system, and dual-targeted therapy are attractive options, by controlling and sustaining drug release from multifunctional, targetable nanocarriers, elastic matrix deposition, and recovery of moderately developed aneurysms.

4.2.2 Peripheral Arterial Disease

Peripheral arterial disease (PAD) is one kind of disease which resulted from lower extremity arterial vessel chronic narrowing; it would lead to loss of limb functional capacity. Traditional PAD therapy mainly include two aspects: noninvasive and invasive therapy. The noninvasive therapy includes the medicinal approaches such as statin therapy, lipid inhibitors, beta-blockers and diuretics, Ramipril administration, and cell-based therapy. Though some efforts were obtained, there were still some shortcomings of these approaches. When administered as single or combination drugs to PAD patients, some terrible complications such as life-threatening bleeding, cardiac complications, and liver problems can occur. Invasive therapies for PAD treatment consist of endovascular intervention, bare metal stents, surgical bypass, and angioplasty balloon catheters. The blood flow was significantly recovered after the invasive therapies were implemented. What limited these approaches was availability of venous vessels, evasion of treating diseased vessels, impaired blood flow at the donor site, and compressive lipid plaques onto the endothelial wall. Besides, the stent implantation was reported due to the problem of stent fractures and the vascular trauma and endothelium denudation. Therefore, it is critical to develop

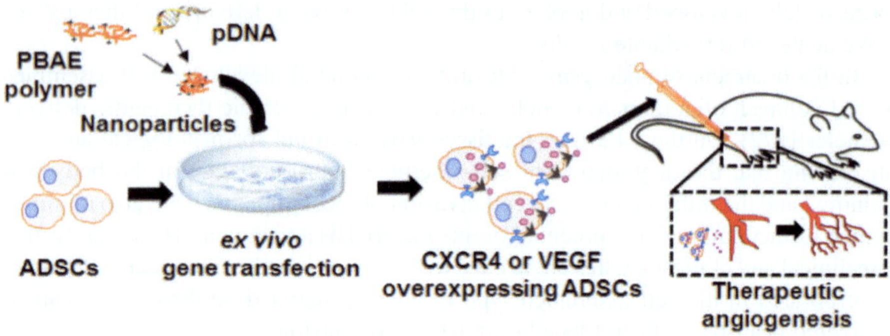

Fig. 4.17 Preparation of CXCR4- and/or VEGF-overexpressing ADSCs using PBAE nanoparticles. Reproduced with permission from [274]

some new methods that overcome limitations of current situation for PAD therapy. In the past decade, numerous materials including synthetic, natural, biodegradable, and biocompatible materials are applied to the treatment of PAD. Moreover, various recent research approaches such as nanoparticles, hydrogel, liposome, functional cells, and so on are being explored to target delivery of various therapeutic agents to treat PAD. Herein, researches for PAD therapy are reviewed in detail in addition to a brief discussion about current treatment modalities.

Polymeric Nanoparticles

An injectable poly (lactide-co-glycolide) (PLGA) microspheres encapsulating WKYMVm peptides were fabricated for the treatment of PAD. WKYMVm peptide has been identified as a novel peptide agonist for the FPR-2 which play a significant role in inflammation and angiogenesis. Injection of microspheres was effective in restoring blood flow from ischemic injury and promoting capillary growth in the hind limb ischemia model [273]. Nanoparticles modified with PEG could extend the circulatory half-lives of protein drugs, however, also reducing their extravasation and retention at the target site. As shown in Fig. 4.17, nanoparticle-induced CXCR4-overexpressing stem cells in a mouse hind limb ischemia model were reported [274]. The system increased stem cell survival, decreased inflammation, and accelerated blood reperfusion, resulting in limb salvage and skeletal muscle repair, markedly outperforming the efficacy of the conventional angiogenic factor VEGF. Importantly, CXCR4 overexpression promotes the change of stem cell phenotype into a pro-angiogenic and anti-inflammatory stem cells. In order to obtain better therapeutic effects of PAD, it must understand the molecular distribution of VEGF and VEGFR because they play a crucial role in the angiogenesis during the recovery of PAD. Yoh Hamada found VEGFR distribution in ischemic tissues increased over a specific period and remained steady at a threefold increase. This several-fold increase of VEGFR at the ischemic site may be effective for controlled angiogenesis which result in increasing the number of sprouting vessels [275].

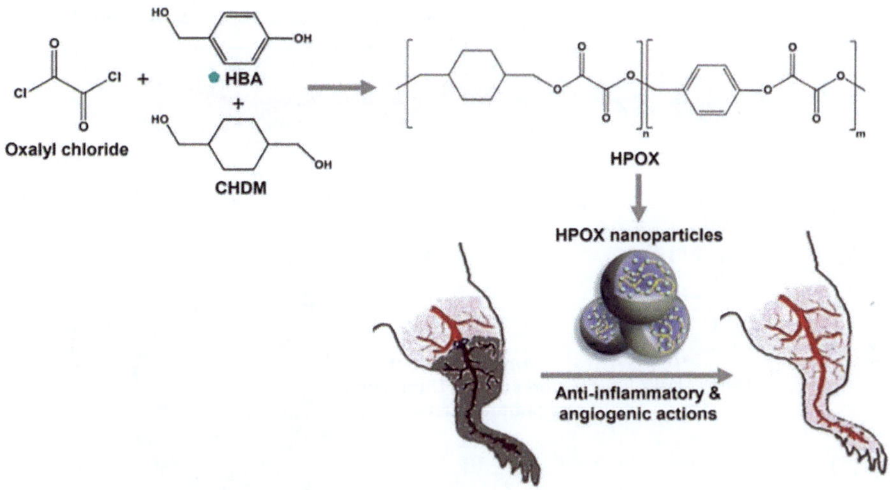

Fig. 4.18 A diagram of therapeutic effects of HPOX particles on ischemic injury. Reproduced with permission from [276]

There are some reports about ROS-responsive materials which were more suitable for the target drug release in the ischemic site due to microenvironment of PAD. Such as PSS, HPOX, PVAX, and so on. Molecularly engineered solid HPOX nanoparticles were designed which could capture and eliminate H_2O_2. Thus HPOX nanoparticles have potential to application in therapeutics for H_2O_2-associated inflammatory diseases. This HPOX delivery system minimizes side effects while targeting release drugs in the high ROS areas to obtain better effect. After that, as depicted in Fig. 4.18, this biodegradable nanoparticle was developed to increase blood flow recovery through angiogenesis [276]. The therapeutic effects were highly correlated with expression of angiogenic inducers, VEGF and Akt/AMPK/eNOS in ischemic muscles.

Another H_2O_2-responsive molecularly engineered polymer nanoparticle with therapeutic and imaging strategies designed with the basic of ROS-responsive was prepared by Dongwon Lee. Gas bubble would generate when the ROS-responsive carriers react with the ROS in pathologic microenvironment. Therefore, the ultrasound imaging and on-demand therapy of peripheral arterial diseases were put forward [277, 278]. Eunkyeong Jung designed a H_2O_2-activated bubble generating anti-inflammatory polymer particles named CUR-PVAX and a stimulus-activatable echogenic maltodextrin nanoparticles as nanotheranostic agents (see Fig. 4.19) for the treatment of peripheral arterial disease. Both delivery systems showed outstanding therapeutic effects due to the improvement of microenvironment which was character with high-level ROS and inflammation. Summary of PAD treatment with polymeric particles can be seen in Table 4.18.

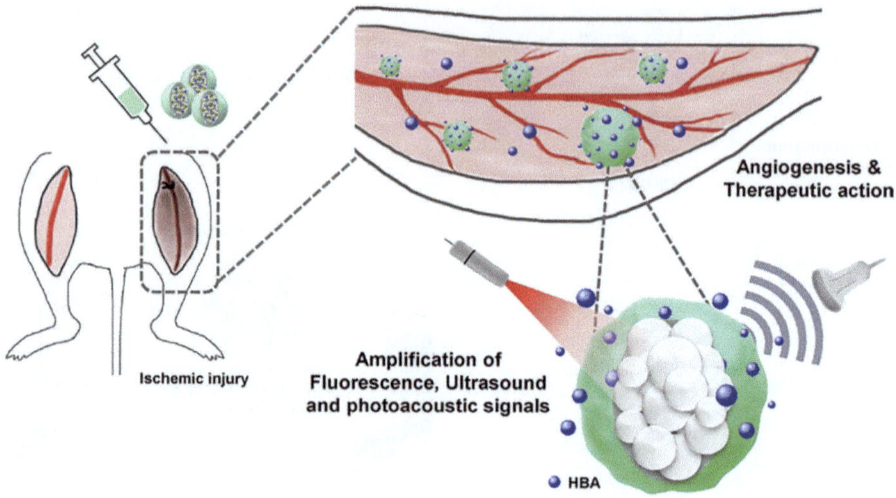

Fig. 4.19 Illustration for therapeutic effects of HPOX particles. Reproduced with permission from [278]

Hydrogel

Hydrogel was widely reported as one kind of the most potential biomaterials. There was a report about the controlled release of deferoxiamine with the hydrogel. The release of deferoxiamine from the hydrogel showed a degradation-dependent manner after adding the collagenase. It was the deferoxamine-activated hypoxia-inducible factor-1a that increased the therapeutic effect of hind limb ischemia [291]. Nitric oxide (NO) was accepted as a promising therapeutic strategy in treatment of PAD. As shown in Fig. 4.20, the NO generation hydrogel was reported by Qiang Zhao with a stable comb-shaped polymer (CSeNO) which contained glycosylated NO compound and chitosan (CS) as the backbone. The CSeNO polymers could form into injectable hydrogel, and the protective effect of CSeNO was observed against limb necrosis [292]. Various growth factors which promote collateral vessel development such as erythropoietin, bFGF, C-CSF, VEGF, and so on are regarded as promising for the treatment of vascular occlusive diseases. Considering that an efficacious delivery system could induce better therapeutic effects, an ionic gelatin-based hydrogels employed to deliver the bFGF was designed. The ionic gelatin-based hydrogels achieve a controlled release of bFGF and exhibited marked reperfusion which was 69.3% higher than control group [293]. Recent clinical trials proved that increased angiogenesis would trigger severe inflammation in a proportionally coupled fashion with treatment of angiogenic growth factors. In order to obtain better therapeutic effect, co-delivery of pro-angiogenic C16 and anti-inflammatory Ac-SDKP peptides was proposed. As expected, the promoted angiogenesis, muscle regeneration, and perfusion recovery were observed after treatment with the co-delivery system [294].

Table 4.18 Summary of PAD treatment with polymeric particles

Drug	Formulations	Characteristic	Application
Pitavastatin	PLGA nanoparticles [279]	Cell-selective delivery of NP into the capillary and arteriolar endothelium	Therapeutic neovascularization
Lead oxide and gelatin	Polyion complex (PIC) micelles [280]	Size-dependent accumulation and retention in the collateral arteries	High accumulation in the lesions
MicroRNA-126	PLGA nanoparticles [281]	Regulation of the VEGF receptor pathway by the in vivo overexpression of miR-126	A murine hind limb ischemia
VEGF gene	Low molecular weight polyethylenimine with 2,6-pyridinedicarboxaldehyde (PDA) as a biodegradable polycationic material (PDAPEI) [282]	Enhanced the transfection performance of VEGF and exerted an excellent therapeutic effect at 40 w/w ratio	Promising gene vector in the treatment of peripheral ischemic artery disease (PAD)
	PLGA nanosphere [283]	High encapsulation efficiency (87%) and sustained release for 11 days	Carrier for skeletal muscle gene delivery applications
	Nanofibrous hydrogel [283]	Thixotropic hydrogel, improve significantly angiogenesis and tissue recovery within 7 days	Rapidly infiltrated by host cells and degraded while promoting the generation of neovessels
	Poly(β-amino esters) (PBAE) nanoparticles [284]	Promoting engraftment of HUVECs into mouse vasculature as well as survival of transplanted HUVECs in ischemic tissues, leading to improved angiogenesis and ischemic limb salvage	Significantly enhanced VEGF expression in HUVECs
VEGF protein	Heparin-functionalized PLGA nanoparticle-fibrin [285]	Sustained release and the enhanced efficacy of VEGF, and fibrin plays a role as a matrix for local delivery, in which endothelial cells can survive and migrate	An effective local and sustained delivery system of VEGF for therapeutic angiogenesis

(continued)

Table 4.18 (continued)

Drug	Formulations	Characteristic	Application
Secretoneurin	S-protected thiolated chitosan nanoparticles [286]	Secretoneurin modified by induction of a cysteine residue to gain higher stability against enzymatic degradation	Restored blood flow in a mouse hind limb ischemia model
Anti-PEG immunoglobulin M antibodies	Mixture of hyaluronic acid (HA) hydrogel [287]	Repeated systemic injections permit "reloading" of the capture depot, allowing the use of complex multi-stage therapies	Injected at a target body site, can capture and retain a variety of systemically injected PEGylated therapeutics at that site
Peptide amphiphiles	Arg-Gly-Asp-Ser peptide nanofiber gel [288]	p38 and p44/p42MAP kinases. An integrin-binding domain of fibronectin improves regenerative efficacy of cell-based strategies in ischemic tissue by enhancing cell survival, retention, and reparative functions	Ischemic tissue repair
Curcumin	Poly(propylene sulfide) microspheres [289]	Hydrophobic PSS would become hydrophilic situation after being oxidized and achieves ROS-demanded drug release	Targeted and eliminated intracellular ROS, reduced in vitro cell death in the presence of cytotoxic levels of ROS, and decreased tissue-level ROS in vivo
Vanillyl alcohol (VA)	Copolyoxalate containing vanillyl alcohol nanoparticles (PVAX) [290]	H_2O_2-responsive peroxalate ester linkages covalently in PVAX backbone	Anti-inflammatory and anti-apoptotic activity

Cell transplantation represents a novel strategy, and with the development of cell therapy, numerous functional cells were applied as a therapeutic option to rescue critical limb ischemia (CLI). Unfortunately, the outcomes have been not so satisfactory. There were rather small number of injected functional cells that would exert therapeutic effect at the ischemic site due to the rapid loss of bioactivity and immunogenicity. Besides, high cost of cell therapy limited its application as well. The current researches have proved the therapeutic contribution of functional cells to vascular regeneration and functional recovery of ischemic tissue; however the

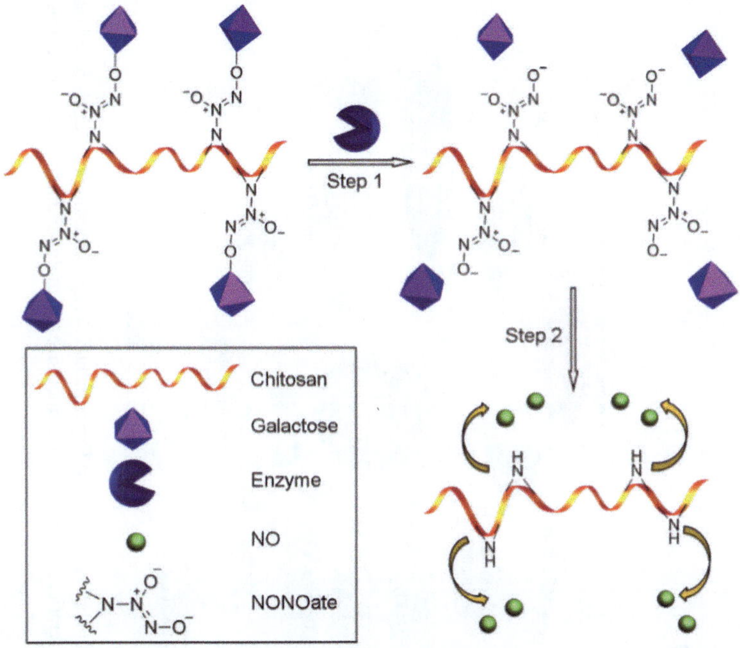

Fig. 4.20 A schematic diagram of the decomposition of CSeNO under the catalysis of enzyme. Reproduced with permission from [292]

application was limited by the short-term cell survival in vivo; therefore it is critical to enhance cell survival. Great efforts were put in how to ameliorate or supply suitable microenvironment for transplanting cells and prolong the survival and therapeutic time in vivo. Thus, multifunction hydrogel researches were widely reported such as PEGylated, catechol-Functionalized, thermos-responsive, nitric oxide release, polypeptide, and so on. Hyun-Ji Park developed a mussel-inspired, catechol-functionalized hyaluronic acid (HA-CA) hydrogel that enabled effective cell transplantation due to the enhanced biocompatibility and tissue adhesiveness; then researcher performed further utility of HA-CA in stem cell therapies for PAD disease [295]. Besides, due to the temperature of ischemia tissues decreases, a reversible thermos-responsive hydrogel based on methylcellulose-salt system was designed to deliver the stem cells for the treatment of critical limb ischemia [296]. The thermos-responsive hydrogel system effectively inhibited the fibrosis and muscular atrophy of ischemic tissue and accelerates the blood reperfusion of hind limb. Meanwhile, there was a report about modulating angiogenic paracrine in stem cells with a shape-controlled tuneable hydrogel (Fig. 4.21), which used naturally occurring extracellular matrix molecules highlighting the need for highly customized matrices with an array of self-assembling proteins that dictate specific cell function resembling the native tissue of interest for repair [297].

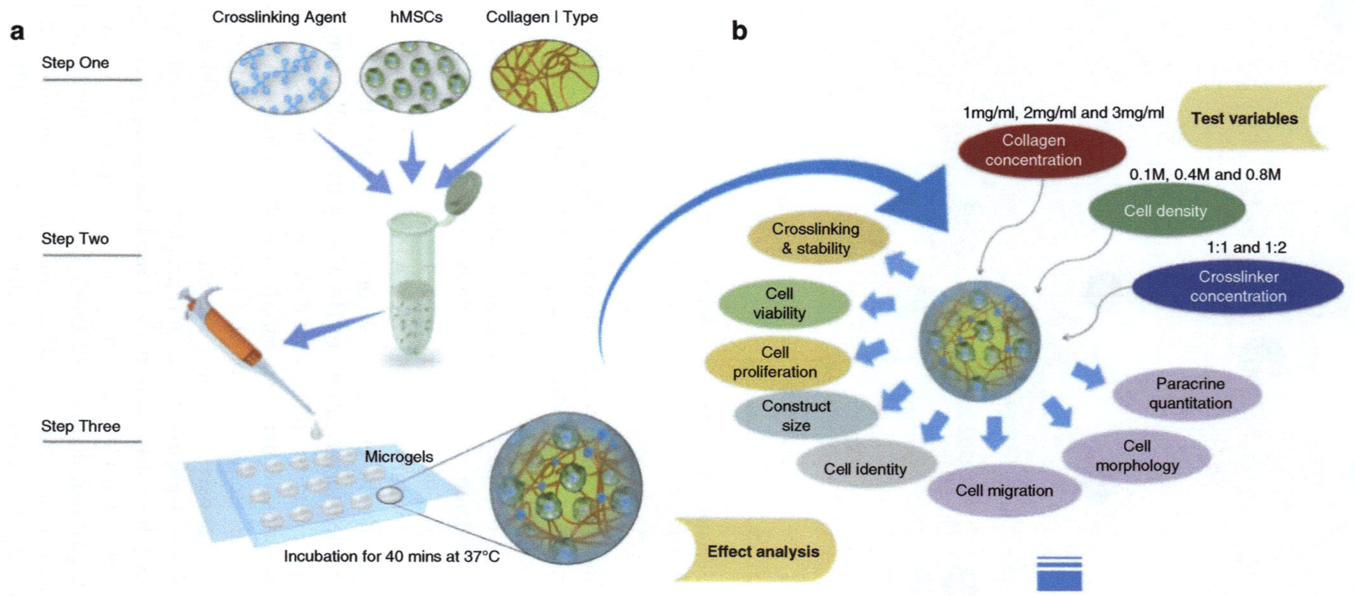

Fig. 4.21 (**a**) Schematic depicting fabrication of hMSC embedded 4S-StarPEG cross-linked microgels. (**b**) Experimental design for testing microgels based on collagen concentration, cell density, and cross-linker ratio on hMSC survival, proliferation, migration, construct size, phenotypic analysis, and paracrine response. Reproduced with permission from [297]

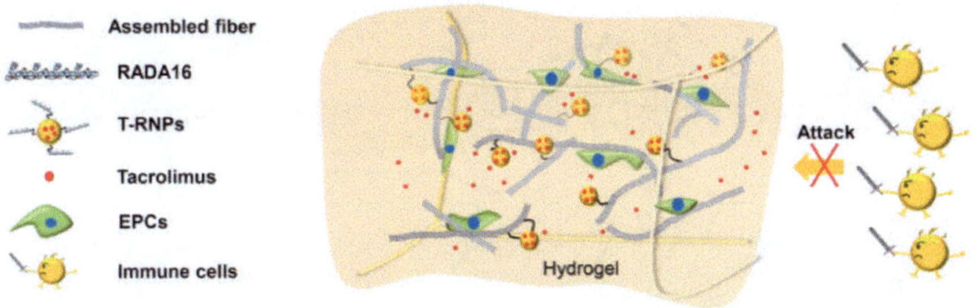

Fig. 4.22 Schematic illustration of a tacrolimus-loaded nanoparticle-anchoring hydrogel. Reproduced with permission from [299]

To seek the better therapeutic effect, there were some new ideas about multifunctional gels integrating highly efficient growth factor delivery with functional cell transplantation. The functional gel could co-deliver cells and protein factors which act as a depot release cytokines and growth factors for a long time, resulting in an improved therapeutic effect. The therapeutic effects of outgrowth endothelial cells (OECs) and growth factors encapsulated within RGD-alginate microgels were formed by electrospraying method. The injectable microgel system containing OECs and growth factors not only decrease the host immune response against the injected materials but also allow for repeated administrations [298].

Besides, as depicted in Fig. 4.22, a sustained release of immunosuppressant by nanoparticle-anchoring hydrogel scaffold was reported to improve the survival of transplanted stems, which showed a significantly promoted blood perfusion (~91% vs. ~36% in control group) [299]. Xiaomin Wang prepared a bioactive hydrogel by mixing chitosan and hyaluronic acid immobilized with C domain peptide of insulin-like growth factor 1 (IGF-1C). As shown in Fig. 4.23, the functional hydrogel could augment stem cell survival and proangiogenic activity of adipose-derived stromal cells (ADSCs). Cotransplantation of hydrogel and ADSCs into ischemic hind limbs of mice effectively improved blood recovery and muscle regeneration, leading to superior limb salvage [300].

Except these treatments with hydrogels for PAD, there are some hydrogels' researches about delivery-engineered functional cells. Genetically modified stem cells offer potential advantages for therapeutically inducing angiogenesis via enhanced paracrine release mechanisms and tuned dynamic responses to environmental stimuli at disease sites. Joan Lee designed a genetically engineered myoblast sheet named C2C12 which formed on Tetronic-tyramine (Tet-TA)-RGD hydrogel for therapeutic angiogenesis. The VEGF plasmid transfection enhanced the VEGF expression and secretion. More importantly, it also augmented the formation of capillaries and arterioles in ischemic muscles, attenuated the muscle necrosis, and prevented ischemic limb loss [301]. Based on cells in three-dimensional (3D) organization maximizing the cell-cell communication and signaling which are critical for cell function, a 3D cell aggregates contained of human umbilical vein endothelial cells (HUVECs) and cord-blood mesenchymal stem cells (cbMSCs)

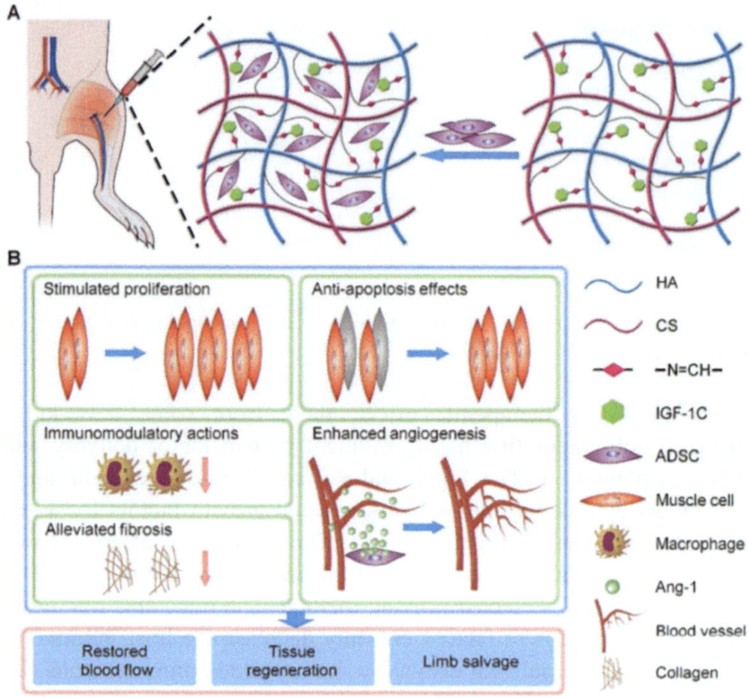

Fig. 4.23 Schematic depiction of the interplay between CS-HA-IGF-1C hydrogel and ADSCs which leads to enhanced cell survival and therapeutic benefits. Reproduced with permission from [300]

were employed for therapeutic neovascularization of critical limb ischemia. The transplanted cell aggregates improved functional vessel formation and protected it from degeneration [302]. After that the 3D cell aggregates were delivered with a methylcellulose hydrogel system. It was the hypoxic microenvironment in the inner cores that activated the hypoxia-inducible factor-1α-dependent angiogenic mechanisms. As expected, the expression of the pro-angiogenic genes was significantly elevated [303]. Delivering cytokine and functional cells with hydrogel in the treatment of PAD can be seen in Table 4.19.

Metal Associate Materials and Stent

According to existing research, the muscles of PAD patients are in a harmful microenvironment characterized by increased oxidative stress and inflammatory infiltration. It is essential to improve the microenvironment along with angiogenesis. Hence, as depicted in Fig. 4.24, a metal-polyphenol capsule (Cu-EGCG) was synthesized through coordination between EGCG and copper ions, which combined the anti-inflammatory and antioxidant activities of EGCG and the angiogenic activity of copper ions. The functional metal-polyphenol capsule not only supplies a

Table 4.19 Delivering cytokine and functional cells with hydrogel in the treatment of PAD

Drug	Formulations	Characteristic	Applications
Erythropoietin	Gelatin hydrogel microsphere [304]	Sustained release EPO	Recovery of blood perfusion in the ischemia site
Basic fibroblast growth factor (bFGF)	Acidic gelatin hydrogel microspheres (AGHMs) [305]	The size of AGHMs is an important factor to achieve pinpoint delivery. 29 μm diameter AGHMs strongly induced functional collateral vessels	Augment functional collateral vessels
hMSCs and desferrioxamine	Chitosan gel [306]	The gel started a thermos-gelation at 33 °C and achieved a sustained, biologically active release of DFO for 7 days	Multimodal proangiogenic therapeutic thermos-responsive gel, the survival, proliferation, and migration of encapsulated hMSCs were increased
Human adipose-derived mesenchymal stem cells (hMSCs)	Gelatin microcryogels (GMs) [307]	The 3D microniche-based therapy only with 10^5 hMSCs realized a better therapeutic effect of ischemic limb than treatment with 10^6 free injected hMSCs	The 3D microniches protected cell during injection and in vivo cell retention, survival, and ultimate therapeutic functions in treatment of critical limb ischemia
Human placenta-derived mesenchymal stem cells (hP-MSCs)	Self-assembled Nap-GFFYK-Thiol peptide hydrogel [308]	Disulfide bonds as cleavable linkers to control the molecular self-assembly	Increasing proangiogenic and anti-apoptotic effects of hP-MSCs Reducing inflammatory cell infiltration and enhancing angiogenesis as well as suppressed collagen deposition
Cell adhesive peptide (RGD) and growth factor (bFGF)	Multifunctional Tetronic-tyramine hydrogels [309]	Thermosensitive, the change in the size of the hydrogels was precisely controlled by a repeated increase and decrease from 37 to 4 °C	The transferred cell sheet was highly viable and maintained robust cell-cell junction. An ideal substrate to modulate the formation of an endothelial cell layer that could potentially be utilized to treat peripheral arterial disease

(continued)

Table 4.19 (continued)

Drug	Formulations	Characteristic	Applications
Human platelet lysate	Fibrinogen-rich platelet lysate hydrogel [310]	This PL hydrogel had comparable storage and loss modulus and behaved as a viscoelastic solid similar to fibrin hydrogels despite having 1/4–1/10th of the fibrin content of standard fibrin gels	Enhance pro-angiogenic activity by promoting hMSC growth and invasion in a 3D environment
FGF-2 and C-CSF	Ionic, gelatin-based hydrogel scaffold which is prepared from the polymerization of gelatin and poly-L-glutamic acid (PLG) [311]	Immobilization and controlled release of human recombinant basic fibroblast growth factor (FGF-2) and human recombinant granulocyte colony-stimulating factor (G-CSF)	FGF-2 and G-CSF were used in combination; there was an 82% increase in endothelial branch point formation and CD31 and α-SMA in vivo
VEGF and EPCs	RADA16 hydrogel [312]	Sustained release functional macromolecules for 4 weeks	Enhancing the proliferation and differentiation of stem cells and neovascularization in a mouse model of hind limb ischemia

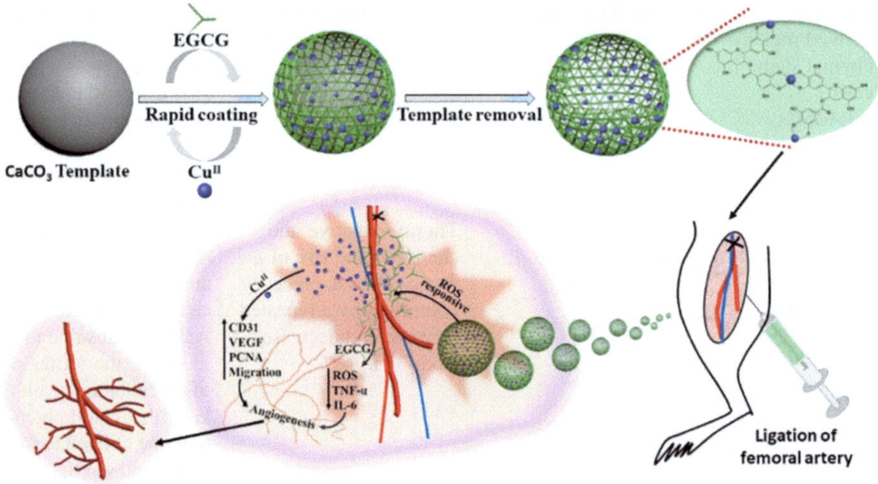

Fig. 4.24 Schematic diagram of Cu-EGCG capsule with therapeutic function on ischemic injury. Reproduced with permission from [313]

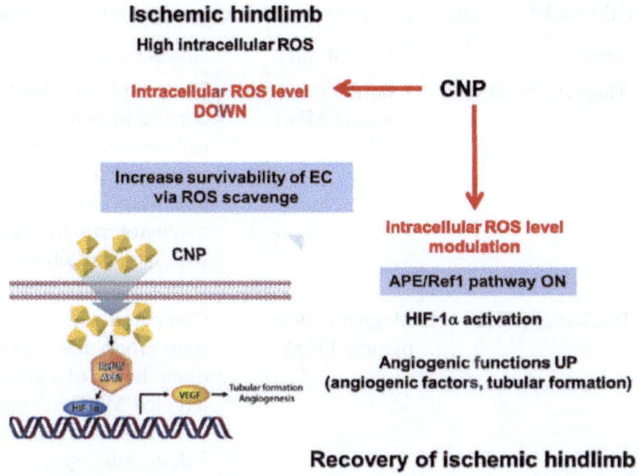

Fig. 4.25 Schematic depicting the proposed roles of CNP in treating CLI. Reproduced with permission from [314]

suitable microenvironment with the antioxidant and anti-inflammatory of EGCG but also induces the therapeutic angiogenesis with the copper ions [313].

As displayed in Fig. 4.25, cerium oxide nanoparticles (CNP) were designed by In-Su Park, and CNP could eliminate the excess ROS and increase the survivability of endothelial cell. Meanwhile, CNP promoted the endothelial cell tubule formation and induced Ref-1/APE1-dependent angiogenesis [314].

Endothelial progenitor cells (EPCs) were widely reported with the effect in augmenting neovascularization to cure the ischemia disease. There was a report about therapeutic angiogenesis with a novel magnetite tissue engineering technology. Masakazu Ishii developed a magnetized MSCs according to magnetic force [315]. The magnetized MSC sheets contained therapeutic mesenchymal stem cell (MSC) sheet for reparative angiogenesis. In this study, apoptosis within the MSC sheets was not observed at 24 h. The survival time of cell in vivo was prolonged, and the therapeutic effect was improved. Besides, stent was reported to the therapy of PAD. An implant composed of spiral magnesium (Mg) and a coating made using poly(lactic-co-glycolic acid) (PLGA) with encapsulated rhbFGF (Mg-PLGA-rhbFGF) was prepared. The incorporation of Mg raised the microclimate pH in the polymer, which strengthen the therapeutic effect due to the stability of rhbFGF was preserved [316]. Summary of other metal associate materials and stent in the treatment of PAD can be seen in Table 4.20.

Liposome

Liposome with no known toxicity are clinically proven for disease therapy as well as PAD. PEGylated liposomes loaded with angiogenic peptides was reported to improve ischemic limb perfusion and promote angiogenic responses [320]. In this study, liposomal angiogenic treatment via intra-arterial infusion resulted in an equally therapeutic efficacy compared to intramuscular injection. Besides, as depicted in Fig. 4.26, glypican-1 nanoliposomes were designed for enhancing FGF-2 activity in therapeutic angiogenesis, which mainly mediated endothelial

Table 4.20 Summary of other metal associate materials and stent in the treatment of PAD

Drug	Formulations	Characteristic	Application
Magnetized EPCs	External magnetic forces (EMFs) [317]	The peripheral blood-derived mononuclear cells were cultured to obtain EPCs and generated Mag-EPCs by a magnetofection method with nanoparticles	Migration capacity of Mag-EPCs with EMFs was 160% more than EPCs or Mag-EPCs without them. Supplying a new therapeutic strategy for patients with CLI
Paclitaxel (PTX)	Magnetic nano-particle [318]	Compared with the bare stent group, the arterial tissue levels of stent-targeted MNPs remained 4- to 10-fold higher over 5 days delivery. A significant inhibition of in-stent restenosis with a relatively low dose of MNP-encapsulated PTX (7.5 μg PTX/stent)	Arterial smooth muscle cells were significantly inhibited with PTX-loaded MNPs. Site-specific drug delivery to implanted magnetizable stents by uniform field controlled targeting of MNPs with efficacy for in-stent restenosis
Recombinant human basic fibroblast growth factor (rhbFGF) and Mg $(OH)_2$	Poly(DL-lactic-co-glycolic acid) (PLGA) millicylindrical implant [319]	Adjust the microclimate pH of PLGA polymer	Reduce the in-stent restenosis Stabilized rhbFGF incorporated in pH modified PLGA millicylinders effectively promotes site-directed in vivo angiogenesis and also enables preservation of ischemic hind limb function

cell proliferation, migration, and tube formation [321]. Co-delivery of glypisomes with FGF-2 markedly increased the recovery of blood and vessel formation in ischemic hind limbs.

Bionic Materials

Nowadays, the increased understanding of various cell surface receptors, their corresponding ligands, and downstream signaling networks appeal researchers to employ bionic strategies to utilize cell membrane, extracellular vesicles, fat extract, and other materials from tissue to strengthen target to the disease site and improve therapeutic effects.

As shown in Fig. 4.27, human adipose-derived stem cell (hASC) membranes were used to prepare the bioengineered stem cell membrane-functionalized nanocarriers (BSMNCs) [322]. BSMNCs contained C-X-C chemokine receptor type 4 (CXCR4) to offer a robust targeting and prolong the retention time in ischemic tissue as well. BSMNCs were prepared and employed in the treatment of hind limb ischemia. The injection of BSMNCs loaded with VEGF into hind limb

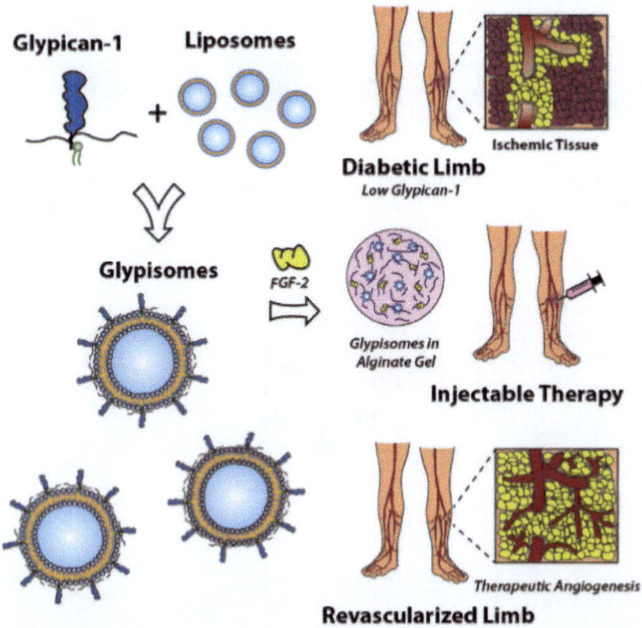

Fig. 4.26 Overall diagram of glypisome-based therapeutics for the treatment of ischemia. Recombinant glypican-1 is embedded in a liposome membrane using detergent extraction. Reproduced with permission from [321]

ischemia mice resulted in enhancement of blood reperfusion, muscle repair, and limb salvage. Other bionic materials employed in the treatment of PAD were shown in Table 4.21.

Last, but not least, despite being a fairly new field, researchers are exploring the use of nanotheranostics for PAD therapy. The reported strategy could be used to create biocompatible and custom-tailored biomimetic nanoparticles with various hybrid functionalities, which may overcome the limitations of current nanoparticle-based therapeutic and imaging platforms.

4.2.3 Vein Thrombosis

Venous thromboembolism (VT) is a common disease, with an annual incidence rate ranging from 1 to 100 in every 10,000 young adults [325]. Among them, deep venous thrombosis (DVT) is the most common and has become one of the most important killers of human health [326]. DVT refers to the abnormal intersection of blood in the deep venous cavity, which connected with the vena cava, resulting in venous reflux disorder, resulting in varying degrees of chronic deep venous

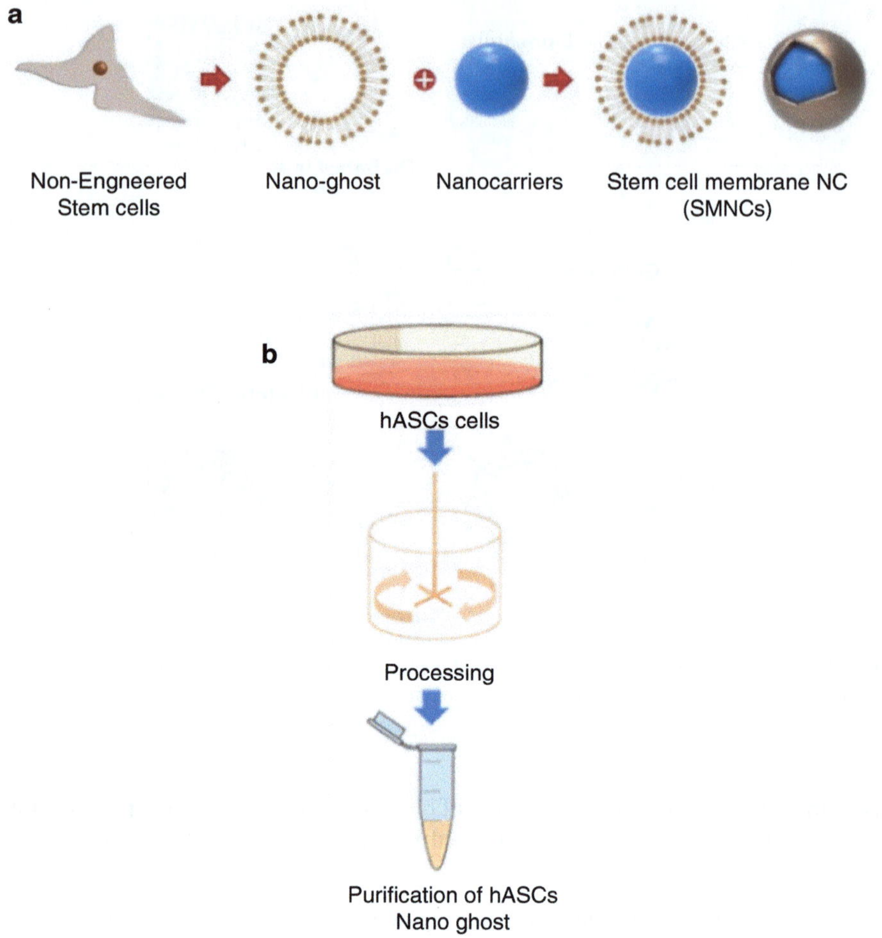

Fig. 4.27 Preparation of stem cell membrane-nanocarrier (SMNCs). (**a**) Schematic diagram details the concept and preparation of SMNCs, (**b**) schematic showing the preparation of hASC membrane-derived nanoghosts. Reproduced with permission from [322]

insufficiency [327, 328]. The most common manifestations of DVT are femoral vein thrombosis, inferior vena cava thrombosis, and lower extremity venous thrombosis, which can cause serious injury to patients. Anticoagulation therapy is the standard treatment for DVT; however, it increases the risk of bleeding [329]. Many drugs, such as aspirin and heparin, approved by the US Food and Drug Administration (FDA), are used for antithrombotic therapy. However, these drugs have serious side effects, such as bleeding risk, gastrointestinal ulcer, drug resistance rate, etc. Therefore, it is urgent to develop new drugs for vein thrombosis.

Table 4.21 Other bionic materials employed in the treatment of PAD

Origin	Formulations	Characteristic	Application
Mesenchymal stem cells (MSCs)	Extracellular vesicles [323]	Improve the recovery of blood perfusion by activating receptors in endothelial cells, such as VEGFR1, VEGFR2, SRC, AKT, and ERK	Cellular proliferation, migration, and neoangiogenesis were strengthened
Human adipose tissue (fat extract (FE))	Cell-free aqueous [324]	Cell-free therapeutic agent FE	Attenuate severe limb loss and improved capillary density and blood flow of the ischemic tissue

Liposome

Thrombomodulin (TM) maintains the antithrombotic ability of endothelium by inactivating thrombin and activating egg white system [330]. Adenovirus-mediated TM gene can successfully promote TM expression. However, recent reports have shown that adenovirus gene therapy can cause serious adverse reactions [331, 332]. Non-viral gene therapy can avoid viral vector-related immune system response, but its low efficiency limits the application of non-viral vector in gene therapy [333]. The efficiency of gene transfer in vitro promoted by ligands has been improved. Moreover, the introduction of transferrin allows cells to absorb the complex rapidly and efficiently at the initial direct contact, resulting in systemic therapeutic results at the level of expressed exogenous cDNA without causing immune response [334]. Noriyuki Tabuchi et al. have demonstrated that the in vivo gene delivery system using ligand-facilitated transfer of cationic liposome could effectively transfer TM gene into rat vein endothelial cells and prevent thrombosis [335].

Polymeric Particle

Nanoparticles are defined as submicron colloidal systems with particle sizes ranging from 10 to 999 nanometers. Nanoparticles are widely used, providing an attractive option for the gradual and continuous delivery of drugs. This drug delivery system has many advantages, such as sustained release of drugs in specific tissues; maintenance of long-term blood drug concentration to reduce the number of drug administration; improving drug bioavailability; reducing drug side effects; and so on [336–338]. Nanoparticles as drug carriers have been widely used in cardiovascular diseases. Low molecular weight heparin (LMWH) is a negative charged oligosaccharide used in the treatment of deep venous thrombosis and pulmonary embolism. It is an anticoagulant drug approved by many countries. Anticoagulation therapy can effectively prevent the occurrence of venous thromboembolism in high-risk patients and reduce the incidence rate and mortality of patients with confirmed thromboembolic diseases. To improve the activity of the anticoagulant and reduce the frequency of administration, it may be beneficial to encapsulate LMWH in nanoparticles such as polymer nanoparticles [339–342].

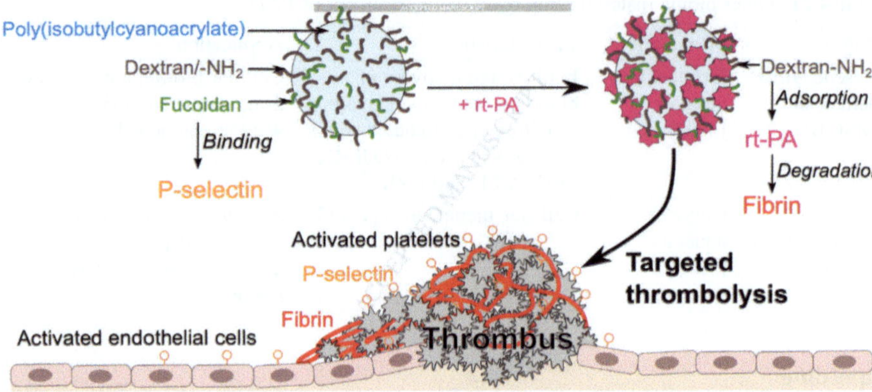

Fig. 4.28 Schematic diagram of preparation and antithrombotic mechanism of RT-PA-Fuco-NPS. Reproduced with permission from [347]

Rt-PA is a serine protease that catalyzes the conversion of plasminogen to plasmin. rt-PA is the standard drug for thrombolysis; however, it shows a risk of bleeding and limited efficiency. Moreover, due to the short half-life of rt-PA, large doses of rt-PA need to be injected, which will produce a cascade reaction in the circulation, leading to serious side effects. Intracranial hemorrhage was observed in 6% of the treated patients, resulting in 50% of the follow-up mortality [343–346]. Therefore, it may be helpful to encapsulate rt-PA in nanoparticles for delivering. Maya Juenet et al. designed a polysaccharide polyisobutyl cyanoacrylate nanoparticles (rt-PA-Fuco-NPs) based on redox radical emulsion polymerization (Fig. 4.28). The results showed that fucoidan functionalized nanoparticles showed high affinity for P-selectin and demonstrated the effectiveness of rt-PA fucoidan NPs in antithrombotic therapy in vivo [347].

Diosgenin is also studied as an antithrombotic drug, which can inhibit thrombosis and anticoagulant function and inhibit platelet aggregation. However, the hydrophobicity of diosgenin and its derivatives is strong, and the understanding of its molecular targets is low, which limits its application [348]. Therefore, it is necessary to develop a drug delivery system to improve the bioavailability of diosgenin, which will improve the efficiency of antithrombotic therapy. Zeliang wei et al. designed a novel prodrug nanocarrier based on polyethylene glycol and diosgenin derivatives to inhibit thrombosis (Fig. 4.29) [349]. The compound was coupled with PEG by Schiff-base bond to form prodrug. Finally, it was found that the precursor micelles in the thrombosis site could be cleaved under acidic conditions, and diosgenin was released rapidly to inhibit the decomposition of thrombus. Diosgenin prodrug micelles with no bleeding risk can prevent thrombosis by inhibiting platelet activation and apoptosis, which is better than aspirin and other clinical antithrombotic drugs. Other treatments of PAD with polymeric particles were shown in Table 4.22.

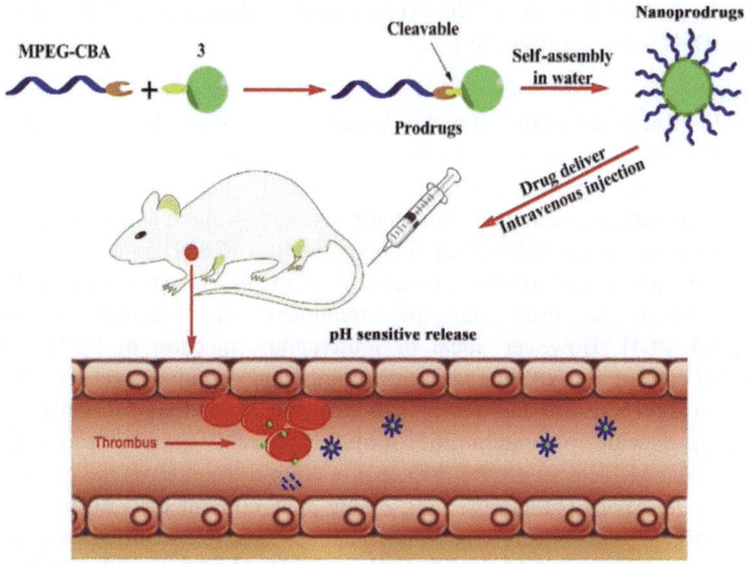

Fig. 4.29 Schematic diagram of MPEG-CBA preparation and in vivo treatment of thrombus. Reproduced with permission from [349]

Inorganic Particle

Metal-organic frameworks (MOFs) have attracted much attention in recent years. Compared with traditional porous materials, MOFs have a variety of advantages, including large surface area, adjustable pore size and shape, adjustable composition and structure, biodegradability, and general functions, which make MOFs a promising platform for drug delivery, molecular imaging, and therapeutic applications [353–357]. Shasha song et al. prepared a luminescent coordination polymer containing Zn (II) metal center, which combined with trimer n donor linker and tricarboxylate in mixed solvent of water and DMF. The compound significantly

Table 4.22 Other treatments of PAD with polymeric particles

Drug	Formulations	Application	Characteristic
Enoxaparin	Biodegradable nanoparticles of poly (epsilon-caprolactone) (PCL) [350]	Anticoagulant and antithrombotic activity in vivo	Showed a sustained release for a greater period and with excellent antithrombotic action
Low molecular weight heparin (LMWH)	Novel polymer lipid hybrid nanoparticles [351]	For oral delivery of LMWH	Excellent performance features by smaller size, acceptable biocompatibility, higher drug encapsulation efficiency, sustained release
LMWH	Drug-dendrimer complex [352]	Enhance pulmonary absorption of LMWH	Positively charged dendrimers increased the relative bioavailability of enoxaparin by 40%

reduced the level of IL-17A and the expression of p-ERK2 and p53. It had protective effect on DVT successfully [358].

Hydrogel
Hydrogel is composed of reticulated polymers dispersed in a large amount of water. It has attracted wide attention in biomedical research for its good formulation stability and biocompatibility [359]. Biodegradable, injectable in situ gel system has many advantages, such as in situ pharmacy store without surgical treatment and easy loading or dosing [360–362]. Pingyangmycin (PYM) is a new hydrophilic glycopeptide anti-tumor antibiotic isolated from the soil *Streptococcus* in Pingyang County. PYM is used in the treatment of vascular malformation (VM) [363, 364]. However, local or intravenous injection of PYM can cause endothelial cell injury, vascular wall thickening, and vascular occlusion. Moreover, PYM has a short half-life and numbers of side effects (such as pulmonary toxicity), which limits its wide application [365, 366]. It is of great significance to carry out the research of PYM local sustained-release preparation to improve the curative effect and reduce the adverse reactions. Ling Zhang et al. combined PYM liposomes with in situ thermo gel successfully to prepare a liposome CS/GP in situ thermo gel which improved the sustained and partial delivery of PYM significantly [365]. In vivo pharmacokinetics and in vivo embolization studies indicated that PYM liposome thermo gel could make drug sustained and local release longer and achieve the purpose of drug-controlled release.

Virus
Nanotechnology is expected to target the disease or injury site. Most of the nanoparticles being developed are spherical carriers. These carriers may not have the best flow and edge characteristics and have the disadvantages of poor specificity and low thrombus permeability. Andrzej s pitek et al. developed a new platform for thrombolytic drugs based on tobacco mosaic virus (TMV). It could be loaded with streptokinase (STK) with density up to 1 mg/mg. The drug delivery system enhanced the interaction with cardiovascular thrombosis and maximized the antithrombotic effect due to its special high aspect ratio and easy aggregation in vascular wall [367].

References

1. Schulz C, Massberg S. Atherosclerosis—multiple pathways to lesional macrophages. Sci Transl Med. 2014;6:239–2.
2. Witztum JL, Steinberg D. The oxidative modification hypothesis of atherosclerosis: does it hold for humans? Trends Cardiovas Med. 2001;11:93–102.
3. Simionescu M. Implications of early structural-functional changes in the endothelium for vascular disease. Arterioscler Thromb Vasc Biol. 2007;27:266–74.
4. Nguyen LTH, Muktabar A, Tang J, Dravid VP, Thaxton CS, Venkatraman S, Ng KW. Engineered nanoparticles for the detection, treatment and prevention of atherosclerosis: how close are we? Drug Discov Today. 2017;22:1438–46.
5. Falk E. Pathogenesis of atherosclerosis. J Am Coll Cardiol. 2006;47:C7–12.

6. Nordestgaard BG, Nicholls SJ, Langsted A, Ray KK, Tybjærg-Hansen A. Advances in lipid-lowering therapy through gene-silencing technologies. Nat Rev Cardiol. 2018;15:261–72.
7. Zhang J, Zu Y, Dhanasekara CS, Li J, Wu D, Fan Z, Wang S. Detection and treatment of atherosclerosis using nanoparticles. Wiley Interdiscip Rev Nanomed Nanobiotechnol. 2017;9: e1412.
8. Torchilin VP. Recent advances with liposomes as pharmaceutical carriers. Nat Rev Drug Discov. 2005;4:145–60.
9. Chono S, Tauchi Y, Deguchi Y, Morimoto K. Efficient drug delivery to atherosclerotic lesions and the antiatherosclerotic effect by dexamethasone incorporated into liposomes in atherogenic mice. J Drug Target. 2005;13:267–76.
10. Poh S, Chelvam V, Ayala-López W, Putt KS, Low PS. Selective liposome targeting of folate receptor positive immune cells in inflammatory diseases. Nanomedicine. 2018;14:1033–43.
11. Almer G, Frascione D, Pali-Schöll I, Vonach C, Lukschal A, Stremnitzer C, et al. Interleukin-10: an anti-inflammatory marker to target atherosclerotic lesions via PEGylated liposomes. Mol Pharmaceut. 2013;10:175–86.
12. Maiseyeu A, Mihai G, Kampfrath T, Simonetti OP, Sen CK, Roy S, et al. Gadolinium-containing phosphatidylserine liposomes for molecular imaging of atherosclerosis. J Lipid Res. 2009;50:2157–63.
13. Homem De Bittencourt PI, Lagranha DJ, Maslinkiewicz A, Senna SM, Tavares AM, Baldissera LP, et al. LipoCardium: endothelium-directed cyclopentenone prostaglandin-based liposome formulation that completely reverses atherosclerotic lesions. Atherosclerosis. 2007;193:245–58.
14. Lobatto ME, Calcagno C, Otten MJ, Millon A, Ramachandran S, Paridaans MP, et al. Pharmaceutical development and preclinical evaluation of a GMP-grade anti-inflammatory nanotherapy. Nanomedicine. 2015;11:1133–40.
15. Van Der Valk FM, Van Wijk DF, Lobatto ME, Verberne HJ, Storm G, Willems MC, et al. Prednisolone-containing liposomes accumulate in human atherosclerotic macrophages upon intravenous administration. Nanomedicine. 2015;11:1039–46.
16. Fang D, Jin Q, Jin Z, Wang F, Huang L, Yang Y, et al. Folate-modified liposomes loaded with telmisartan enhance anti-atherosclerotic potency for advanced atherosclerosis in ApoE(-/-) mice. J Biomed Nanotechnol. 2019;15:42–61.
17. Li X, Xiao H, Lin C, Sun W, Wu T, Wang J, et al. Synergistic effects of liposomes encapsulating atorvastatin calcium and curcumin and targeting dysfunctional endothelial cells in reducing atherosclerosis. Int J Nanomed. 2019;14:649–65.
18. Zhang Y, Li Q, Welsh WJ, Moghe PV, Uhrich KE. Micellar and structural stability of nanoscale amphiphilic polymers: implications for anti-atherosclerotic bioactivity. Biomaterials. 2016;84:230–40.
19. Wu T, Chen X, Wang Y, Xiao H, Peng Y, Lin L, et al. Aortic plaque-targeted andrographolide delivery with oxidation-sensitive micelle effectively treats atherosclerosis via simultaneous ROS capture and anti-inflammation. Nanomedicine. 2018;14:2215–26.
20. Hou X, Lin H, Zhou X, Cheng Z, Li Y, Liu X, et al. Novel dual ROS-sensitive and CD44 receptor targeting nanomicelles based on oligomeric hyaluronic acid for the efficient therapy of atherosclerosis. Carbohyd Polym. 2020;232:115787.
21. Gao W, Sun Y, Cai M, Zhao Y, Cao W, Liu Z, et al. Copper sulfide nanoparticles as a photothermal switch for TRPV1 signaling to attenuate atherosclerosis. Nat Commun. 2018;9: 231.
22. Zhang L, Tian XY, Chan CKW, Bai Q, Cheng CK, Chen FM, et al. Promoting the delivery of nanoparticles to atherosclerotic plaques by DNA coating. ACS Appl Mater Interfaces. 2019;11:13888–904.
23. Qu G, Wu Q, Zhao B, Miao J, Su L. The promotion effect of novel magnetic nanoparticles on atherosclerotic plaque vulnerability in apolipoprotein E(-/-) mice. Toxicology. 2019;419:24–31.

24. Zhu MT, Wang B, Wang Y, Yuan L, Wang HJ, Wang M, et al. Endothelial dysfunction and inflammation induced by iron oxide nanoparticle exposure: risk factors for early atherosclerosis. Toxicol Lett. 2011;203:162–71.
25. Suzuki Y, Tada-Oikawa S, Ichihara G, Yabata M, Izuoka K, Suzuki M, et al. Zinc oxide nanoparticles induce migration and adhesion of monocytes to endothelial cells and accelerate foam cell formation. Toxicol Lett. 2014;278:16–25.
26. Kamaly N, Fredman G, Fojas JJ, Subramanian M, Choi WI, Zepeda K, et al. Targeted Interleukin-10 nanotherapeutics developed with a microfluidic chip enhance resolution of inflammation in advanced atherosclerosis. ACS Nano. 2016;10:5280–92.
27. Chmielowski RA, Abdelhamid DS, Faig JJ, Petersen LK, Gardner CR, Uhrich KE, et al. Athero-inflammatory nanotherapeutics: ferulic acid-based poly(anhydride-ester) nanoparticles attenuate foam cell formation by regulating macrophage lipogenesis and reactive oxygen species generation. Acta Biomater. 2017;57:85–94.
28. Ul Ain Q, Chung H, Chung JY, Choi JH, Kim YH. Amelioration of atherosclerotic inflammation and plaques via endothelial adrenoceptor-targeted eNOS gene delivery using redox-sensitive polymer bearing l-arginine. J Control Release. 2017;262:72–86.
29. Alaarg A, Senders ML, Varela-Moreira A, Pérez-Medina C, Zhao Y, Tang J, et al. A systematic comparison of clinically viable nanomedicines targeting HMG-CoA reductase in inflammatory atherosclerosis. J Control Release. 2017;262:47–57.
30. Gao M, Liang C, Song X, Chen Q, Jin Q, Wang C, Liu Z. Erythrocyte-membrane-enveloped perfluorocarbon as nanoscale artificial red blood cells to relieve tumor hypoxia and enhance cancer radiotherapy. Adv Mater. 2017;29:1701429.
31. Wang Y, Zhang K, Qin X, Li T, Qiu J, Yin T, et al. Biomimetic nanotherapies: red blood cell based core-shell structured nanocomplexes for atherosclerosis management. Adv Sci. 2019;6: 1900172.
32. Dehaini D, Wei X, Fang RH, Masson S, Angsantikul P, Luk BT, et al. Erythrocyte-platelet hybrid membrane coating for enhanced nanoparticle functionalization. Adv Mater. 2017;29: 1606209.
33. Huang Y, Li T, Gao W, Wang Q, Li X, Mao C, et al. Platelet-derived nanomotor coated balloon for atherosclerosis combination therapy. J Mater Chem B. 2020;8:5765–75.
34. Matuszak J, Lutz B, Sekita A, Zaloga J, Alexiou C, Lyer S, Cicha I. Drug delivery to atherosclerotic plaques using superparamagnetic iron oxide nanoparticles. Int J Nanomed. 2018;13:8443–60.
35. Zhang S, Xu W, Gao P, Chen W, Zhou Q. Construction of dual nanomedicines for the imaging and alleviation of atherosclerosis. Artif Cells Nanomed Biotechnol. 2020;48:169–79.
36. Mishra S, Bedja D, Amuzie C, Foss CA, Pomper MG, Bhattacharya R, et al. Improved intervention of atherosclerosis and cardiac hypertrophy through biodegradable polymer-encapsulated delivery of glycosphingolipid inhibitor. Biomaterials. 2015;64:125–35.
37. Hong Z, Xu Y, Yin JF, Jin J, Jiang Y, Du Q. Improving the effectiveness of (-)-epigallocatechin gallate (EGCG) against rabbit atherosclerosis by EGCG-loaded nanoparticles prepared from chitosan and polyaspartic acid. J Agr Food Chem. 2014;62:12603–9.
38. Nakashiro S, Matoba T, Umezu R, Koga J, Tokutome M, Katsuki S, et al. Pioglitazone-incorporated nanoparticles prevent plaque destabilization and rupture by regulating monocyte/macrophage differentiation in ApoE-/- mice. Arterioscler Thromb Vasc Biol. 2016;36:491–500.
39. Stigliano C, Ramirez MR, Singh JV, Aryal S, Key J, Blanco E, Decuzzi P. Methotraxate-loaded hybrid nanoconstructs target vascular lesions and inhibit atherosclerosis progression in ApoE(-/-) Mice. Adv Healthc Mater. 2017;6:1601286.
40. Boada C, Zinger A, Tsao C, Zhao P, Martinez JO, Hartman K, et al. Rapamycin-loaded biomimetic nanoparticles reverse vascular inflammation. Circ Res. 2020;126:25–37.
41. Gao C, Huang Q, Liu C, Kwong CHT, Yue L, Wan JB, et al. Treatment of atherosclerosis by macrophage-biomimetic nanoparticles via targeted pharmacotherapy and sequestration of proinflammatory cytokines. Nat Commun. 2020;11:2622.

42. Boas U, Heegaard PM. Dendrimers in drug research. Chem Soc Rev. 2004;33:43–63.
43. Li MH, Choi SK, Thomas TP, Desai A, Lee KH, Kotlyar A, et al. Dendrimer-based multivalent methotrexates as dual acting nanoconjugates for cancer cell targeting. Eur J Med Chem. 2012;47:560–72.
44. Bhadra D, Bhadra S, Jain NK. PEGylated peptide dendrimeric carriers for the delivery of antimalarial drug chloroquine phosphate. Pharm Res. 2006;23:623–33.
45. Poh S, Putt KS, Low PS. Folate-targeted dendrimers selectively accumulate at sites of inflammation in mouse models of ulcerative colitis and atherosclerosis. Biomacromolecules. 2017;18:3082–8.
46. Spyropoulos-Antonakakis N, Sarantopoulou E, Trohopoulos PN, Stefi AL, Kollia Z, Gavriil VE, et al. Selective aggregation of PAMAM dendrimer nanocarriers and PAMAM/ZnPc nanodrugs on human atheromatous carotid tissues: a photodynamic therapy for atherosclerosis. Nanoscale Res Lett. 2015;10:210.
47. He H, Lancina MG 3rd, Wang J, Korzun WJ, Yang H, Ghosh S. Bolstering cholesteryl ester hydrolysis in liver: a hepatocyte-targeting gene delivery strategy for potential alleviation of atherosclerosis. Biomaterials. 2017;130:1–13.
48. Ficker M, Theeuwen MJM, Janaszewska A, Gorzkiewicz M, Svenningsen SW, Klajnert-Maculewicz B, Christensen JB. Complexes of indomethacin with 4-carbomethoxy-pyrrolidone PAMAM dendrimers show improved anti-inflammatory properties and temperature-dependent binding and release profile. Mol Pharmaceut. 2018;15:3573–82.
49. He H, Yuan Q, Bie J, Wallace RL, Yannie PJ, Wang J, et al. Development of mannose functionalized dendrimeric nanoparticles for targeted delivery to macrophages: use of this platform to modulate atherosclerosis. Transl Res. 2018;193:13–30.
50. Yao SY, Shen ML, Li SJ, Wu XD, Zhang MM, Ma LN, Li YP. Application of a mechanically responsive, inflammatory macrophage-targeted dual-sensitive hydrogel drug carrier for atherosclerosis. Colloids Surf B. 2020;186:110718.
51. Xue Y, Wu Y, Wang Q, Xue L, Su Z, Zhang C. Cellular vehicles based on neutrophils enable targeting of atherosclerosis. Mol Pharmaceut. 2019;16:3109–20.
52. Yi S, Allen SD, Liu YG, Ouyang BZ, Li X, Augsornworawat P, et al. Tailoring nanostructure morphology for enhanced targeting of dendritic cells in atherosclerosis. ACS Nano. 2016;10: 11290–303.
53. Thygesen K, Alpert JS, Jaffe AS, Chaitman BR, Bax JJ, Morrow DA, et al. Fourth universal definition of myocardial infarction. Circulation. 2018;138:e618–51.
54. Neri M, Riezzo I, Pascale N, Pomara C, Turillazzi E. Ischemia/reperfusion injury following acute myocardial infarction: a critical issue for clinicians and forensic pathologists. Mediators Inflamm. 2017;2017:7018393.
55. Cheng B, Toh EK, Chen KH, Chang YC, Hu CJ, Wu HC, et al. Biomimicking platelet-monocyte interactions as a novel targeting strategy for heart healing. Adv Healthc Mater. 2016;5:2686–97.
56. Zhang N, Li C, Zhou D, Ding C, Jin Y, Tian Q, et al. Cyclic RGD functionalized liposomes encapsulating urokinase for thrombolysis. Acta Biomater. 2018;70:227–36.
57. Dasa SSK, Suzuki R, Gutknecht M, Brinton LT, Tian Y, Michaelsson E, et al. Development of target-specific liposomes for delivering small molecule drugs after reperfused myocardial infarction. J Control Release. 2015;220:556–67.
58. Lipinski MJ, Albelda MT, Frias JC, Anderson SA, Luger D, Westman PC, et al. Multimodality imaging demonstrates trafficking of liposomes preferentially to ischemic myocardium. Cardiovasc Revasc Med. 2016;17:106–12.
59. Leuschner F, Dutta P, Gorbatov R, Novobrantseva TI, Donahoe JS, Courties G, et al. Therapeutic siRNA silencing in inflammatory monocytes in mice. Nat Biotechnol. 2011;29: 1005–10.
60. Dasa SSK, Suzuki R, Mugler E, Chen L, Jansson-Löfmark R, Michaëlsson E, et al. Evaluation of pharmacokinetic and pharmacodynamic profiles of liposomes for the cell type-specific delivery of small molecule drugs. Nanomedicine. 2017;13:2565–74.

61. Fraccarollo D, Thomas S, Scholz CJ, Hilfiker-Kleiner D, Galuppo P, Bauersachs J. Macrophage mineralocorticoid receptor is a pleiotropic modulator of myocardial infarct healing. Hypertension. 2019;73:102–11.
62. Pawlowski CL, Li W, Sun M, Ravichandran K, Hickman D, Kos C, et al. Platelet microparticle-inspired clot-responsive nanomedicine for targeted fibrinolysis. Biomaterials. 2017;128:94–108.
63. Yu H, Chen PL, Zhao Y, Gu X, He W, Gong Z. Effect of sphingosine-1-phosphate and myoblast transplantation on rat acute myocardial infarction. Genet Mol Res. 2015;14:13843–51.
64. Sosunov EA, Anyukhovsky EP, Sosunov AA, Moshnikova A, Wijesinghe D, Engelman DM, et al. pH (low) insertion peptide (pHLIP) targets ischemic myocardium. Proc Natl Acad Sci U S A. 2013;110:82–6.
65. Bejarano J, Navarro-Marquez M, Morales-Zavala F, Morales JO, Garcia-Carvajal I, Araya-Fuentes E, et al. Nanoparticles for diagnosis and therapy of atherosclerosis and myocardial infarction: evolution toward prospective theranostic approaches. Theranostics. 2018;8:4710–32.
66. Aizik G, Grad E, Golomb G. Monocyte-mediated drug delivery systems for the treatment of cardiovascular diseases. Drug Deliv Transl Res. 2018;8:868–82.
67. Tokutome M, Matoba T, Nakano Y, Okahara A, Fujiwara M, Koga J-I, et al. Peroxisome proliferator-activated receptor-gamma targeting nanomedicine promotes cardiac healing after acute myocardial infarction by skewing monocyte/macrophage polarization in preclinical animal models. Cardiovasc Res. 2019;115:419–31.
68. Zhu K, Li J, Wang Y, Lai H, Wang C. Nanoparticles-assisted stem cell therapy for ischemic heart disease. Stem Cells Int. 2016;2016:1384658.
69. Blocki A, Beyer S, Dewavrin JY, Goralczyk A, Wang Y, Peh P, et al. Microcapsules engineered to support mesenchymal stem cell (MSC) survival and proliferation enable long-term retention of MSCs in infarcted myocardium. Biomaterials. 2015;53:12–24.
70. Chang Y, Lee E, Kim J, Kwon YW, Kwon Y, Kim J. Efficient in vivo direct conversion of fibroblasts into cardiomyocytes using a nanoparticle-based gene carrier. Biomaterials. 2019;192:500–9.
71. Margulis K, Neofytou EA, Beygui RE, Zare RN. Celecoxib nanoparticles for therapeutic angiogenesis. ACS Nano. 2015;9:9416–26.
72. Oduk Y, Zhu W, Kannappan R, Zhao M, Borovjagin AV, Oparil S, Zhang JJ. VEGF nanoparticles repair the heart after myocardial infarction. Am J Physiol Heart Circ Physiol. 2018;314:H278–h284.
73. Kandzari DE, Koolen JJ, Doros G, Massaro JJ, Garcia-Garcia HM, Bennett J, et al. Ultrathin bioresorbable polymer sirolimus-eluting stents versus thin durable polymer everolimus-eluting stents. J Am Coll Cardiol. 2018;72:3287–97.
74. Zanjanizadeh Ezazi N, Ajdary R, Correia A, Mäkilä E, Salonen J, Kemell M, et al. Fabrication and characterization of drug-loaded conductive poly(glycerol sebacate)/nanoparticle-based composite patch for myocardial infarction applications. ACS Appl Mater Interfaces. 2020;12:6899–909.
75. Li Z, Zhang J, Li Y, Zhao S, Zhang P, Zhang Y, et al. Carbon dots based photoelectrochemical sensors for ultrasensitive detection of glutathione and its applications in probing of myocardial infarction. Biosens Bioelectron. 2018;99:251–8.
76. Grimaldi AM, Forte E, Infante T, Cavaliere C, Salvatore M, Cademartiri F. Future perspectives of nanoparticle-based contrast agents for cardiac magnetic resonance in myocardial infarction. Nanomedicine. 2019;17:329–41.
77. Bejerano T, Etzion S, Elyagon S, Etzion Y, Cohen S. Nanoparticle delivery of miRNA-21 mimic to cardiac macrophages improves myocardial remodeling after myocardial infarction. Nano Lett. 2018;18:5885–91.

78. Fujiwara M, Matoba T, Koga JI, Okahara A, Funamoto D, Nakano K, et al. Nanoparticle incorporating Toll-like receptor 4 inhibitor attenuates myocardial ischaemia-reperfusion injury by inhibiting monocyte-mediated inflammation in mice. Cardiovasc Res. 2019;115:1244–55.
79. Keliher EJ, Ye YX, Wojtkiewicz GR, Aguirre AD, Tricot B, Senders ML, et al. Polyglucose nanoparticles with renal elimination and macrophage avidity facilitate PET imaging in ischaemic heart disease. Nat Commun. 2017;8:14064.
80. Katsuki S, Matoba T, Koga JI, Nakano K, Egashira K. Anti-inflammatory nanomedicine for cardiovascular disease. Front Cardiovasc Mede. 2017;4:87.
81. Matoba T, Koga JI, Nakano K, Egashira K, Tsutsui H. Nanoparticle-mediated drug delivery system for atherosclerotic cardiovascular disease. J Cardiol. 2017;70:206–11.
82. Sager HB, Dutta P, Dahlman JE, Hulsmans M, Courties G, Sun Y, et al. RNAi targeting multiple cell adhesion molecules reduces immune cell recruitment and vascular inflammation after myocardial infarction. Sci Transl Med. 2016;8:342ra380.
83. Somasuntharam I, Yehl K, Carroll SL, Maxwell JT, Martinez MD, Che PL, et al. Knockdown of TNF-alpha by DNAzyme gold nanoparticles as an anti-inflammatory therapy for myocardial infarction. Biomaterials. 2016;83:12–22.
84. Buehler A, Van Zandvoort MA, Stelt BJ, Hackeng TM, Schrans-Stassen BH, Bennaghmouch A, et al. cNGR: a novel homing sequence for CD13/APN targeted molecular imaging of murine cardiac angiogenesis *in vivo*. Arterioscler Thromb Vasc Biol. 2006;26: 2681–7.
85. Hu B, Zeng M, Chen J, Zhang Z, Zhang X, Fan Z, Zhang X. External magnetic field-induced targeted delivery of highly sensitive iron oxide nanocubes for MRI of myocardial infarction. Small. 2016;12:4707–12.
86. Zhao K, Tang M, Wang H, Zhou Z, Wu Y, Liu S. Simultaneous detection of three biomarkers related to acute myocardial infarction based on immunosensing biochip. Biosens Bioelectron. 2019;126:767–72.
87. Rufaihah AJ, Seliktar D. Hydrogels for therapeutic cardiovascular angiogenesis. Adv Drug Deliv Rev. 2016;96:31–9.
88. Fan C, Shi J, Zhuang Y, Zhang L, Huang L, Yang W, et al. Myocardial-infarction-responsive smart hydrogels targeting matrix metalloproteinase for on-demand growth factor delivery. Adv Mater. 2019;31:e1902900.
89. Mihic A, Cui Z, Wu J, Vlacic G, Miyagi Y, Li SH, et al. A conductive polymer hydrogel supports cell electrical signaling and improves cardiac function after implantation into myocardial infarct. Circulation. 2015;132:772–84.
90. Carlini AS, Gaetani R, Braden RL, Luo C, Christman KL, Gianneschi NC. Enzyme-responsive progelator cyclic peptides for minimally invasive delivery to the heart post-myocardial infarction. Nat Commun. 2019;10:1735.
91. Mihalko E, Huang K, Sproul E, Cheng K, Brown AC. Targeted treatment of ischemic and fibrotic complications of myocardial infarction using a dual-delivery microgel therapeutic. ACS Nano. 2018;12:7826–37.
92. Yang H, Qin X, Wang H, Zhao X, Liu Y, Wo HT, et al. An in vivo miRNA delivery system for restoring infarcted myocardium. ACS Nano. 2019;13:9880–94.
93. Dorsey SM, Mcgarvey JR, Wang H, Nikou A, Arama L, Koomalsingh KJ, et al. MRI evaluation of injectable hyaluronic acid-based hydrogel therapy to limit ventricular remodeling after myocardial infarction. Biomaterials. 2015;69:65–75.
94. Bao R, Tan B, Liang S, Zhang N, Wang W, Liu W. A π-π conjugation-containing soft and conductive injectable polymer hydrogel highly efficiently rebuilds cardiac function after myocardial infarction. Biomaterials. 2017;122:63–71.
95. Liang S, Zhang Y, Wang H, Xu Z, Chen J, Bao R, et al. Paintable and rapidly bondable conductive hydrogels as therapeutic cardiac patches. Adv Mater. 2018;30:e1704235.
96. Liang W, Chen J, Li L, Li M, Wei X, Tan B, et al. Conductive hydrogen sulfide-releasing hydrogel encapsulating ADSCs for myocardial infarction treatment. ACS Appl Mater Interfaces. 2019;11:14619–29.

97. Peña B, Maldonado M, Bonham AJ, Aguado BA, Dominguez-Alfaro A, Laughter M, et al. Gold nanoparticle-functionalized reverse thermal gel for tissue engineering applications. ACS Appl Mater Interfaces. 2019;11:18671–80.
98. Dong D, Hao T, Wang C, Zhang Y, Qin Z, Yang B, et al. Zwitterionic starch-based hydrogel for the expansion and "stemness" maintenance of brown adipose derived stem cells. Biomaterials. 2018;157:149–60.
99. Hao T, Li J, Yao F, Dong D, Wang Y, Yang B, Wang C. Injectable fullerenol/alginate hydrogel for suppression of oxidative stress damage in brown adipose-derived stem cells and cardiac repair. ACS Nano. 2017;11:5474–88.
100. Pitek AS, Park J, Wang Y, Gao H, Hu H, Simon DI, Steinmetz NF. Delivery of thrombolytic therapy using rod-shaped plant viral nanoparticles decreases the risk of hemorrhage. Nanoscale. 2018;10:16547–55.
101. García-Olloqui P, Rodriguez-Madoz JR, Di Scala M, Abizanda G, Vales Á, Olagüe C, et al. Effect of heart ischemia and administration route on biodistribution and transduction efficiency of AAV9 vectors. J Tissue Eng Regen M. 2020;14:123–34.
102. Gallego-Colon E, Villalba M, Tonkin J, Cruz F, Bernal JA, Jimenez-Borregureo LJ, et al. Intravenous delivery of adeno-associated virus 9-encoded IGF-1Ea propeptide improves post-infarct cardiac remodelling. NPJ Regen Med. 2016;1:16001.
103. Ruozi G, Bortolotti F, Falcione A, Dal Ferro M, Ukovich L, Macedo A, et al. AAV-mediated in vivo functional selection of tissue-protective factors against ischaemia. Nat Commun. 2015;6:7388.
104. Tani H, Sadahiro T, Ieda M. Direct cardiac reprogramming: a novel approach for heart regeneration. Int J Mol Sci. 2018;19:2629.
105. Falk E, Fuster V. Angina pectoris and disease progression. Circulation. 1995;92:2033–5.
106. Rousan TA, Mathew ST, Thadani U. Drug therapy for stable angina pectoris. Drugs. 2017;77: 265–84.
107. Taylor SH. Alpha- and beta-blockade in angina pectoris. Drugs. 1984;28(Suppl 2):69–87.
108. Fan Y, Jiang Y, Fu X, Cai J, Wang Y, Li W, et al. Effects of liposomal prostaglandin E1 on periprocedural myocardial injury in patients with unstable angina undergoing an elective percutaneous coronary intervention. Coronary Artery Dis. 2015;26:671–7.
109. Chonkar AD, Rao JV, Managuli RS, Mutalik S, Dengale S, Jain P, Udupa N. Development of fast dissolving oral films containing lercanidipine HCl nanoparticles in semicrystalline polymeric matrix for enhanced dissolution and ex vivo permeation. Eur J Pharm Biopharm. 2016;103:179–91.
110. Khan AA, Abdulbaqi IM, Abou Assi R, Murugaiyah V, Darwis Y. Lyophilized hybrid nanostructured lipid carriers to enhance the cellular uptake of verapamil: statistical optimization and in vitro evaluation. Nanoscale Res Lett. 2018;13:323.
111. Chadha R, Bhandari S, Kataria D, Gupta S, Singh Jain D. Exploring the potential of lecithin/chitosan nanoparticles in enhancement of antihypertensive efficacy of hydrochlorothiazide. J Microencapsul. 2012;29:805–12.
112. Repkova MN, Levina AS, Seryapina AA, Shikina NV, Bessudnova EV, Zarytova VF, Markel AL. Toward gene therapy of hypertension: experimental study on hypertensive ISIAH rats. Biochemistry. 2017;82:454–7.
113. Ahn BN, Kim SK, Shim CK. Preparation and evaluation of proliposomes containing propranolol hydrochloride. J Microencapsul. 1995;12:363–75.
114. Arzani G, Haeri A, Daeihamed M, Bakhtiari-Kaboutaraki H, Dadashzadeh S. Niosomal carriers enhance oral bioavailability of carvedilol: effects of bile salt-enriched vesicles and carrier surface charge. Int J Nanomed. 2015;10:4797–813.
115. Kapoor H, Aqil M, Imam SS, Sultana Y, Ali A. Formulation of amlodipine nano lipid carrier: formulation design, physicochemical and transdermal absorption investigation. J Drug Deliv Sci Tec. 2019;49:209–18.

116. Qumbar M, Ameeduzzafar ISS, Ali J, Ahmad J, Ali A. Formulation and optimization of lacidipine loaded niosomal gel for transdermal delivery: in-vitro characterization and in-vivo activity. Biomed Pharmacother. 2017;93:255–66.
117. Mohuczy D, Gelband CH, Phillips MI. Antisense inhibition of AT1 receptor in vascular smooth muscle cells using adeno-associated virus-based vector. Hypertension. 1999;33:354–9.
118. Wang T, Li H, Zhao C, Chen C, Li J, Chao J, et al. Recombinant adeno-associated virus-mediated kallikrein gene therapy reduces hypertension and attenuates its cardiovascular injuries. Gene Ther. 2004;11:1342–50.
119. Yan JT, Wang T, Wang DW. Recombinant adeno-associated virus-mediated human kallikrein gene therapy protects against hypertensive target organ injuries through inhibiting cell apoptosis. Acta Pharmacol Sin. 2009;30:1253–61.
120. Schillinger KJ, Tsai SY, Taffet GE, Reddy AK, Marian AJ, Entman ML, et al. Regulatable atrial natriuretic peptide gene therapy for hypertension. Proc Natl Acad Sci U S A. 2005;102:13789–94.
121. Miller WH, Brosnan MJ, Graham D, Nicol CG, Morecroft I, Channon KM, et al. Targeting endothelial cells with adenovirus expressing nitric oxide synthase prevents elevation of blood pressure in stroke-prone spontaneously hypertensive rats. Mol Ther. 2005;12:321–7.
122. Wang H, Katovich MJ, Gelband CH, Reaves PY, Phillips MI, Raizada MK. Sustained inhibition of angiotensin I-converting enzyme (ACE) expression and long-term antihypertensive action by virally mediated delivery of ACE antisense cDNA. Circ Res. 1999;85:614–22.
123. Phillips MI. Somatic gene therapy for hypertension. Braz J Med Biol Res. 2000;33:715–21.
124. Arnold AS, Tang YL, Qian KP, Shen LP, Valencia V, Phillips MI, Zhang YC. Specific beta (1)-adrenergic receptor silencing with small interfering RNA lowers high blood pressure and improves cardiac function in myocardial ischemia. J Hypertens. 2007;25:197–205.
125. Wu WH, Hu CP, Chen XP, Zhang WF, Li XW, Xiong XM, Li YJ. MicroRNA-130a mediates proliferation of vascular smooth muscle cells in hypertension. Am J Hypertens. 2011;24:1087–93.
126. Li Y, Liu F, Su C, Yu B, Liu D, Chen HJ, et al. Biodegradable therapeutic microneedle patch for rapid antihypertensive treatment. ACS Appl Mater Interfaces. 2019;11:30575–84.
127. Venishetty VK, Chede R, Komuravelli R, Adepu L, Sistla R, Diwan PV. Design and evaluation of polymer coated carvedilol loaded solid lipid nanoparticles to improve the oral bioavailability: a novel strategy to avoid intraduodenal administration. Colloids Surf B. 2012;95:1–9.
128. Liu D, Yu S, Zhu Z, Lyu C, Bai C, Ge H, et al. Controlled delivery of carvedilol nanosuspension from osmotic pump capsule: in vitro and in vivo evaluation. Int J Pharm. 2014;475:496–503.
129. Rizwan M, Aqil M, Azeem A, Talegaonkar S, Sultana Y, Ali A. Enhanced transdermal delivery of carvedilol using nanoemulsion as a vehicle. J Exp Nanosci. 2010;5:390–411.
130. Taveira SF, Varela-Garcia A, Dos Santos SB, Marreto RN, Martin-Pastor M, Concheiro A, Alvarez-Lorenzo C. Cyclodextrin-based poly(pseudo)rotaxanes for transdermal delivery of carvedilol. Carbohyd Polym. 2018;200:278–88.
131. Ge H, Hu Y, Jiang X, Cheng D, Yuan Y, Bi H, Yang C. Preparation, characterization, and drug release behaviors of drug nimodipine-loaded poly(epsilon-caprolactone)-poly(ethylene oxide)-poly(epsilon-caprolactone) amphiphilic triblock copolymer micelles. J Pharm Sci. 2002;91:1463–73.
132. Quan P, Shi K, Piao H, Piao H, Liang N, Xia D, Cui F. A novel surface modified nitrendipine nanocrystals with enhancement of bioavailability and stability. Int J Pharm. 2012;430:366–71.
133. Geeva Prasanth A, Sathish Kumar A, Sai Shruthi B, Subramanian S. Kinetic study and in vitro drug release studies of nitrendipine loaded arylamide grafted chitosan blend microspheres. Mater Res Express. 2020;6:125427.

134. Ansari KA, Pagar KP, Anwar S, Vavia PR. Design and optimization of self-microemulsifying drug delivery system (SMEDDS) of felodipine for chronotherapeutic application. Braz J Pharm Sci. 2014;50:203–12.
135. Sahu BP, Das MK. Nanosuspension for enhancement of oral bioavailability of felodipine. Appl Nanosci. 2013;4:189–97.
136. Detroja C, Chavhan S, Sawant K. Enhanced antihypertensive activity of candesartan cilexetil nanosuspension: formulation, characterization and pharmacodynamic study. Sci Pharm. 2011;79:635–51.
137. Alexa IF, Ignat M, Popovici RF, Timpu D, Popovici E. In vitro controlled release of antihypertensive drugs intercalated into unmodified SBA-15 and MgO modified SBA-15 matrices. Int J Pharm. 2012;436:111–9.
138. Beg S, Swain S, Singh HP, Patra CN, Rao MEB. Development, optimization, and characterization of solid self-nanoemulsifying drug delivery systems of valsartan using porous carriers. AAPS Pharm Sci Tech. 2012;13:1416–27.
139. Chhabra G, Chuttani K, Mishra AK, Pathak K. Design and development of nanoemulsion drug delivery system of amlodipine besilate for improvement of oral bioavailability. Drug Dev Ind Pharm. 2011;37:907–16.
140. Mao YJ, Koga JI, Tokutome M, Matoba T, Ikeda G, Nakano K, Egashira K. Nanoparticle-mediated delivery of pitavastatin to monocytes/macrophages inhibits left ventricular remodeling after acute myocardial infarction by inhibiting monocyte-mediated inflammation. Int. Heart J. 2017;58:615–23.
141. Cohen JE, Purcell BP, Macarthur JW, Mu AB, Shudo Y, Patel JB, et al. A Bioengineered hydrogel system enables targeted and sustained intramyocardial delivery of neuregulin, activating the cardiomyocyte cell cycle and enhancing ventricular function in a murine model of ischemic cardiomyopathy. Circ Heart Fail. 2014;7:619–26.
142. Mentkowski KI, Lang JK. Exosomes engineered to express a cardiomyocyte binding peptide demonstrate improved cardiac retention in vivo. Sci Rep. 2019;9:10041.
143. Maurice JP, Hata JA, Shah AS, White DC, Mcdonald PH, Dolber PC, et al. Enhancement of cardiac function after adenoviral-mediated in vivo intracoronary beta2-adrenergic receptor gene delivery. J Clin Invest. 1999;104:21–9.
144. Miyamoto MI, Del Monte F, Schmidt U, Disalvo TS, Kang ZB, Matsui T, et al. Adenoviral gene transfer of SERCA2a improves left-ventricular function in aortic-banded rats in transition to heart failure. Proc Natl Acad Sci U S A. 2000;97:793–8.
145. Del Monte F, Harding SE, Dec GW, Gwathmey JK, Hajjar RJ. Targeting phospholamban by gene transfer in human heart failure. Circulation. 2002;105:904–7.
146. Hoshijima M, Ikeda Y, Iwanaga Y, Minamisawa S, Date MO, Gu Y, et al. Chronic suppression of heart-failure progression by a pseudophosphorylated mutant of phospholamban via in vivo cardiac rAAV gene delivery. Nat Med. 2002;8:864–71.
147. Szatkowski ML, Westfall MV, Gomez CA, Wahr PA, Michele DE, Dellorusso C, et al. In vivo acceleration of heart relaxation performance by parvalbumin gene delivery. J Clin Invest. 2001;107:191–8.
148. Byrne MJ, Power JM, Preovolos A, Mariani JA, Hajjar RJ, Kaye DM. Recirculating cardiac delivery of AAV2/1SERCA2a improves myocardial function in an experimental model of heart failure in large animals. Gene Ther. 2008;15:1550–7.
149. Most P, Pleger ST, Völkers M, Heidt B, Boerries M, Weichenhan D, et al. Cardiac adenoviral S100A1 gene delivery rescues failing myocardium. J Clin Invest. 2004;114:1550–63.
150. El-Armouche A, Singh J, Naito H, Wittkopper K, Didie M, Laatsch A, et al. Adenovirus-delivered short hairpin RNA targeting PKCalpha improves contractile function in reconstituted heart tissue. J Mol Cell Cardiol. 2007;43:371–6.
151. Grubb DR, Gao XM, Kiriazis H, Matsumoto A, Mcmullen JR, Du XJ, Woodcock EA. Expressing an inhibitor of PLCbeta1b sustains contractile function following pressure overload. J Mol Cell Cardiol. 2016;93:12–7.

152. Lai NC, Roth DM, Gao MH, Tang T, Dalton N, Lai YY, et al. Intracoronary adenovirus encoding adenylyl cyclase VI increases left ventricular function in heart failure. Circulation. 2004;110:330–6.
153. Rengo G, Lymperopoulos A, Zincarelli C, Donniacuo M, Soltys S, Rabinowitz JE, Koch WJ. Myocardial adeno-associated virus serotype 6-betaARKct gene therapy improves cardiac function and normalizes the neurohormonal axis in chronic heart failure. Circulation. 2009;119:89–98.
154. Bernardo BC, Weeks KL, Pongsukwechkul T, Gao X, Kiriazis H, Cemerlang N, et al. Gene delivery of medium chain acyl-coenzyme A dehydrogenase induces physiological cardiac hypertrophy and protects against pathological remodelling. Clin Sci. 2018;132:381–97.
155. Tenhunen O, Soini Y, Ilves M, Rysa J, Tuukkanen J, Serpi R, et al. p38 Kinase rescues failing myocardium after myocardial infarction: evidence for angiogenic and anti-apoptotic mechanisms. FASEB J. 2006;20:1907–9.
156. Wang Y, Ding N, Guan G, Liu G, Huo D, Li Y, et al. Rapid delivery of Hsa-miR-590-3p using targeted exosomes to treat acute myocardial infarction through regulation of the cell cycle. J Biomed Nanotechnol. 2018;14:968–77.
157. Leotta E, Patejunas G, Murphy G, Szokol J, Mcgregor L, Carbray J, et al. Gene therapy with adenovirus-mediated myocardial transfer of vascular endothelial growth factor 121 improves cardiac performance in a pacing model of congestive heart failure. J Thorac Cardiovasc Surg. 2002;123:1101–13.
158. Bonci D, Cittadini A, Latronico MV, Borello U, Aycock JK, Drusco A, et al. 'Advanced' generation lentiviruses as efficient vectors for cardiomyocyte gene transduction in vitro and in vivo. Gene Ther. 2003;10:630–6.
159. Woo YJ, Panlilio CM, Cheng RK, Liao GP, Atluri P, Hsu VM, et al. Therapeutic delivery of cyclin A2 induces myocardial regeneration and enhances cardiac function in ischemic heart failure. Circulation. 2006;114:I206–1213.
160. Turnbull IC, Eltoukhy AA, Fish KM, Nonnenmacher M, Ishikawa K, Chen J, et al. Myocardial delivery of lipidoid nanoparticle carrying modRNA induces rapid and transient expression. Mol Ther. 2016;24:66–75.
161. Suckau L, Fechner H, Chemaly E, Krohn S, Hadri L, Kockskamper J, et al. Long-term cardiac-targeted RNA interference for the treatment of heart failure restores cardiac function and reduces pathological hypertrophy. Circulation. 2009;119:1241–52.
162. Sundararaman S, Miller TJ, Pastore JM, Kiedrowski M, Aras R, Penn MS. Plasmid-based transient human stromal cell-derived factor-1 gene transfer improves cardiac function in chronic heart failure. Gene Ther. 2011;18:867–73.
163. Shentu WH, Yan CX, Liu CM, Qi RX, Wang Y, Huang ZX, et al. Use of cationic microbubbles targeted to P-selectin to improve ultrasound-mediated gene transfection of hVEGF165 to the ischemic myocardium. J Zhejiang Univ Sci B. 2018;19:699–707.
164. Walker MJA. Antiarrhythmic drug research. Br J Pharmacol. 2009;147:S222–31.
165. Dorian P, Cass D, Schwartz BS, Cooper RM, Gelaznikas R, Barr A. Amiodarone as compared with lidocaine for shock-resistant ventricular fibrillation. New Engl J Med. 2002;346:884–90.
166. Vassallo P, Trohman RG. Prescribing Amiodarone: an evidence-based review of clinical indications. JAMA. 2007;298:1312–22.
167. Nikalje AP. Nanotechnology and its applications in medicine. Med Chem. 2015;5:1–9.
168. Elgart A, Cherniakov I, Aldouby Y, Domb AJ, Hoffman A. Improved oral bioavailability of BCS Class 2 compounds by self nano-emulsifying drug delivery systems (SNEDDS): the underlying mechanisms for amiodarone and talinolol. Pharm Res. 2013;30:3029–44.
169. Akbarzadeh A, Rezaei-Sadabady R, Davaran S, Joo SW, Zarghami N, Hanifehpour Y, et al. Liposome: classification, preparation, and applications. Nanoscale Res Lett. 2013;8:102.
170. Takahama H, Shigematsu H, Asai T, Matsuzaki T, Sanada S, Fu H, et al. Liposomal amiodarone augments anti-arrhythmic effects and reduces hemodynamic adverse effects in an ischemia/reperfusion rat model. Cardiovasc Drug Ther. 2013;27:125–32.

171. Wang Y, Su W, Li Q, Li C, Wang H, Li Y, et al. Preparation and evaluation of lidocaine hydrochloride-loaded TAT-conjugated polymeric liposomes for transdermal delivery. Int J Pharmaceut. 2013;441:748–56.
172. Takahama H, Minamino T, Asanuma H, Fujita M, Asai T, Wakeno M, et al. Prolonged targeting of ischemic/reperfused myocardium by liposomal adenosine augments cardioprotection in rats. J Am Coll Cardiol. 2009;53:709–17.
173. Dey B, Nath L, Mohanti B, Bhowmik B. Development and evaluation of propranolol hydrochloride transdermal patches by using hydrophilic and hydrophobic polymer. Indian J Pharm Educ. 2007;41:388–93.
174. Chonn A, Cullis PR. Recent advances in liposomal drug-delivery systems. Curr Opin Biotech. 1995;6:698–708.
175. Allen TM, Cullis PR. Liposomal drug delivery systems: from concept to clinical applications. Adv Drug Deliver Rev. 2013;65:36–48.
176. Benedini L, Antollini S, Fanani ML, Palma S, Messina P, Schulz P. Study of the influence of ascorbyl palmitate and amiodarone in the stability of unilamellar liposomes. Mol Membr Biol. 2014;31:85–94.
177. Xiong F, Liu K, Liu S, Chen J, Liu J, Wang H, Gu N. Safety, heart specificity, and therapeutic effect evaluation of Guanfu base A-loaded solid nanolipids in treating arrhythmia. Drug Deliv Transl Re. 2018;8:1471–82.
178. Guo X, Zhu X, Liu D, Gong Y, Sun J, Dong C. Continuous delivery of propranolol from liposomes-in-microspheres significantly inhibits infantile hemangioma growth. Int J Nanomed. 2017;12:6923–36.
179. Kazemzadeh-Narbat M, Reid M, Brooks M, Ghanem A. Chitosan nanoparticles as adenosine carriers. J Microencapsul. 2015;32:1–7.
180. Abdel Mouez M, Zaki NM, Mansour S, Geneidi AS. Bioavailability enhancement of verapamil HCl via intranasal chitosan microspheres. Eur J Pharm Sci. 2014;51:59–66.
181. Ahmed M, Rodell C, Hulsmans M, Kohler R, Aguirre A, Nahrendorf M, Weissleder R. A supramolecular nanocarrier for delivery of amiodarone anti-arrhythmic therapy to the heart. Bioconjugate Chem. 2019;30:733–40.
182. Pritchard EM, Szybala C, Boison D, Kaplan DL. Silk fibroin encapsulated powder reservoirs for sustained release of adenosine. J Control Release. 2010;144:159–67.
183. Labhasetwar V, Underwood T, Gallagher M, Murphy G, Langberg J, Levy RJ. Sotalol controlled-release systems for arrhythmias: in vitro characterization, in vivo drug disposition, and electrophysiologic effects. J Pharm Sci. 1994;83:156–64.
184. Chourasiya V, Bohrey S, Pandey A. Formulation, optimization, characterization and in-vitro drug release kinetics of atenolol loaded PLGA nanoparticles using 33 factorial design for oral delivery. Mater Discov. 2016;5:1–13.
185. Garcia J, Campbell P, Kumar G, Langberg J, Cesar L, Deppen J, et al. Minimally invasive delivery of hydrogel-encapsulated amiodarone to the epicardium reduces atrial fibrillation. Circ-Arrhythmia Elec. 2018;11:e006408.
186. Bolderman R, Hermans J, Rademakers L, De Jong M, Bruin P, Dias A, et al. Epicardial application of an amiodarone-releasing hydrogel to suppress atrial tachyarrhythmias. Int J Cardiol. 2011;149:341–6.
187. Bongianino R, Priori S. Gene therapy to treat cardiac arrhythmias. Nat Rev Cardiol. 2015;12: 531–46.
188. Farraha M, Chong J, Kizana E. Therapeutic prospects of gene therapy for atrial fibrillation. Heart Lung Circ. 2016;25:808–13.
189. Ailawadi S, Wang X, Gu H, Fan G-C. Pathologic function and therapeutic potential of exosomes in cardiovascular disease. BBA-Mol Basis Dis. 2015;1852:1–11.
190. Johnsen KB, Gudbergsson JM, Skov MN, Pilgaard L, Moos T, Duroux M. A comprehensive overview of exosomes as drug delivery vehicles—endogenous nanocarriers for targeted cancer therapy. BBA-Rev Cancer. 2014;1846:75–87.

191. Roos-Hesselink JW, Kerstjens-Frederikse WS, Meijboom FJ, Pieper PG. Inheritance of congenital heart disease. Neth Heart J. 2005;13:88–91.
192. Parvin Nejad S, Blaser MC, Santerre JP, Caldarone CA, Simmons CA. Biomechanical conditioning of tissue engineered heart valves: too much of a good thing? Adv Drug Deliv Rev. 2016;96:161–75.
193. Kolwicz SC Jr, Odom GL, Nowakowski SG, Moussavi-Harami F, Chen X, Reinecke H, et al. AAV6-mediated cardiac-specific overexpression of ribonucleotide reductase enhances myocardial contractility. Mol Ther. 2016;24:240–50.
194. Chen J, Fan J, Wang S, Sun Z. Secreted klotho attenuates inflammation-associated aortic valve fibrosis in senescence-accelerated mice P1. Hypertension. 2018;71:877–85.
195. Franz W-M, Müller OJ, Katus HA. Cardiomyopathies: from genetics to the prospect of treatment. Lancet. 2001;358:1627–37.
196. Prondzynski M, Mearini G, Carrier L. Gene therapy strategies in the treatment of hypertrophic cardiomyopathy. Pflugers Arch. 2019;471:807–15.
197. Mcnally EM, Mestroni L. Dilated cardiomyopathy: genetic determinants and mechanisms. Circ Res. 2017;121:731–48.
198. Shimizu-Motohashi Y, Komaki H, Motohashi N, et al. Restoring dystrophin expression in duchenne muscular dystrophy: current status of therapeutic approaches. J Pers Med. 2019;9:1.
199. Keeler CE. Gene therapy. J Hered. 1947;38:294–8.
200. Fishbein I, Chorny M, Levy R. Site-specific gene therapy for cardiovascular disease. Curr Opin Drug Disc. 2010;13:203–13.
201. Huynh K, Bernardo BC, Mcmullen JR, Ritchie RH. Diabetic cardiomyopathy: mechanisms and new treatment strategies targeting antioxidant signaling pathways. Pharmacol Therapeut. 2014;142:375–415.
202. Morilla M, Romero E. Nanomedicines against Chagas disease: an update on therapeutics, prophylaxis and diagnosis. Nanomedicine. 2015;10:465–81.
203. Chamberlain K, Riyad JM, Weber T. Cardiac gene therapy with adeno-associated virus-based vectors. Curr Opin Cardiol. 2017;32(3):275–82.
204. Jiang J, Wakimoto H, Seidman JG, Seidman CE. Allele-specific silencing of mutant Myh6 transcripts in mice suppresses hypertrophic cardiomyopathy. Science. 2013;342:111–4.
205. Merkulov S, Chen X, Chandler MP, Stelzer JE. In vivo cardiac myosin binding protein C gene transfer rescues myofilament contractile dysfunction in cardiac myosin binding protein C null mice. Circ Heart Fail. 2012;5:635–44.
206. Martino TA, Liu P, Sole MJ. Viral infection and the pathogenesis of dilated cardiomyopathy. Circ Res. 1994;74:182–8.
207. Zhang H, Su Y, Guo S, Yuan J, Lim T, Liu J, et al. Targeted delivery of anti-coxsackievirus siRNAs using ligand-conjugated packaging RNAs. Antivir Res. 2009;83:307–16.
208. Kim D, Nam J. Application of attenuated coxsackievirus B3 as a viral vector system for vaccines and gene therapy. Hum Vaccines. 2011;7:410–6.
209. Joanne P, Kitsara M, Boitard S, Naemetalla H, Vanneaux V, Pernot M, et al. Nanofibrous clinical-grade collagen scaffolds seeded with human cardiomyocytes induces cardiac remodeling in dilated cardiomyopathy. Biomaterials. 2016;80:157–68.
210. Duan D. Challenges and opportunities in dystrophin-deficient cardiomyopathy gene therapy. Hum Mol Genet. 2006;15:R253–61.
211. Shin JH, Nitahara-Kasahara Y, Hayashita-Kinoh H, Ohshima-Hosoyama S, Kinoshita K, Chiyo T, et al. Improvement of cardiac fibrosis in dystrophic mice by rAAV9-mediated microdystrophin transduction. Gene Ther. 2011;18:910–9.
212. Bostick B, Shin J-H, Yue Y, Duan D. AAV-microdystrophin therapy improves cardiac performance in aged female mdx mice. Mol Ther. 2011;19:1826–32.
213. Duan D. Systemic AAV micro-dystrophin gene therapy for duchenne muscular dystrophy. Mol Ther. 2018;26:2337–56.

214. Kizhner T, Azulay Y, Hainrichson M, Tekoah Y, Arvatz G, Shulman A, et al. Characterization of a chemically modified plant cell culture expressed human α-Galactosidase-A enzyme for treatment of Fabry disease. Mol Genet Metab. 2015;114:259–67.
215. Dai B, Li H, Fan J, Zhao Y, Yin Z, Nie X, et al. MiR-21 protected against diabetic cardiomyopathy induced diastolic dysfunction by targeting gelsolin. Cardiovasc Diabetol. 2018;17:123.
216. Jia C, Chen H, Wei M, Chen X, Zhang Y, Cao L, et al. Gold nanoparticle-based miR155 antagonist macrophage delivery restores the cardiac function in ovariectomized diabetic mouse model. Int J Nanomed. 2017;12:4963–79.
217. Karuppagounder V, Arumugam S, Giridharan V, Sreedhar R, Bose R, Vanama J, et al. Tiny molecule, big power: multi-target approach for curcumin in diabetic cardiomyopathy. Nutrition. 2017;34:47–54.
218. Maradana MR, Thomas R, O'sullivan BJ. Targeted delivery of curcumin for treating type 2 diabetes. Mol Nutr Food Res. 2013;57:1550–6.
219. Zhao Y, Zhang M, Wong H, Tian X, Zheng L, Yu X, et al. Prevent diabetic cardiomyopathy in diabetic rats by combined therapy of aFGF-loaded nanoparticles and ultrasound-targeted microbubble destruction technique. J Control Release. 2016;223:11–21.
220. Zhao Y, Tian X, Zhang M, Cai L, Ru A, Shen X, et al. Functional and pathological improvements of the hearts in diabetes model by the combined therapy of bFGF-loaded nanoparticles with ultrasound-targeted microbubble destruction. J Control Release. 2014;186:22–31.
221. Meloni M, Descamps B, Caporali A, Zentilin L, Floris I, Giacca M, Emanueli C. Nerve growth factor gene therapy using adeno-associated viral vectors prevents cardiomyopathy in type 1 diabetic mice. Diabetes. 2012;61:229–40.
222. Barry M, Wang Q, Jones K, Heffernan M, Buhaya M, Beaumier C, et al. A therapeutic nanoparticle vaccine against Trypanosoma cruzi in a BALB/c mouse model of Chagas disease. Hum Vacc Immunother. 2016;12:976–87.
223. Dumonteil E, Bottazzi ME, Zhan B, Heffernan MJ, Jones K, Valenzuela JG, et al. Accelerating the development of a therapeutic vaccine for human Chagas disease: rationale and prospects. Expert Rev Vaccines. 2012;11:1043–55.
224. Park S, Yoon J, Bae S, Park M, Kang C, Ke Q, et al. Therapeutic use of H_2O_2-responsive antioxidant polymer nanoparticles for doxorubicin-induced cardiomyopathy. Biomaterials. 2014;35:5944–53.
225. Patil RR, Guhagarkar SA, Devarajan PV. Engineered nanocarriers of doxorubicin: a current update. Crit Rev Ther Drug Carrier Syst. 2008;25:1–61.
226. Kalaria DR, Sharma G, Beniwal V, Ravi Kumar MNV. Design of biodegradable nanoparticles for oral delivery of doxorubicin: in vivo pharmacokinetics and toxicity studies in rats. Pharm Res. 2009;26:492–501.
227. Borer JS, Sharma A. Drug therapy for heart valve diseases. Circulation. 2015;132:1038–45.
228. Van Der Ven C, Wu P, Tibbitt M, Van Mil A, Sluijter J, Langer R, Aikawa E. In vitro 3D model and miRNA drug delivery to target calcific aortic valve disease. Clin Sci. 2017;131: 181–95.
229. Wong F, Ho M, Yamagami M, Lam M, Grande-Allen K, Suh J. Effective gene delivery to valvular interstitial cells using adeno-associated virus serotypes 2 and 3. Tissue Eng Part C-Me. 2015;21:808–15.
230. Imazio M, Gaita F, Lewinter M. Evaluation and treatment of pericarditis: a systematic review. JAMA. 2015;314:1498–506.
231. Rahman A, Saraswat A. Pericarditis. Aust Fam Physician. 2017;46:810–4.
232. Ji YL, Li RZ, Xue LF, Li P, Liang XH, Dong L, et al. Therapeutic effects of 5-fluorouracil sustained-release particles in 81 malignant pericardial effusion patients. Kaohsiung J Med Sci. 2015;31:96–101.
233. Clough RE, Nienaber CA. Management of acute aortic syndrome. Nat Rev Cardiol. 2015;12: 103–14.

234. Oricco S, Perego M, Poggi M, Tursi M, Biasato I, Santilli RA. Aortic dissection in four cats: clinicopathological correlations. J Vet Cardiol. 2019;25:52–60.
235. Mussa FF, Horton JD, Moridzadeh R, Nicholson J, Trimarchi S, Eagle KA. Acute aortic dissection and intramural hematoma: a systematic review. JAMA. 2016;316:754–63.
236. Sbarouni E, Georgiadou P. MicroRNAs in acute aortic dissection. J Thorac Dis. 2018;10: 1256–7.
237. Vacante F, Denby L, Sluimer JC, Baker AH. The function of miR-143, miR-145 and the MiR-143 host gene in cardiovascular development and disease. Vascul Pharmacol. 2019;112: 24–30.
238. Yeh YT, Wei J, Thorossian S, Nguyen K, Hoffman C, Del Álamo JC, et al. MiR-145 mediates cell morphology-regulated mesenchymal stem cell differentiation to smooth muscle cells. Biomaterials. 2019;204:59–69.
239. Jia L-X, Zhang W-M, Li T-T, Liu Y, Piao C-M, Ma Y-C, et al. ER stress dependent microparticles derived from smooth muscle cells promote endothelial dysfunction during thoracic aortic aneurysm and dissection. Clin Sci. 2017;131:1287–99.
240. Shao Y, Li G, Huang S, Li Z, Qiao B, Chen D, et al. Effects of extracellular matrix softening on vascular smooth muscle cell dysfunction. Cardiovasc Toxicol. 2020;20:548–56.
241. Bishop CJ, Kozielski KL, Green JJ. Exploring the role of polymer structure on intracellular nucleic acid delivery via polymeric nanoparticles. J Control Release. 2015;219:488–99.
242. Wagner E. Polymers for nucleic acid transfer-an overview. Adv Genet. 2014;88:231–61.
243. Xu C, Zhang Y, Xu K, Nie JJ, Yu B, Li S, et al. Multifunctional cationic nanosystems for nucleic acid therapy of thoracic aortic dissection. Nat Commun. 2019;10:3184.
244. Li N, Xu J, Feng R, Hu TL, Bu XH. Governing metal-organic frameworks towards high stability. Chem Commun. 2016;52:8501–13.
245. Fan Y, Zhang Y, Zhao Q, Xie Y, Luo R, Yang P, Weng Y. Immobilization of nano Cu-MOFs with polydopamine coating for adaptable gasotransmitter generation and copper ion delivery on cardiovascular stents. Biomaterials. 2019;204:36–45.
246. Liu Y, Feng F, Lv G-X, Gao Z-S. Fluorescent Tb(III)-MOF for Fe(III) Ion detection and treatment effect in aortic dissection by reducing β-receptor expression and d-dimer production. J Fluoresc. 2020;30:793–9.
247. He Y, Xu J, Sun X, Ren X, Maharjan A, York P, et al. Cuboidal tethered cyclodextrin frameworks tailored for hemostasis and injured vessel targeting. Theranostics. 2019;9:2489–504.
248. Durlach J. A possible advance in arterial gene therapy for aortic complications in the Marfan syndrome by local transfer of an antisense Mg-dependent hammerhead ribozyme. Magnes Res. 2001;14:65–7.
249. Neve RL, Neve KA, Nestler EJ, Carlezon WA Jr. Use of herpes virus amplicon vectors to study brain disorders. Biotechniques. 2005;39:381–91.
250. Wang Y, Miao Y, Gong K, Cheng X, Chen Y, Zhao M-H. Plasma complement protein C3a level was associated with abdominal aortic calcification in patients on hemodialysis. J Cardiovasc Transl. 2019;12:496–505.
251. Liu O, Xie W, Qin Y, Jia L, Zhang J, Xin Y, et al. MMP-2 gene polymorphisms are associated with type A aortic dissection and aortic diameters in patients. Medicine. 2016;95:e5175.
252. Maguire EM, Pearce SWA, Xiao R, Oo AY, Xiao Q. Matrix metalloproteinase in abdominal aortic aneurysm and aortic dissection. Pharmaceuticals. 2019;12:118.
253. Alexander H, Dowlati E, Mcgowan JE, Mason RB, Anaizi A. C2-C3 spinal fracture subluxation with ligamentous and vascular injury: a case report and review of management. Spinal Cord Ser Cases. 2019;5:4.
254. Gu H, Fisher AJ, Mickler EA, Duerson F 3rd, Cummings OW, Peters-Golden M, et al. Contribution of the anaphylatoxin receptors, C3aR and C5aR, to the pathogenesis of pulmonary fibrosis. FASEB J. 2016;30:2336–50.

255. Ren W, Liu Y, Wang X, Piao C, Ma Y, Qiu S, et al. The complement C3a–C3aR axis promotes development of thoracic aortic dissection via regulation of MMP2 expression. J Immunol. 2018;200:1829–38.
256. Baxter BT, Pearce WH, Waltke EA, Littooy FN, Hallett JW Jr, Kent KC, et al. Prolonged administration of doxycycline in patients with small asymptomatic abdominal aortic aneurysms: report of a prospective (Phase II) multicenter study. J Vasc Surg. 2002;36:1–12.
257. Sivaraman B, Ramamurthi A. Multifunctional nanoparticles for doxycycline delivery towards localized elastic matrix stabilization and regenerative repair. Acta Biomater. 2013;9:6511–25.
258. Camardo A, Seshadri D, Broekelmann T, Mecham R, Ramamurthi A. Multifunctional, JNK-inhibiting nanotherapeutics for augmented elastic matrix regenerative repair in aortic aneurysms. Drug Deliv Transl Res. 2018;8:964–84.
259. Venkataraman L, Sivaraman B, Vaidya P, Ramamurthi A. Nanoparticulate delivery of agents for induced elastogenesis in three-dimensional collagenous matrices. J Tissue Eng Regen M. 2016;10:1041–56.
260. Sivaraman B, Swaminathan G, Moore L, Fox J, Seshadri D, Dahal S, et al. Magnetically-responsive, multifunctional drug delivery nanoparticles for elastic matrix regenerative repair. Acta Biomater. 2017;52:171–86.
261. Rouer M, Xu BH, Xuan HJ, Tanaka H, Fujimura N, Glover KJ, et al. Rapamycin limits the growth of established experimental abdominal aortic aneurysms. Eur J Vasc Endovasc Surg. 2014;47:493–500.
262. Shirasu T, Koyama H, Miura Y, Hoshina K, Kataoka K, Watanabe T. Nanoparticles effectively target rapamycin delivery to sites of experimental aortic aneurysm in rats. PLoS One. 2016;11:e0157813.
263. Cheng J, Zhang R, Li C, Tao H, Dou Y, Wang Y, et al. A targeting nanotherapy for abdominal aortic aneurysms. J Am Coll Cardiol. 2018;72:2591–605.
264. Nosoudi N, Nahar-Gohad P, Sinha A, Chowdhury A, Gerard P, Carsten CG, et al. Prevention of abdominal aortic aneurysm progression by targeted inhibition of matrix metalloproteinase activity with batimastat-loaded nanoparticles. Circ Res. 2015;117:e80–90.
265. Sylvester A, Sivaraman B, Deb P, Ramamurthi A. Nanoparticles for localized delivery of hyaluronan oligomers towards regenerative repair of elastic matrix. Acta Biomater. 2013;9:9292–302.
266. Nosoudi N, Chowdhury A, Siclari S, Karamched S, Parasaram V, Parrish J, et al. Reversal of vascular calcification and aneurysms in a rat model using dual targeted therapy with EDTA- and PGG-loaded nanoparticles. Theranostics. 2016;6:1975–87.
267. Nosoudi N, Chowdhury A, Siclari S, Parasaram V, Karamched S, Vyavahare N. Systemic delivery of nanoparticles loaded with pentagalloyl glucose protects elastic lamina and prevents abdominal aortic aneurysm in rats. J Cardiovasc Transl Res. 2016;9:445–55.
268. Farrell K, Simmers P, Mahajan G, Boytard L, Camardo A, Joshi J, et al. Alterations in phenotype and gene expression of adult human aneurysmal smooth muscle cells by exogenous nitric oxide. Exp Cell Res. 2019;384:111589.
269. Jennewine B, Fox J, Ramamurthi A. Cathepsin K-targeted sub-micron particles for regenerative repair of vascular elastic matrix. Acta Biomater. 2017;52:60–73.
270. Zehtabi F, Ispas-Szabo P, Djerir D, Sivakumaran L, Annabi B, Soulez G, et al. Chitosan-doxycycline hydrogel: an MMP inhibitor/sclerosing embolizing agent as a new approach to endoleak prevention and treatment after endovascular aneurysm repair. Acta Biomater. 2017;64:94–105.
271. Yu M, Dong A, Chen C, Xu S, Cao Y, Liu S, et al. Thermosensitive hydrogel containing doxycycline exerts inhibitory effects on abdominal aortic aneurysm induced by pancreatic elastase in mice. Adv Healthc Mater. 2017;6:1700671.
272. Yrineo AA, Adelsperger AR, Durkes AC, Distasi MR, Voytik-Harbin SL, Murphy MP, Goergen CJ. Murine ultrasound-guided transabdominal para-aortic injections of self-assembling type I collagen oligomers. J Control Release. 2017;249:53–62.

273. Choi YH, Heo SC, Kwon YW, Kim HD, Kim SH, Jang IH, et al. Injectable PLGA microspheres encapsulating WKYMVM peptide for neovascularization. Acta Biomater. 2015;25: 76–85.
274. Deveza L, Choi J, Lee J, Huang N, Cooke J, Yang F. Polymer-DNA nanoparticle-induced CXCR4 overexpression improves stem cell engraftment and tissue regeneration in a mouse hindlimb ischemia model. Theranostics. 2016;6:1176–89.
275. Hamada Y, Gonda K, Takeda M, Sato A, Watanabe M, Yambe T, et al. In vivo imaging of the molecular distribution of the VEGF receptor during angiogenesis in a mouse model of ischemia. Blood. 2011;118:e93–e100.
276. Cho BR, Ryu DR, Lee KS, Lee DK, Bae S, Kang DG, et al. p-Hydroxybenzyl alcohol-containing biodegradable nanoparticle improves functional blood flow through angiogenesis in a mouse model of hindlimb ischemia. Biomaterials. 2015;53:679–87.
277. Jung E, Noh J, Kang C, Yoo D, Song C, Lee D. Ultrasound imaging and on-demand therapy of peripheral arterial diseases using H2O2-activated bubble generating anti-inflammatory polymer particles. Biomaterials. 2018;179:175–85.
278. Jung E, Lee J, Jeong L, Park S, Lee M, Song C, Lee D. Stimulus-activatable echogenic maltodextrin nanoparticles as nanotheranostic agents for peripheral arterial disease. Biomaterials. 2019;192:282–91.
279. Kubo M, Egashira K, Inoue T, Koga J, Oda S, Chen L, et al. Therapeutic neovascularization by nanotechnology-mediated cell-selective delivery of pitavastatin into the vascular endothelium. Arterioscler Thromb Vasc Biol. 2009;29:796–801.
280. Suhara M, Miura Y, Cabral H, Akagi D, Anraku Y, Kishimura A, et al. Targeting ability of self-assembled nanomedicines in rat acute limb ischemia model is affected by size. J Control Release. 2018;286:394–401.
281. Tsumaru S, Masumoto H, Minakata K, Izuhara M, Yamazaki K, Ikeda T, et al. Therapeutic angiogenesis by local sustained release of microRNA-126 using poly lactic-co-glycolic acid nanoparticles in murine hindlimb ischemia. J Vasc Surg. 2018;68:1209–15.
282. Liu G, Fang Z, Yuan M, Li W, Yang Y, Jiang M, et al. Biodegradable carriers for delivery of VEGF plasmid DNA for the treatment of critical limb ischemia. Front Pharmacol. 2017;8:528.
283. Kumar VA, Liu Q, Wickremasinghe NC, Shi S, Cornwright TT, Deng Y, et al. Treatment of hind limb ischemia using angiogenic peptide nanofibers. Biomaterials. 2016;98:113–9.
284. Cho SW, Yang F, Son SM, Park HJ, Green JJ, Bogatyrev S, et al. Therapeutic angiogenesis using genetically engineered human endothelial cells. J Control Release. 2012;160:515–24.
285. Chung YI, Kim SK, Lee YK, Park SJ, Cho KO, Yuk SH, et al. Efficient revascularization by VEGF administration via heparin-functionalized nanoparticle-fibrin complex. J Control Release. 2010;143:282–9.
286. Albrecht-Schgoer K, Barthelmes J, Schgoer W, Theurl M, Nardin I, Lener D, et al. Nanoparticular delivery system for a secretoneurin derivative induces angiogenesis in a hind limb ischemia model. J Control Release. 2017;250:1–8.
287. Wu JP, Cheng B, Roffler SR, Lundy DJ, Yen CY, Chen P, et al. Reloadable multidrug capturing delivery system for targeted ischemic disease treatment. Sci Transl Med. 2016;8: 365ra160.
288. Tongers J, Webber MJ, Vaughan EE, Sleep E, Renault MA, Roncalli JG, et al. Enhanced potency of cell-based therapy for ischemic tissue repair using an injectable bioactive epitope presenting nanofiber support matrix. J Mol Cell Cardiol. 2014;74:231–9.
289. Poole KM, Nelson CE, Joshi RV, Martin JR, Gupta MK, Haws SC, et al. ROS-responsive microspheres for on demand antioxidant therapy in a model of diabetic peripheral arterial disease. Biomaterials. 2015;41:166–75.
290. Lee D, Bae S, Hong D, Lim H, Yoon JH, Hwang O, et al. H_2O_2-responsive molecularly engineered polymer nanoparticles as ischemia/reperfusion-targeted nanotherapeutic agents. Sci Rep. 2013;3:2233.
291. Saito T, Tabata Y. Hypoxia-induced angiogenesis is increased by the controlled release of deferoxiamine from gelatin hydrogels. Acta Biomater. 2014;10:3641–9.

292. Zhao Q, Zhang J, Song L, Ji Q, Yao Y, Cui Y, et al. Polysaccharide-based biomaterials with on-demand nitric oxide releasing property regulated by enzyme catalysis. Biomaterials. 2013;34:8450–8.
293. Layman H, Spiga MG, Brooks T, Pham S, Webster KA, Andreopoulos FM. The effect of the controlled release of basic fibroblast growth factor from ionic gelatin-based hydrogels on angiogenesis in a murine critical limb ischemic model. Biomaterials. 2007;28:2646–54.
294. Zachman AL, Wang X, Tucker-Schwartz JM, Fitzpatrick ST, Lee SH, Guelcher SA, et al. Uncoupling angiogenesis and inflammation in peripheral artery disease with therapeutic peptide-loaded microgels. Biomaterials. 2014;35:9635–48.
295. Park HJ, Jin Y, Shin J, Yang K, Lee C, Yang HS, Cho SW. Catechol-functionalized hyaluronic acid hydrogels enhance angiogenesis and osteogenesis of human adipose-derived stem cells in critical tissue defects. Biomacromolecules. 2016;17:1939–48.
296. Wang R, Yao X, Li T, Li X, Jin M, Ni Y, Yuan W. Reversible thermoresponsive hydrogel fabricated from natural biopolymer for the improvement of critical limb ischemia by controlling release of stem cells. Adv Healthc Mater. 2019;8:e1900967.
297. Thomas D, Fontana G, Chen X, Sanz-Nogues C, Zeugolis DI, Dockery P, et al. A shape-controlled tuneable microgel platform to modulate angiogenic paracrine responses in stem cells. Biomaterials. 2014;35:8757–66.
298. Kim PH, Yim HG, Choi YJ, Kang BJ, Kim J, Kwon SM, et al. Injectable multifunctional microgel encapsulating outgrowth endothelial cells and growth factors for enhanced neovascularization. J Control Release. 2014;187:1–13.
299. Li R, Liang J, He Y, Qin J, He H, Lee S, et al. Sustained release of immunosuppressant by nanoparticle-anchoring hydrogel scaffold improved the survival of transplanted stem cells and tissue regeneration. Theranostics. 2018;8:878–93.
300. Wang X, Zhang J, Cui W, Fang Y, Li L, Ji S, et al. Composite hydrogel modified by IGF-1C domain improves stem cell therapy for limb ischemia. ACS Appl Mater Interfaces. 2018;10: 4481–93.
301. Lee J, Jun I, Park HJ, Kang TJ, Shin H, Cho SW. Genetically engineered myoblast sheet for therapeutic angiogenesis. Biomacromolecules. 2014;15:361–72.
302. Chen DY, Wei HJ, Lin KJ, Huang CC, Wang CC, Wu CT, et al. Three-dimensional cell aggregates composed of HUVECs and cbMSCs for therapeutic neovascularization in a mouse model of hindlimb ischemia. Biomaterials. 2013;34:1995–2004.
303. Huang CC, Chen DY, Wei HJ, Lin KJ, Wu CT, Lee TY, et al. Hypoxia-induced therapeutic neovascularization in a mouse model of an ischemic limb using cell aggregates composed of HUVECs and cbMSCs. Biomaterials. 2013;34:9441–50.
304. Li L, Okada H, Takemura G, Esaki M, Kobayashi H, Kanamori H, et al. Sustained release of erythropoietin using biodegradable gelatin hydrogel microspheres persistently improves lower leg ischemia. J Am Coll Cardiol. 2009;53:2378–88.
305. Hosaka A, Koyama H, Kushibiki T, Tabata Y, Nishiyama N, Miyata T, et al. Gelatin hydrogel microspheres enable pinpoint delivery of basic fibroblast growth factor for the development of functional collateral vessels. Circulation. 2004;110:3322–8.
306. Hastings CL, Kelly HM, Murphy MJ, Barry FP, O'brien FJ, Duffy GP. Development of a thermoresponsive chitosan gel combined with human mesenchymal stem cells and desferrioxamine as a multimodal pro-angiogenic therapeutic for the treatment of critical limb ischaemia. J Control Release. 2012;161:73–80.
307. Li Y, Liu W, Liu F, Zeng Y, Zuo S, Feng S, et al. Primed 3D injectable microniches enabling low-dosage cell therapy for critical limb ischemia. Proc Natl Acad Sci USA. 2014;111:13511–6.
308. Huang A, Liu D, Qi X, Yue Z, Cao H, Zhang K, et al. Self-assembled GFFYK peptide hydrogel enhances the therapeutic efficacy of mesenchymal stem cells in a mouse hindlimb ischemia model. Acta Biomater. 2019;85:94–105.
309. Kim SJ, Jun I, Kim DW, Lee YB, Lee YJ, Lee JH, et al. Rapid transfer of endothelial cell sheet using a thermosensitive hydrogel and its effect on therapeutic angiogenesis. Biomacromolecules. 2013;14:4309–19.

310. Robinson ST, Douglas AM, Chadid T, Kuo K, Rajabalan A, Li H, et al. A novel platelet lysate hydrogel for endothelial cell and mesenchymal stem cell-directed neovascularization. Acta Biomater. 2016;36:86–98.
311. Layman H, Sacasa M, Murphy AE, Murphy AM, Pham SM, Andreopoulos FM. Co-delivery of FGF-2 and G-CSF from gelatin-based hydrogels as angiogenic therapy in a murine critical limb ischemic model. Acta Biomater. 2009;5:230–9.
312. Li R, Pang Z, He H, Lee S, Qin J, Wu J, et al. Drug depot-anchoring hydrogel: a self-assembling scaffold for localized drug release and enhanced stem cell differentiation. J Control Release. 2017;261:234–45.
313. Duan J, Chen Z, Liang X, Chen Y, Li H, Tian X, et al. Construction and application of therapeutic metal-polyphenol capsule for peripheral artery disease. Biomaterials. 2020;255: 120199.
314. Park IS, Mahapatra C, Park JS, Dashnyam K, Kim JW, Ahn JC, et al. Revascularization and limb salvage following critical limb ischemia by nanoceria-induced Ref-1/APE1-dependent angiogenesis. Biomaterials. 2020;242:119919.
315. Isner JM. Arterial gene transfer of naked DNA for therapeutic angiogenesis: early clinical results. Adv Drug Deliv Rev. 1998;30:185–97.
316. Bao H, Lv F, Liu T. A pro-angiogenic degradable Mg-poly(lactic-co-glycolic acid) implant combined with rhbFGF in a rat limb ischemia model. Acta Biomater. 2017;64:279–89.
317. Koiwaya H, Sasaki K, Ueno T, Yokoyama S, Toyama Y, Ohtsuka M, et al. Augmented neovascularization with magnetized endothelial progenitor cells in rats with hind-limb ischemia. J Mol Cell Cardiol. 2011;51:33–40.
318. Chorny M, Fishbein I, Yellen BB, Alferiev IS, Bakay M, Ganta S, et al. Targeting stents with local delivery of paclitaxel-loaded magnetic nanoparticles using uniform fields. Proc Natl Acad Sci U S A. 2010;107:8346–51.
319. Zhong Y, Zhang L, Ding AG, Shenderova A, Zhu G, Pei P, et al. Rescue of SCID murine ischemic hindlimbs with pH-modified rhbFGF/poly(DL-lactic-co-glycolic acid) implants. J Control Release. 2007;122:331–7.
320. Hwang H, Kim HS, Jeong HS, Rajasaheb BT, Kim M, Oh PS, et al. Liposomal angiogenic peptides for ischemic limb perfusion: comparative study between different administration methods. Drug delivery. 2016;23:3619–28.
321. Monteforte AJ, Lam B, Das S, Mukhopadhyay S, Wright CS, Martin PE, et al. Glypican-1 nanoliposomes for potentiating growth factor activity in therapeutic angiogenesis. Biomaterials. 2016;94:45–56.
322. Bose RJ, Kim BJ, Arai Y, Han IB, Moon JJ, Paulmurugan R, et al. Bioengineered stem cell membrane functionalized nanocarriers for therapeutic targeting of severe hindlimb ischemia. Biomaterials. 2018;185:360–70.
323. Gangadaran P, Rajendran RL, Lee HW, Kalimuthu S, Hong CM, Jeong SY, et al. Extracellular vesicles from mesenchymal stem cells activates VEGF receptors and accelerates recovery of hindlimb ischemia. J Control Release. 2017;264:112–26.
324. Yu Z, Cai Y, Deng M, Li D, Wang X, Zheng H, et al. Fat extract promotes angiogenesis in a murine model of limb ischemia: a novel cell-free therapeutic strategy. Stem Cell Res Ther. 2018;9:294.
325. Mannucci PM, Poller L. Venous thrombosis and anticoagulant therapy. Br J Haematol. 2001;114:258–70.
326. Sarica-Kucukoglu R, Akdag-Kose A, Kayabal IM, Yazganoglu KD, Disci R, Erzengin D, Azizlerli G. Vascular involvement in Behcet's disease: a retrospective analysis of 2319 cases. Int J Dermatol. 2006;45:919–21.
327. Asbeutah AM, Riha AZ, Cameron JD, Mcgrath BP. Five-year outcome study of deep vein thrombosis in the lower limbs. J Vasc Surg. 2004;40:1184–9.
328. Markel A. Origin and natural history of deep vein thrombosis of the legs. Semin Vasc Med. 2005;5:65–74.

329. Beyer-Westendorf J, Lensing AW, Arya R, Bounameaux H, Cohen AT, Wells PS, et al. Choosing wisely: the impact of patient selection on efficacy and safety outcomes in the EINSTEIN-DVT/PE and AMPLIFY trials. Thromb Res. 2017;149:29–37.
330. Mook-Kanamori BB, Valls Seron M, Geldhoff M, Havik SR, Van Der Ende A, Baas F, et al. Thrombin-activatable fibrinolysis inhibitor influences disease severity in humans and mice with pneumococcal meningitis. J Thromb Haemost. 2015;13:2076–86.
331. Kim JA, Hedrick CC, Xie D, Fisher MJ. Adenoviral-mediated transfer of tissue plasminogen activator gene into brain capillary endothelial cells in vitro. Angiology. 2001;52:627–34.
332. Mcneil MM, Paradowska-Stankiewicz I, Miller ER, Marquez PL, Seshadri S, Collins LC Jr, Cano MV. Adverse events following adenovirus type 4 and type 7 vaccine, live, oral in the Vaccine Adverse Event Reporting System (VAERS), United States, October 2011-July 2018. Vaccine. 2019;37:6760–7.
333. Gallego I, Villate-Beitia I, Martinez-Navarrete G, Menendez M, Lopez-Mendez T, Soto-Sanchez C, et al. Non-viral vectors based on cationic niosomes and minicircle DNA technology enhance gene delivery efficiency for biomedical applications in retinal disorders. Nanomedicine. 2019;17:308–18.
334. Zhang B, Zhang Y, Yu D. Lung cancer gene therapy: transferrin and hyaluronic acid dual ligand-decorated novel lipid carriers for targeted gene delivery. Oncol Rep. 2017;37:937–44.
335. Tabuchi N, Shichiri M, Shibamiya A, Koyama T, Nakazawa F, Chung J, et al. Non-viral in vivo thrombomodulin gene transfer prevents early loss of thromboresistance of grafted veins. Eur J Cardiothorac Surg. 2004;26:995–1001.
336. De Jong WH, Borm PJ. Drug delivery and nanoparticles: applications and hazards. Int J Nanomed. 2008;3:133–49.
337. Banerjee A, Qi J, Gogoi R, Wong J, Mitragotri S. Role of nanoparticle size, shape and surface chemistry in oral drug delivery. J Control Release. 2016;238:176–85.
338. Wang Z, Wu Z, Liu J, Zhang W. Particle morphology: an important factor affecting drug delivery by nanocarriers into solid tumors. Expert Opin Drug Del. 2018;15:379–95.
339. Jiang H, Meng J, Guo T, Zhao JN, Wang YC, Wang J, et al. Comparison of apixaban and low molecular weight heparin in preventing deep venous thrombosis after total knee arthroplasty in older adults. Yonsei Med J. 2019;60:626–32.
340. Robinson R, Wirt TC, Barbosa C, Amidi A, Chen S, Joseph RM, Fleischer AE. Routine use of low-molecular-weight heparin for deep venous thrombosis prophylaxis after foot and ankle surgery: a cost-effectiveness analysis. J Foot Ankle Surg. 2018;57:543–51.
341. Xia ZN, Zhou Q, Zhu W, Weng XS. Low molecular weight heparin for the prevention of deep venous thrombosis after total knee arthroplasty: a systematic review and meta-analysis. Int J Surg. 2018;54:265–75.
342. Zee AA, Van Lieshout K, Van Der Heide M, Janssen L, Janzing HM. Low molecular weight heparin for prevention of venous thromboembolism in patients with lower-limb immobilization. Cochrane Database Syst Rev. 2017;8:Cd006681.
343. Tang JC, Li S, Mcgray P, Vecchio A. Current status of activity assays for tissue plasminogen activator. Ann N Y Acad Sci. 1984;434:536–40.
344. Nolin WB, Emmetsberger J, Bukhari N, Zhang Y, Levine JM, Tsirka SE. tPA-mediated generation of plasmin is catalyzed by the proteoglycan NG2. Glia. 2008;56:177–89.
345. Merino P, Diaz A, Manrique LG, Cheng L, Yepes M. Urokinase-type plasminogen activator (uPA) promotes ezrin-mediated reorganization of the synaptic cytoskeleton in the ischemic brain. J Biol Chem. 2018;293:9234–47.
346. Merino P, Diaz A, Jeanneret V, Wu F, Torre E, Cheng L, Yepes M. Urokinase-type plasminogen activator (uPA) binding to the uPA receptor (uPAR) promotes axonal regeneration in the central nervous system. J Biol Chem. 2017;292:2741–53.
347. Juenet M, Aid-Launais R, Li B, Berger A, Aerts J, Ollivier V, et al. Thrombolytic therapy based on fucoidan-functionalized polymer nanoparticles targeting P-selectin. Biomaterials. 2018;156:204–16.

348. Zhang R, Huang B, Du D, Guo X, Xin G, Xing Z, et al. Anti-thrombosis effect of diosgenyl saponins in vitro and in vivo. Steroids. 2013;78:1064–70.
349. Wei Z, Xin G, Wang H, Zheng H, Ji C, Gu J, et al. The diosgenin prodrug nanoparticles with pH-responsive as a drug delivery system uniquely prevents thrombosis without increased bleeding risk. Nanomedicine. 2018;14:673–84.
350. Pazzini C, Marcato PD, Prado LB, Alessio AM, Hoehr NF, Montalvao S, et al. Polymeric nanoparticles of enoxaparin as a delivery system: in vivo evaluation in normal rats and in a venous thrombosis rat model. J Nanosci Nanotechnol. 2015;15:4837–43.
351. Hallan SS, Kaur V, Jain V, Mishra N. Development and characterization of polymer lipid hybrid nanoparticles for oral delivery of LMWH. Artif Cells Nanomed Biotechnol. 2017;45: 1631–9.
352. Bai S, Thomas C, Ahsan F. Dendrimers as a carrier for pulmonary delivery of enoxaparin, a low-molecular weight heparin. J Pharm Sci. 2007;96:2090–106.
353. Cui Y, Li B, He H, Zhou W, Chen B, Qian G. Metal-organic frameworks as platforms for functional materials. Acc Chem Res. 2016;49:483–93.
354. Carrasco S. Metal-organic frameworks for the development of biosensors: a current overview. Biosensors. 2018;8:92.
355. Drake T, Ji P, Lin W. Site isolation in metal-organic frameworks enables novel transition metal catalysis. Acc Chem Res. 2018;51:2129–38.
356. Yuan S, Feng L, Wang K, Pang J, Bosch M, Lollar C, et al. Stable metal-organic frameworks: design, synthesis, and applications. Adv Mater. 2018;30:e1704303.
357. Vaitsis C, Sourkouni G, Argirusis C. Metal organic frameworks (MOFs) and ultrasound: a review. Ultrason Sonochem. 2019;52:106–19.
358. Song S, Song W, Li P, Cheng Y. Zn(II)-based coordination polymer for detection of picric acid in aqueous media and treatment effect against deep vein thrombosis by reducing P-ERK2 and p53 expression. J Fluoresc. 2020;30:717–23.
359. Ahmed EM. Hydrogel: preparation, characterization, and applications: a review. J Adv Res. 2015;6:105–21.
360. Alizadeh B, Bahari Javan N, Akbari Javar H, Khoshayand MR, Dorkoosh F. Prolonged injectable formulation of Nafarelin using in situ gel combination delivery system. Pharm Dev Technol. 2018;23:132–44.
361. Mei L, Xie Y, Huang Y, Wang B, Chen J, Quan G, et al. Injectable in situ forming gel based on lyotropic liquid crystal for persistent postoperative analgesia. Acta Biomater. 2018;67:99–110.
362. Zhang K, Zhou L, Chen F, Chen Y, Luo X. Injectable gel self-assembled by paclitaxel itself for in situ inhibition of tumor growth. J Control Release. 2019;315:197–205.
363. Zheng JW, Yang XJ, Wang YA, He Y, Ye WM, Zhang ZY. Intralesional injection of Pingyangmycin for vascular malformations in oral and maxillofacial regions: an evaluation of 297 consecutive patients. Oral Oncol. 2009;45:872–6.
364. Li J, Qiao JB, Liu QY. Pingyangmycin pretreatment influences the biological behavior of ocular venous malformation and relates with galectin-3 expression. Chinese Med J. 2017;130: 1804–9.
365. Zhang L, Chen F, Zheng J, Wang H, Qin X, Pan W. Chitosan-based liposomal thermogels for the controlled delivery of pingyangmycin: design, optimization and in vitro and in vivo studies. Drug Deliv. 2018;25:690–702.
366. Stuepp RT, Scotti FM, Melo G, Munhoz EA, Modolo F. Effects of sclerosing agents on head and neck hemangiomas: a systematic review. J Clin Exp Dent. 2019;11:e1033–44.
367. Pitek AS, Wang Y, Gulati S, Gao H, Stewart PL, Simon DI, Steinmetz NF. Elongated plant virus-based nanoparticles for enhanced delivery of thrombolytic therapies. Mol Pharmaceut. 2017;14:3815–23.

Chapter 5
Manufacturing Methods of Materials for Cardiac Implant

Yufeng Zheng and Zhenyuan Yin

5.1 Manufacturing Methods of Metallic Cardiac Materials for Implant

Up to now, most of the cardiac metallic implants on the market are produced through traditional metal processing techniques, including smelting, forging, rolling, extrusion, drawing, heat treatment, mechanical processing, polishing, and so forth; these approaches are summarized in Fig. 5.1a. In cardiac interventions, coronary stents are recognized as the major applications of metal materials. During the case of stents fabrication, the metal materials are processed by a series of complex procedures as mentioned above, which aims to endow the raw materials with enhanced mechanical properties. The obtained bulk materials are machined to billet, followed by drawing and heat treatment cycle; the billets are made into thin-wall microtubes that are directly used for laser cutting. Since the laser-cut surfaces are relatively rough and covered with molten slag, the stents are treated with acid pickling and electrochemical polishing cleaning to achieve a desired surface finish of the implanted materials. Currently, laser cutting is recognized as the most mainstream technology in the manufacturing of commercial stents. However, the conventional laser cutting technique requires complicated manufacturing procedures, and the obtained stents cannot achieve personalized customization. Therefore, additive manufacturing (AM) technology has been investigated in the fabrication of implantable devices such as powder metallurgy (PM), binder jetting (BJT), directed energy deposition

Y. Zheng (✉)
School of Materials Science and Engineering, Peking University, Beijing, China
e-mail: yfzheng@pku.edu.cn

Z. Yin
School of Materials Science and Engineering, Peking University, Beijing, China

Academy for Advanced Interdisciplinary Studies (AAIS), Peking University, Beijing, China
e-mail: yinzhenyuan@pku.edu.cn

Y.-D. Tang et al. (eds.), *Application of Biomaterials in the Treatment of Cardiovascular Diseases*, https://doi.org/10.1007/978-981-19-7712-1_5

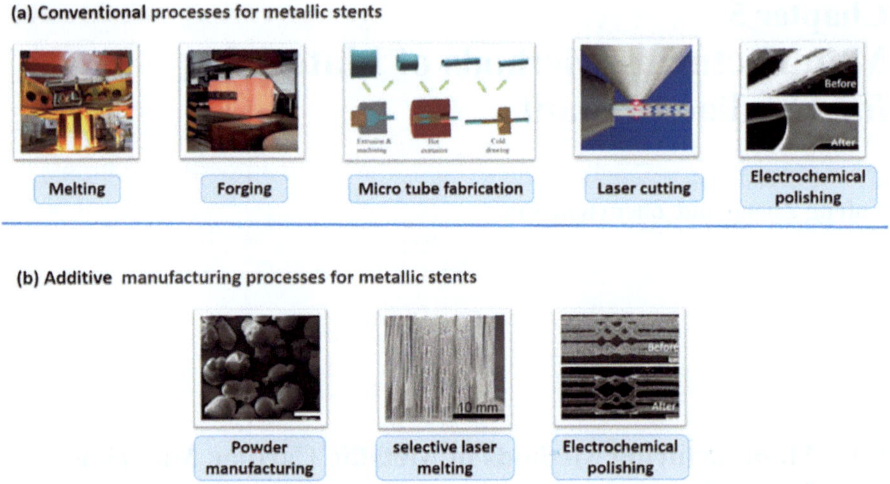

Fig. 5.1 Two types of metallic stent manufacturing chain [11–13]: (**a**) conventional processes and (**b**) AM processes

(DED), and powder bed fusion (PBF). On account of hyperfine dimensional accuracy, selective laser melting (SLM) based on PBF is utilized to produce patient-specific implants like orthopedic prosthesis [1–7]. Although only a few AM studies focused on cardiac devices, it could be a potentially revolutionary manufacturing technology to replace complicated steps for conventional modeling approaches by selectively fusing metallic powders layer by layer [8]. For instance, different from laser micro-cutting, the SML process is based on a powder bed fusion where powder size is between 20 and 50 μm to guarantee effective melting and flowing. In SLM technology, the laser beam selectively melting the metal powder layer by layer to complete the prototype of stents according to the 3D model, which exhibited superior efficiency compared with the conventional fabrication techniques (Fig. 5.1). The metallic powders as the feedstock material directly influence performance of devices, such as density, surface quality, mechanical properties, and corrosion resistance [9]. Existing powder processes include gas, water, plasma, and centrifugal atomization technologies. In view of application, the gas-atomized powder is considered as the most commonly used feedstock in SLM processing as it produces spherical powders with optimal cost compared to irregular gas-atomized powders and high cost of plasma-atomized powders [10]. However, rapid heating and cooling would inevitably cause some defects in prototypes, such as pores and unmelted powder. Therefore, heat treatment, surface cleaning, and electrochemical polishing are executed to acquire stents with the desired properties.

5.1.1 Stainless Steel

The stainless steel is commonly fabricated via smelting techniques; the ingredient needed in such process mainly includes scrap steel, molten iron, chromium ore, nickel ore, stainless slags, and dusts. The smelting processes of stainless steel has undergone three stages of development, one-step method, two-step method, and three-step method, among them the two-step method accounts for about 70% in smelting processes. One-step method is gradually phased out due to excessive carbon content without any other decarburization processes in electric-arc furnace (EAF). On the contrary, a refining process is involved into two-step method, in which the steels are melted by the argon oxygen decarburization (AOD) or vacuum oxygen decarburization (VOD) processes after EAF, producing all kinds of stainless steels except ultra-low carbon. Unlike the two-step method, the three-step method combines AOD and VOD methods, accomplishing ultra-low carbon and nitrogen content control. Compared with the nitrogen-free stainless steel, high nitrogen stainless steels (HNS) exhibit attractive mechanical properties, good corrosion resistance, and biocompatibility, which can be produced by normal smelting processes below 0.4% nitrogen as well. To achieve higher nitrogen content and avoid denitriding during solidification, high-pressure melting-pouring and addition of nitralloy are required, such as pressurized electroslag remelting (PESR), hot isostatic pressing (HIP), pressurized plasma furnace melting, and pressurized induction furnace melting [14–17].

316L stainless steel (316L) is one of mainstream medical metallic materials, which has been widely used to fabricate dental, orthopedic, and cardiac implants [18–21]. This alloy is usually forged from a temperature of 1150–1260 °C depending upon the severity of deformation. Heat treatment is required either in the manufacturing processes or after fabrication depending on the processing technique to remove the effects of cold forming or to dissolve precipitated chromium carbides resulting from hot working. Solution annealing for 316L is generally accomplished by heating at the temperature of 1040 to 1120 °C, followed by air cooling or a water quenching in order to prevent the formation of undesirable intermetallic phase caused by long exposure at 650 to 870 °C and provide optimum corrosion resistance [22].

Heat-treated feedstock is then used to fabricate microtubes via the following steps: (1) producing solid rods by hot working like forging and extrusion, (2) deeply drilling a hole in the rod, and (3) reducing the diameter and wall thickness of tubes by multi-pass cold drawing and intermediate annealing.

The quality of stents after laser cutting directly affects its effectiveness and safety, such as surface roughness, kerf width, recast layer, heat affected zone (HAZ), and dross adherence. Ulteriorly, laser beam parameters (e.g., pulse frequency, pulse width, laser beam power) and process parameters (e.g., gas pressure and cutting speed) are further optimized to fully meet the quality requirements [23]. Since the properties and performances of stents are closely related to the surface roughness, a stent with smoother surface exhibits higher biocompatibility and longer fatigue life [24–27]. Chemical polishing can effectively reduce the surface roughness, but it is difficult to uniformly control the thickness of struts. The experimental results showed that the surface roughness of stents became higher with the longer interaction time between the laser beam and the 316L microtubes, which is caused by the increase of pulse width and cutting speed [28]. Ensuring the stable kerf width of stents is a precondition for dimensional accuracy requirements. With the

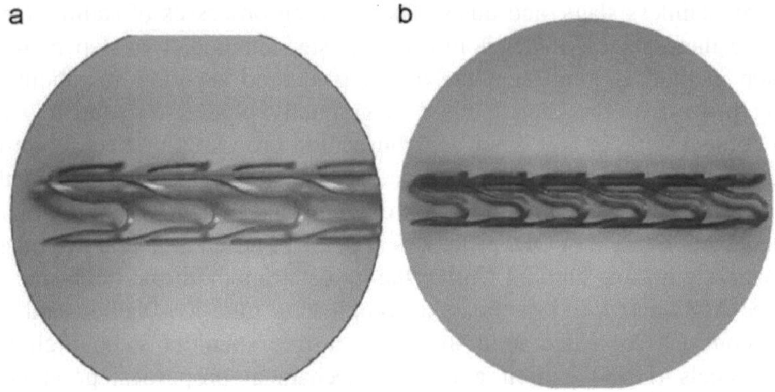

Fig. 5.2 The 316L cardiovascular stent: (**a**) fiber laser cutting, (**b**) YAG laser cutting [31]

precondition, narrow kerf width means less fusion of material and thinner recast layer, which is produced by low laser power and low cutting speed, but some microcracks were observed at the kerf surfaces [29]. During the cutting process, melted and evaporated materials solidify on the kerf surface again, forming a recast layer. A portion of rest heat transfers to the surrounding areas, leading to distortion and forming HAZ [28]. Due to the existing of microcracks, coarse grains, and hardening caused by the recast layer and HAZ, the intrinsic properties of material may get worse. In order to reduce heat accumulation, shorter plus width and wet cutting are usually applied to produce narrow HAZ [29]. Splashes of dross resulting from the melted materials not only increase the roughness of kerf surface but also cause back damage when they attach to inner wall of stent. High-pressure assist gas and ultra-short pulse width laser effectively reduce the amount of dross [30]. Besides, water-assisted laser cutting produce less back wall damage due to the confining high plasma pressure and quick removal of debris. In recent years, researchers have continuously optimized cutting quality through coming up with original techniques and adjusting process parameters. Meng et al. [31] conducted 316L cardiovascular stents based on gas-assisted fiber laser system (Fig. 5.2). As the kerf width size was less than 20 μm, the stents exhibited low roughness and the HAZ that showed a sharp contrast with the outcome by Nd:YAG laser in the same parameters (laser output power 7 W, pulse length 0.15 ms, repeat frequency 1500 Hz, cutting speed 8 mm/s, and assisting oxygen pressure 0.3 MPa). Muhammad et al. [32] compared wet cutting of 316L tubes by flowing water with gas-assisted dry cutting, which indicated that the wet cutting produced higher comprehensive qualities with lower surface roughness, narrow kerf, thinner HAZ, and less dross and back wall damages. García-López et al. [33] combined appropriate process parameters (peak power 160 W, pulse length 0.18 ms, repeat frequency 900 Hz, cutting speed 250 mm/min, pulse overlapping factor 77.74%, pulse energy 28.8 mJ, and assisting nitrogen pressure 10.34 bar) to achieve less than 1 μm in surface roughness at the laser-cut edge and less than 3.5% dross deposits at the back wall of the miniature tubes with fiber laser. Sun et al. [28] investigated the relationship between process parameters (pulse width, laser power, frequency, cutting speed) and surface integrity indexes

Table 5.1 The parameters of 316L stent acid pickling

Composition	Conditions	References
3 ml HF (48%) + 9 ml HNO_3 (65%) + 88 ml H_2O	1. Immersion in pickling solution at 40 °C for 40 min 2. Ultrasonic agitation bath in distilled water for 15 min and dried in air	[32]
70% HNO_3	1. Acid bath in ultrasonic cleaner at 50–60 °C for 3–5 h 2. Washed with deionized water and deionized water	[50]
1 ml HF + 9 ml HNO_3 + 90 ml H_2O	1. At 45 °C for 15 min by ultrasonic cleaner 2. Cleaned ultrasonically using deionized water and dried by hot air-dryer	[51]

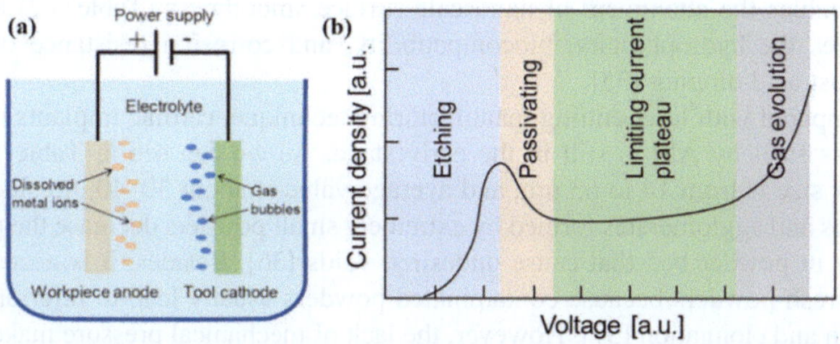

Fig. 5.3 Schematic illustration of (**a**) electrochemical polishing processes and (**b**) current density-voltage curve [11]

(surface topography, surface roughness, HAZ, recast layer, dross, subsurface nano-hardness) with fiber laser wet cutting of 316L stents. In conclusion, the short pulse width, medium cutting speed, small power, and high frequency are considered as the optimum single-factor parameters, respectively.

Acid pickling is then applied to remove dross, burr, oil contamination, and oxide scale. Generally, stents are immersed in an acid aqueous solution for tens of minutes followed by ultrasonic cleaning and drying. As shown in Table 5.1, due to unqualified surface smoothness after pickling, electrochemical polishing process is applied to achieve anodic leveling and brightening based on anodic polarization [11]. In the schematic diagram of electrochemical polishing processes (Fig. 5.3a), the 316L stent is connected to the positive terminal of DC power as the anode immersed in the electrolyte, while the cathode is usually made of stainless steel sheet. When the current flows through the stent, surface metal begins to dissolve. Because the peaks have extremely higher charge concentration on them than that on the valleys of the stents, and thus the material of peaks will dissolve faster than that of the valleys, the desire level of surface smoothness is reached [34]. According to the current density-voltage curve (Fig. 5.3b) for electrochemical polishing, the current of stent varies nonlinearly with the applied voltage, which shows etching, passivating, polishing, and gas evolution region. For the anode potentials in the etching region, continuing dissolve occurs as the surface is etched in the active stage. Hence, electrochemical polishing enhances the pickling effect partly. In the latter half passivating, the anodic polarization occurs that current density decreases with the increase of anode potential region, in which the dissolution rate of surface decreases and the oxide film begins to form. In the polishing region, the current in this range is usually utilized in polishing process because a stable current density is achieved as the voltage increases. When the current density continues to rise and reaches a sufficiently high state, the passivation layer will be peeled off, and the surface defects will appear such as pits and stains [11]. Electrochemical polishing is regarded as an optimal technique to polish complex shapes, and it is also the most widely used polishing process for stents. The phosphoric acid and sulfuric acid mixed system, commonly employed as a polishing electrolyte of stainless steel stents, not only enables the attainment of nanoscale surface smoothness (Table 5.2) but also improves the hydrophilicity, biocompatibility, and corrosion resistance of stents with post acid dipping [35].

Compared with laser cutting manufacturing technique, cardiac implants made of stainless steel by AM is still in the early stage. As we can see in Table 5.3, the powder size is from 14 to 63 μm, and average value is about 30–40 μm since large particles and agglomerates formed by extremely small powders decrease the packing density in powder bed that cause undesired voids [36]. Besides, it is necessary to adopt fresh powders because contaminated powders usually lead to inferior tensile strength and elongation [37]. However, the lack of mechanical pressure makes SLM process to expose more defects than conventional counterpart like pores and cracks [38]. Thus, post heat treatment is recognized as a common way to achieve the high density that can substantially reduce the porosity as the heat treatment temperature increases because of the homogenized structure [39]. On the other hand, Z. Sun et al. [40] combined fast scanning speeds (625–3000 mm/s) with high power laser (380 W) to fabricate 316L samples that the density values of >99% and overall build rate were enhanced by 72% as compared to commonly used processing parameters (Fig. 5.4). The samples manufactured by SLM are significantly anisotropic as variable building direction and scanning strategy, coupled with layer by layer stacking. Suryawanshi et al. [41] evaluated and compared single melt (SM) and checker board (CB) laser scanning strategies with 316L. According to experimental results, substantial sub-grain microstructure in SLM sample resulted in significantly higher yield strength (536.4 MPa) than that of conventional manufacturing sample (220–207 MPa) while ductility markedly reduced. CB scanning strategy produced samples with better mechanical properties than that of SM whatever the tensile stress was parallel or perpendicular to scanning-transverse directions plane. Moreover, better properties were obtained in parallel direction than that in perpendicular direction. Shrestha et al. [42] investigated how layer orientation impacted the mechanical properties and fatigue life of 316L, and then they found that horizontally built specimens performed optimal fatigue resistance, followed by vertically built specimens, while the lowest fatigue resistance was obtained for diagonal specimens (Fig. 5.5). Sander et al. [43] found that anisotropy was also present in the corrosion

Table 5.2 The parameters of 316L stent electrochemical polishing

Electrolyte	Potential (V)	Current	Time	Temperature (°C)	Surfaces condition	References
40 ml H_2SO_4 (95–97%) + 45 ml H_3PO_4 (85%) + 14 ml H_2O	3.5	0.4 A	0.5 min	75	N/A N/A	[52]
150 ml H_2SO_4 (98%) + 390 ml H_3PO_4 (88%) + 1.5 g $Na_5O_{10}P_3$ + 60 ml H_2O	8	0.4 A	80 s	80–90	N/A N/A	[50]
50% H_2SO_4 + 50% H_3PO_4	9.5	0.44 A	180 s	75	N/A N/A	[51]
75 ml H_2SO_4 (96%) + 195 ml H_3PO_4 (88%) + 0.75 g $Na_5O_{10}P_3$	N/A	0.65 A	75 s	40	N/A	[53]
	N/A	0.5 A	60 s	75	N/A	
20% H_2SO_4 + 50% H_3PO_4 + 30% H_2O	N/A	8 A/cm^2	180 s	68	Surface roughness < 0.59 μm	[34]
35% H_3PO_4 (85%) + 50% $C_3H_8O_3$ + 15% H_2O	N/A	0.75 A		N/A	Surface roughness = 3–180 nm	[54]
20% H_2SO_4 (95–97%) + 60% H_3PO_4 (85%) + 10% $C_3H_8O_3$ 10% H_2O	4	N/A	3 min	65–70	Surface roughness = 77 ± 4 nm	[55]
40% H_2SO_4 + 60% H_3PO_4	5	0.25 A/cm^2	20	50	Surface roughness = 5.05 ± 0.28 nm	[35]

Table. 5.3 The process parameters and properties of SLM 316L

Powder size (μm)	Sample size	SLM process parameters				Mechanical properties			Corrosion behavior	References
		Laser power (W)	Wavelength (nm)	Scan speed (mm/s)	Layer thickness (μm)	YS (MPa)	UTS (MPa)	EL (%)		
10–53	5 mm × 10 mm × 50 mm	205	1070	960	40	577 (horizontal) 520 (vertical) 536 (diagonal) 311 (wrought)	688 (horizontal) 620 (vertical) 660 (diagonal) 641 (wrought)	65 (horizontal) 75 (vertical) 62 (diagonal) 70 (wrought)	icorr (μA/ cm^2): 0.019 (horizontal), 0.027 (vertical), 0.056 (diagonal), 0.059 (wrought) Epit (V): 1.01 (horizontal), 0.87 (vertical), 0.90 (diagonal), 0.25 (wrought)	[43]
22.5–56.4 42.83 μm (average)	Thickness is 2 mm	150	1090	600	40	488.9 (fresh powder) 436.3 (contaminated powder)	664.7 (fresh powder) 499.8 (contaminated powder)	29.3 (fresh powder) 13.7 (contaminated powder)	N/A	[37]
20–50	Plate dimensions of 89.5 × 89.5 mm^2	90	N/A	1000	30	511.6 (SS-SM-∥) 430.4 (SS-SM-⊥) 536.4(SS-CB-∥) 448.5 (SS-CB -⊥)	621.7(SS-SM-∥) 509.0 (SS-SM-⊥) 668.4(SS-CB-∥) 527.9 (SS-CB -⊥)	20.4(SS-SM-∥) 12.4 (SS-SM-⊥) 24.7(SS-CB-∥) 11.6 (SS-CB -⊥)	N/A	[41]

35–50	Circular disks with diameter of 15.8 cm and thickness of 0.5 cm	200	1060	N/A	30	N/A	N/A	N/A	The pitting potential values of AM 316L ss were −138.4, −96.2, −252.3 mV in serum, PBS, 0.9 M NaCl, respectively	[44]
20–100	N/A	70	N/A	417	30	~480	~600	~25	N/A	[56]

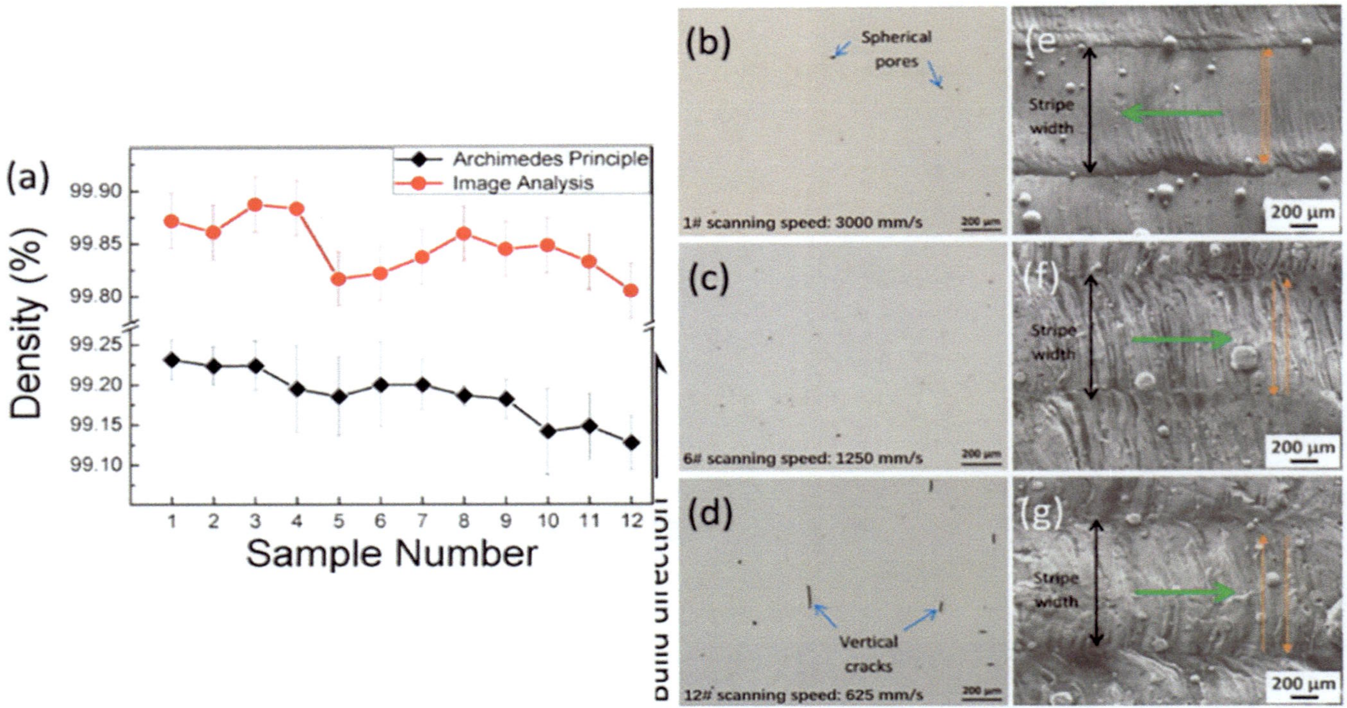

Fig. 5.4 The relationship between scanning speed and 316L ss sample's density with high power laser [40]: (**a**) Density graphs obtained from Archimedes and image analysis methods. The OM images of (**b**) 1#(**c**) 6# (**d**) 12# showing the pores. SEM images showing the topography of (**e**) 1# (**f**) 6# (**g**) 12#

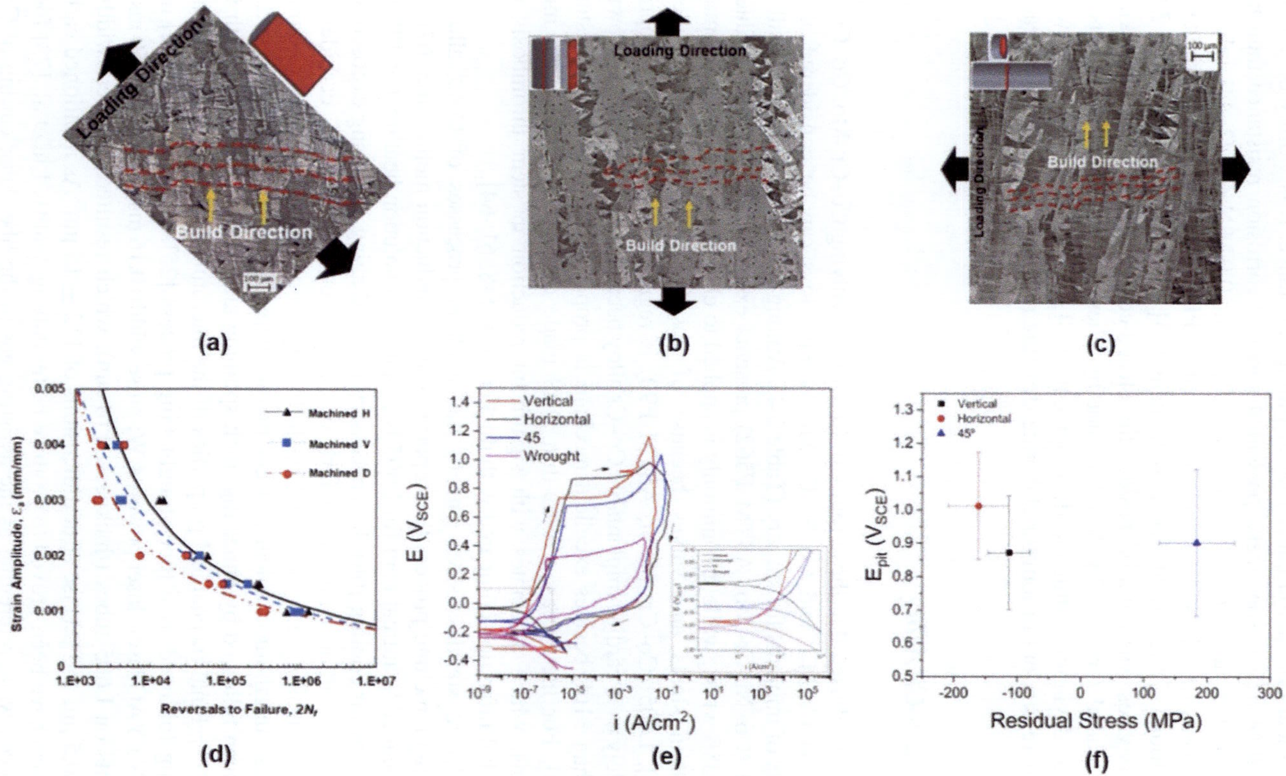

Fig. 5.5 The effect of build orientation on the fatigue and corrosion behaviors of SLM 316L [42, 43]: Optical micrographs showing microstructural features for (**a**) diagonal, (**b**) vertical, and (**c**) horizontal specimens. (**d**) Strain-life fatigue data, (**e**) cyclic-potentiodynamic polarization curves and for (**f**) pitting potential as a function of residual stress diagonal, vertical, and horizontal samples

process, which indicated that horizontal build orientation produced samples with the highest resistance against pitting but nanosegregation on the sub-cellular boundaries of the vertical orientation caused lower pitting resistance. It is worthy of note that though SLM 316 L specimens presented superior corrosion resistance than their wrought counterpart respectively in serum, PBS, and 0.9 M NaCl [43, 44], relatively high amounts of released Cr were observed in PBS+BSA probably attributed to protein-bound Cr for SLM 316 L specimens [45]. Up to now, a few SLM 316L cardiac devices was reported besides the closed-cell design stent (strut thickness, 200 μm; radial forces, 2.38–7.11 N) made by Wiesent et al. [46]. Size effects and geometric deviations limit the development of SLM 316L stents [47–49], which means there is a lot of room for elaborate structure research and clinical application.

5.1.2 Co–Cr Alloy

ASTM has advised that these four Co–Cr alloys involving Co–Cr–Mo, Co–Cr–W–Ni, Co–Ni–Cr–Mo, and Co–Ni–Cr–Mo–W–Fe alloys can be considered as the materials of implants for surgery (Table 5.4). Among these, for Co–Cr-based alloys, MP35N (Co–Ni–Cr–Mo, ASTM F562) presents excellent SCC resistance due to adding 35%wt Ni, while it commonly is applied to electrodes of cardiac pacemaker rather than stents and valves because of possible increasing nickel allergy [57, 58]. L605 (Co–Cr–W–Ni, ASTM F90) is regarded as the most appropriate and widely used cardiac implantable Co–Cr alloy because its Ni content (~10%wt) is lower than 316L besides excellent mechanical properties and corrosion resistance [59, 60]. For these reason, ultra-thin DES made of L605 have attracted much attentions, which associates with significant reduction in stent thrombosis and myocardial infarction compared with thicker struts [18, 61–64].

Similar to stainless steel, one of the smelting processes of CoCr alloys is electroslag remelting process that replaces vacuum induction melting [65–67]. The recommended heat treatment for L605 is to anneal or solution heat treat at 1200 °C followed by air cooling [68]. It is necessary to avoid aging and being exposed long time below solution temperature for CoCr alloys; otherwise excess carbides will precipitate, making adverse effect on following processes [69]. Besides common multi-pass thermomechanical process of fine wires and microtubes [70], such as cold-drawn followed by annealing, melt spinning also can produce wires with fine grains [71]. Final mechanical properties of the microtubes are achieved by vacuum annealing followed by a tube-straightening process [25]. Sudheer et al. [72] used pulsed Nd:YAG laser (laser power, 4 W; pulse width, 0.16 ms; dwell time, 50 ms) to cut stents on L605 tubes (thickness, 110 μm), which resulted in a kerf width of 22.5 ± 0.5 μm and precise strut dimensions of 112 ± 15 μm. As mentioned earlier, fiber lasers have been proved to produce better cutting quality. Liu et al. [73] found that oxygen was preferential as the assisting gas in the fiber laser cutting of CoCr stents under fiber laser microprofiling of CoCr tubes (peak power, 20 W; repetition rate, 6000 Hz; pulse width, 0.1 ms; speed, 4 mm/s). According to Table 5.5, like the

Table 5.4 Chemical composition and mechanical properties of biomedical Co–Cr-based alloys [84]

	Chemical composition (wt%)									Mechanical properties		
Material	Co	Cr	Mo	W	Ni	Mn	Si	Fe	C	YS (MPa)	UTS (MPa)	Fatigue strength (MPa)
Co–Cr–Mo (F75)	58.9–69.5	27.0–30.0	5.0–7.0	0.2	2.5	1.0	1.0	0.75	0.3	448–517	655–899	207–310
Co–Cr–W–Ni (F90)	45.5–56.2	19.0–21.0	/	14.0–16.0	9.0–11.0	2.0	0.4	3.0	0.15	448–1606	951–1896	586
Co–Ni–Cr–Mo (F562)	29.0–38.8	19.0–21.0	9.0–10.5	/	33.0–37.0	0.15	0.15	1.0	0.025	965–1500	1206–1795	500–793
Co–Ni–Cr–Mo–W–Fe (F563)	27.9–54.9	18.0–22.0	3.0–4.0	3.0–4.0	15.0–25.0	1.0	0.5	4.0–6.0	0.05	276–1310	600–1586	N/A

Table 5.5 The process of CoCr stent polishing

Acid pickling		Electrochemical polishing					
Solution	Processes	Electrolyte	Potential (V)	Current (A)	Time (s)	Temperature (°C)	Ref.
$HF + HNO_3$	Ultrasonic agitation	$HClO_4$: CH_3COOH: H_2O= 2:7:1	25	0.8	60 s	40–50	[72]
HF:HNO_3:H_2O = 2:15:83	Etching for 3 h at room temperature						[73]
HF:HNO_3:H_2O = 3:9:38	Ultrasonic bath at 55 °C for 1 h						[85]
	Ultrasonic agitation	H_2SO_4:H_3PO_4:H_2O = 9:10:1	0–15	1–1.5	15–30	55	[86]
$HF + HNO_3 + H_2O$	Ultrasonic bath at 5 °C for 200 min	HF:HNO_3:H_2O = 3:20:77	9	0.24	300	55	[87]
		0.3 M $(NH_4)_2SO_4$ aqueous solution with a pH of 5.3	450		480	750	[74]

pickling process used for stainless steel stents, such a kind of electrolyte containing HF and HNO_3 also can be considered as the mainstream candidate. In addition to ordinary electrochemical polishing processes, plasma electrolytic polishing is a potential polishing technique for CoCr stent because of simple electrolyte [74].

As mentioned earlier, the properties of SLM samples are influenced by various factors. Combining laser power, scanning velocity, distance between two consequent laser scans, and layer thickness, laser energy density (LED) is directly related to some of the features involved in the densification process such as density, porosity content, and defect formation mechanism [75, 76]. Tonelli et al. [77] found that LED well correlated to surface quality, microstructural features, and hardness and 150–200 J/mm^3 was the optimized range for high-quality Co28Cr6Mo SLM parts (Fig. 5.6). Li et al. [78] concluded that the powder size had a little effect on the microstructural and mechanical properties of the as-built SLM Co–Cr–W–Si parts while an apparent influence on those of the heat-treated ones. Post heat treatment, like solution and aging, generally causes a drastic reduction of UTS and YS and a remarkable increase of ductility [78, 79]. On the other hand, post heat treatment parts usually lead to lower corrosion resistance and higher ion release than as-built counterparts because of precipitates and more internal stresses generated from the water quenching [79–81]. Demir et al. [82] compared two different scanning strategies and found that concentric scanning had demonstrated lower average surface roughness in low peak power condition compared with hatching scanning (Fig. 5.7). Omar et al. [83] adopted hot isostatic pressing (HIP) as posttreatment, which enhanced the stiffness by 10% compared with as-built samples. However, the strut thickness about 200 μm did not bring CoCr alloy's superiority into full play. Finazzi et al. [12] reduced the strut thickness approximately to 90 μm though optimizing stent mesh and SLM parameters. The strength of mini tensile specimens with the same thickness to stents was comparable overall to the conventional L605 alloy. Moreover, SLM CoCr stents are promising in terms of their mechanical properties as their balloon expandability has been confirmed expanded from 0 to 8 bar, making SLM CoCr stents closer to conventional counterparts. Notably, although electrochemical polishing is effective in reducing the surface roughness, it could be further reduced to fit the requirements of cardiac devices.

5.1.3 *Shape Memory and Superelastic TiNi Alloy*

TiNi (Ti–50.0% at Ni) alloys are used in stents, occluders, and guidewires because of their irreplaceable shape memory effect and superelasticity, as well as good biocompatibility, radio-impermissibility, and MR compatibility [88–91]. TiNi alloys can realize reversible shape changes by heating and cooling between the shape of the parent phase of austenite at high temperature and that of martensite at low temperature, called two-way shape memory effect [92]. One-way shape memory effect is the recovery of large strains (up to 8%) mechanically created in the low temperature range through reorientation/detwinning [93]. Firstly, the shape of device is fixed

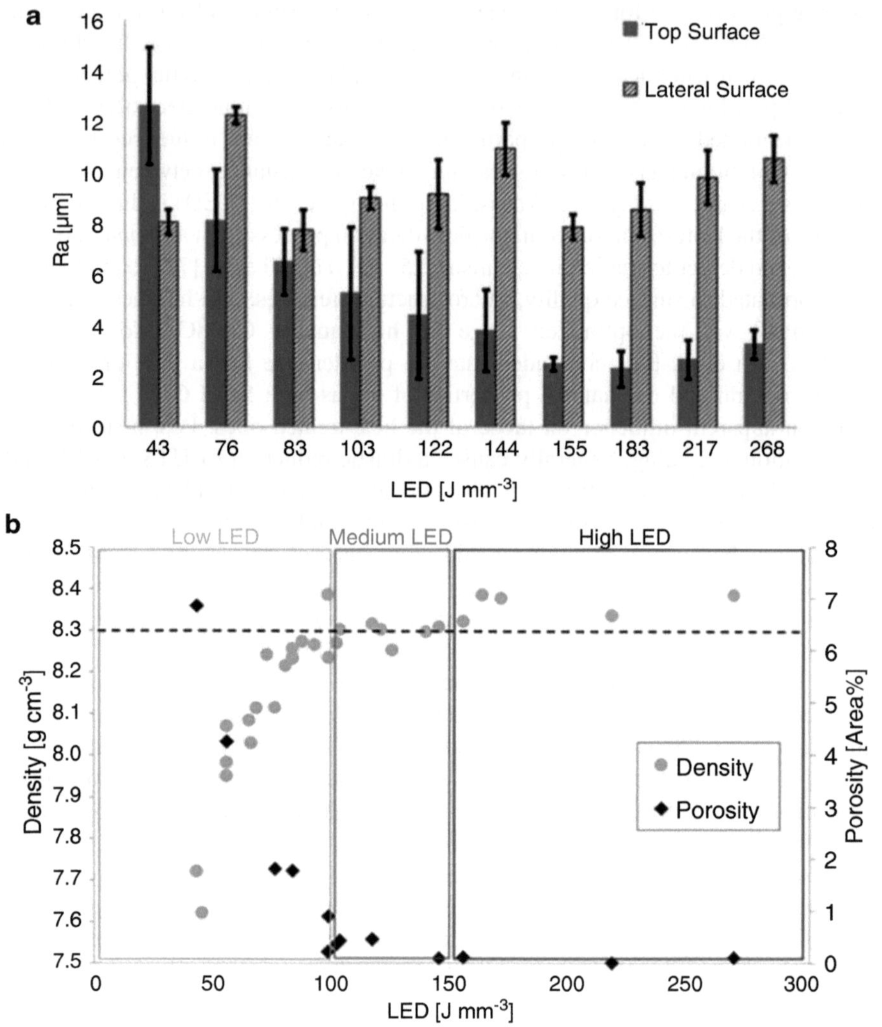

Fig. 5.6 (**a**) Average surface roughness (Ra), (**b**) density and internal porosity of SLM samples with increasing LED values [77]

above the austenite finish temperature (A_f). Cooling to the start of the martensitic transition temperature (M_s), thermoelastic martensitic transformation occurs, and the secondary forming is applied until the temperature is reduced to the martensite finish temperature (M_f). The recovery starts when the temperature rises to that for the start of austenite formation (A_s). These transition temperatures depend on the chemical composition and manufacturing processes. For example, a shift about 1% in the amount of either Ni or Ti in the alloy will result in a 100 °C change in the alloy's transformation temperature [94]. According to Ti–Ni phase diagram (Fig. 5.8), the presence of single-phase TiNi is recognized at near the equi-atomic composition, but

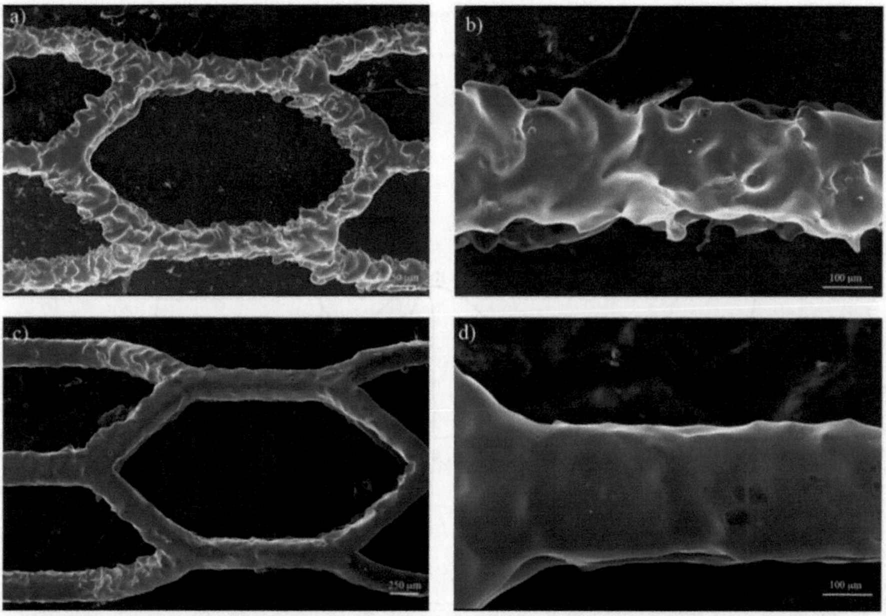

Fig. 5.7 SEM images of prototype stents after electrochemical polishing [82]. Stent realized with (**a**, **b**) hatching $P_{peak} = 180$ W and (**c**, **d**) concentric scanning, $P_{peak} = 40$ W

other precipitates like Ni_4Ti_3, Ni_3Ti, and Ni_3Ti_2 will alter the transformation temperatures [95]. The superelasticity allows TiNi alloys to recover virtually any shape as long as the deformations are below 10% above A_f, while surgical steel and Ti are limited to 1% elastic deformations [96].

From raw materials to devices, TiNi alloys conventionally entail arc or induction melting followed by a hot working process and machining to obtain the final shape. As already mentioned, the shape memory effect and superelasticity of TiNi are sensitive to components; hence the smelting process needs to focus on the control of impurity content to meet the medical standards (ASTM 2063). Two major methods of melting commonly used are vacuum induction melting (VIM), vacuum arc remelting (VAR), and electron beam melting (EBM) [93]. The VIM processes generally prefer graphite crucibles due to their easy handling and inexpensiveness, where the carbon content in the TiNi ingot lies between 200 and 500 ppm as a source of TiC contamination [97]. The VAR procedure obtains high homogeneity through multiple melting cycles without any crucible, while these cycles bring more oxygen and carbon if vacuum leakage occurs. Therefore, VIM and VAR are combined to refining. EBM process can reduce carbon content as low as 0.007% with a water-cooled copper crucible, but it is difficult to control the nominal chemical composition due to the high vacuum operation and heating temperature [98]. The respective value of tensile strength and elongation of TiNi ingots could not be lower than 551 MPa and 10% after annealing at 800 °C followed by rapid cooling [99].

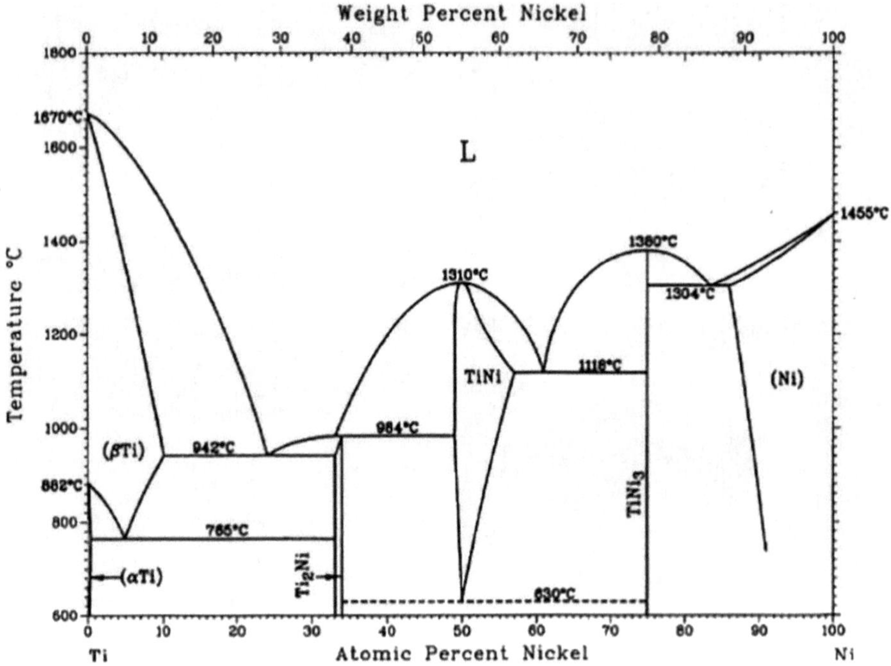

Fig. 5.8 Phase diagram of the Ni–Ti system [93]

The TiNi alloys in hot-deformation processes, like hot forging, rolling, and extrusion, are performed at temperature range of 600 °C to 900 °C, which are typically used to reduce the ingots' cross-sectional to less than 15 cm, because commercial TiNi ingots are generally 50 cm in diameter and weigh in excess of 2500 kg [100]. Dynamic recovery is dominant since the grains are obviously elongated at 600 °C, and then almost complete dynamic recrystallization occurs at 700 °C. When the temperature beyond 800 °C, TiNi alloys exhibit complete dynamic recrystallization, which are sufficient to lower the flow stresses to allow relatively large deformation steps [101]. Semi-finished products, such as wires, tubes, and sheets, are produced via cold working processes that almost determine the final shapes and refined microstructures. Among them, TiNi wires are the most widely available profiles through cold drawing. TiNi alloys work-harden quite severely, so interannealing at 600–800 °C during the drawing process is necessary [100, 102]. Compared to metallurgy processes, Saito et al. [103] adopted clad-chip extrusion (CCE) procedure and obtained almost 30 μm TiNi single-phase wires by 1173 °C heat treatment, but its elongation was lower (Fig. 5.9). Wires and tubes usually need continuous strand straightening at 450–550 °C, leading to 2–3% strain during the straightening process due to the shape memory effect [104]. The wires can be knitted directly into occluders and woven-type stents, and further laser cutting was needed for the tubes [105, 106]. Before the fabrication of devices, mechanical polishing is applied to wires and tubes. After laser cutting or weaving, complicated

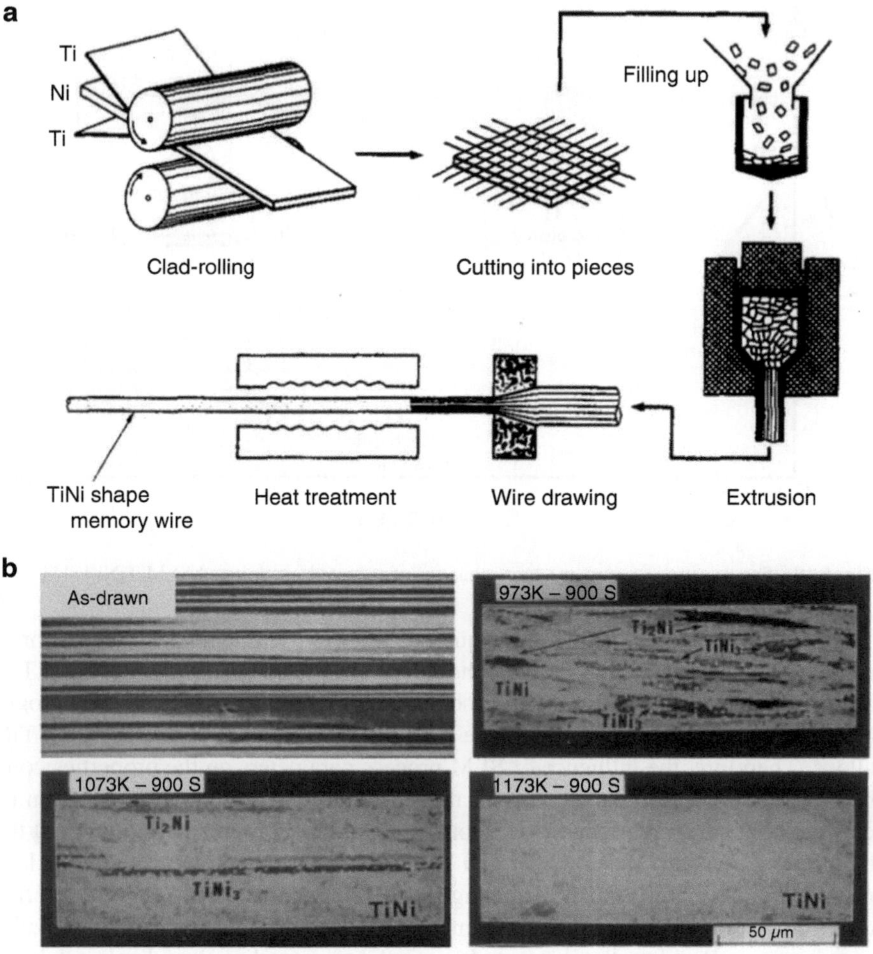

Fig. 5.9 Drawn fabrication process of TiNi wire by the CCE method. (**a**) Schematic diagram of CCE procedure. (**b**) Phase composition via different heat treatment [103]

structures need to be treated with pickling and electrochemical polishing to achieve smooth and clean surfaces. The mixed aqueous solution of hydrofluoric acid and nitric acid is often used for pickling and higher concentrations and shorter reaction time, performing better effect within a certain range [107–109]. Zhao et al. [110] selected an electrolyte containing 5 mL perchloric acid (70%) and 100 mL acetic acid (99.8%) and optimized electrochemical polishing parameters for TiNi stents (20 V; 0.15 A; 1.5–2 min). Lopes et al. [111] investigated the effect of a 3.5 mol•L^{-1} methanolic H_2SO_4 electrolyte on superelastic and shape memory TiNi alloys and found that electrochemical polishing effectively reduced the surface roughness of both superelasticity and shape memory effect of TiNi alloys and improved corrosion resistance by removing superficial nickel-rich layers.

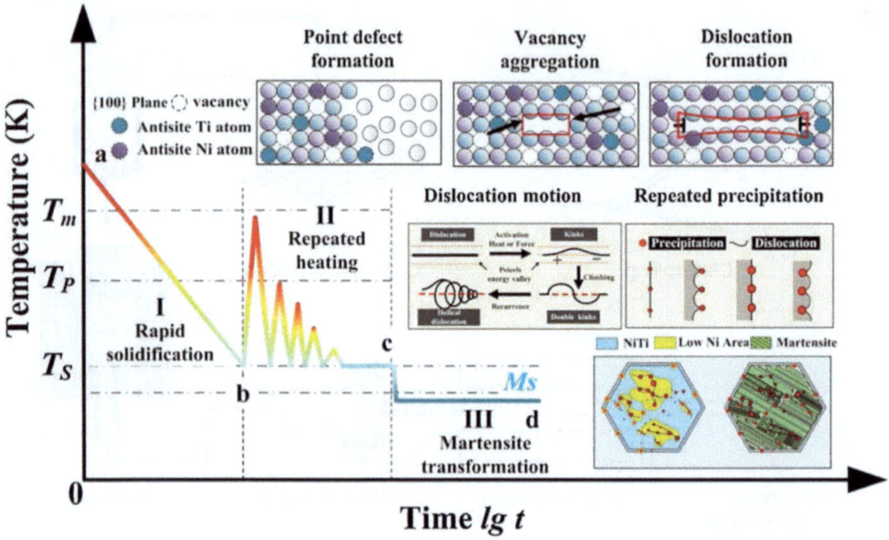

Fig. 5.10 The process of microstructures changes of repeated precipitated SLM TiNi [115]

Up to now, although SLM technique has not yet been applied in fabrication of cardiac TiNi alloy devices, the researches on properties and parameters of SLM TiNi alloys have also been discussed. As mentioned before, the processing technologies have a dramatic impact on the shape memory effect and superplasticity of TiNi alloys. Therefore, the influence of SLM process parameters on the properties could be considered because the metallurgical process of SLM is quite different from the conventional smelting processes. Dadbakhsh et al. [112] found a curious result that high laser parameters (laser power 250 W, scanning speed 160 mm/s) and low laser parameters (laser power 40 W, scanning speed 110 mm/s) made no significant effect on composition, but high laser parameters lowered the austenite transformation temperatures. Besides, they reported vertical building direction displayed better shape memory effect than that of horizontal and incline samples, implying that different scanning directions will affect grain orientation [113]. Moreover, Qiu et al. [114] also proved that vertical building samples had higher corrosion resistance, because smaller grain was beneficial to the formation of uniform passivation film. Zhang et al. [115] employed a stripe rotation scanning strategy to make distinctive microstructures (Fig. 5.10), which fabricated high ductility (15.2 ± 0.8%) parts with 98% strain recovery when pre-deformed to 4% (laser power 120 W, scanning speed 500 mm/s, hatch spacing 80 μm) (Fig. 5.10). Moghaddam et al. [116] conducted the SLM TiNi samples whose strain recovery rate reached 98% under 5.62% pre-deformation without heat treatment (laser power 250 W, scanning speed 1250 mm/s, hatch spacing 80 μm). However, there is still a lot of room for improvement in shape memory effect and superelasticity of SLM TiNi devices compared to conventional parts.

5.1.4 Noble Metals

The noble metals of gold, silver, and platinum are often used to fabricate physiological electrodes and radiopaque markers because of their excellent corrosion resistance, electrical conductivity, radiation resistance, and easy processing [117–121]. The processing technology of noble metal equipment is the same as that of ordinary biomedical metal, which is also manufactured through smelting, heat treatment, pressure processing, and other methods [122, 123]. However, the extraction method of noble metals is very complicated because of rare reserves and the stable chemical properties. Taking gold for example, the average concentration of gold in the world is about 0.005 g/t which means that it is usually required to increase by a factor of 3000–4000 to reach commercial concentrations during gold ore formation [124]. The following paragraph will take gold as an example to introduce the extraction of noble metals.

There are many methods to extract gold, such as gravity concentration, flotation, cyanidation, pre-treatment, non-cyanidation leaching and amalgam, and so forth [125]. The specific method to choose depends on the chemical composition of the ore, occurrence state of gold, and the requirements for the product. For free milling of gold ores, cyanidation is recognized as the most mainstream wet gold extraction method, the principle of which is that Au element is first oxidized to Au^+ complexing with CN^- into the solution and finally be precipitated out of gold from the leaching with zinc powder or electrodeposition [126]. For example, hydrogen peroxide assistant leaching method promotes the gold leaching rate more than 98% by adding hydrogen peroxide as the feeding source, which shortens leaching time and reduces the consumption of cyanide [127]. In addition to cyanidation, gravity concentration processes and grinding have also been used in gold extraction. Gravity concentration processes perform separation based on the difference in density of mineral particles in the ore, which does not consume chemical agents but only for coarse separation because of its small processing capacity and poor sorting ability [128]. Flotation method produces gold minerals attaching to air bubbles and separated from other minerals by adding chemical reagents followed by intense agitation and air sparging [129]. With the large-scale mining, the rich ore and the easily treated ore decrease year by year, and the refractory ore has become the main ore resources of the gold industry [130]. Refractory gold ore generally refers to the ore of which the direct gold leaching rate is lower than 80% when gold is extracted by conventional cyanidation method, owing to the preg-robbing of dissolved Au species ($Au(CN)_2^-$) by carbonaceous matter and encapsulation of the gold in sulfide minerals [131]. Therefore, pre-treatment is needed to improve the gold leaching rate [132]. The pre-treatment methods of gold ore are mainly roasting oxidation method, biological oxidation method, and chemical oxidation method [133]. The roasting method exposes the gold by heating the ore at high temperatures to decompose the arsenic and sulfides and inactivate the carbon-containing materials. The fabrication process is relatively simple, but what followed is the production of serious environmental pollution. Microwave and additives are introduced to improve the high

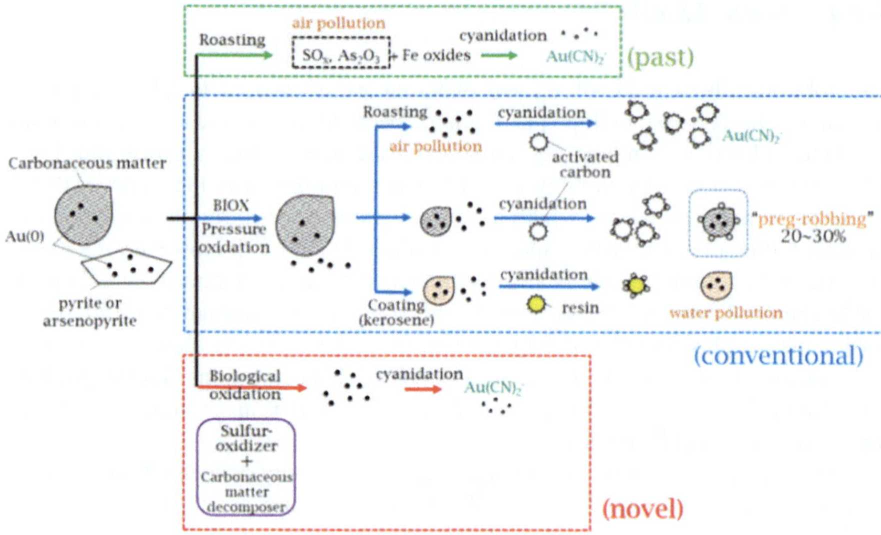

Fig. 5.11 Environmentally friendly biological treatment for gold extraction [137]

thermal efficiency and exposure area of gold in the roasting method [134, 135]. The chemical oxidation method oxidizes and removes impurity elements through oxidants under normal pressure or pressure conditions, and oxidants include ozone, peroxide, permanganate, chlorine, perchlorate, hypochlorite, Fe^{3+}, oxygen, etc. [130]. In acidic medium, the gold leaching rate increased from 19.5% to 96.5% by ozone and Fe^{3+} oxidation [136]. Chemical methods inevitably produce harmful substances containing arsenic and sulfur. Hence, it is necessary to develop green and low-cost technologies for extracting gold from refractory minerals. Biological treatment has become a hot spot, because it has the following characteristics: (1) mediating oxidation the dissolution of sulfide matrix, making gold easier to be leached by chemical impregnation; (2) stabilizing gold by secreting ligands to form gold-rich complexes and/or colloids; (3) producing potential alternative gold leaching agents; and (4) precipitating gold by gold-bound ligands or biosorption [137]. Compared with traditional treatment methods, biological treatment methods cause almost no air and water pollution (Fig. 5.11) [137]. *Acidithiobacillus ferrooxidans* produced a bio-oxidation stage that showed gold extractions four times as large (85%) as that of the initial samples [138]. Thermophilic archaeon *Acidianus brierleyi* combining white-rot fungus *Phanerochaete chrysosporium* significantly improved the gold recovery from 25% to 92% (Fig. 5.12), in which *Acidianus brierleyi* oxidized sulfides and *Phanerochaete chrysosporium* decomposed carbonaceous matters [139]. Microorganisms producing cyanide and thiosulfate are regarded as potential bioleaching agents, such as *Pseudomonas fluorescens*, *Pseudomonas aeruginosa*, *Alkaliphilic bacillus* spp., and *Streptomyces fradiae* [140]. Notably, it is necessary that screening the targeted microorganisms for different types of minerals and process links to shorten the biological treatment cycle

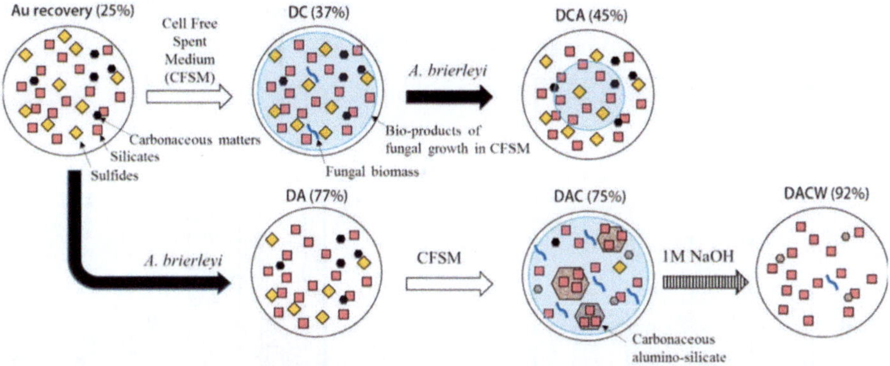

Fig. 5.12 Environmentally friendly biological treatment for gold extraction [139]

and improve gold recovery rate [141]. In order to avoid the toxicity and environmental pollution caused by cyanidation, cyanide-free gold leaching agents are being developed, including thiourea leaching, thiosulfate leaching, chlorination, polysulfide method, bromination, iodization, and sulfur mixture method [142–145].

5.2 Manufacturing Methods of Inorganic Cardiac Materials for Implant

Inorganic materials such as pyrolytic carbon and ceramics are often used in the surface coatings of cardiac devices because of their high hardness and wear resistance, leading to machining difficulties. Although commercial fully pyrolytic carbon artificial valves and single-crystal alumina disk are available, inorganic materials are mainly used in cardiac devices in the form of coating. The commonly used deposition techniques mainly include physical vapor deposition and chemical vapor deposition. The specific deposition methods will be described in detail later. According to the different material characteristics, the preparation process of pyrolytic carbon and diamond-like carbon (DLC) films will be introduced as follows.

5.2.1 Manufacturing Methods of Pyrolytic Carbon

Pyrolytic carbon is a kind of carbon material formed by dehydrogenation of gaseous hydrocarbons on the surface of hot matrix, divided into granular carbons, laminar carbons, and isotropic carbons according to their structures. Low-temperature isotropic carbon (LTIC) is directly used to make the part or coating of mechanical heart valves through chemical vapor deposition (CVD) because of its good compatibility with blood and proteins. CVD can introduce the vapor containing gaseous or liquid

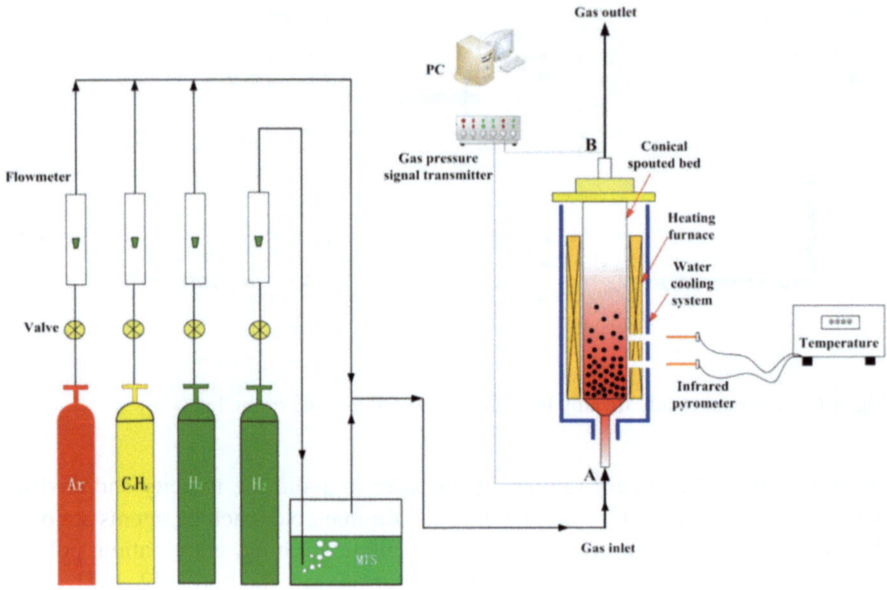

Fig. 5.13 Schematic diagram of FBCVD system [152]

reactant which constitutes the target film layer element and other gases needed for the reaction into the reaction chamber to form the film through chemical reaction on the surface of the devices. This technology has been widely used in the preparation of inorganic functional films due to many advantages, for instance, the produced film is dense and uniform with high stability, the strong adhesive force can be obtained between film and substrate, film composition can be easily controlled, the deposition speed is fast, and it is easy to achieve mass production. Fluidized bed chemical vapor deposition (BFCVD) is the most commonly used technology to prepare pyrolytic carbon (Fig.5.13). Droplet model is considered as the most commonly used model that formed by the condensation of large hydrocarbon molecules [146]. Raw materials for pyrolytic carbon include methane, propane, acetylene, benzene, liquefied petroleum gas, or liquid hydrocarbons, which can be used alone or in combination. In addition, inert nitrogen is needed as a carrier gas to remove the hydrogen gas produced in the reaction. According to the different status of materials, coating or bulk, researchers have improved the process parameters (Table 5.6). Serino et al. [147] found that setting a slower deposition rate (0.021 μm/s) during the first hour of deposition can produce highly homogeneous layers on the graphite substrates. Sheikholeslami et al. [148] utilized camphor ($C_{10}H_{16}O$) as a natural and renewable carbon precursor to manufacture nanopyrolytic carbon coating on smooth graphite and quartz in a low temperature range from 650 to 800 °C. This coating compared to St. Jude commercial mechanical heart valve (coated with pyrolytic carbon) a smaller number of platelets sticked to it. Xu et al. [149] introduced boron into the precursor and found boron-doped pyrolytic carbon has good isotropic microstructure. Tan

Table 5.6 The process parameters of producing pyrolytic carbon by CVD method

Deposition temperature (°C)	Gases composition	Flow rates	Time (h)	References
1300	Propane, methyltrichlorosilane, and nitrogen	N/A	8	[147]
650–800	Camphor and nitrogen	Camphor: 15 ml/min	5	[148]
1250	Methane, boron bromide, hydrogen, and argon	Methane: 200 ml/min boron bromide: 125 ml/min hydrogen: 400 ml/min	1-2	[149]
1550	Propylene, hydrogen, and argon	$V_{Prolylene}$:$V_{Hydrogen}$:V_{Argon} = 1:15:10	30	[150]

et al. [150] created a "litchi" hierarchical structure pyrolytic carbon that possessed superhydrophobic properties and resistance to *E. coli* and protein adhesion. After the deposition, Al_2O_3 and SiC were used to remove excess parts and polish the surface. To remove residual polish or contaminants, a cleaning process was required before the final assembly of the heart valve components. Smith et al. [151] introduced a typical cleaning process:

1. Ultrasonic cleaning in water-soluble detergent for 5 min at 60–65 °C.
2. Ultrasonic stirring in Aerosol OT for 5 min, followed by enhanced water rinsing. The flyers were rinsed by ultrasonic stirring for 5 min at room temperature.
3. Deionized water irrigation at 75–80 °C.
4. Ultrasonic stirring with isopropanol for 5 min at room temperature followed by air-drying. Cardiac valve was sterilized at 120° for half an hour after assembly.

5.2.2 Manufacturing Methods of DLC Film

DLC film has been applied to commercial stents and valve coatings. Although DLC materials have similar properties to diamond, such as high hardness, low friction coefficient, and high biological stability, the structures of DLC materials are usually amorphous, where two carbon-carbon bonds, sp^3 that occurs in diamond and sp^2 form found in graphite, are both contained carbon-carbon bonds. Therefore, the performance of DLC film can be adjusted flexibly by doping Si, O, N, Cr, and Ti to change the sp^3:sp^2 ratio [153]. Various techniques have been developed for DLC deposition, such as plasma-enhanced chemical vapor deposition (PECVD), magnetron sputtering, cathodic-arc deposition [154], etc. PECVD performed better adhesion than other CVD techniques through initial plasma surface treatment. Cho et al. [155] found that when PECVD bias power was 30 V with a film thickness of 150 nm, the corrosion rate (8.05 nm/year) could be significantly improved without affecting the biocompatibility. Jo et al. [156] deposited Ti-DLC on the Ti platen by

PVD/DC-PECVD. The bonding force increased between Ti-DLC and the metal substrate with the increased Ti content because of improvement of the film's electrical conductivity. Magnetron sputtering is a type of physical vapor deposition that does not require precursors. As long as a target can be fabricated, almost materials can be deposited. Non-conductive materials can be sputtered by means of radio-frequency (RF) power. Bociaga et al. [157] sputtered Si-DLC on the 316L multi-target DC-RF magnetron sputtering method. The increase of Si concentration inhibited platelet adhesion and decreased platelet activation level. Patnaik et al. [158] utilized DC magnetron sputtering prepared Ag-doped Cr/DLC coatings on 316L. Alloying with Ag can enhance elongation rate of the film close to 33%, facilitating the release of internal stress. Filtered cathodic vacuum arc (FCVA) method, as a kind of plasma assisted deposition techniques, is suitable to produce ultra-hard films, like virtually hydrogen-free DLC coatings [154]. Xi Rao et al. [159] found that the faster N_2 flowed, the higher C–C sp^2/sp^3ratio (1.09–2.66) was in FCVA system. Orrit-Prat et al. [160] fabricated bactericidal Ag (2 at.%)-DLC on CoCrMo substrates by pulsed filtered cathodic arc co-deposition using pure graphite and silver targets, showing some fullerene-like structures or carbon nano-onions.

5.3 Manufacturing Methods of Organic Cardiac Materials for Implant

Organic materials are indispensable to both non-implanted and implantable devices for cardiac field. For those natural polymer materials that are derived from animals and plants, the immunogenicity needs to be eliminated, and the mechanical properties need to be improved in the manufacturing procedure. However, synthetic polymers are firstly polymerized into polymer materials using chemical monomer and then processed into medical devices by polymer processing technology. According to the structural changes from monomer to polymer, polymerization reactions are classified into addition polymerization and condensation polymerization [161]. For addition polymerization, monomers form polymers through addition reactions without loss of small molecules. Many non-degradable polymers are made by addition polymerization, because plenty of heat would be released when π-bond of the monomer is converted to a sigma bond of the polymer, promoting the rapid synthesis of high molecular weight chain [162]. For condensation polymerization, functional groups contained in the monomer are connected with each other to form condensation polymers. In addition, there are by-products such as water, alcohol, ammonia, or hydrogen chloride. The monomer used for condensation polymerization has two or more functional groups, such as –OH, $–NH_2$, or –COOH, rather than double bonds. Generally, the two functional groups in the monomer are condensed to form chain polymers, while the monomer containing more than two functional groups is used to produce cross-linked polymers [161]. Polymerization methods can be roughly divided into homogeneous polymerization and heterogeneous polymerization. In

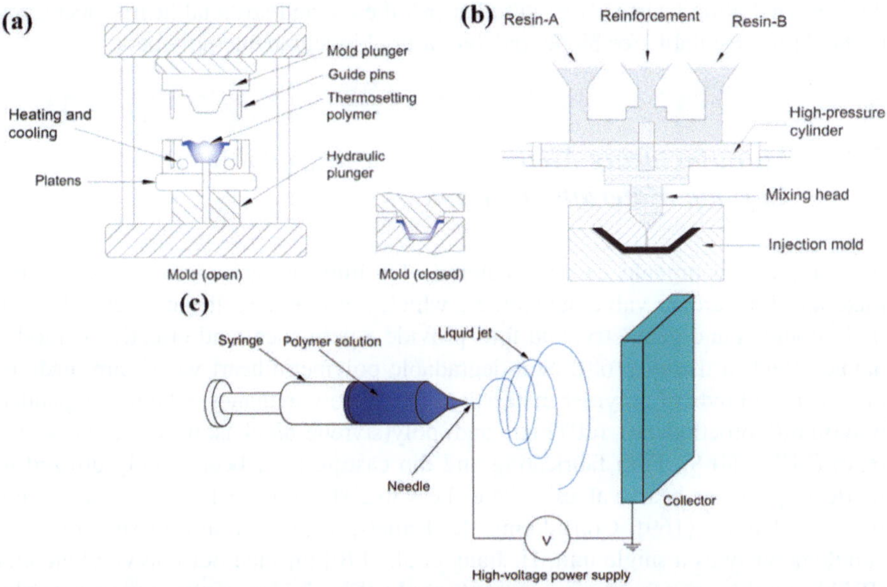

Fig. 5.14 Common polymer processing technology: (**a**) compression molding, (**b**) injection molding, (**c**) electrospinning [162]

the former, the reactants and products always exist in the same phase including bulk polymerization and solution polymerization, while in the latter, the polymer generated new phase and precipitates, such as suspension polymerization, emulsion polymerization, and precipitation polymerization [163]. Different types of biomedical devices and their components are suitable for different processing methods. Biomedical polymers are generally shaped through compression, extrusion, blow, injection molding, and spinning processes (Fig. 5.14), whatever biomedical films or shaped objects [162]. The basic compression process is to heat a thermosetting polymer material in a closed cavity under high pressure until the polymer is cured into a desired shape [164]. Extrusion molding is an economical method of producing polymer profiles, where the polymer materials are fed into a heating cylinder and then the melted polymers are further extruded or introduced into the mold through the rotating screw [165]. Blow molding, by injecting compressed gas into a molten polymer parison, makes polymer parison expand to contact with mold wall, and the devices will form after subsequent cooling process [166]. Injection molding is a method of high-speed manufacturing of thermoplastic and thermosetting polymer devices. The polymer powder is melted and injected into the die cavity through a nozzle with the aid of pressure, forming after cooling or solidification [166]. The spinning process is used to produce implants made of polymer fibers, which firstly sprays out polymer fibers and then moves a nozzle to print the desired shape [166, 167]. The following will introduce the production of organic heart repair devices, including heart valves, vascular prostheses, and artificial hearts, according to the devices suitable for biodegradable or nonbiodegradable polymers.

5.3.1 Manufacturing Methods of Non-degradable Organic Cardiac Materials for Implant

Non-degradable organic cardiac materials for implants are regarded as suitable materials for cardiac valve substitutes, which are closer to the natural valves in both modulus and geometry, and thus provide a permanent and effective hemodynamic effect in theory [168]. Non-degradable polymeric heart valves are made of one or more kinds of polymer materials, such as polyurethanes, silicone, expanded poly(tetrafluoroethylene) (ePTFE), and poly(styrene-*block*-isobutylene-*block*-styrene) (SIBS) [168]. Film fabricating and dip casting have been widely utilized to fabricate polymer heart valves, where the valve stents were fabricated separately from the leaflets [169]. Considering the leaflets, stent, and sewing ring function simultaneously as a single unit, H. Jiang et al. [170] injected hot polyvinyl alcohol (PVA) solution into the designed cavity mold (Fig. 5.15a) of the static assembly, which was sealed and clamped followed by immersion in the water bath. Upon removal from the mold after controlled freezing (at −20 °C) and thawing (at +20 °C) cycles, the small remnants of material were removed and the leaflets separated as shown in Fig. 5.15b. Cavallo et al. [171] fabricated a minimally invasive aortic valve replacement consisting of an external metallic stent and a valve leaflet using thermoplastic silicone-polycarbonate-urethane. The fabrication of valve leaflets process adopted a molding method with combination of dip coating and spray techniques that comprises the following steps (Fig. 5.16): The auxiliary molding stent was positioned on the tubular mold and dipped in 3% CarboSil solution of tetrahydrofuran/dioxane 1:1 for 10 times. The polymer was deposited into a tri-leaflet tubular valve by spraying a 2.5% CarboSil solution in tetrahydrofuran/dioxane 1:1 solution onto a tubular mold (400 cycles, flow rate of 2 ml/min, pressure 12 psi, translation

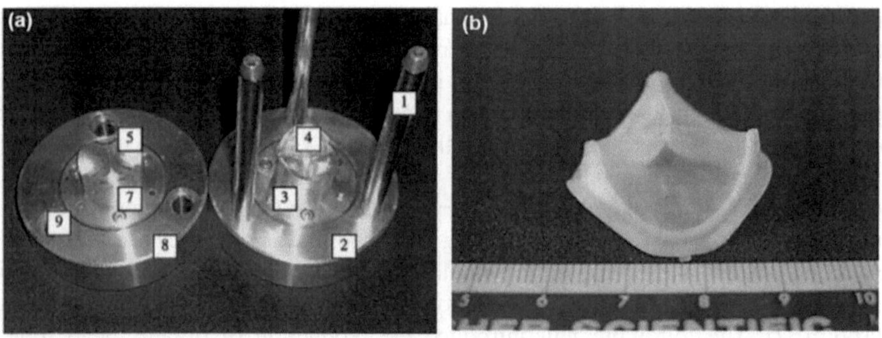

Fig. 5.15 Integrated PVA valve fabrication: (**a**) cavity mold, (**b**) PVA prototype [170]

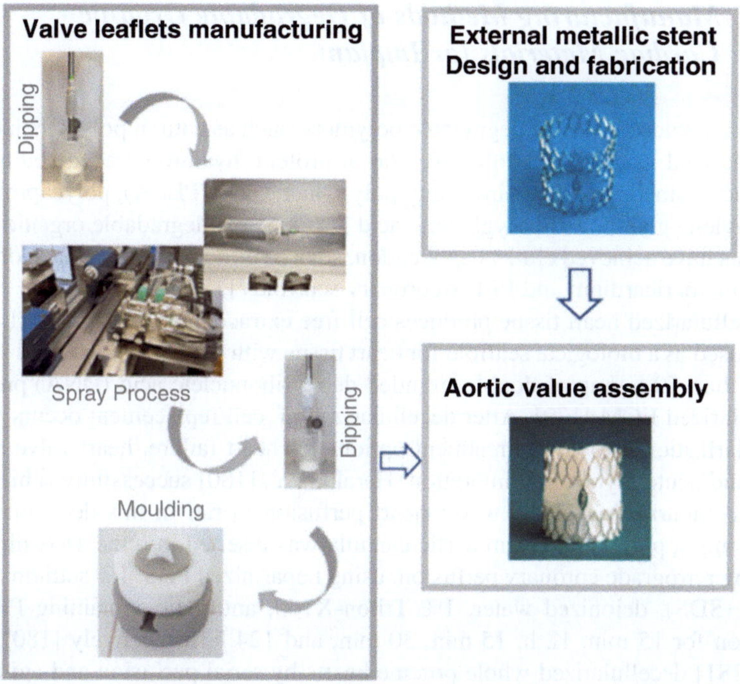

Fig. 5.16 The dipping-spraying-molding manufacturing processes of nonbiodegradable polymeric valve [171]

speed of 23.3 cm/min, and rotation speed of 88 RPM). After 2 h, the valve was dipped again in 3% CarboSil solution for 10 times. Finally, cutting the stent base ring, the valve was consolidated by molding in 40 °C distilled water and assembled with the metal stent after removing the excess material on the leaflets by thermal precision cutting equipment at 110 °C.

Non-degradable polymer materials are more suitable for manufacturing vessels with large diameter, while the fabrication of small diameter vessels also has been explored in recent years [172]. Seifalian et al. [173] found that when poly(carbonate-urea) urethane polymer solution is extruded into coagulant with a rotating mandrel head, the microporous structure was generated. In order to prepare the bio-functionalized ePTFE vascular prosthesis, Wang et al. [174] prepared a PTFE small-diameter vascular prosthetic through the paste extrusion and expansion processing, where fully mixed bio-functional molecules with PTFE powder were compressed into a preformed mold and extruded followed by expanding three times under 900 mm/min stretching (expansion) rate at 40 °C for 5 min. He et al. [175] fabricated a small-diameter polyurethane vascular prosthesis by electrospinning (voltage 22 kV, tip-mandrel distance 17 cm, solution flow rate 1 ml/h, electrospinning time 1–4 h, and mandrel rotation speeds 500–1500 r/min), which showed significantly higher cell proliferation rate than the PTFE group.

5.3.2 Manufacturing Methods of Degradable Organic Cardiac Materials for Implant

There are a wide variety of degradable polymers, such as natural polymers including decellularized matrix, silk, collagen, fibroin protein, hyaluronic acid, sodium alginate, and synthetic polymers including poly-L-lactic acid (PLLA), polycaprolactone, polyethylene glycol, and polyglycolic acid [176]. Some degradable organic cardiac materials have achieved clinical application, such as biological valve made of bovine or porcine pericardium and PLLA coronary scaffolds [177, 178].

Decellularized heart tissue produces cell-free extracellular matrix (ECM), which can be used as a biological scaffold for heart tissue with the widely accepted standard of less than 50 ng/mg of double-stranded deoxyribonucleic acid (DNA) present in decellularized ECM [179]. After decellularization, cell replacement occurs through recellularization, providing treatment options for heart failure, heart valve replacement, and acute myocardial infarction. Harald et al. [180] successfully achieved the first decellularization of the whole-heart perfusion in rat. In this decellularization processing, a prefilled 1.8 mm aortic cannula was inserted into the ascending aorta allowing retrograde coronary perfusion, using heparinized PBS, 1% sodium dodecyl sulfate (SDS), deionized water, 1% Triton-X100, antibiotic-containing PBS, and perfusion for 15 min, 12 h, 15 min, 30 min, and 124 h, respectively [180]. Methe et al. [181] decellularized whole porcine hearts by serial perfusion and agitation of hypotonic solution, repeating eight times in this order −4% SDS, agitation with SDS, 1% Triton X-100, and perfusion with Triton X-100. Guyette et al. [182] decellularized human hearts with only 0.95% DNA content remaining by a chemical perfusion with 1% SDS for 168 h, deionized water for 24 h, 1% Triton X-100 for 24 h, and final PBS wash for 168 h.

At present, electrospinning technique is widely used to fabricate biodegradable small-diameter vascular prostheses to promote the adhesion, proliferation, and growth of cells, owing to the similarity to the construction of extracellular matrix (ECM) for native blood vessel [183]. Ju et al. [184] fabricated bilayer vascular prostheses by electrospinning a polymer blend of PCL and type I collagen (Fig. 5.17). In this electrospinning processing [184], the inner layer was made of 5% w/v PCL/collagen blend solution with electrospinning parameters as an 18 gauge blunt tip at 3 mL/h flow rate and 10 cm distance, while the outer layer was subsequently electrospun through a 16 gauge blunt tip at 10 mL/h flow rate and 15 cm distance, with 750 μL of 18% w/v PCL/collagen blend solution. After the electrospinning procedure, the PCL/collagen vascular prostheses were cross-linked in vapor of a 2.5% glutaraldehyde solution at room temperature for 8–10 h. Kuang et al. [185] combined conjugate electrospinning with freeze-dry technology to produce poly(l-lactide-co-caprolactone) (PLCL)-based vascular prostheses (Fig. 5.18). In conjugated electrospinning equipment, two bundles of PLCL nanofibers carrying opposite charges were sprayed out from two needles that were applied to high voltage static current of opposite polarity, and the fibers were intertwined in the air because of electrostatic interaction the. In order to mimic the

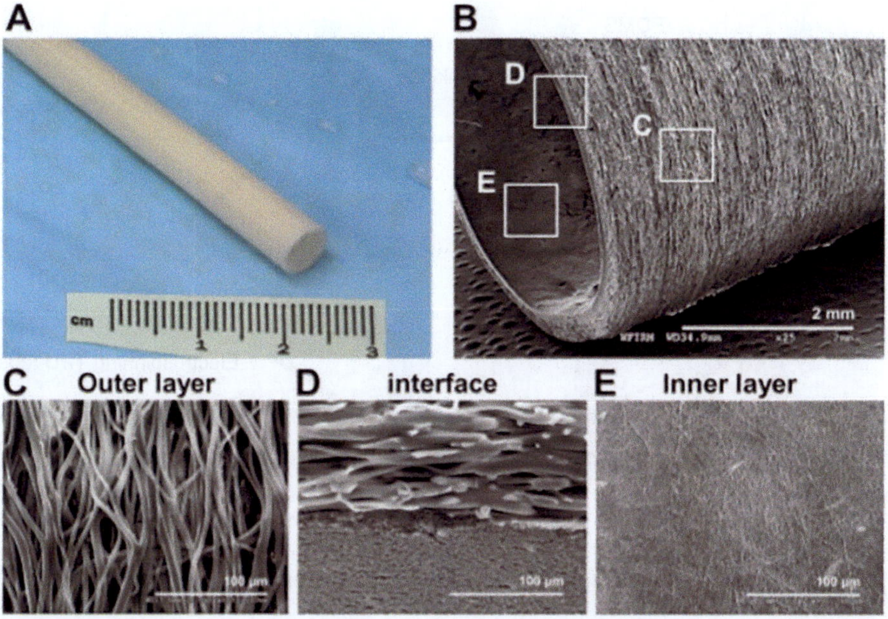

Fig. 5.17 The bi-layered PCL/collagen vascular prostheses fabricated by electrospinning [184]

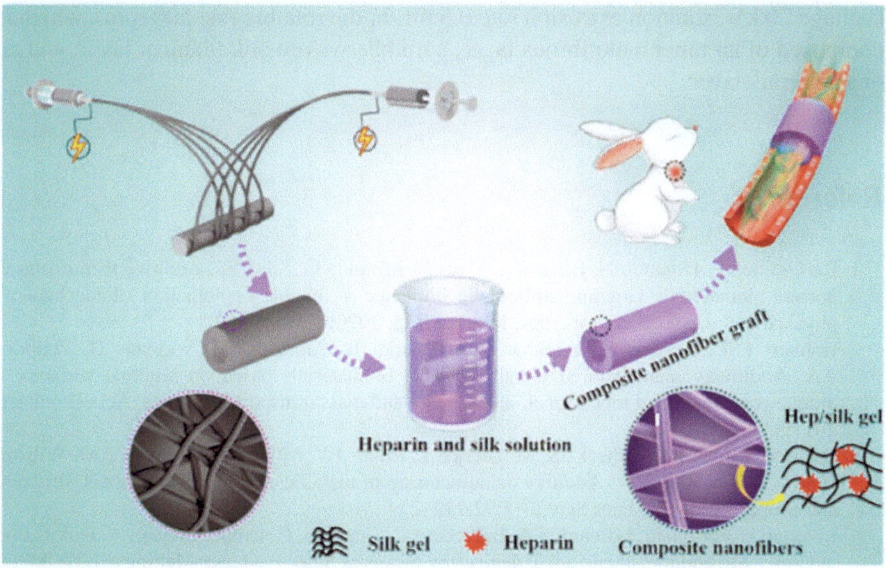

Fig. 5.18 The schematic diagram of conjugate electrospinning with freeze-dry technology [185]

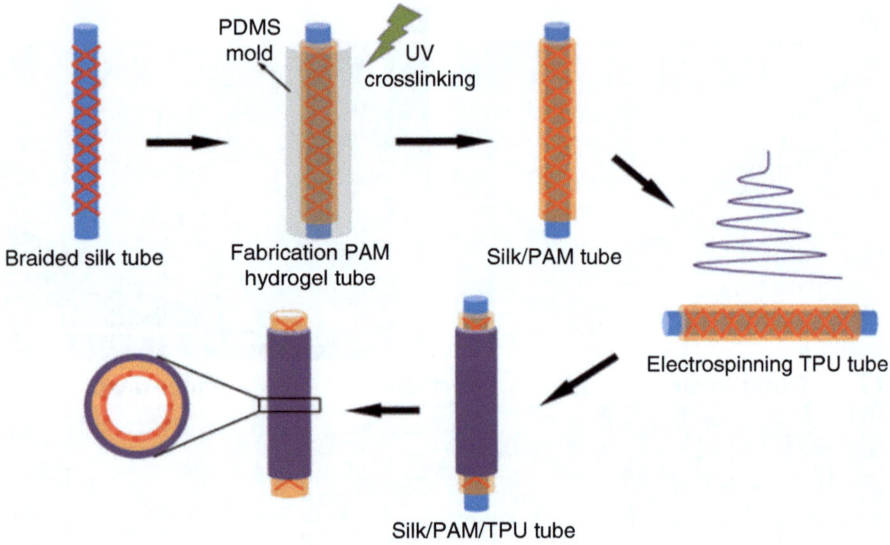

Fig. 5.19 The braiding-electrospinning procedure for triple-layered small-diameter vessel [186]

structure of native blood vessels, Mi et al. [186] fabricated triple-layered small-diameter vascular prostheses (Fig. 5.19) by combining braiding and electrospinning (voltage 20 kV, solution extrusion rate 0.5 mL/h, the rotating rate 500 rpm), which is composed of an inner nanofibrous layer, a middle woven-silk filament layer, and an outer porous layer.

References

1. Bartolomeu F, Dourado N, Pereira F, Alves N, Miranda G, Silva FS. Additive manufactured porous biomaterials targeting orthopedic implants: a suitable combination of mechanical, physical and topological properties. Mat Sci Eng. 2020;C-Mater 107
2. Bobbert FSL, Lietaert K, Eftekhari AA, Pouran B, Ahmadi SM, Weinans H, Zadpoor AA. Additively manufactured metallic porous biomaterials based on minimal surfaces: a unique combination of topological, mechanical, and mass transport properties. Acta Biomater. 2017;53:572–84.
3. DebRoy T, Wei HL, Zuback JS, Mukherjee T, Elmer JW, Milewski JO, Beese AM, Wilson-Heid A, De A, Zhang W. Additive manufacturing of metallic components - process, structure and properties. Prog Mater Sci. 2018;92:112–224.
4. Liverani E, Fortunato A, Leardini A, Belvedere C, Siegler S, Ceschini L, Ascari A. Fabrication of Co-Cr-Mo endoprosthetic ankle devices by means of Selective Laser Melting (SLM). Mater Design. 2016;106:60–8.
5. Ponader S, von Wilmowsky C, Widenmayer M, Lutz R, Heinl P, Korner C, Singer RF, Nkenke E, Neukam FW, Schlegel KA. In vivo performance of selective electron beam-melted Ti-6Al-4V structures. J Biomed Mater Res A. 2010;92a(1):56–62.

6. Qin Y, Wen P, Guo H, Xia DD, Zheng YF, Jauer L, Poprawe R, Voshage M, Schleifenbaum JH. Additive manufacturing of biodegradable metals: current research status and future perspectives. Acta Biomater. 2019;98:3–22.
7. Sharma M, Soni M. Direct metal laser sintering of TI6Al4V alloy for patient-specific temporo mandibular joint prosthesis and implant - ScienceDirect. Materials today: proceedings; 2020.
8. Valentina F, Ali G, Demir C, Alberto B, Claudio C. Design rules for producing cardiovascular stents by selective laser melting: geometrical constraints and OpportuTiNies - ScienceDirect. Procedia Struct Integ. 2019;15(C):16–23.
9. Vock S, Klöden B, Kirchner A, Weißgärber T, Kieback B. Powders for powder bed fusion: a review. Prog Add Manuf. 2019;4(4):383–97. https://doi.org/10.1007/s40964-019-00078-6.
10. Wallner S. Powder production technologies. BHM Berg- und Hüttenmännische Monatshefte. 2019;164(3):108–11. https://doi.org/10.1007/s00501-019-0832-2.
11. Han W, Fang F. Fundamental aspects and recent developments in electropolishing. Int J Mach Tools Manuf. 2019;139:1–23.
12. Finazzi V, Demir AG, Biffi CA, Migliavacca F, Petrini L, Previtali B. Design and functional testing of a novel balloon-expandable cardiovascular stent in CoCr alloy produced by selective laser melting. J Manuf Process. 2020;55:161–73.
13. Wang LX, Fang G, Qian LY, Leeflang S, Duszczyk J, Zhou J. Forming of magnesium alloy microtubes in the fabrication of biodegradable stents. Prog Nat Sci-Mater. 2014;24(5):500–6.
14. Katada Y, Sagara M, Kobayashi Y, Kodama T. Fabrication of high strength high nitrogen stainless steel with excellent corrosion resistance and its mechanical properties. Mater Manuf Process. 2004;19(1):19–30.
15. Naumenko VV, Shlyamnev AP, Filippov GA. Nitrogen in AusteTiNic stainless steels of different alloying systems. Metallurgist+. 2011;55(5-6):410–8.
16. Li HB, Jiang ZH, Feng H, Ma QF, Zhan DP. Aging precipitation behavior of 18Cr-16Mn-2Mo-1. 1N high nitrogen AusteTiNic stainless steel and its influences on mechanical properties. J Iron Steel Res Int. 2012;19(8):43–51.
17. Li HB, Jiang ZH, Ma QF, Li WM. Manufacturing high nitrogen AusteTiNic stainless steels by pressurized induction furnace. Appl Mech Mater. 2011;52–54:1687–91.
18. Mani G, Feldman MD, Patel D, Agrawal CM. Coronary stents: a materials perspective. Biomaterials. 2007;28(9):1689–710.
19. Tonmukayakul U, Martin R, Clark R, Brownbill J, Manton D, Hall M, Armfield J, Smith M, Shankumar R, Sivasithamparam K, Martin-Kerry J, Calache H. Protocol for the Hall Technique study: a trial to measure clinical effectiveness and cost-effectiveness of stainless steel crowns for dental caries restoration in primary molars in young children. Contemp Clin Trials. 2015;44:36–41.
20. Ren YB, Zhao HC, Yang K, Zhang Y. Biomechanical compatibility of high strength nickel free stainless steel bone plate under lightweight design. Mat Sci Eng C-Mater. 2019;101:415–22.
21. Inamdar VV, Fitzpatrick E, Alferiev I, Nagaswami C, Spruce LA, Fazelinia H, Bratinov G, Seeholzer SH, Levy RJ, Fishbein I, Stachelek SJ. Stability and bioactivity of pepCD47 attachment on stainless steel surfaces. Acta Biomater. 2020;104:231–40.
22. Padilha AF, Plaut RL, Rios PR. Stainless steels heat treatment. Steel Heat Treatment Handbook. 2007;2:695–739.
23. GarcíaLópez E, Siller HR. Rodríguez CA Development of AISI 316L stainless steel coronary stent. Laser-based Micro-& Nanoprocessing XII; 2018.
24. Suresh S, Ritchie RO. A geometric model for fatigue crack closure induced by fracture surface-roughness. Metall Trans A. 1982;13(9):1627–31.
25. Marrey RV, Burgermeister R, Grishaber RB, Ritchie RO. Fatigue and life prediction for cobalt-chromium stents: a fracture mechanics analysis. Biomaterials. 2006;27(9):1988–2000.
26. Rimondini L, Fare S, Brambilla E, Felloni A. The effect of surface roughness on early in vivo plaque colonization on titanium. J Periodontol. 1997;68(6):556–62.

27. Milleret V, Hefti T, Hall H, Vogel V, Eberli D. Influence of the fiber diameter and surface roughness of electrospun vascular grafts on blood activation. Acta Biomater. 2012;8(12): 4349–56.
28. Sun X, Wei X, Li Z, Zhang J, Wang Y. Study on process technology and surface integrity of 316L cardiovascular stents by fiber laser wet cutting. Int J Adv Manuf Technol. 2019;105(11): 4517–26.
29. Yilbas BS, Shaukat MM, Ashraf F. Laser cutting of various materials: Kerf width size analysis and life cycle assessment of cutting process. Opt Laser Technol. 2017;93:67–73.
30. Garcia-Lopez E, Ibarra-Medina JR, Siller HR, Lammel-Lindemann JA, Rodriguez CA. Surface finish and back-wall dross behavior during the fiber laser cutting of AZ31 magnesium alloy. Micromachines-Basel. 2018;9(10)
31. Meng HG, Liao JH, Zhou YH, Zhang QM. Laser micro-processing of cardiovascular stent with fiber laser cutting system. Opt Laser Technol. 2009;41(3):300–2.
32. Muhammad N, Whitehead D, Boor A, Li L. Comparison of dry and wet fibre laser profile cutting of thin 316L stainless steel tubes for medical device applications. J Mater Process Tech. 2010;210(15):2261–7.
33. García-López E, Medrano-Tellez AG, Ibarra-Medina JR, Siller HR, Rodriguez CA. Experimental study of back wall dross and surface roughness in fiber laser microcutting of 316l miniature tubes. Micromachines-Basel. 2018;9(1):4.
34. Mohammad A, Wang DW. Electrochemical mechanical polishing technology: recent developments and future research and industrial needs. Int J Adv Manuf Technol. 2016;86(5-8): 1909–24.
35. Latifi A, Imani M, Khorasani MT, Joupari MD. Electrochemical and chemical methods for improving surface characteristics of 316L stainless steel for biomedical applications. Surf Coat Tech. 2013;221:1–12.
36. Tan JH, Wong WLE, Dalgarno KW. An overview of powder granulometry on feedstock and part performance in the selective laser melting process. Addit Manuf. 2017;18:228–55.
37. Yang L, Yang Y, Mai S, Di W, Song C. Investigation into spatter behavior during selective laser melting of AISI 316L stainless steel powder. Mater Design. 2015;87:797–806.
38. Kruth J-P, Badrossamay M, Yasa E, Deckers J, Thijs L, Van Humbeeck J. Part and material properties in selective laser melting of metals. In: Proceedings of the 16th International Symposium on Electromachining (ISEM XVI); 2010. p. 3–14.
39. Kaynak Y, Karabulut Y, Tascioglu E. Influence of heat treatment temperature on the microstructural, mechanical, and wear behavior of 316L stainless steel fabricated by laser powder bed additive manufacturing. Int J Adv Manuf Technol. 2020;
40. Sun Z, Tan X, Tor SB, Yeong WY. Selective laser melting of stainless steel 316L with low porosity and high build rates. Mater Design. 2016;104:197–204. https://doi.org/10.1016/j.matdes.2016.05.035.
41. Suryawanshi J, Prashanth KG, Ramamurty U. Mechanical behavior of selective laser melted 316L stainless steel. Mat Sci Eng A-Struct. 2017;696:113–21.
42. Shrestha R, Simsiriwong J, Shamsaei N. Fatigue behavior of additive manufactured 316L stainless steel parts: effects of layer orientation and surface roughness. Addit Manuf. 2019;28: 23–38.
43. Sander G, Babu AP, Gao X, Jiang D, Birbilis N. On the effect of build orientation and residual stress on the corrosion of 316L stainless steel prepared by selective laser melting. Corros Sci. 2021;179.
44. Lodhi MJK, Deen KM, Greenlee-Wacker MC, Haider W. Additively manufactured 316L stainless steel with improved corrosion resistance and biological response for biomedical applications. Addit Manuf. 2019;27:8–19.
45. Atapour M, Wang XY, Farnlund K, Wallinder IO, Hedberg Y. Corrosion and metal release investigations of selective laser melted 316L stainless steel in a synthetic physiological fluid containing proteins and in diluted hydrochloric acid. Electrochim Acta. 2020;354.

46. Wiesent L, Schultheiss U, Lulla P, Noster U, Schratzenstaller T, Schmid C, Nonn A, Spear A. Computational analysis of the effects of geometric irregularities and post-processing steps on the mechanical behavior of additively manufactured 316L stainless steel stents. Plos One. 2020;15(12)
47. Lei H, Li C, Meng J, Zhou H, Liu Y, Zhang X, Wang P, Fang D. Evaluation of compressive properties of SLM-fabricated multi-layer lattice structures by experimental test and μ-CT-based finite element analysis. Mater Design; 2019.
48. Barba D, Alabort C, Tang YT, Viscasillas MJ, Reed RC, Alabort E. On the size and orientation effect in additive manufactured Ti-6Al-4V. Mater Design. 2020;186.
49. Ravari MRK, Kadkhodaei M. A computationally efficient modeling approach for predicting mechanical behavior of cellular lattice structures. J Mater Eng Perform. 2015;24(1):245–52.
50. Raval A, Choubey A, Engineer C, Kothwala D. Development and assessment of 316LVM cardiovascular stents. Mat Sci Eng A-Struct. 2004;386(1-2):331–43.
51. Sojitra P, Engineer C, Kothwala D, Raval A, Kotadia H, Mehta G. Electropolishing of 316LVM stainless steel cardiovascular stents: an investigation of material removal, surface roughness and corrosion behavior. Trends Biomater Artificial Organs. 2010;23(3):115–21.
52. Zhao H, Van Humbeeck J, Sohier J, De Scheerder I. Electrochemical polishing of 316L stainless steel slotted tube coronary stents. J Mater Sci-Mater M. 2002;13(10):911–6.
53. Gellér ZE, Albrecht K, Dobránszky J. Electropolishing of coronary stents. Mater Sci Forum. 2008;589:367–72.
54. Haidopoulos M, Turgeon S, Sarra-Bournet C, Laroche G, Mantovani D. Development of an optimized electrochemical process for subsequent coating of 316 stainless steel for stent applications. J Mater Sci-Mater M. 2006;17(7):647–57.
55. Habibzadeh S, Li L, Shum-Tim D, Davis EC, Omanovic S. Electrochemical polishing as a 316L stainless steel surface treatment method: towards the improvement of biocompatibility. Corros Sci. 2014;87:89–100.
56. Bartolomeu F, Buciumeanu M, Pinto E, Alves N, Carvalho O, Silva FS, Miranda G. 316L stainless steel mechanical and tribological behavior—a comparison between selective laser melting, hot pressing and conventional casting. Addit Manuf. 2017;81–9.
57. Younkin CN. Multiphase Mp35n alloy for medical implants. J Biomed Mater Res. 1974;8(3): 219–26.
58. Honari G, Ellis SG, Wilkoff BL, Aronica MA, Svensson LG, Taylor JS. Hypersensitivity reactions associated with endovascular devices. Contact Dermatitis. 2008;59(1):7–22.
59. Feng QM, Zhang DY, Xin CH, Liu XD, Lin WJ, Zhang WQ, Chen S, Sun K. Characterization and in vivo evaluation of a bio-corrodible nitrided iron stent. J Mater Sci-Mater M. 2013;24(3): 713–24.
60. Pound BG. Corrosion behavior of metallic materials in biomedical applications. II. Stainless steels and Co-Cr alloys. Corrosion Rev. 2014;32(1–2):21–41. https://doi.org/10.1515/corrrev-2014-0008.
61. Suttorp MJ, Stella PR, Dens J, McKenzie JM, Park KS, Frambach P, Investigators S. Ultra-thin strut cobalt chromium bare metal stent usage in a complex real-world setting. (SOLSTICE registry). Neth Heart J. 2015;23(2):124–9.
62. von Birgelen C, Zocca P, Buiten RA, Jessurun GAJ, Schotborgh CE, Roguin A, Danse PW, Benit E, Aminian A, van Houwelingen KG, Anthonio RL, Stoel MG, Somi S, Hartmann M, Linssen GCM, Doggen CJM, Kok MM. Thin composite wire strut, durable polymer-coated (Resolute Onyx) versus ultrathin cobalt-chromium strut, bioresorbable polymer-coated (Orsiro) drug-eluting stents in allcomers with coronary artery disease (BIONYX): an international, single-blind, randomised non-inferiority trial. Lancet. 2018;392(10154):1235–45.
63. Kandzari DE, Mauri L, Koolen JJ. Ultrathin, bioresorbable polymer sirolimus-eluting stents versus thin, durable polymer everolimus-eluting stents in patients undergoing coronary revascularisation (BIOFLOW V): a randomised trial. Lancet. 2017;390(10105):1832.
64. Iantorno M, Lipinski MJ, Garcia-Garcia HM, Forrestal BJ, Rogers T, Gajanana D, Buchanan KD, Torguson R, Weintraub WS, Waksman R. Meta-analysis of the impact of strut thickness

on outcomes in patients with drug-eluting stents in a coronary artery. Am J Cardiol. 2018;122 (10):1652–60.
65. Zhang EL, Liu C. A new antibacterial Co-Cr-Mo-Cu alloy: preparation, biocorrosion, mechanical and antibacterial property. Mat Sci Eng C-Mater. 2016;69:134–43.
66. Keyvani M, Garcin T, Militzer M, Fabregue D. Laser ultrasonic measurement of recrystallization and grain growth in an L605 cobalt superalloy. Mater Character. 2020;167:110465. https://doi.org/10.1016/j.matchar.2020.110465.
67. Lu Y, Ren L, Xu X, Yang Y, Wu S, Luo J, Yang M, Liu L, Zhuang D, Yang K, Lin J. Effect of Cu on microstructure, mechanical properties, corrosion resistance and cytotoxicity of CoCrW alloy fabricated by selective laser melting. J Mech Behav Biomed. 2018;81:130–41. https://doi.org/10.1016/j.jmbbm.2018.02.026.
68. Haynes Alloy No. 25 (L-605). In: Simmons W, Cross H (eds). ASTM International, West Conshohocken, PA; 1954. p. 133–142. https://doi.org/10.1520/STP46128S
69. Poncin P, Gruez B, Missillier P, Comte-Gaz P, Proft J, Metallurgical Solutions FC. L605 precipitates and their effects on stent applications. Med Device Mater. 2006;Iii:85.
70. Li CL, Oh JM, Choi SW, Hong JK, Yeom JT, Mei XM, Mei QS, Yu ZT, Park CH. Study on microstructure and mechanical property of a biomedical Co-20Cr-15W-10Ni alloy during multi-pass thermomechanical processing. Mat Sci Eng A-Struct. 2020;785.
71. Chiba A, Nomura N, Ono Y. Microstructure and mechanical properties of biomedical Co–29Cr–8Mo alloy wire fabricated by a modified melt-spinning process. Acta Mater. 2007;55(6): 2119–28. https://doi.org/10.1016/j.actamat.2006.10.060.
72. Sudheer S, Kothwala D, Prathibha S, Engineer C, Raval A, Kotadia H. Laser microfabrication of L605 cobalt-chromium cardiovascular stent implants with modulated pulsed Nd: YAG laser. J Micro/Nanolithography, MEMS, MOEMS. 2008;7(3):033012.
73. Liu L, Li D, Tong Y, Zhu Y. The influences of assisting gas type and process parameters on the fiber laser microprofiling of thin CoCr tubes for vascular stents. Appl Sci. 2017;7(6):608.
74. Seo B, Park HK, Kim HG, Kim WR, Park K. Corrosion behavior of additive manufactured CoCr parts polished with plasma electrolytic polishing. Surf Coat Tech. 2021;406.
75. de Terris T, Andreau O, Peyre P, Adamski F, Koutiri I, Gorny C, Dupuy C. Optimization and comparison of porosity rate measurement methods of selective laser melted metallic parts. Addit Manuf. 2019;28:802–13.
76. Herzog D, Seyda V, Wycisk E, Emmelmann C. Additive manufacturing of metals. Acta Mater. 2016;117:371–92.
77. Tonelli L, Fortunato A, Ceschini L. CoCr alloy processed by selective laser melting (SLM): effect of laser energy density on microstructure, surface morphology, and hardness. J Manuf Process. 2020;52:106–19. https://doi.org/10.1016/j.jmapro.2020.01.052.
78. Li KF, Mao XH, Khanlari K, Song KK, Shi Q, Liu X. Effects of powder size distribution on the microstructural and mechanical properties of a Co-Cr-W-Si alloy fabricated by selective laser melting. J Alloy Compd. 2020;825.
79. Lu YJ, Wu SQ, Gan YL, Zhang SY, Guo S, Lin JJ, Lin JX. Microstructure, mechanical property and metal release of As-SLM CoCrW alloy under different solution treatment conditions. J Mech Behav Biomed. 2016;55:179–90.
80. Lashgari HR, Zangeneh S, Hasanabadi F, Saghafi M. Microstructural evolution during isothermal aging and strain-induced transformation followed by isothermal aging in Co-Cr-Mo-C alloy: a comparative study. Mat Sci Eng A-Struct. 2010;527(16-17):4082–91.
81. Zhang XQ, Li YP, Tang N, Onodera E, Chiba A. Corrosion behaviour of CoCrMo alloys in 2 wt% sulphuric acid solution. Electrochim Acta. 2014;125:543–55.
82. Demir AG, Previtali B. Additive manufacturing of cardiovascular CoCr stents by selective laser melting. Mater Design. 2017;119:338–50.
83. Omar MA, Baharudin BTHT, Sulaiman S, Ismail MIS, Omar MA. Evaluation of chemical composition, heat treatment, mechanical properties and electro chemical polishing for additively manufactured stent using ASTM F75 cobalt based superalloy (CoCrMo) by selective

laser melting (SLM) technology. Adv Mater Process Technol. 2020;1–20. https://doi.org/10.1080/2374068X.2020.1860598
84. Mani G. Chapter 1b metallic biomaterials: Cobalt-Chromium Alloys. In: Murphy W, Black J, Hastings G, editors. Handbook of biomaterial properties. New York: Springer; 2016. p. 159–66. https://doi.org/10.1007/978-1-4939-3305-1_15.
85. Kumar A, Bhatnagar N. Finite element simulation and testing of cobalt-chromium stent: a parametric study on radial strength, recoil, foreshortening, and dogboning. Comput Methods Biomech Biomed Engin. 2020;1-15 https://doi.org/10.1080/10255842.2020.1822823.
86. Kaufmann C, Mani G, Marton D, Johnson D, Agrawal CM. Long-term stability of self-assembled monolayers on electropolished L605 cobalt chromium alloy for stent applications. J Biomed Mater Res Part B: Appl Biomater. 2011;98(2):280–9.
87. Sojitra P, Engineer C, Rval A, Kothwala D, Jariwala A, Kotadia H, Adeshara S, Mehta G. Surface enhancement and characterization of L-605 cobalt alloy cardiovascular stent by novel electrochemical treatment. Artificial Organs (India). 2009;20081210:31.
88. Verschuur EML, Homs MYV, Steyerberg EW, Haringsma J, Wahab PJ, Kuipers EJ, Siersema PD. A new esophageal stent design (TiNi-S stent) for the prevention of migration: a prospective study in 42 patients. Gastrointest Endosc. 2006;63(1):134–40.
89. Plant SD, Grant DM, Leach L. Behaviour of human endothelial cells on surface modified TiNi alloy. Biomaterials. 2005;26(26):5359–67.
90. Shao AL, Cheng Y, Zhou Y, Li N, Xi TF, Zheng YF, Wei SC, Zhang DY. Electrochemistry properties of multilayer TiN/Ti coatings on TiNi alloy for cardiac occluder application. Surf Coat Tech. 2013;228:S257–61.
91. Loger K, de Miranda RL, Engel A, Marczynski-Buhlow M, Lutter G, Quandt E. Fabrication and evaluation of TiNinol thin film heart valves. Cardiovasc Eng Techn. 2014;5(4):308–16.
92. Quandt E, Halene C, Holleck H, Feit K, Kohl M, Schlossmacher P, Skokan A, Skrobanek KD. Sputter deposition of TiNi, TiNiPd and TiPd films displaying the two-way shape-memory effect. Sensor Actuat A-Phys. 1996;53(1-3):434–9.
93. Elahinia MH, Hashemi M, Tabesh M, Bhaduri SB. Manufacturing and processing of TiNi implants: a review. Prog Mater Sci. 2012;57(5):911–46.
94. Otsuka K, Ren XB. Recent developments in the research of shape memory alloys. Intermetallics. 1999;7(5):511–28.
95. Shabalovskaya SA. On the nature of the biocompatibility and on medical applications of TiNi shape memory and superelastic alloys. Bio-medical Mater Eng. 1996;6(4):267–89.
96. Levi DS, Kusnezov N, Carman GP. Smart materials applications for pediatric cardiovascular devices. Pediatric Res. 2008;63(5):552–8.
97. Otubo J, Rigo O, Neto C, Kaufman M, Mei P. TiNi shape memory alloy ingot production by EBM. J Phys IV. 2003;112:813–20.
98. Otubo J, Rigo OD, Moura Neto C, Kaufman MJ, Mei PR. Low carbon content TiNi shape memory alloy produced by electron beam melting. Mater Res. 2004;7(2):263–7.
99. International A. ASTM F2063. Standard specification for wrought nickel-titanium shape memory alloys for medical devices and surgical implants. West Conshohocken, PA: ASTM International; 2018.
100. Pelton AR, Russell SM, DiCello J. The physical metallurgy of TiNinol for medical applications. Jom-J Min Met Mat S. 2003;55(5):33–7.
101. Jiang S-y, Zhang Y-q, Zhao Y-N. Dynamic recovery and dynamic recrystallization of TiNi shape memory alloy under hot compression deformation. T Nonferr Metal Soc. 2013;23(1):140–7.
102. Wu SK, Lin HC, Yen YC. A study on the wire drawing of TiNi shape memory alloys. Mat Sci Eng A-Struct. 1996;215(1–2):113–9.
103. Saito S, Wachi T, Hanada S. A new fabrication process of Tini shape memory wire. Mat Sci Eng A-Struct. 1993;161(1):91–6.
104. Pelton AR, Dicello J, Miyazaki S. Optimisation of processing and properties of medical grade TiNinol wire. Minimally Invasive Therapy Allied Technol. 2000;9(2):107–18.

105. Teng PT, Chang FY, Chang YT, Liang PC, Huang KW. Micro cutting of biliary stent with nanosecond fiber laser system. In: Advanced materials research. Trans Tech Publ; 2014. p. 209–13.
106. Liu L, Li DB, Tong YF, Zhu YF. Fiber laser micromachining of thin TiNi tubes for shape memory vascular stents. Appl Phys A. 2016;122(7):1–9.
107. Zhao H, Stalmans R, Van Humbeeck J, De Scheerder I. Pickling of laser-cut TiNi slotted tube stents: effect on surface morphology, dimension changes and mechanical behaviour. In: Journal de Physique IV (Proceedings). EDP sciences; 2003. p. 1125–8.
108. Anikeev S, Hodorenko V, Gunther V, Chekalkin T, Kang JH, Kang SB. The effect of mechano-chemical treatment on structural properties of the drawn TiNi-based alloy wire. Mater Res Express. 2018;5(1)
109. Nespoli A, Biffi CA, Previtali B, Villa E, Tuissi A. Laser and surface processes of TiNi shape memory elements for micro-actuation. Metallurgical Mater Trans A. 2014;45(4):2242–9.
110. Zhao H, Van Humbeeck J, De Scheerder I. Surface conditioning of nickel–titanium alloy stents for improving biocompatibility. Surface Eng. 2001;17(6):451–8.
111. Lopes NIA, Silva LÁO, Santos LA, Buono VTL. Surface characterization of TiNi superelastic and shape memory alloys after electrolytic polishing. Mater Res. 2017;20:572–9.
112. Dadbakhsh S, Speirs M, Kruth JP, Schrooten J, Luyten J, Van Humbeeck J. Effect of SLM parameters on transformation temperatures of shape memory nickel titanium parts. Adv Eng Mater. 2014;16(9):1140–6.
113. Dadbakhsh S, Vrancken B, Kruth JP, Luyten J, Van Humbeeck J. Texture and anisotropy in selective laser melting of TiNi alloy. Mat Sci Eng a-Struct. 2016;650:225–32.
114. Qiu P, Gao P, Wang S, Li Z, Yang Y, Zhang Q, Xiong Z, Hao S. Study on corrosion behavior of the selective laser melted TiNi alloy with superior tensile property and shape memory effect. Corros Sci. 2020;175:108891.
115. Zhang QQ, Hao SJ, Liu YT, Xiong ZW, Guo WQ, Yang Y, Ren Y, Cui LS, Ren LQ, Zhang ZH. The microstructure of a selective laser melting (SLM)-fabricated TiNi shape memory alloy with superior tensile property and shape memory recoverability. Appl Mater Today. 2020;m 19.
116. Moghaddam NS, Saedi S, Amerinatanzi A, Hinojos A, Ramazani A, Kundin J, Mills MJ, Karaca H, Elahinia M. Achieving superelasticity in additively manufactured TiNi in compression without post-process heat treatment. Sci Rep-UK. 2019;9.
117. Rattfält L, Lindén M, Hult P, Berglin L, Ask P. Electrical characteristics of conductive yarns and textile electrodes for medical applications. Med Biol Eng Comput. 2007;45(12):1251–7.
118. Corti C, Holliday R. Gold: science and applications. CRC Press; 2009.
119. Aubry J-F, Beaulieu L, Girouard L-M, Aubin S, Tremblay D, Laverdière J, Vigneault E. Measurements of intrafraction motion and interfraction and intrafraction rotation of prostate by three-dimensional analysis of daily portal imaging with radiopaque markers. Int J Radiat Oncol* Biol* Phys. 2004;60(1):30–9.
120. Cowley A. A healthy future: platinum in medical applications. Platinum Metals Rev. 2011;55 (2):98–107.
121. Brummer SB, Turner M. Electrochemical considerations for safe electrical stimulation of the nervous system with platinum electrodes. IEEE Trans Biomed Eng. 1977;1:59–63.
122. O'Brien BJ, Stinson JS, Larsen SR, Eppihimer MJ, Carroll WM. A platinum-chromium steel for cardiovascular stents. Biomaterials. 2010;31(14):3755–61.
123. Menown IBA, Noad R, Garcia EJ, Meredith I. The Platinum Chromium element stent platform: from alloy, to design, to clinical practice. Adv Ther. 2010;27(3):129–41.
124. Eugene WWL, Mujumdar AS. Gold extraction and recovery processes. Minerals, Metals, and Materials Technology Centre, National University of Singapore; 2009.
125. Prasad MS, Mensahbiney R, Pizarro RS. Modern trends in gold processing - overview. Minerals Eng. 1991;4(12):1257–77.
126. Prasad M, Mensah-Biney R, Pizarro R. Modern trends in gold processing—overview. Minerals Eng. 1991;4(12):1257–77.

127. Eksteen J, Oraby E. The leaching and adsorption of gold using low concentration amino acids and hydrogen peroxide: effect of catalytic ions, sulphide minerals and amino acid type. Minerals Eng. 2015;70:36–42.
128. Gül A, Kangal O, Sirkeci AA, Önal G. Beneficiation of the gold bearing ore by gravity and flotation. Int J Minerals Metallurgy Mater. 2012;19(2):106–10.
129. Valderrama L, Rubio J. High intensity conditioning and the carrier flotation of gold fine particles. Int J Mineral Process. 1998;52(4):273–85.
130. Zhang Y, Wang S, Zhou J. Review of the gold extracting methods. J Shangrao Normal College. 2005;3
131. Marsden J, House I. The chemistry of gold extraction. SME; 2006.
132. Labrooy SR, Linge HG, Walker GS. Review of gold extraction from ores. Minerals Eng. 1994;7(10):1213–41.
133. Fraser KS, Walton RH, Wells JA. Processing of refractory gold ores. Minerals Eng. 1991;4 (7–11):1029–41.
134. Nanthakumar B, Pickles CA, Kelebek S. Microwave pretreatment of a double refractory gold ore. Minerals Eng. 2007;20(11):1109–19.
135. Liu XL, Li Q, Zhang Y, Jiang T, Yang YB, Xu B, He YH. Improving gold recovery from a refractory ore via Na2SO4 assisted roasting and alkaline Na2S leaching. Hydrometallurgy. 2019;185:133–41.
136. Li Q, Li D, Qian F. Pre-oxidation of high-sulfur and high-arsenic refractory gold concentrate by ozone and ferric ions in acidic media. Hydrometallurgy. 2009;97(1-2):61–6.
137. Konadu KT, Mendoza DM, Huddy RJ, Harrison STL, Kaneta T, Sasaki K. Biological pretreatment of carbonaceous matter in double refractory gold ores: a review and some future considerations. Hydrometallurgy. 2020;196.
138. de Carvalho LC, da Silva SR, Giardini RMN, de Souza LFC, Leao VA. Bio-oxidation of refractory gold ores containing stibnite and gudmundite. Environ Technol Inno. 2019;15.
139. Konadu KT, Huddy RJ, Harrison STL, Osseo-Asare K, Sasaki K. Sequential pretreatment of double refractory gold ore (DRGO) with a thermophilic iron oxidizing archeaon and fungal crude enzymes. Minerals Eng. 2019;138:86–94.
140. Reith F, Lengke MF, Falconer D, Craw D, Southam G. The geomicrobiology of gold. Isme J. 2007;1(7):567–84.
141. Kaksonen AH, Mudunuru BM, Hackl R. The role of microorganisms in gold processing and recovery-a review. Hydrometallurgy. 2014;142:70–83.
142. Aylmore MG, Muir DM. Thiosulfate leaching of gold - a review. Minerals Eng. 2001;14(2): 135–74.
143. Ahtiainen R, Lundström M. Cyanide-free gold leaching in exceptionally mild chloride solutions. J Cleaner Prod. 2019;234:9–17.
144. Sun C-b, Zhang X-l, Kou J, Xing Y. A review of gold extraction using noncyanide lixiviants: fundamentals, advancements, and challenges toward alkaline sulfur-containing leaching agents. Int J Minerals Metallurgy Mater. 2020;27(4):417–31.
145. Gökelma M, Birich A, Stopic S, Friedrich B. A review on alternative gold recovery re-agents to cyanide. J Mater Sci Chem Eng. 2016;4(08):8.
146. López-Honorato E, Meadows P, Xiao P. Fluidized bed chemical vapor deposition of pyrolytic carbon–I. Effect of deposition conditions on microstructure. Carbon. 2009;47(2):396–410.
147. Serino G, Gusmini M, Audenino AL, Bergamasco G, Ieropoli O, Bignardi C. Multiscale characterization of isotropic pyrolytic carbon used for mechanical heart valve production. Processes. 2021;9(2):338.
148. Sheikholeslami ZS, Yousefi M, Imani M, Joupari MD. Low-temperature, chemical vapor deposition of thin-layer pyrolytic carbon coatings derived from camphor as a green precursor. J Mater Sci. 2018;53(2):959–76.
149. Xu L, Wu J, Bai S. Fabrication and microstructure of boron-doped isotropic pyrolytic carbon. Carbon. 2012;50(12):4705–10.

150. Tan R, Xie H, She J, Liang J, He H, Li J, Fan Z, Liu B. A new approach to fabricate superhydrophobic and antibacterial low density isotropic pyrocarbon by using catalyst free chemical vapor deposition. Carbon. 2019;145:359–66.
151. Smith K, Black K. Characterization of the treated surfaces of silicon alloyed pyrolytic carbon and SiC. J Vacuum Sci Technol A: Vacuum, Surfaces Films. 1984;2(2):744–7.
152. Liu R, Liu M, Shao Y, Chen X, Ma J, Liu B. A novel coated-particle design and fluidized-bed chemical vapor deposition preparation method for fuel-element identification in a nuclear reactor. Particuology. 2017;31:35–41.
153. Tyagi A, Walia R, Murtaza Q, Pandey SM, Tyagi PK, Bajaj B. A critical review of diamond like carbon coating for wear resistance applications. Int J Refractory Metals Hard Mater. 2019;78:107–22.
154. Dearnaley G, Arps JH. Biomedical applications of diamond-like carbon (DLC) coatings: a review. Surface Coatings Technol. 2005;200(7):2518–24.
155. Cho Y-S, Liao L-K, Hsu C-H, Hsu Y-H, Wu W-Y, Liao S-C, Chen K-H, Lui P-W, Zhang S, Lien S-Y. Effect of substrate bias on biocompatibility of amorphous carbon coatings deposited on Ti6Al4V by PECVD. Surf Coatings Technol. 2019;357:212–7.
156. Jo YJ, Zhang TF, Son MJ, Kim KH. Synthesis and electrochemical properties of Ti-doped DLC films by a hybrid PVD/PECVD process. Appl Surf Sci. 2018;433:1184–91.
157. Bociaga D, Kaminska M, Sobczyk-Guzenda A, Jastrzebski K, Swiatek L, Olejnik A. Surface properties and biological behaviour of Si-DLC coatings fabricated by a multi-target DC–RF magnetron sputtering method for medical applications. Diam Relat Mater. 2016;67:41–50.
158. Patnaik L, Maity SR, Kumar S. Effect of lubricated sliding wear against CFRPEEK on the nanomechanical properties of Ag alloyed Cr/DLC thin film. J Mech Behav Biomed. 2021;118: 104478.
159. Rao X, Yang J, Chen Z, Yuan Y, Chen Q, Feng X, Qin L, Zhang Y. Tuning C–C sp2/sp3 ratio of DLC films in FCVA system for biomedical application. Bioact Mater. 2020;5(2):192–200.
160. Orrit-Prat J, Bonet R, Rupérez E, Punset M, Ortiz-Hernández M, Guillem-Marti J, Lousa A, Cano D, Díaz C, Fuentes GG. Bactericidal silver-doped DLC coatings obtained by pulsed filtered cathodic arc co-deposition. Surf Coatings Technol. 2021;411:126977.
161. Odian G. Principles of polymerization. Wiley; 2004.
162. Kariduraganavar MY, Kittur AA, Kamble RR. Polymer synthesis and processing. In: Natural and synthetic biomedical polymers. Elsevier; 2014. p. 1–31.
163. Hiemenz PC, Lodge TP. Polymer chemistry. CRC; 2007.
164. Tatara RA. Compression molding. In: Applied plastics engineering handbook. Elsevier; 2017. p. 291–320.
165. Lafleur PG, Vergnes B. Polymer extrusion. Wiley Online Library; 2014.
166. Vlachopoulos J, Strutt D. Polymer processing. Mater Sci Technol. 2003;19(9):1161–9.
167. Hu J, Chen S. A review of actively moving polymers in textile applications. J Mater Chem. 2010;20(17):3346–55.
168. Bezuidenhout D, Williams DF, Zilla P. Polymeric heart valves for surgical implantation, catheter-based technologies and heart assist devices. Biomaterials. 2015;36:6–25.
169. Mackay T, Wheatley D, Bernacca G, Fisher A, Hindle C. New polyurethane heart valve prosthesis: design, manufacture and evaluation. Biomaterials. 1996;17(19):1857–63.
170. Jiang H, Campbell G, Boughner D, Wan W-K, Quantz M. Design and manufacture of a polyvinyl alcohol (PVA) cryogel tri-leaflet heart valve prosthesis. Med Eng Phys. 2004;26(4): 269–77.
171. Cavallo A, Gasparotti E, Losi P, Foffa I, Al Kayal T, Vignali E, Celi S, Soldani G. Fabrication and in-vitro characterization of a polymeric aortic valve for minimally invasive valve replacement. J Mech Behav Biomed. 2021;115
172. Hiob MA, She S, Muiznieks LD, Weiss AS. Biomaterials and modifications in the development of small-diameter vascular grafts. ACS Biomater Sci Eng. 2017;3(5):712–23.
173. Seifalian AM, Salacinski HJ, Tiwari A, Edwards A, Bowald S, Hamilton G. In vivo biostability of a poly(carbonate-urea)urethane graft. Biomaterials. 2003;24(14):2549–57.

174. Wang D, Xu Y, Lin Y-J, Yilmaz G, Zhang J, Schmidt G, Li Q, Thomson JA, Turng L-S. Biologically functionalized expanded polytetrafluoroethylene blood vessel grafts. Biomacromolecules. 2020;21(9):3807–16.
175. He W, Hu Z, Xu A, Liu R, Yin H, Wang J, Wang S. The preparation and performance of a new polyurethane vascular prosthesis. Cell Biochem Biophys. 2013;66(3):855–66.
176. Song R, Murphy M, Li CS, Ting K, Soo C, Zheng Z. Current development of biodegradable polymeric materials for biomedical applications. Drug Des Dev Ther. 2018;12:3117–45.
177. Bloomfield P. Choice of heart valve prosthesis. Heart. 2002;87(6):583–9.
178. Li C, Guo C, Fitzpatrick V, Ibrahim A, Zwierstra MJ, Hanna P, Lechtig A, Nazarian A, Lin SJ, Kaplan DL. Design of biodegradable, implantable devices towards clinical translation. Nat Rev Mater. 2020;5(1):61–81.
179. Tang-Quan KR, Mehta NA, Sampaio LC, Taylor DA. Whole cardiac tissue bioscaffolds. Cardiac Extracellular Matrix. 2018;85–114.
180. Ott HC, Matthiesen TS, Goh SK, Black LD, Kren SM, Netoff TI, Taylor DA. Perfusion-decellularized matrix: using nature's platform to engineer a bioartificial heart. Nat Med. 2008;14(2):213–21.
181. Methe K, Bäckdahl H, Johansson BR, Nayakawde N, Dellgren G, Sumitran-Holgersson S. An alternative approach to decellularize whole porcine heart. BioResearch open access. 2014;3 (6):327–38.
182. Guyette JP, Charest JM, Mills RW, Jank BJ, Moser PT, Gilpin SE, Gershlak JR, Okamoto T, Gonzalez G, Milan DJ. Bioengineering human myocardium on native extracellular matrix. Circ Res. 2016;118(1):56–72.
183. Awad NK, Niu HT, Ali U, Morsi YS, Lin T. Electrospun fibrous scaffolds for small-diameter blood vessels: a review. Membranes-Basel. 2018;8(1).
184. Ju YM, Ahn H, Arenas-Herrera J, Kim C, Abolbashari M, Atala A, Yoo JJ, Lee SJ. Electrospun vascular scaffold for cellularized small diameter blood vessels: a preclinical large animal study. Acta Biomater. 2017;59:58–67.
185. Kuang HZ, Wang Y, Shi Y, Yao WC, He X, Liu XZ, Mo XM, Lu SY, Zhang P. Construction and performance evaluation of Hep/silk-PLCL composite nanofiber small-caliber artificial blood vessel graft. Biomaterials. 2020;259.
186. Mi HY, Jiang YC, Jing X, Enriquez E, Li H, Li Q, Turng LS. Fabrication of triple-layered vascular grafts composed of silk fibers, polyacrylamide hydrogel, and polyurethane nanofibers with biomimetic mechanical properties. Mat Sci Eng C-Mater. 2019;98:241–9.

Chapter 6
Stent and Perfusion Balloon

Yong Zeng and Xiliang Zhao

6.1 Introduction

Percutaneous coronary intervention (PCI) refers to the use of cardiac catheterization technology to dilate stenosed or diseased vessels to improve myocardial blood circulation, including percutaneous coronary angioplasty and intravascular stent implantation [1]. With the development of PCI, a variety of new interventional devices emerge in endlessly, among which the most common devices are balloon and stent [2].

6.2 Stent

Stent is a commonly used medical instrument in cardiac interventional surgery. The main materials are stainless steel, nickel titanium alloy, or cobalt chromium alloy. It first appeared in the 1980s and experienced the development process of bare mental stent (BMS), drug-eluting stent (DES), and bioabsorbable vascular scaffolds (BVS). Each technological change overcomes the limitations of the previous generation technology to a certain extent, and the drug-eluting stent is the biggest progress and highlight in the process of interventional medicine (Table 6.1).

Y. Zeng (✉) · X. Zhao
Beijing Anzhen Hospital, Capital Medical University, Beijing, China

Y.-D. Tang et al. (eds.), *Application of Biomaterials in the Treatment of Cardiovascular Diseases*, https://doi.org/10.1007/978-981-19-7712-1_6

Table 6.1 Overview of principal characteristics of selected commercially available DES

Stent characteristics	Xience Sierra (EES)	Resolute Onyx (ZES)	Synergy (EES)	Orsiro (SES)	Ultimaster (SES)	BioFreedom (BCS)	EluNIR
Company	Abbott	Metronic	Boston Scientific	Biotronik	Terumo	Eurocor Biosensors	Medinol
Platform	MULTI-LINK8	Onyx	Premier	PRO-Kinetic Energy	Kamame	Juno	–
Material	Co-Cr	Co-Ni + Pt-Ir core	Pt-Cr	Co-Cr	Co-Cr	Stainless steel	Co-Cr
Strut thickness (μm)	81	81	74	60	80	112	87
Drug type	Everolimus	Zotarolimus	Everolimus	Sirolimus	Sirolimus	Biolimus A9	Ridaforolimus
Drug dose	100 μg/cm^2	1.6 μg/mm^2	100 μg/cm^2	1.4 μg/mm^2	3.9 μg/mm	15.6 μg/mm	1.1 μg/mm^2
Polymer materials	PBMA-PVDF-HFP	BioLinx	PLGA	PLLA, silicon carbide	Poly(D,L-lactide-co-caprolactone)	No polymer	PBMA CarboSil 20 55D
Coating type	Circumferential	Circumferential	Abluminal	Circumferential	Abluminal	Microstructured; abluminal	Circumferential
Coating thickness (μm)	7–8 per side	6 per side	4	Conformal (3.5/7.5)	14	NA	7
Strut+ coating thickness (μm)	95–97	93	78	71	94	112	104
Strut design	Open cell	Open cell	Open cell	Open cell	Open cell	Open cell	Open cell
Link pattern	Peak to valley	Peak to valley	Offset peak to peak	Midstrut connectors	Peak to valley	Peak to valley	Peak to valley

BCS, Biolimus A9-coated stent; EES, everolimus-eluting stent; NA, not applicable; PBMA, poly(n-butyl methacrylate); PLGA, poly(lactic-co-glycolic acid); PLLA, poly(L-lactic acid); PVDF-HFP, poly(vinylidene fluoride-co-hexafluoropropylene); SES, sirolimus-eluting stent; ZES, zotarolimus-eluting stent

6.2.1 Classification

There are many ways to classify stents, such as self-expanding stent and balloon expandable stent [3]. Self-expanding stent is limited in clinical application because of its high metal coverage, high restenosis rate, high rate of subacute coronary artery occlusion, and great influence on the side branch after implantation. Balloon expandable stent can be further classified into wound, tubular, annular, and special stent according to its structural characteristics [4]. At present, coronary stents are mostly tubular and annular stents made by seamless metal laser cutting. Clinically, there are also stent bifurcation, stent side branch, and stent coronary stent graft, which are special purpose stents designed according to different purposes.

Other classification methods include mesh size after stent expansion (open-loop stent, closed-loop stent), stent material (316 L medical stainless steel, tantalum metal, alloy stent), and special coating treatment of stent surface (bare metal stent, drug-coated stent).

6.2.2 Performance and Selection of Support

The choice of stents is mainly based on the characteristics of target vessel lesions, on which the operator can make a comprehensive judgment according to the understanding of stent performance. In interventional therapy, we think that the ideal stent should have the following characteristics: flexibility, good tracer, small profile, X-ray imperceptibility, strong visibility, antithrombotic, good biocompatibility, reliable expansion performance, good supporting force, good coverage, and small surface area, in line with hydrodynamics. These excellent characteristics are the guarantee for the success of interventional therapy, but, in fact, none of the stents can fully meet all the above characteristics. Each stent has its own characteristics, so clinical interventional physicians are faced with complex stent selection problems. Familiarity with the characteristics of various stents determines the success of interventional therapy and the reduction of postoperative complications to the greatest extent.

6.2.3 Application

Cardiac stents can be used in a variety of cardiovascular diseases [5], such as coronary heart disease, as well as congenital heart diseases, such as tetralogy of Fallot, patients with abnormal coronary artery anatomy, and percutaneous right ventricular outflow tract stent implantation is less risk than simple surgery. There are also some complicated congenital heart diseases such as severe pulmonary artery stenosis or pulmonary atresia, which are suitable for percutaneous placement of stents [6].

Compared with drug therapy alone, revascularization is more effective, including PCI and CABG [7]. However, the risk of revascularization is inevitable. We should carefully weigh the benefits and risks of various treatment options. In addition to paying attention to the vascular anatomy, we also need to consider the clinical situation, technical factors, and realistic conditions of patients. Based on evidence-based medicine, we should formulate the best individualized treatment plan for individual patients and flexibly refer to the guidelines.

6.2.4 The Future Development Trend of Stent

With the development of interventional technology, the rate of restenosis and the incidence of stent thrombosis have been reduced [8]. As a permanent implanted metal support, the mainstream permanent scaffold still has a series of problems due to its nonspecific inhibition on endothelial cells. Over the years, we have been committed to the research and development of new antiproliferative drugs, biodegradable polymer coating, directed drug delivery system, etc. At present, another research hotspot of DES is to promote endothelial healing stent [9]. The design is no longer aimed at drug inhibition of proliferation, but to improve the wound healing speed of patients. A series of special designs are aimed at promoting the functional healing of vascular intima, so as to reduce the cost of traditional DES. The long-term catch-up effect of implantation induced accelerated neointimal atherosclerosis and the incidence of cardiac events [10].

Another hot topic is the concept of "intervening in non-implantable." Under this background, the emerging BVS emerges as the times require [11]. Its principle is only as a kind of "temporary stent," which provides short-term target vessel support in the necessary stage of vascular healing. After complete degradation, it not only solves the negative problems such as metal residue, long-term stent rupture, stent thrombosis and restenosis, and long-term antithrombotic treatment but also does not affect long-term re-stent and surgical treatment [12]. It is no longer limited by nuclear magnetic resonance and other examinations. It is the fourth revolution of coronary intervention medicine and is the research frontier in the field of vascular stent. It can be seen from the above that BVS is more suitable for young people, middle anterior descending artery, and multivessel diffuse stenosis, which not only realizes non-implantable treatment but also benefits the long-term prognosis and treatment of such patients. However, its biggest limitation lies in the large diameter of stent and the potential risk of fracture, which is easy to cause adverse events such as reduction of support force, blood vessel retraction, and interference of blood flow. The ideal absorbable stent can provide support for blood vessels in the early stage of implantation; it is gradually degraded and absorbed in the later stage to avoid inflammatory reaction caused by foreign body and reduce the occurrence of late stent thrombosis events; with the complete disappearance of the stent, the natural shape and vasomotor function of blood vessels are finally restored. Therefore, how to improve the mechanical properties is the key problem in the development of BVS.

6.3 Perfusion Balloon

In 1977, the world's first epoch-making percutaneous coronary angioplasty (PTCA) opened a new era in the interventional treatment of coronary heart disease. PTCA is the delivery of balloon catheter through the peripheral artery to the vascular lesion, through balloon dilation to relieve vascular stenosis and restore coronary blood flow [13]. The development history of balloon has experienced from the early fixed guide wire balloon to the over the wire system (OTW), rapid exchange system (Rx) and monorail system, as well as the later cutting balloon, double guide wire balloon, and drug-coated balloon, which promoted the development of PCI and brought great gospel for the interventional treatment of various complex lesions.

6.3.1 Classification and Performance

The performance of the balloon needs to pass a comprehensive evaluation of a number of indicators, including the following:

(1) The outer diameter, which is the minimum outer diameter of the marked part of the balloon or the clearance sheet, is the standard for PCI to select the balloon model.
(2) Flexibility (compliance), which determines the ability to go through tortuous lesions
(3) Tracking: the ability to pass through the target lesion along the guide wire.
(4) Compliance: the ability to increase or elongate the balloon diameter with the increase of inflation pressure.

According to the characteristics of balloon materials, it can be divided into compliant, semi compliant, and noncompliant balloon. According to its design characteristics, it can be divided into ordinary balloon, cutting balloon, double guide wire balloon, drug-coated balloon, etc. The common balloon can compress and displace the stenosis plaque and dilate the lumen. It is mainly divided into compliant, semi compliant, and noncompliant balloon. After dilation, it is easy to cause serious damage of vascular intima and promote the occurrence of in stent restenosis (ISR) and can lead to acute vascular occlusion and thrombosis and other serious events. The dilation effect of stent restenosis, calcification, and other special lesions is not ideal. Cutting balloon is a special kind of balloon, which perfectly combines micro cutting operation with balloon dilation mechanism. When the balloon began to expand, it was not fully opened, and the blade on the surface of the balloon gradually appeared on the surface of the balloon. The blade on the cutting balloon radially cut the fiber cap, elastic fiber, and smooth muscle along the longitudinal axis of the blood vessel in advance to form a geometric model of expansion. When the balloon was completely expanded, the filling pressure of the balloon was reduced, the elastic retraction degree of the blood vessel was reduced, and the blood pressure was reduced in order to reduce the incidence of restenosis and reduce the vascular injury caused by simple balloon PTCA. At the same time, this method of cutting before expanding can reduce the severe and irregular tear of intima through controlled injury, which not only has good immediate effect but also reduces the loss of lumen diameter in the long term, so as to effectively reduce the restenosis. The cutting balloon can achieve the purpose of lumen expansion by dilating and compressing the plaque, rather than the passive dilation of the blood vessels, which proved that the cutting balloon can reduce the damage to the vascular media and intima compared with the common balloon. It further confirmed that the cutting balloon had less trauma to the vascular wall, so it had less intimal hyperplasia than the common balloon group. Most of them are used in coronary artery disease, calcification, small vessel disease, and in stent restenosis. However, due to its structural characteristics, cutting balloon in the treatment of calcified lesions and high flexion lesions, the operation is difficult, so we should pay attention to it.

In the process of interventional therapy, the classic procedure is balloon pre expansion-stent release-balloon post expansion if necessary. Most of the stent platforms are not suitable for direct stenting [14]. DES wall does not fit well with the vascular wall, and there are complications such as acute thrombosis and in stent restenosis. Balloon ensures the efficacy of DES, especially in complex lesions [15].

6.3.2 *Drug-Coated Balloons*

At present, the imbalance of drug release between bone and bone intersection of DES and the stimulation of carrier polymer to vascular wall after drug depletion promote the in stent restenosis rate. In 2014, the guideline of the European Society of Cardiology recommended DCB and new generation DES as the first choice strategy for ISR treatment [16]. Compared with traditional stent, DCB has many theoretical advantages. More and more clinical studies have confirmed its advantages in the treatment of small vessel disease, bifurcation disease, chronic total occlusion disease, and diffuse disease [17] (Table 6.2).

Firstly, because DCB has no metal skeleton, it can avoid the imbalance of drug release, make the drug evenly distributed in the specific vascular wall area, at the same time preserve the original vascular anatomy, and avoid the influence on the blood flow pattern in the treatment of small vascular lesions and bifurcation lesions; secondly, it can avoid the double-layer stent to reduce the vascular lumen in the treatment of stent stenosis, and many studies have confirmed that ISR is the preferred choice for DCB indication [18]. In addition, DCB has no polymer carrier, which can reduce chronic inflammatory reaction and late thrombosis, so it can shorten the time of dual antiplatelet therapy (only 1–3 months after) [19]. In addition to the above

Table 6.2 Overview of principal characteristics of selected commercially available DCBs

Drug and device	Company	Additive	Substance class	Dose (μg/mm^2)
Paclitaxel				
Agent	Boston Scientific	Acetyl tributyl citrate	Plasticizer	2
Elutax SV	Aachen Resonance			2.2
Danubio	Minvasys	n-Butyryl tri-n-hexyl citrate	Plasticizer	2.5
SeQuent Please	B. Braun	Iopromide	X-ray contrast medium	3
Pantera Lux	Biotronik	n-Butyryl tri-n-hexyl citrate	Plasticizer	3
RESTORE	Cardionovum	Shellac	Varnish	3
AngioSculptX	Spectranetics	Nordihydroguaiaretic acid	Antioxidant	3
Chocolate touch	QT Vascular	Undisclosed		3
Dior II, BioStream	Eurocor Biosensors	Shellac	Varnish	3
Essential	iVascular	Undisclosed		3
IN.PACT (Admiral, Pacific, Falcon)	Medtronic Vascular	Urea	Endogenous metabolite	3.5
Sirolimus				
Selution	Med Alliance	Biodegradable polymer	Microreservoirs	
Virtue	Caliber Therapeutics	Biodegradable polyester-based polymers	Submicrometer nanoparticles	
Magic Touch	Concept Medical		Phospholipids	
Sequent Please SCB	B. Braun		Crystalline sirolimus	4

population, DCB can also be applied to patients with high risk of bleeding, such as hemophilia, previous bleeding history, gastric ulcer, and severe renal failure; patients with oral anticoagulants or recent surgery, such as with atrial fibrillation; patients with prosthetic heart valve replacement; patients with vascular endothelial dysfunction or history of subacute stent thrombosis; and patients who refuse to implant foreign bodies.

DCB plays an important role in the field of interventional therapy. However, there are still some problems to be solved. DCB-only percutaneous coronary intervention (PCI) is the concept of treating coronary stenoses while limiting the need for permanent or semipermanent implants to those lesions at high risk for acute vessel closure or unfavorable long-term results [20]. Potentially, every PCI should aim at using the DCB-only strategy, which implies a similar approach for every lesion preparation.

6.4 Future Perspective

The continuous innovation of technology is the process of human struggle with disease. Since the successful implementation of the first PTCA, cardiac interventional therapy has become the first choice for countless patients with incomparable advantages of surgery. The invention of each new type of instrument has gathered the great efforts of scientific researchers and clinicians. There is no end to medicine. With the development of technology, it is believed that it will be renewed in the future. However, it is worth mentioning that interventional therapy is not the end point, and it is more important to do a good job of "heart protection" treatment after operation, and drug treatment is still the cornerstone.

References

1. Landau C, Lange RA, Hillis LD. Percutaneous transluminal transluminal coronary angioplasty. N Engl J Med. 1994;330(14):981–93.
2. Shengshou H, Runlin G, Lisheng L, et al. Summary of Chinese cardiovascular disease report 2018. Chin J Circ. 2019;3:209–20.
3. Laham RJ, Carrozza JP, Berger C, et al. Long-term (4-to 6-year) outcome of Palmaz-Schatz stenting: paucity of late clinical stent-related problems. J Am Coll Cardiol. 1996;28(4):820–6.
4. Zhe Q. Analysis of risk factors for adverse cardiac events after percutaneous coronary intervention. J Chengdu Med Coll. 2013;1:95–7.
5. Yeh RW, Mauri L, Wolf RE, et al. Population trends in rates of coronary revascularization. JAMA Intern Med. 2015;175(3):454–6.
6. Wang PL, Zhang L, Wang SL, et al. Long—term follow-up of Chinese herbal medicines combined with conventional treatment in patients with acute coronary syndrome after per—cutaneous coronary intervention: a multicenter randomized controlled trial. Chin J Integr Med. 2017;10:22–8.
7. Min F. Study on nursing measures for complications after coronary angiography and intracoronary stent implantation. J Pract Clin Med. 2016;20(4):4–7.
8. S-SY, LT, G-HG, et al. Meta-analysis of the long term effects of different interventions on chronic total coronary occlusions. Eur Rev Med Pharmacol Sci. 2013;17(12)
9. Alfonso F, Byrne RA, Rivero F, et al. Current treatment of in-stent restenosis. J Am Coll Cardiol. 2014;63(24):2659–73.
10. Chenfeng G, Qiuyu F, Jianxiong L. Research progress of coronary drug-eluting stent restenosis. Prog Cardiol. 2017;38(1):70–3.
11. Mindrescu C, Brener SJ, Guerchicoff A, et al. Impact of scheduled angiographic follow up in patients treated with primary percutaneous coronary intervention for ST-segment elevation myocardial infarction. J Interv Cardiol. 2013;26(4):319–24.
12. Yates DJ, Savage ML, Walters DL, et al. Retrospective study of first generation drug eluting stents, second generation drug-eluting stents and non-drug eluting stent methods Ln the

treatment of native vessel in-stent restenosis in real world clinical practice. Heart Lung Circ. 2015;25(4):342–51.
13. Members WG, Mozaffarian D, Benjamin EJ, et al. Heart disease and stroke statistics-2016 update: a report from the American heart association. Circulation. 2016;133(4):e38 e360.
14. Ningfu W, Shuang Z, Feng F, et al. Effect of simple balloon angioplasty on coronary stent restenosis. Prev Treatment Cardiovasc Cerebrovasc Dis. 2005;5(1):23–4. https://doi.org/10.3969/j.issn.1009-816X.2005.01.009.
15. De Labriolle A, Pakala R, Bonello L, et al. Paclitaxel-eluting balloon: from bench to bed. Catheter Cardiovasc Interv. 2009;73(5):643–52.
16. Windecker S, Kolh P, Alfonso F, et al. 2014 ESC / EACTS guidelines on myocardial revascularization: the task force on myocardial revascularization of the European society of cardiology (ESC) and the European Association for Cardio-Thoracic Surgery (EACTS) developed with the special contribution of the European Association of Percutaneous Cardiovascular Interventions (EAPCI). Eur Heart J. 2014;35(37):2541–619.
17. Sergie Z, Dangas GD. In-stent restenosis of bifurcation lesions: experience with drug-eluting balloons. Catheter Cardiovasc Interv. 2012;79(3):397–8.
18. Kufner S, Cassese S, Valeskini M, et al. Long-term efficacy and safety of paclitaxel-eluting balloon for the treatment of drug- eluting stent restenosis: 3-year results of a randomized controlled trial. JACC Cardiovasc Interv. 2015;8(7):877–84.
19. Cremers B, Toner JL, Schwartz LB, et al. Inhibition of neointimal hyperplasia with a novel zotarolimus coated ballooncatheter. Clin Res Cardiol. 2012;101(6):469–76.
20. Kongyong C, Shuzheng LV, Xiantao S, et al. Research progress of drug-eluting balloon in the treatment of coronary artery disease other than in stent restenosis. Chin J Cardiovasc Dis. 2017;45(1):70–3.

Chapter 7
Cardiac Valves

Yida Tang

7.1 Transcatheter Treatments of Aortic Valve Disease

7.1.1 Introduction and Etiology

Degenerative aortic stenosis, according to available epidemiological data, has a prevalence of 1.3% in people aged 60–69 years, 3.9% in people aged 70–79 years, and 9.8% in people aged 80–89 years. Patients with severe aortic stenosis, whether symptomatic or not, have a poor prognosis. Patients with asymptomatic severe aortic stenosis have a 2-year survival rate of 56–63% and a 5-year survival rate of 25–33%. Patients with symptomatic aortic stenosis have a 1-year morbidity and mortality rate of 32–51%. Of these, approximately 20–40% of patients have to undergo minimally invasive endoluminal treatment because of high surgical risk [1–3].

The population prevalence of congenital aortic valve malformation is 0.5–2.0%, with congenital bicuspid aortic valve (BAV) stenosis accounting for approximately 49%. However, patients with congenital aortic stenosis usually complete aortic valve replacement before age 65, and congenital aortic stenosis accounts for only 15–33% of all patients with minimally invasive aortic valve treatment [4–7].

The incidence of clinically significant aortic regurgitation (AR) increases with age, typically peaking in the fourth to sixth decade of life. It is more common in men than women. The prevalence of AR in the Framingham study was reported to be 4.9%, with regurgitation of moderate or greater severity occurring in 0.5% [8, 9].

Y. Tang (✉)
Department of Cardiology and Institute of Vascular Medicine, Peking University Third Hospital, Beijing, China
e-mail: tangyida@bjmu.edu.cn

Y.-D. Tang et al. (eds.), *Application of Biomaterials in the Treatment of Cardiovascular Diseases*, https://doi.org/10.1007/978-981-19-7712-1_7

7.1.2 Anatomy and Pathophysiology of Aortic Valve Disease

7.1.2.1 Anatomy and Pathophysiology of Aortic Stenosis

An aortic valve area of l cm or less is considered severe AS (normal, 3–4 cm), and the hemodynamic severity of AS is best characterized by transaortic peak velocity ≧4 m/s or mean pressure gradient ≧40 mmHg. Patients with aortic stenosis may have a combination of widened aortic internal diameter and thickened and enhanced echogenicity. In severe aortic stenosis, thickening of the interventricular septum and posterior left ventricular wall may occur, and the ascending aorta may be dilated after stenosis.

Aortic stenosis can be congenital or acquired. Rheumatic heart disease used to account for the majority of AS. In recent years, the incidence of degenerative aortic valve disease has increased significantly. In rheumatic lesions, there are varying degrees of valve leaflet thickening, echogenic enhancement, valve deformation, stiffness, and a significant reduction in opening amplitude. Ultrasound shows that the aortic valve loses its normal "Y" shape during diastolic closure, and the opening area becomes smaller, deformed, and asymmetrical in a "plum-shaped" manner.

Another acquired cause is atherosclerosis, which leads to degenerative lesions such as aortic valve fibrosis and calcification, with calcification occurring mainly at the root of the valve leaflets and the annulus. Echo enhancement of the annulus and root of the leaflets, thickening of one or more leaflets (>2 mm), and stiffening of the motion of one or more leaflets may be seen on ultrasound in cases of degeneration, and in severe cases the body and cusps may be involved.

Congenital aortic stenosis can be unicuspid, bicuspid, tricuspid, or quadracuspid, with bicuspid malformations being more common. Due to the malformation of the valve leaflets, the opening and closing activities after birth may cause damage to the valve leaflets, fibrosis, and calcification, which eventually lead to stenosis. Patients with aortic stenosis may also have widened aortic internal diameter and thickened and enhanced echogenicity. In severe stenosis, thickening of the ventricular septum and posterior left ventricular wall may occur [10].

7.1.2.2 Anatomy and Pathophysiology of Aortic Regurgitation

Rheumatic disease remains the most common cause of AR in developing countries; however, in Western Europe and North America, the leading causes of AR are congenital (especially due to bicuspid valves) or degenerative diseases, including annular aortic ectasia.

AR results in left ventricular (LV) volume overload. In severe AR, regurgitant flow may equal or even exceed the effective ejection volume. The increase in LV end-diastolic volume is the primary compensatory mechanism for maintaining normal effective stroke volume. This significantly enlarged perivalvular structure of the aortic valve, as well as the relatively mild or no calcification of the valve, may be detrimental to the fixation of the prosthesis and present a greater risk of perivalvular leakage.

The left ventricular ejection fraction is initially normal, but, over time, the left ventricular end-diastolic volume continues to increase further, and the ejection fraction decreases; these changes may actually precede the onset of clinical symptoms. Acute AR can be life-threatening because lung expansion and other compensatory mechanisms do not develop rapidly enough to avoid hemodynamic deterioration. Regurgitant volumes that are well tolerated in chronic AR can lead to a significant increase in LV end-diastolic pressure and a decrease in effective beat volume, resulting in pulmonary edema, hypotension, and even cardiogenic shock [9].

7.1.3 Indications and Contraindications for TAVR

In general, aortic valve interventions are recommended for patients with severe AS and direct objective imaging evidence or surrogate symptomatic measures of left ventricular dysfunction. In 2020 ACC/AHA Guideline for the Management of Patients with Valvular Heart Disease [11], it is stated that, for symptomatic and asymptomatic patients with severe AS with any indication of AVR who are <65 years of age or have a life expectancy >20 years, SAVR is recommended. For symptomatic patients with severe AS who are 65–80 years of age and have no anatomic contraindication to transfemoral TAVR, either SAVR or transfemoral TAVR is recommended after shared decision-making about the balance between expected patient longevity and valve durability. For symptomatic patients with severe AS who are >80 years of age or for younger patients with a life expectancy <10 years and no anatomic contraindication to transfemoral TAVR, transfemoral TAVR is recommended in preference to SAVR. For symptomatic patients of any age with severe AS and a high or prohibitive surgical risk, TAVR is recommended if predicted post TAVI survival is >12 months with an acceptable quality of life.

When referring to isolated chronic AR, TAVR is still challenging because of dilation of the aortic annulus and aortic root and, in many patients, lack of sufficient leaflet calcification. Risks of TAVR for treatment of AR include transcatheter valve migration and significant paravalvular leak. TAVR is rarely feasible and then only in carefully selected patients with severe AR and HF who have a prohibitive surgical risk and in whom valvular calcification and annular size are appropriate for a transcatheter approach.

The early pivotal TAVR trials excluded patients with BAV. Initial studies using early-generation valves suggested a higher rate of paravalvular leak in the BAV population. Data from the STS/ACC Transcatheter Valve Therapies Registry, which includes all consecutive TAVR procedures performed in the United States, suggest that with the use of newer-generation prosthetic valves, the rate of paravalvular leak is no different in patients with a BAV than in patients with a TAV. This registry also showed no difference in mortality rate at 30 days and 1 year between the BAV and

TAV groups. However, the stroke rate at 30 days was higher in the BAV group. Other considerations are the younger age of patients with a BAV, for whom the risk-benefit ratio of TAVR versus SAVR needs careful consideration. RCTs are needed to obtain full clarity on the optimal use of TAVR in this population, as well as long-term outcomes [12].

7.1.4 Devices for TAVR

In 2002, Alain Cribier and his colleagues pioneered the use of TAVR in a 57-year-old man who was unfit for surgery because of severe AS and multiple comorbidities leading to cardiogenic shock [13]. After this groundbreaking success, the concept of TAVR was accepted, the technology continued to advance, and the indications gradually expanded. The development of the TAVR device concept was divided into two distinct routes.

One route was based on the same concept as the first valve and was based on the balloon dilation technique previously used for stents. This route was dominated by successive improved models developed by Edwards Life Sciences (Sapien, Sapien XT, Sapien 3; see Fig. 7.1), known as balloon-expandable valves. A few years later, CoreValve developed another valve with a nickel-titanium alloy backbone, which has a unique feature: it contracts at low temperatures and expands at 37 °C, hence the name of this valve as a self-expandable valve. This technological route is followed up mainly by several modified models from Medtronic (CoreValve, Evolut R, Evolut PRO) [14, 15].

Several other companies have also developed balloon-expandable or self-expandable prostheses. Recently, a large network meta-analysis showed similar clinical study results for balloon-expandable and self-expandable prostheses [16]. The main technical trends in the development of TAVR over the past decade have been the gradual reduction in the use of other alternative access methods (including transapical access) in favor of transfemoral access, the use of smaller

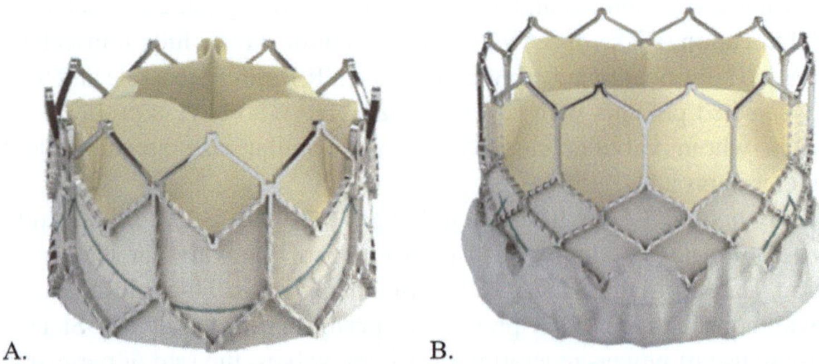

Fig. 7.1 Overview of Edwards Lifesciences balloon-expandable valves. (**a**) Edwards SAPIEN XT, (**b**) Edwards SAPIEN3

catheters and valves of different sizes tailored to anatomic needs, the design of retrievable devices to reposition the valve when needed, the use of an outer skirt to reduce paravalvular leakage (PVL), or improved delivery systems.

7.1.4.1 Sapien Balloon-Expandable Valve

Edwards' initial valve was the Cribier-Edwards THV (Edwards Lifesciences, Irvine, CA, USA), which consisted of a stainless steel frame and equine pericardial leaflets. The subsequent modification was the Edwards SAPIEN THV, which featured a higher sealing cuff and bovine pericardial leaflets, among others (Fig. 7.1a). The SAPIEN THV was followed by the SAPIEN XT THV (Fig. 7.1b), which still has the bovine pericardial leaflet, but the frame was upgraded from stainless steel to cobalt-chromium alloy that provides a high radial strength for circularity and optimal hemodynamics. The latest generation of Edwards balloon-expandable valves is the SAPIEN 3 (S3) THV, which still uses a three-piece valve tissue made of bovine pericardium and a cobalt-chromium alloy frame, but with a lower frame height and open cell geometry to preserve access for coronary interventions, and a polyethylene terephthalate (PET) skirt wrapped around the bottom outer edge of the valve to reduce valve paravalvular leakage (PVL) (Fig. 7.1c). The S3 THV is currently available in four sizes for the operator: 20, 23, 26, and 29 mm. The choice of size needs to be based on the results of the CT evaluation, and S3 can treat a wide range of aortic annulus diameters, from 18.6 to 29.5 mm.

The Sapien series valves are positioned according to the marker in the center of the dilating balloon. The S3 THV can be delivered via a 14-French (Fr) (20, 23, or 26 mm S3 THV) or 16-Fr (29 mm S3 THV) expandable sheath (eSheath; Edwards Lifesciences). The outer diameter of the expandable sheath is 18 Fr (for 20–26 mm S3 THV) and 20 Fr (for 29 mm S3 THV), which is characterized by an increase to 24 Fr (for 20–26 mm S3 THV) and 27 Fr (for 29 mm S3 THV) during THV passage. Considering that the artery has some compliance, the minimum vessel diameters recommended for the transfemoral approach are 5.5 mm (for 14-Fr eSheath) and 6 mm (for 16-Fr eSheath), respectively. However, some caution is required when there is round vessel calcification in the arterial pathway [17].

S3 THV has received CE (Communaute europeenne) mark approval in Europe in January 2014 and FDA approval in the United States in June 2015. Clinical results from the PARTNER II trial confirm the promising results of the S3 THV system: the all-cause mortality was 32.7% and aortic valve re-replacement was required in 1.9% at 3 years. Mean transaortic gradient was 35.0 mmHg at baseline, decreasing to 17.8 mmHg at 30-day follow-up and 16.6 mmHg at 3-year follow-up [18]. Longer follow-up results showed l-year mortality rates of 17.7% for inoperable patients and 12.7% for high-risk patients. The PARTNER III trial, which expanded the study population to include low-risk patients, showed an incidence of the primary endpoint of 8.5% for TAVR and 15.1% for SAVR, confirming the noninferiority and superiority of the TAVR group in patients with severe aortic stenosis and low surgical risk. The PARTNER III trial provided evidence that the use of TAVR can achieve at least comparable results to SAVR in the treatment of low-risk surgical candidates [17].

7.1.4.2 CoreValve

The representative of the self-expandable valve is the Evolut series of valves designed and developed by CoreValve. The valve is composed of a nickel-titanium frame and a tricuspid valve made from porcine pericardium. The frame is a shape with a narrow waist and wide ends, and its portion in the left ventricular outflow tract (LVOT) provides high radial forces in the aortic annulus, so the valve is stabilized and fixed. In addition, the 12 mm porcine pericardial "skirt" plays an important role in preventing regurgitation and paravalvular leakage. The central portion of the skeleton has a tricuspid valve that is designed in such a way as to avoid blockage of the coronary artery orifice. Available valves include 26, 29, and 31 mm, which can cover 20–29 mm of the aortic annulus. With the addition of the CoreValve Evolut (23 mm), however, its length is reduced by 10%, and 18–20 mm of the aortic annulus is covered. The CoreValve is the only available TAVR valve that covers 18–29 mm of the aortic annulus. The Evolut R and EnVeo R are new prosthetic valves and delivery systems from Medtronic. The valve is 10% less in length than the original CoreValve, allowing implantation of the valve in angled anatomic regions and in noncircular or asymmetric calcified aortic annulus. To reduce perivalvular leakage after valve implantation, its "skirt" extends toward the LVOT (Fig. 7.2).

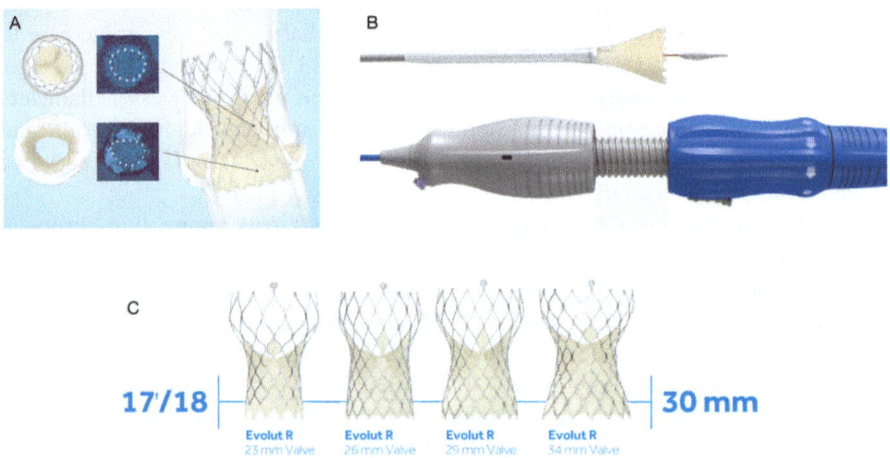

Fig. 7.2 Evolut R valve. (**a**) The supra-annular valve design of the nitinol frame maximizes the valve orifice and leaflet coaptation. (**b**) The EnVeo PRO delivery system assists in accurate positioning of the valve and the option to recapture and reposition. (**c**) Evolut R valve has expanded the annulus range from 23 to 34 mm

CoreValve has some major advantages, such as its larger range of annular sizes that can be covered and the lower profile of the inlet device, which can reduce vascular events. However, CoreValve also has a vital disadvantage of a high implantation rate of permanent pacemakers, a problem that has been addressed in the newer valves.

The primary clinical evidence for CoreValve comes from the ADVANCE, SURTAVI, and Extreme Risk US Pivotal trials.

The ADVANCE trial enrolled patients who could not undergo the procedure ($n = 1015$). Three years after TAVR, the rate of major adverse cardiac/cerebrovascular events was 38.5%, all-cause mortality was 33.7%, and cardiovascular mortality was 22.3%. Average mean aortic valve gradient remained ≤10 mmHg, and 12.6% of patients had moderate and none had severe paravalvular regurgitation. For the two groups older and younger than 75 years, there was no difference in mortality at 2 years (23.6 vs. 26%, $p = 0.448$), indicating a benefit of TAVR for both groups of patients (older and younger than 75 years) [19]. The SURTAVI trial randomized 1660 patients, and the primary 2-year endpoint of all-cause mortality or disabling strokes for the first 1400 patients was 12.7% in the TAVR group and 12.6% in the surgery group, which confirmed the noninferiority, with respect to the frequency of all-cause mortality or disabling stroke, of TAVR to surgery [20].

Results from the second leg of the Extreme Risk US Pivotal Trial showed that at 1 year, patients who underwent TAVR had a lower mortality rate (14.2 vs. 19.1%, $p < 0.04$), fewer cerebrovascular events, and lower bleeding rates compared to SAVR patients. However, patients with TAVR had more complications (6.2 vs. 2%, $p < 0.004$) and permanent pacemaker implantation (22.3 vs. 11.3%, $p < 0.001$), in addition to a high late mortality in patients with severe aortic regurgitation after TAVR [3, 21].

7.1.4.3 Lotus and ACURATE

The Lotus™ Valve System (Boston Scientific, Natick, MA, USA) consists of three bovine pericardial valves supported on a woven nitinol frame (Fig. 7.3). A radiopaque marker is located in the vertical center of the implant to aid in positioning. An adaptive seal around the ventricular portion of the device, designed to conform to the irregular surface of the local anatomy, has the potential to reduce paravalvular aortic regurgitation. The system's most prominent features utilize an intuitive design and a high degree of user control to facilitate safe and accurate device placement. To simplify handling, preparation, and loading, the Lotus valve is pre-attached to the delivery system, requiring no special tools or clamps. The delivery handle enables a controlled and accurate deployment process through two controls: a control knob and a release collar. By turning the control knob counterclockwise, the valve can be unsheathed, deployed, and locked in a smooth and progressive manner. The system has a retrieval feature that allows the device to be advanced or retracted as needed if positioning is not ideal, and the valve can be fully retracted into the delivery sheath as long as the final release is not completed. In addition, the device can be safely

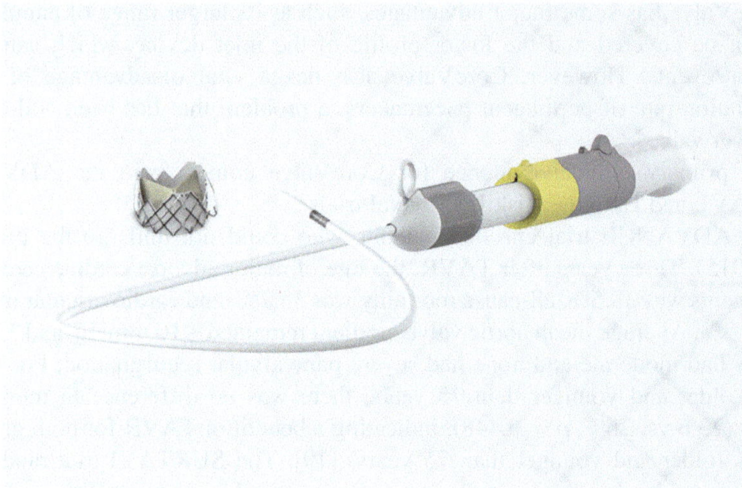

Fig. 7.3 The J-Valve™ system

retracted during the procedure if it is decided not to implant. The valve is released by sliding the release ring forward and turning the release mechanism. The valve is ready to start working when the first half of the release process is completed, ensuring stable hemodynamics during the procedure and allowing for adequate operator consideration and control [22].

In 2017, the RESponD trial published initial prospective data utilizing the Lotus valve system. In this prospective, open-label, single-arm study, 1014 patients were enrolled in the post-marketing study. In terms of device safety, the 30-day all-cause mortality rate was 2.6% (95% CI 1.7, 3.8, $p < 0.001$), and the rate of disabling stroke was 2.2%, which were comparable to those from the PARTNER II trial and the CoreValve SURTAVI trial. Although the incidence of more than moderate perivalvular leaks was significantly lower (0.3%), it came at the cost of a high permanent pacemaker implantation rate of 34.6%. The Depth Guard technique was subsequently introduced with the aim of anchoring and reducing the depth of the valve frame earlier in the deployment process, thereby reducing LVOT interactions. In the ResponD extension study, 50 patients underwent TAVR with the Lotus valve with Depth Guard, and only 8 (16.0%) required permanent pacemaker implantation, significantly lower than previous REPRISE data.

The most recent data from the REPRISE II trial represent the first head-to-head comparison data between the self-expandable CoreValve TAVR and the mechanical Lotus Edge valve. The Lotus system met the 1-year primary effective endpoint of noninferiority and superiority. Paravalvular leakage was seen in 6.9% of CoreValve TAVR patients compared with 0.9% in the Lotus TAVR population. Patients with Lotus TAVR remained at 30 days (35.5 vs. 19.6%) and 1 year (41.4 vs. 23.0%1) had higher rates of permanent pacemaker implantation [23, 24].

All unused LOTUS Edge™ Aortic Valve System was recalled on Nov. 17, 2020, due to complexities associated with the product delivery system.

7.1.4.4 J-Valve

Transapical TAVR is also an option for patients with pure aortic regurgitation and for patients for whom neither transfemoral nor other peripheral arterial approaches are appropriate. A representative of transapical TAVR devices is the J-Valve system (JCMedical, Inc., Burlingame, CA, USA). The J-Valve™ system is composed of a porcine aortic valve attaching to a self-expandable nitinol stent with three U-shaped graspers encircling the valve stent. This unique design greatly simplifies the implantation process and leads to more consistent results, as the operator does not need to actively determine the optimal valve position. Also, by embracing the native valve with three U-shaped graspers, the "J-Valve" system has the advantage of achieving better anchoring, a reduced risk of coronary obstruction due to native valve cusp calcification, and a reduced risk of conduction block (less off-axis fixation and subsequent radial expansion forces). Unlike other second-generation TAVR devices, the J-Valve system's U-shaped graspers increase the interface between the aortic valve leaflet and the valve, which will provide more anchoring force (axial fixation) after valve deployment. In addition, the system's unique two-stage release design (the graspers are fully released first and then the valve is released) can also largely facilitate accurate adjustment of the gripper position prior to valve deployment. In addition, this U-shaped gripper minimizes the risk of native valve perforation [25].

In a study published in 2018 enrolling 107 patients with high-risk AS or AR, the success rate of the J-Valve procedure was 91.6% (98/107). At 1 year, all-cause mortality was 5.0%, stroke rate was 2%, new pacemaker rate was 5.0%, and only mild paravalvular leaks were reported. Event-free survival rates for major adverse cardiovascular events at 1 year were similar for either the AS and AR or the BAV and TAV [26].

7.1.5 Issues with TAVR Devices

The major complications of TAVR therapy include permanent pacemaker implantation, perivalvular leak and aortic regurgitation, coronary risk, and risk of embolism and stroke. These complications are more prevalent in first-generation prosthetic valve devices. The risk of these complications has gradually decreased with our improved understanding of valve structure and operative procedures, refinement of preoperative evaluation, improvements in instrumentation, and the use of appropriate protective devices.

7.1.5.1 Permanent Pacemaker Implantation

Currently, the need for postoperative implantation of a permanent pacemaker (PPM) is one of the most common complications of TAVR [27]. For first-generation devices, a landmark meta-analysis of 32 studies (5258 patients) found a 15% incidence of PPM (25.8% for the CoreValve self-expandable valve and 6.5% for the Sapien balloon-expandable valve) [28]. Another review of 49 studies (16,063 patients) reported an overall probability of 13% of patients requiring PPM, which was five times higher with CoreValve than with Sapien valves [29, 30]. The incidence of PPM was highly variable in newer-generation prostheses (ranging from 2.3 to 26.1%) [31]. The higher incidence of PPM in self-expandable valves appears to be due to the increased radial forces on the left ventricular outflow tract from the expanding nitinol device [32].

The impact of PPM on the prognosis of patients with TAVR is unclear. However, a meta-analysis did not find a negative effect of PPM on 1-year survival [33]. Subsequent results from a large US TVT registry showed that PPM implantation resulted in longer hospital and ICU stays and increased 1-year mortality (24.1 vs. 19.6%) [34]. The impact of PPM on long-term prognosis has not been evaluated. Probably because PVL was earlier shown to be significantly associated with higher mortality, while the prognostic impact of PPM remains unclear, the direction of improvement in the new generation of balloon and self-expandable prostheses has focused on reducing paravalvular leakage (PVL), either with a sealing skirt or a repositionable and retractable device, but not much consideration has been given to the need for PPM reduction [35]. The incidence of PPM use has not decreased and has even increased with newer-generation valves. The Sapien 3 valve has a PPM-resistant outer skirt that may exert greater radial forces on the valve, with PPM rates between 10.1% and 14.1% found in the Sapien 3 multicenter registry, more than twice that of its predecessor, the Sapien/Sapien XT valve [36–39].

A possible solution comes from selecting a higher implantation location, meaning that the portion of the ventricle that protrudes into the ventricle is <25% of the ventricular portion, but this requires further data to confirm and still requires consideration of the risk of valve migration or PVL. A repositionable valve may be useful in reducing PPM by helping to select a higher release position and avoiding excessive penetration into the ventricle. The new generation of self-expandable valves, the Evolut R, has a PPM rate of 16.4% and the Evolut PRO, 11.8%, whereas the new Edwards Centera self-expandable valve has an encouraging PPM rate of 4.9%, with a lower degree of protrusion of the frame into the outflow tract [40–42]. The new Symetis Acurate Neo/TF self-expandable flap used a PPM rate of only 9.9% [43]. Risk factors predicting the use of PPM in TAVR, in addition to self-expandable valves, are preexisting right bundle branch block, male gender, calcification, implant depth (some valve-specific cutoffs have been proposed), first-degree AV block, and aortic annulus/left ventricular outflow tract overstretch [44].

Postoperative pacemaker dependence is not long-lasting in some patients treated with TAVR. In cohort studies of patients treated with various TAVR treatments, long-term dependence ranged from 27 to 68.4% [45, 46]. A study of patients receiving CoreValve self-expandable valves showed a long-term dependence of 46.7%, and 57% and 38% of patients treated with Lotus valves had pacemaker dependence at 30 days and 1 year, respectively [47, 48]. Inflammation, edema, ischemia, or direct injury may cause transient pacemaker dependence, but some patients can subsequently recover. Continuous rhythm monitoring prior to PPM implantation after TAVR may be necessary to reduce unnecessary PPM implantation [49].

7.1.5.2 Paravalvular Leak

PVL is one of the most common complications of TAVR and is currently the main driver for device design improvements. Moderate-to-severe PVL is associated with higher short- and long-term complications and mortality [50–54]. The overall incidence of mild PVL for first-generation valves ranged from 20 to 80%, and the incidence of moderate-to-severe PVL ranged from 5 to 22% [55]. A meta-analysis of 45 studies (11,244 patients) reported an 11.7% incidence of PVL, with CoreValve having a significantly higher risk of moderate-to-severe PVL compared with Sapien (16.0 vs. 9.1%) [53]. Another meta-analysis (20 studies, 11,244 patients) also showed that the risk of PVL was significantly higher with self-expandable valves than with balloon-expandable valves [56, 57]. Although self-expandable valves had a higher rate of more than mild PVL, a direct comparison of the two techniques showed no effect on mortality.

The incidence of moderate-to-severe PVL decreases substantially with newer-generation valves. The incidence of PVL was reported to be 2% or 3.4% with the Edwards Sapien 3 [36, 58]. Early findings showed that patients receiving the CoreValve Evolut PRO device did not experience more than mild PVL within 30 days (0%); moderate PVL occurred in only 1% of Lotus valves, and no severe PVL occurred [59]. Recent trials in low-risk patients have demonstrated very low rates of PVL. In the PARTNER 3 trial (next-generation Sapien 3 balloon-expandable valve), the 30-day rate of moderate-to-severe PVL was 0.8% in TAVR versus SAVR, and in the Evolut low-risk trials (CoreValve, Evolut R, or Evolut PRO self-expandable valves), the 30-day PVL rates were 3.5% and 0.5%, respectively [60, 61]. Such findings confirm both the reduced risk of new-generation valves and the high risk of self-expandable prostheses.

Reasons for the reduced incidence of PVL in recent years include increased operator experience in valve positioning, improvements in design, and refinement of preoperative imaging assessment of valve dimensions. The design of newer-generation valves has improved considerably. Manufacturers have added a cuff (Sapien, Lotus) or an external sealing system (Evolut or Portico) to the inflow portion of the valve. In addition, retrievable systems have been effective in reducing PVL [55, 62]. However, as previously mentioned, there are some concerns about the potential trade-off between PVL and PPM complications, considering that implantation of new-generation valves has not reduced PPM or even increased it.

Minimal or mild PVL usually does not require corrective measures. If more than mild PVL is detected after TAVR, immediate correction is recommended by post-balloon dilation [63, 64] or valve-in-valve manipulation [65, 66]. However, because of the persistent dilation effect of the metal skeleton in some types of valves, it has been reported that undergoing SAPIEN 3 or CoreValve devices have shown a decrease in PVL severity after a period of follow-up [67, 68].

7.1.5.3 Stroke and Embolic Protection

Stroke is one of the most feared complications after aortic valve replacement. A meta-analysis of 25 studies (1225 patients) found that among patients undergoing TAVR, brain magnetic resonance imaging detected new perioperative lesions in up to 77.5% (95% CI, 71.7–83.3%), and 59.5% had multiple lesions [69]. Fortunately, most such lesions are so-called resting strokes and are not associated with clinical symptoms; the incidence of clinically significant 30-day strokes is approximately 2.8%, and the risk of stroke is decreasing with advances in experience and technology [70]. However, because perioperative stroke is associated with greatly increased in-hospital and 1-year mortality, as well as decreased quality of life and severe disability in survivors, there is still a need for methods to prevent neurological embolic complications [71].

A number of embolic protection devices have been developed and are in clinical use. These mesh devices are designed to prevent surgically generated debris, including blood clots, valve calcification/tissue, and arterial wall debris, from entering and occluding the cerebral vasculature. There are currently two main types. One is a device that deflects debris, including the Embrella Embolic Deflector (Edwards Lifesciences, CA, USA) and TriGuard (Keystone Heart, Israel); the other is a filter-type device that intercepts embolic material, thereby reducing the risk of peripheral embolism. These include Embol-X (Edwards Lifesciences, CA, USA) and Sentinel (Boston Scientific Corporation, CA USA) (formerly known as Montage, Claret Medical), among others [71].

Although such devices have been shown to be safe and effective in filtering debris and reducing the volume of ischemic lesions, however, no difference could be demonstrated in reducing clinically significant strokes [72]. A randomized controlled trial using the Sentinel device to assess mortality did not find that such cerebral protection devices reduced mortality and the primary composite endpoint including all major adverse cardiac and cerebrovascular events. Currently, the quality of evidence supporting clinical efficacy in this area of research remains low, and the volume of lesions remains small [73]. Further evidence on this issue is urgently needed.

7.1.5.4 TAVR and Cardiac Ischemic Disease

TAVR treatment may result in coronary artery acute occlusion (CAO), which is more common in self-expandable valves. There are several anatomic features that predict coronary risk during TAVR, including the height of the coronary ostium, the size of the Valsalva sinus, rheumatic valve disease, and thickening or calcification of the tips of leaflets. In addition, larger sized valves also put the risk of coronary occlusion at increased risk. Because of the potentially catastrophic consequences, once a high risk of CAO is considered, proactive coronary protection measures should be performed rather than remedial maneuvers. There are a number of methods of coronary protection, including guidewire protection, balloon protection, guidezilla protection, and pre-positioned stents. The pre-positioned stent approach has a high reliability but may cause stent derotation during the withdrawal of the stent for postoperative confirmation that no coronary protection is required. Chimney stenting is a commonly used technique for the treatment of CAO. A total of 60 (approximately 0.5%) patients underwent chimney stenting, of which 44 were preemptively coronary protected, 25 were used to treat CAO that had already occurred, and 35 were prophylactically used in patients at high risk for CAO. The results showed that the absence of prior coronary protection was the only independent risk factor for the combined endpoint of death, cardiogenic shock, or myocardial infarction. Two stent failures (one in-stent restenosis and one possible late stent thrombosis) were reported at a median follow-up time of 612 days, after 157 and 374 days, respectively. [74].

Bioprosthetic or native aortic scallop intentional laceration to prevent iatrogenic coronary artery obstruction during TAVR (BASILICA) is a transcatheter procedure performed immediately before TAVR aimed to protect coronary artery from CAO. The target aortic leaflet is split using focused radiofrequency energy directed by catheters and guidewires. The BASILICA trial enrolled 30 subjects. Primary success was met in 28 (93%) subjects. BASILICA traversal and laceration was successful in 35 of 37 (95%) attempted leaflets. There was 100% freedom from coronary obstruction and reintervention. Primary safety was met in 21 (70%), driven by 6 (20%) major vascular complications related to TAVR but not BASILICA [75].

Because degenerative AS and atherosclerosis share many risk factors and pathogenic mechanisms, coronary artery disease (CAD) is frequently comorbid in patients with severe AS (40–75%) [76]. A recent meta-analysis that included 8013 patients from 15 studies showed a significant adverse effect of CAD on 1-year mortality [77]. Another meta-analysis showed (6 studies, 3107 patients) that residual SYNTAX scores after coronary revascularization predicted higher mortality, using 8 as a cutoff value [78]. With the expansion of the indications for TAVR, especially the increase in younger and lower-risk patients with longer life expectancy, there is a need to adequately manage CAD and to preserve the opportunity for interventions that may be required later in the coronary arteries of these patients, especially for self-expandable valves [79].

7.1.5.5 Durability

The longevity of bioprosthetic valves is always an issue to consider, both for TAVR and SAVR, especially as the indications for TAVR gradually reach younger, lower-risk patients with longer life expectancy. Significant degeneration of the surgical bioprosthetic flap usually begins about 10 years after surgery and increases dramatically in the second decade after surgery [80]. Because TAVR is a relatively new technology, there is a lack of long-term data to assess the durability of porcine and bovine flaps. In addition, the low 5-year survival rate of early clinical studies, which were limited to high-risk patients, also limits the assessment of valve durability [81]. Potential factors affecting durability include physical factors and clinical/metabolic factors [82–84].

Results were equally favorable in the 5-year results of the PARTNER II trial, the 3-year results of CoreValve's high-risk pivotal trial, the 5-year results of the ADVANCE trial, and the 5-year results of an Italian multicenter registry which were reassuring [85–88]. A meta-analysis of 13 observational studies (8914 patients) with a follow-up period of 1.6–5 years reported only 0–1.34 valve deteriorations per 100 patients. In contrast, the UK registry reported a deterioration rate of 8.7% after a mean follow-up of 6.1 years [89].

In 2017, the European Association for Percutaneous Cardiovascular Interventions (EAPCI), with the support of the European Society of Cardiology (ESC) and the European Association for Cardiothoracic Surgery (EACTS), published a consensus statement defining the concepts and grades of "hemodynamic SVD" and "morphologic SVD," and in 2018, the Valve International Data (VIVID) group defined structural valve deterioration (SVD) lack of consistency was addressed. Using such a new definition, the rates of severe SVD and moderate/severe SVD were 2.5% and 13.3%, respectively, over 5 years in the FRANCE 2 registry and 2.4% and 8.3%, respectively, over 8 years in the local Italian REPLACE registry [90, 91]. Thus, the interim follow-up data are reassuring, but we eagerly await the 10- and 15-year data to inform our current decisions on the use of TAVR in low-risk patients.

7.1.6 Conclusions

The advent of TAVR in the last decade has led to a major revolution in the treatment of severe AS and indeed in the field of structural heart disease in adults. While current device outcomes are still accompanied by many complications and device design is not perfect, PVL problems appear to be diminishing with current improvements in device design, and PPM implantation requirements are a problem that must be addressed with the new generation of prostheses. The incidence of stroke is low, but its adverse effects make it a continuing concern; the clinical performance of existing embolic protection devices still needs further demonstration. The role of

TAVR in patients with comorbid ischemic heart disease should also be better defined. In the coming years, the availability of durability data will be critical to determine the role of current prostheses in low-risk patients with longer life expectancy or the need to develop additional devices. In addition, experience will accumulate in off-label field including BAV and pure aortic regurgitation.

7.2 Transcatheter Mitral Valve Treatment

7.2.1 Introduction

Mitral valve disease is an important type of heart valve disease. Rheumatic mitral stenosis is no longer the predominant type of mitral valve disease, which has been replaced by mitral regurgitation (MR), the incidence of which is increasing with an aging population, and severe MR significantly decreases patient survival. In addition to optimal medicine therapy, conventional surgery is the mainstay of treatment for MR. In recent years, interventional techniques for MR have flourished, and significant breakthroughs have been achieved. In this article, we will systematically present the current status and progress of interventional treatment of transcatheter mitral valve disease (mainly mitral regurgitation).

7.2.2 Anatomy and Imaging

The mitral valve (MV) is located in the left atrioventricular groove, adjacent to the aortic valve and the left ventricular outflow tract (LVOT). The core structures of the mitral valve are the anterior and posterior leaflets, which are adjacent to each other to form the "mitral smile" (Fig. 7.5) and the saddle-shaped mitral annulus. With the left auricular side as the lateral side, the leaflets are divided into three regions (A1, A2, and A3 and P1, P2, and P3, respectively) from the lateral to the medial. The anterior and posterior leaflets meet at the lateral and medial commissure, forming a junction area. In addition, the subvalvular apparatus of a MV device consists of chordae tendineae, papillary muscle, and left ventricular wall [92]. Not only appropriate size of the annulus and intact structure of leaflets, a healthy mitral valve also requires contraction of papillary muscle to pull chordae tendineae and normal ventricular morphology and function. Any exception would cause MR.

The structure of the MV is dynamic, asymmetric, and heterogeneous, which is significantly different from other valves. The MV has four distinct anatomic features:

1. The mitral annulus has a "saddle-shaped" spatial structure, with fibrous tissue anterior to the annulus, continuous with cardiac fibrous tissue through the left and

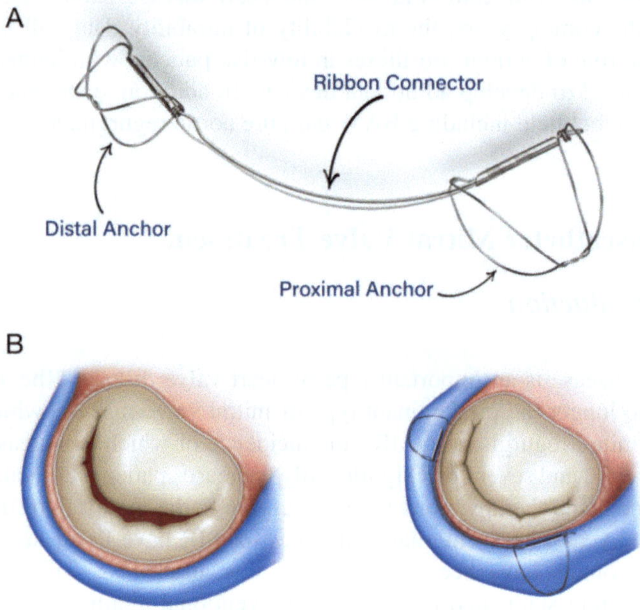

Fig. 7.5 The Carillon Mitral Contour System. A. The distal and proximal anchors of the Carillon device are connected by a shaping ribbon that utilizes the heart's venous anatomy to cinch the mitral apparatus. B. Before and after Carillon implant

right fibrous triangle and movable in horizontal and vertical planes (dynamic saddle-shaped structure)

2. The valve leaflet's irregular geometry (lesions of the anterior and posterior valves)
3. Large annular diameter, without calcification in most cases
4. Complex subvalvular anatomy and adjacent tissue structures

Advancements in cardiac imaging allow visualization and quantification of the mitral valve in three dimensions.

7.2.3 Mitral Regurgitation (MR)

7.2.3.1 Etiology and Epidemiology

Mitral regurgitation (MR) has become the most prevalent valvular heart disease along with the aging and growth of the population. According to an epidemiological study based on community in the United States, the overall prevalence of MR (moderate and above) was 1.7%, rising to 6.4 and 9.3% among people aged 65–74 years and over 75 years, which was four times higher than that of aortic stenosis

(1.7 vs. 0.4%) [93]. In China, epidemiological data on the prevalence of MR are still lacking, but the results of some studies based on hospitalized populations remain suggestive. Two studies, both involving approximately 1.4 million people undergoing ultrasound examinations, showed a detection rate of 0.68–1.44% for severe MR in hospitalized patients [94, 95]. Among hospitalized patients over 65 years of age, the prevalence of mild MR was as high as 23%, and moderate and severe MR was as high as 2.2% [96], indicating that severe MR is common in the hospitalized population.

The causes of MR are generally classified as ischemic (MR directly due to coronary artery disease) and nonischemic (degenerative mitral valve disease, endocarditis, rheumatic, other reasons like cardiomyopathy, inflammatory diseases, and drug-induced). In developed countries, degenerative mitral valve disease, accounting for 60–70%, is the main cause of MR, followed by ischemic MR (20%) and other reasons (rheumatic represents 2–5%). MR can be broadly categorized as degenerative MR (DMR) or functional MR (FMR) based on the mechanism of mitral valve dysfunction. DMR is intrinsic valvular disease involving annulus dilatation, leaflet prolapse, or chordae tendineae rupture, while the main cause of FMR is dilatation of the annulus secondary to left ventricular dysfunction and dilatation, which results in mitral valve insufficiency. The Carpentier classification is commonly used to divide MR into three types according to leaflet movement—Type I: normal movement with annular dilatation or leaflet perforation; Type II: excessive movement caused by leaflet prolapse or flail; and Type III: restricted movement, in which III A is caused by leaflet thickening, often seen in rheumatic or inflammatory changes, and III B is caused by dilatation of left ventricular or left atrium.

Patients with severe MR, when left untreated, have a mortality rate of 50% within 5 years. Clinical studies have shown that drug therapy can only relieve the symptoms of patients with DMR somehow but could not delay the progression of the disease, prolong the timing of surgery, and improve prognosis. It is estimated that there are 4.1 million cases of MR in the United States, of which 1.67 million (severe MR) require surgical treatment [97], but actually about 30,000, only 2% of patients, underwent mitral valve surgery every year. This is due to high risk of surgery and low diagnosed rate.

The total number of patients with moderate-to-severe MR in China is estimated to be 16.8 million, of which 6.84 million are symptomatic patients with severe MR requiring surgical treatment. But the number of mitral valve surgeries in China is 40,000 (less than 1%) per year. This means that worldwide, most MR patients are not receiving effective treatment.

7.2.3.2 Surgical Versus Transcatheter Mitral Valve Intervention

Clinical studies have shown that drug therapy can only improve the symptoms of DMR patients but cannot improve the prognosis and cannot prolong the survival time or the surgical opportunity [98]. Surgical valve repair or replacement is considered the standard treatment for MR and has been shown to relieve symptoms and prolong survival in patients [12]. According to different etiology (DMR, FMR), different stages of disease progression (advanced stage, severe lesions), and specific indicators and related symptoms, regular surgical valve surgery was selected. There are two types of mitral valve surgical procedures: valve repair and valve replacement. The specific techniques for surgical repair of the mitral valve vary depending on the location of the mitral valve lesion. A series of traditional techniques for posterior leaflet prolapse include rectangular resection, triangular resection, sliding repair, folding technique, et al. General techniques for bilateral leaflet prolapse include artificial chordae tendineae implantation, edge-to-edge repair, and flap ring forming. Flap repair is often assisted by the implantation of a forming ring to reduce recurrence. Mitral valve replacement surgery is used for patients with poor mitral valve structure and difficult to repair. Compared with replacement, repair showed higher safety in intraoperative, postoperative, and long-term prognosis.

Conventional surgery is the standard treatment for severe MR, but it is highly invasive, requires a state of cardiac arrest and extracorporeal circulation, and is risky for patients with low cardiac function, high comorbidity, frailty, and old age. As a result, only about 50% of patients can undergo surgery, which means that about 50% of MR patients lack effective surgical treatment. Since the late twentieth century, minimally invasive valve surgery has been developed. This surgical approach, which reduces trauma by performing tissue separation through small incisions without damaging the bone, has the advantages of less postoperative pain, faster recovery, less bleeding, and smaller scars but still requires cardiopulmonary bypass and cardiac arrest [97].

In recent years, the rapid development of mitral valve interventions has brought new hope to patients with MR. Percutaneous valve interventions, in which transcatheter mitral valve repair (TMVr) or transcatheter mitral valve replacement (TMVR) is performed through apical or peripheral vascular puncture into the heart, avoiding the need for chest opening, cardiopulmonary bypass, and cardiac arrest, can significantly reduce operative mortality and postoperative complications in high-risk patients undergoing conventional procedures [99]. In the 2020 guidelines, MitraClip® has been included in the treatment protocol for mitral regurgitation, marking a gradual shift from surgical to catheter-based mitral valve treatment [12].

7.2.3.3 Transcatheter Mitral Valve Repair

The mechanism of transcatheter mitral valve repair is similar to the surgical operation, but the repair devices are always delivered to the mitral valve site through the minimally invasive approach such as transfemoral or transapical, which furthest preserves the original structure of the mitral valve, with various efficacy and safety. According to technical mechanism, transcatheter mitral valve repair can be divided into the following categories: transcatheter edge-to-edge mitral valve repair, mitral valve artificial chordae tendineae implantation, transcatheter mitral annuloplasty, mitral valve artificial chordae tendineae implantation, and so on.

Transcatheter Edge-to-Edge Repair (TEER)

The mechanism of this technique stems from the edge-to-edge repair surgery pioneered by the Italian surgeon Ottavio Alfieri, in which the mid segments of the anterior and posterior leaflets are sutured together, transforming the mitral valve from one large to two small orifices to reduce regurgitation [100, 101]. Devices to repair mitral valve edges, such as MitraClip, PASCAL, ValveClamp, and DragonFly, are currently in clinical use or undergoing clinical studies. Of these, the MitraClip is the most widely used device with the strongest evidence.

MitraClip

MitraClip (Abbott Medical) has been used in more than 100,000 cases worldwide. The operation is usually performed under general anesthesia, with the insertion of the 24Fr sheath into the left atrium through femoral vein and atrial septal puncture (Fig. 7.6). Clips are used to clamp the anterior and posterior leaflets of the mitral valve under the guidance of transesophageal echocardiography (TEE) and X-ray to reduce regurgitation. The clip can be withdrawn, repositioned, and released. Two or more clips can be placed until the ideal result is achieved and the femoral access is less invasive to the patient resulting in a faster postoperative recovery. EVEREST I study confirmed the feasibility, safety, and effectiveness of MitraClip in the treatment of functional or degenerative MR patients [102]. In a subsequent multicenter

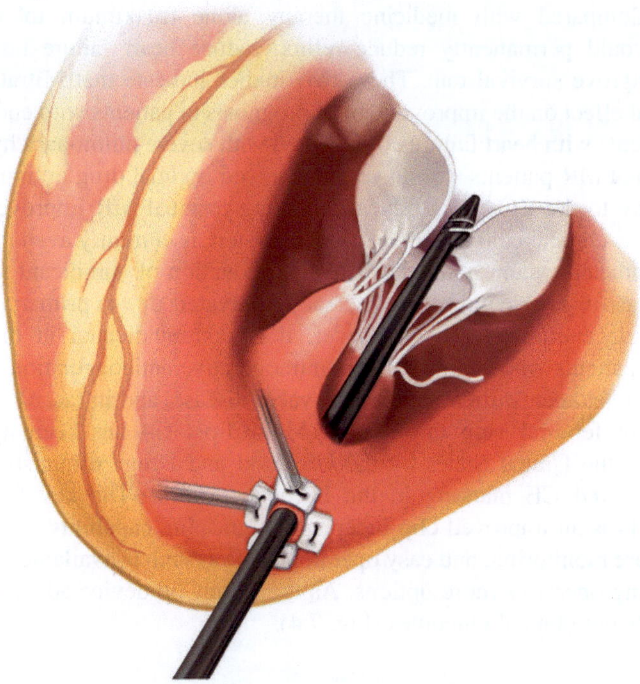

Fig. 7.6 Transapical NeoChord mitral valve repair

randomized controlled trial EVEREST II, 184 patients received MitraClip and 95 patients received surgical mitral valve repair or replacement. Results showed that MitraClip was safer than surgery and had a lower incidence of major adverse events (15 vs. 48%, $P < 1$), mainly due to the high proportion of blood transfusion required in the surgery group. However, the primary endpoint was inferior to surgery (55 vs. 73%, $p = 0.000\ 7$), and the difference in mortality was not statistically significant, mainly because about 20% of patients needed to undergo surgery again due to mitral valve dysfunction [103]. The 5-year follow-up results of EVEREST II study further confirmed the long-term safety and efficacy of MitraClip: the mortality of MitraClip group and surgical group were 20.8% and 26.8%, respectively ($p = 0.4$). The proportion of patients requiring reoperation for mitral valve dysfunction was lower in both groups during the 1–5-year postoperative follow-up period.

There are currently two randomized controlled trials (Mitra-FR and COAPT) that illustrate the efficacy of MitraClip in treating functional MR [104, 105]. In both studies, patients with FMR were randomized to the optimized medication therapy or combined with MitraClip. In the 1-year follow-up of the Mitra-FR study, there was no statistically significant difference in the compound event rate of death or rehospitalization between the two groups. The main reason may be that the enrolled patients had more severe disease, low left ventricular ejection fraction, and poor myocardial activity. Even though the reflux condition was improved, the cardiomyopathy was still progressing. For these patients, it may be too late to correct MR. The 2-year follow-up results of the COAPT trial showed that MitraClip was safe and effective. Compared with medicine therapy alone (maximum tolerated dose), MitraClip could permanently reduce reflux, reduce heart failure hospitalization rate, and improve survival rate. These two studies indicate that MitraClip therapy has a limited effect on the improvement of prognosis in patients with end-stage heart failure, patients with heart failure complicated with severe pulmonary hypertension, and functional MR patients who have not received optimal drug optimization.

According to the COAPT study, minimizing residual MR improves prognosis and maintains clinical benefit. The MitraClip system is currently available in China. This transcatheter system is indicated for the reduction of significant symptomatic MR (moderate-to-severe mitral insufficiency or greater) due to primary mitral valve anomalies (degenerative MR). However, it is contraindicated in the following conditions: patients who cannot tolerate intraoperative anticoagulation or postoperative antiplatelet agents, rheumatic mitral valve disease, and intracardiac or inferior vena cava or femoral vein thrombosis. MitraClip's G4 has been approved for marketing in the United States by the US Food and Drug Administration (FDA) and has received CE marking in the European Union. The key feature of the MitraClip G4 is an improved clip-independent grasping capability, continuous left atrial pressure monitoring, and easy operation. The system is available in four clamp sizes, offering operators more options. As the MitraClip device advances, we look forward to better clinical outcomes (Fig. 7.4).

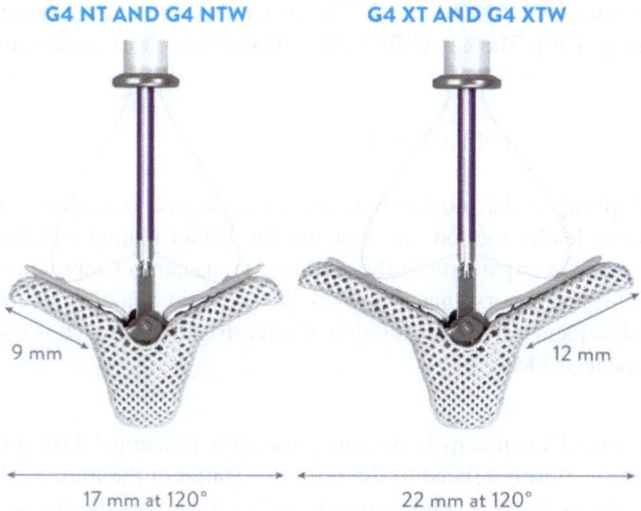

Fig. 7.4 The fourth-generation (G4) MitraClip offers more clip sizes for tailored repair. It also has a new leaflet grasping technology called Controlled Gripper Actuation™, which allows physicians to grasp leaflets simultaneously or independently to confirm and optimize leaflet insertion

PASCAL

The PASCAL (Edwards Lifesciences) system is based on the same mechanism as the MitraClip system, which is to puncture the atrial septum transfemorally and clip the mitral valve into double hole structure [106]. Compared with MitraClip, the clamping arms of PASCAL are wider and longer. The two clamping arms can be operated independently and more simply, allowing for wider range of anatomical application. The CLASP study (NCT 03170349) involved 62 patients treated with the single-arm design PASCAL, of whom 56% had FMR. The results revealed successful deployment as high as 95%, and 98% of the patients had MR ≤2 during the 30-day follow-up. In addition, 6-min walking test results and the quality of life were also significantly improved [107]. A retrospective observational study with a small sample (50 cases) also confirmed the safety and effectiveness of PASCAL, with successful implantation in 98% and 78% (36/46) of patients with MR ≤1 during 1-month follow-up [108]. The PASCAL system received CE certification from the European Union. The large-scale randomized controlled trial CLASP II D/IIF (NCT03706833), which directly compares the safety and efficacy of two edge-to-edge repair devices MitraClip and PASCAL for DMR and FMR, was approved by the FDA.

ValveClamp

The ValveClamp (Hanyu Medical Co., LTD) was developed to clamp the prolapsed mitral valve leaf transapically (Fig. 7.6). It has the advantages of larger capture space, ease in operation, smaller delivery system, and complete sizes. However, the device is currently limited to DMR for A2 and P2 prolapse. The first human experimental study (CLAMP 1) on DMR has validated the safety and effectiveness of ValveClamp [109]. The CLAMP 2 (NCT03869164) EF is underway.

Transcatheterization Mitral Annuloplasty

Mitral annuloplasty is the most commonly used surgical repair for MR. Not only does it preserve leaflet motion and increase the leaflet mating surface, but it also prevents progressive expansion of the annulus. Transcatheter mitral annuloplasty is based on a surgical theory that preserves the intrinsic anatomy of the valve and allows for subsequent valve implantation. Currently, this approach is used primarily for the treatment of FMR.

Carillon

Carillon (Cardiac Dimension) is primarily used for functional MR (Fig. 7.7). The coronary sinus is closely related to the posterior leaflet of the mitral valve. Carillon permanently places the distal and proximal anchor devices in the coronary sinus and then brings them close to each other through traction, so as to reduce the perimeter of the mitral valve ring, promote the petal involution, and produce indirect ring forming effect [110]. As early as 2011, Carillon has already obtained the European Union CE

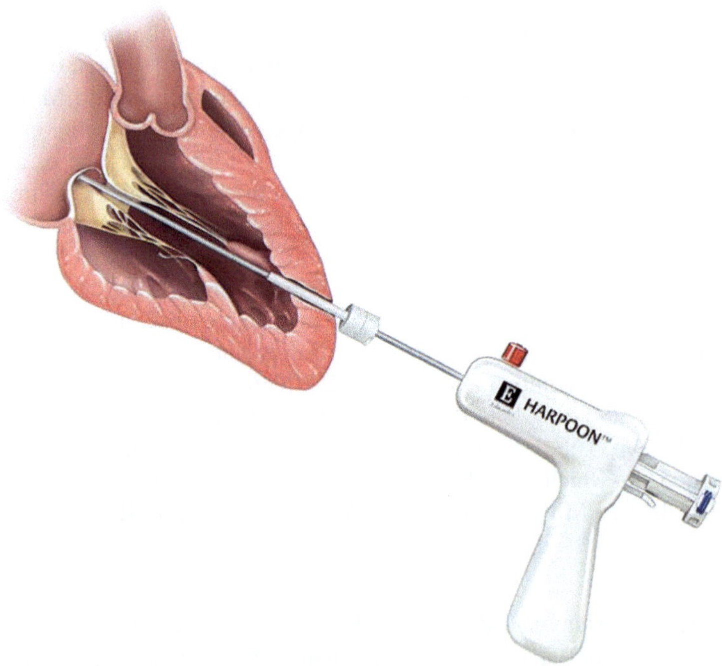

Fig. 7.7 Small-footprint ePTFE chords with proprietary self-forming knots are easily deployed and securely anchored using the preloaded HARPOON delivery system

certification. In the AMADEUS study, Carillon was successfully implanted in 30 of 48 patients (62.5%), and reflux was improved, but the incidence of coronary artery compromise was observed in 17% implants [111]. The subsequent TITAN study included 53 patients with functional MR (64% ischemic), among whom 36 patients (68%) were successfully implanted with Carillon, and 6 and 12 months of follow-up showed significant improvement in MR severity, left ventricular remodeling, cardiac function, and quality of life [112]. The randomized controlled study REDUCE-FMR found that, compared with the control group, the implantation device group had significantly reduced reflux and significantly reversed left ventricular remodeling, but there was no statistically significant difference in the occurrence of clinical endpoints between the two groups [113]. Due to the low success rate of the operation and the risk of coronary artery occlusion caused by compression of the circumfluence branch, its popularization is slow. Due to the limitations of indirect transcoronary mitral annuloplasty, direct transcutaneous mitral annuloplasty was performed and achieved good clinical results (Fig. 7.5).

Cardioband

The Cardioband (Edwards Lifesciences, USA) is an adjustable, catheter-delivered, suture-free, direct valvular ring forming device via atrial septal puncture. In 2015, we obtained the European CE certification. Early clinical experience results of Cardioband were good. Thirty-one patients were included in this study, and 22 of 25 patients (88%) had MR grade ≤2+ during 30-day follow-up, with significant improvement in cardiac function and 6-min walk test results [114]. Follow-up results of the trial suggested further improvement in MR at 24 months of follow-up, with 23 of the 25 patients (92%) having an MR grade ≤2+ and 77% having a New York Heart Association (NYHA) cardiac function grade I/II. The effectiveness and safety of the device was demonstrated in a subsequent multicenter study (NCT01841554) with 1-year follow-up results: the 1-year survival rate was 87%, 61% of patients had moderate MR or less, and patients' cardiac function, quality of life, and exercise capacity were significantly improved [115].

Mitralign

Mitralign (American Mitralign) placed mass on the medial and lateral sides of the mitral valve and pulled each pair of mass together to reduce mitral valve ring and reduce regurgitate [116]. The procedure has the advantages of less trauma and faster recovery, and Mitralign received the European Union CE certification in 2016. The early clinical trial (NCT01852149) included 71 patients with functional MR, 70.4% of whom were successfully implanted with the system. All-cause mortality at 30 days and 6 months was 4.4% and 12.2%, respectively. At 6 months, the patient's cardiac function grade and left ventricular remodeling were significantly improved [117]. However, the scope of the instrument for mitral annulus reduction is limited, and its application is limited.

Transcatheter Mitral Valve Artificial Chordae Tendineae Implantation

NeoChord
The NeoChord system (NeoChord, USA) is transapically implanted by suturing the captured flail leaflet to the artificial chordae tendineae, which is then fixed to the apex to reduce regurgitation by adjusting the length of the artificial chordae tendineae. The device is mainly used for MV prolapse, not suitable for functional MR. Colli et al. reported the results of 62 patients with mitral valve prolapse treated with the NeoChord system, which confirmed the safety and validity, although the recovery is slower compared with the peripheral vascular approach [118]. NeoChord system received CE certification of Europe in 2012. The international registry study included 213 patients with severe MR caused by mitral valve prolapse, and 96.7% [119] of patients were successfully repaired, with 1-year follow-up results showing that the operation was safe and effective [120] (Fig. 7.6).

Harpoon
Harpoon (Edwards Lifesciences, USA) is also a transapical artificial chordae tendineae implantation device that consists of a hemostatic introducer, a delivery system, a 21-gauge needle, and artificial cords. Artificial ePTFE cords are anchored on the targeted segment of the prolapsed MV leaflet, resulting in a great reduction in mitral annular dimension. A recent study suggested that 89% (24/27) of subjects had low MR improvement at 30 days and 85% (22/26) of patients only had mild MR or less at 6 months [121]. It shows that the device has a promising application and safety profile (Fig. 7.7).

MitralStitch
MitralStitch (Hangzhou Dejin Medical Co., LTD., China) is a mitral valve repair device not only implanting artificial chordae tendineae for the treatment of DMR but also allowing for edge-to-edge repair for FMR [122]. The early clinical trial included ten patients with 100% successful MitralStitch implantation. The 30-day follow-up results indicated that all patients had MR grade ≤2, which was safe and effective. Its multicenter clinical study (NCT04061837) is underway.

7.2.3.4 Transcatheter Mitral Valve Replacement (TMVR)

Transcatheter mitral valve replacement (TMVR) is widely used in the treatment of high-risk patients with thin, short, fragile, or calcified MV diseases, especially in those with anatomies not ideally suited for transcatheter mitral valve repair. However, there are a variety of the challenges in developing safe and effective TMVR devices at present. We will discuss current TMVR technology and future directions.

Intrepid Valve
The Intrepid valve (Medtronic) is a retrievable self-expandable stent system which is delivered transapically through a 35-F delivery system. Fifty patients were included in the early trial, with a successful implantation rate of 96% (48/50). There is a

significant improvement in symptom class and Minnesota Heart Failure Questionnaire with the 30-day mortality rate of 14% [123], 1-year survival was 76.5%. A large randomized controlled study, APOLLO (NCT03242642), is currently undergoing.

Tendyne

Tendyne (Abbott Medical) is composed of two self-expandable nitinol stents housing a tri-leaflet porcine pericardial valve delivered transapically through a 34-F delivery system. It has currently the most extensive clinical data and relatively more successful outcomes to date. It is shown that the included 100 patients had a mortality rate of 6% at 30 days after valve replacement and an all-cause mortality rate of 26% at 1-year follow-up. The cardiac function and quality of life were significantly improved, and the results were significant in reducing MR and improving symptoms, with acceptable safety [124]. Tendyne has been granted CE certification in the European Union in 2020. The SUMMIT trial (NCT03433274) is approved and ongoing in the United States.

Tiara

Tiara (Neovasc, Canada) is a D-shaped device consisting of three bovine pericardial lobules with self-expandable nitinol frames, with two sizes—35 and 40 mm—available, delivered transapically through 32- and 36-F delivery system. The valve was fixed to the mitral annulus by means of three fixation points on the ventricle side and one fixation point in the atrium. A total of 33 patients were included in the early feasibility trial, with an implantation success rate of 90% and an early mortality rate of 12% [125]. A multicenter international feasibility trial called Tiara II is currently in development.

SAPIEN M3

The SAPIEN M3 (Edwards Lifesciences, USA) is a transseptal balloon dilated valve modified from the SAPIEN 3 valve for aortic valves. The valve consists of a memory nitinol stent and a trilobal bovine pericardial valve with an added hemline to minimize perivalvular leakage. This valve is mainly used in patients with severe MR caused by failure of mitral annuloplasty or decay after biological valve replacement [126]. Early reports showed that the success rate of implantation was 86.7% (13/15), and the MR of 93.3% (14/15) of patients decreased to mild degree without death, left ventricular outflow tract obstruction, and hemolysis [127].

MitraFix and Mi-thos

MitraFix and Mi-thos (Shanghai, China) are transcatheter mitral valve replacement systems developed by Chinese Yixin and New Pulse Medical in Shanghai, China, respectively. MitraFix has the unique nitinol frame design that combines fixing and leakproof features to reduce valve size and compliance after compression. Results of the first human trial showed significant improvement in MR after 6 months of follow-up [128].

Mi-thos is a transcatheter mitral valve replacement system developed by Shanghai New Pulse Medical in China, which is implanted through the apical approach. The D-type design was adopted and the atrial surface was woven with nitinol alloy wires. The valve was composed of three bovine pericardial lobules. In 2019, the team of Wang Chunsheng of Zhongshan Hospital, Fudan University, successfully completed the first mitral valve replacement by transcatheter Mi-thos in Asia.

In addition, various devices like Cardiovalve, Cephea, and EVOQUE are also in clinical trials. However, the results of early clinical trials of some TMVR devices were not satisfactory enough. For example, the Fortis valve had an all-cause mortality of 38.5%, and valve thrombosis occurred at 30 days after implantation. At 30 days after CardiAQ valve implantation, the all-cause mortality was 53.8%.

7.2.4 *Mitral Stenosis (MS)*

7.2.4.1 Etiology and Epidemiology

Mitral stenosis (MS) is a prevalent valvular disorder, which can be classified into two groups—rheumatic MS (RMS) and degenerative mitral stenosis (DMS). RMS accounts for approximately 40% of rheumatic heart disease. Although the incidence of rheumatic heart disease is gradually declining, it is still relatively common in developing countries, and the prevalence rate is reported to be 0.186% in China, 0.454% in India, and 0.014% in Japan [129]. Nonrheumatic MS contains various pathologies including degenerative mitral stenosis (DMS) and radiation-induced and congenital MS. With the increasing number of aging people, there is a rise in the incidence of mitral annular/valve calcification (MAC), which causes DMS. MAC is estimated to affect 9% of women and 3% of men above the age of 60 years, especially common in patients with diabetes and chronic kidney disease [130].

7.2.4.2 Percutaneous Balloon Valvuloplasty

For patients with moderate-to-severe MS with symptoms, when oral-drug therapy was invalid, surgical options including closed or open valvotomy and mitral valve replacement were the only solutions, while since the advent in 1984, percutaneous balloon mitral valvuloplasty (PMBV) becomes the mainstay of treatment when indicated in anatomically eligible MR patients, due to its safety, effectiveness, and improvement in short- and long-term prognosis.

The study concluded that patients meeting the following characteristics—age <65 years (some 45 years), sinus rhythm, Wilkins score ≤8, absence of calcification under fluoroscopy, and preoperative mitral regurgitation ≤1+ (Sellers class)—are more suitable candidates for PMBV treatment with better immediate and long-term postoperative results. A study that included 202 patients who met these criteria showed a procedure success rate of 81% and a survival rate of 97% at 61 months of follow-up [12, 131].

MS is a common disorder that causes substantial morbidity worldwide, and PMBV has broad application prospects. Guidelines also emphasize that valve replacement should be considered only when there are contraindication of PMBV

[12], and more multicenter researches should be carried out further to provide data for the long-term prognosis of patients after PMBV in China.

7.2.5 *Conclusion*

The safety of chordae tendineae implantation is high, but the scope of application is limited, and more research data are needed to confirm. Direct or indirect annular valvuloplasty is mainly used for functional MR. Large sample studies are needed to confirm the effectiveness and durability. It is expected to be combined with valve repair in the future to further improve MR. TMVR is still in small clinical trials. At present, the anticoagulation regimen after interventional therapy has not been concluded. Anticoagulation therapy is recommended for patients with atrial fibrillation/flutter, and antiplatelet drugs are recommended for the remaining patients. The specific regimen needs to be guided by the results of further studies. It is expected that more and more transcatheter mitral valve repair and replacement technologies will be available to further promote transcatheter mitral valve interventional therapy [132, 133].

7.3 Transcatheter Treatments for Pulmonic Valve Disease

7.3.1 *Introduction*

Pulmonary valve (PV) may be the least valued and poorly studied heart valve. The poststernal position makes it difficult to evaluate by standard transthoracic echocardiography, which is the most common method. This aspect is particularly worrying for young patients, in which early diagnosis of abnormal right ventricular outflow tract (RVOT) can save lives. The incidence rate of congenital PV is significantly higher than acquired PV. As more and more patients with congenital heart disease survive and thrive into adulthood, clinicians should have a strong understanding of the pathophysiology of PV disease and be able to identify when the problem occurs.

Disorders of the RVOT, PV, and suprapulmonic region can disrupt the normal process of PV function and lead to significant hemodynamic compromise. These disorders can present as isolated problems or as part of multisystemic/multichamber heart diseases. They are divided into three main categories: atresia, stenosis, and regurgitation.

Interventional treatment of PV usually leaves partial stenosis and/or regurgitation, which needs regular follow-up and may be the future intervention [134, 135]. Although cardiac surgery has been the gold standard for the treatment of PV for a long time, the availability of transcatheter approach is getting higher and higher, which provides a minimally invasive approach for more and more complex patients. In this review, we will focus on pulmonary valve stenosis (also "pulmonic stenosis," PS) and pulmonary valve regurgitation (also "pulmonic regurgitation," PR) and review their pathophysiology, diagnosis, and treatment.

7.3.2 *Anatomy and Pathophysiology*

7.3.2.1 Pulmonary Valve Stenosis

PS is mainly a congenital disease. Acquired cases also occur, but are not common. The causes of the latter include infective endocarditis, carcinoid heart disease, rheumatic heart disease, and iatrogenic causes [134, 136]. Stenosis may occur at PV level, above or below PV. PS typically occurs as an isolated lesion, but it can also coexist with other congenital heart diseases. Importantly, it is one of the four major features of tetralogy of Fallot (TOF). PS is also part of many syndromes that affect the intrauterine development of PV, including Noonan, 22q11 deletion, and Williams syndrome [137–139]. Severe pulmonary artery stenosis may be accompanied by right ventricular hypertrophy, eventually leading to right ventricular and right atrial enlargement. In addition, the detection of other related malformations is also very important for the evaluation of patients with pulmonary valve stenosis.

PS can be further divided into three types with clinical significance: fornix, dysplasia, and bicuspid/multicuspid. The characteristics of these three forms are that the orifice of the valve is narrowed, resulting in the formation of pressure gradient throughout the valve, and the RV bears pressure load [140].

Dome-shaped PV morphology accounts for 40–60% of valvular PS. It is characterized by bowing of the PV leaflets into the pulmonary trunk caused by incomplete separation of valve cusps, coupled with preserved valvular motion. The leaflets have varying degrees of fibrous thickening and are fused along their commissures. Dilation of the pulmonary artery is frequently present due to the eccentric flow resulting from the stenosis, as well as possible intrinsic connective tissue abnormalities within the pulmonary trunk itself. Abnormal PV morphology accounted for 20% of valve PS and was most common in patients with Noonan syndrome. The thickening of the PV lobule, minimal or no commissural fusion, and restricted activity were observed. [141, 142]. Other segments along the RVOT may also narrow. In this case, pulmonary vein stenosis is often very serious and needs early intervention in children. Mitral and pulmonary venous malformations are considered rare and are often associated with other congenital cardiac abnormalities, such as TOF [143, 144]. Pulmonary regurgitation does occur frequently in this form [145, 146].

Patients with isolated mild PS are usually asymptomatic, and their long-term survival probability is similar to that of the general population [147]. With the aggravation of stenosis, the decrease of cardiac output may lead to the onset of symptoms. The most common symptoms include dyspnea and fatigue when tired. If the stenosis is not relieved, it will lead to right ventricular hypertrophic, accompanied by mild stenosis, diastolic function, and ultimate systolic dysfunction [148].

7.3.2.2 Pulmonary Valve Regurgitation

Pulmonary valve regurgitation is mainly caused by enlargement of pulmonary valve ring and main pulmonary artery. About 75% of normal people can detect a mild physiological pulmonary regurgitation without hemodynamic significance. Pathological pulmonary regurgitation is not common and often accompanied by abnormal valve structure. The most common cause is pulmonary hypertension. Secondary pulmonary regurgitation is the most common in patients with pulmonary hypertension, but the regurgitation is usually small and the leaflet structure is normal. The causes of primary PR overlap with those of PS, including iatrogenic reflux, congenital malformation, infective endocarditis, rheumatic disease, and carcinoid heart disease [149–151]. The main manifestations of primary pulmonary regurgitation were abnormal pulmonary valve leaflets and pulmonary valve undergoing balloon angioplasty due to stenosis. Severe primary pulmonary regurgitation can lead to right ventricular enlargement, but right ventricular function is usually preserved. Chronic severe pulmonary regurgitation can lead to right ventricular dysfunction.

PR is generally better tolerated than aortic regurgitation, and patients may be asymptomatic for many years [152]. To some extent, this can be attributed to the low resistance of the pulmonary vascular system, which is conducive to the contraction of upstream blood flow to the pulmonary capillary bed. In addition, in patients with severe PR and restrictive RV physiology, echo or MRI often showed pulmonary artery late diastolic forward flow, accompanied by each atrial contraction [153]. This suggests that right atrial contraction helps to maintain forward blood flow in these patients. Therefore, even in patients with severe PR, the percentage of reflux is usually less than 50% [154]. This may be further affected by right ventricular afterload, right ventricular compliance, size of regurgitation orifice, and diastolic duration [155]. However, over time, RV eventually succumbs to excessive volume loading and expands to maintain cardiac output [156]. Progressive RV dilatation eventually leads to RV dysfunction, decreased exercise tolerance, and atrial and/or ventricular arrhythmias and may lead to sudden cardiac death [157–160]. Therefore, moderate PR in patients with TOF or PV repair/replacement is the strongest predictor of short-term and long-term adverse outcomes [161, 162].

7.3.3 Balloon Pulmonary Valvuloplasty

In 1913, the first attempt to treat PS occurred in Paris. A tenatome was introduced into the RV to remove the narrow PV, but unfortunately, the patient died the day after surgery. Then it took 30 years to reintroduce new technologies to expand PV. The first transcatheter pulmonary valvuloplasty occurred in 1979. At present, balloon pulmonary valvuloplasty (BPV) has become one of the main methods for the treatment of pulmonary stenosis.

7.3.3.1 Indication and Contradiction of BPV

BPV is the treatment of choice for the vast majority of patients due to its less invasive nature and higher success rate [163]. In addition, a recent study showed that BPV was performed safely and effectively with only echocardiographic guidance, thus avoiding the risk of radiation and contrast agents if necessary [164]. This is particularly useful when pregnant patients have significant hemodynamic PS.

BPV is currently recommended for asymptomatic patients with domed valve shape and peak echo gradient >60 mmHg (mean > 40), accompanied by ≤ moderate PR and symptomatic patients with peak echo gradient >50 mmHg (mean > 30), accompanied by ≤ moderate PR [165]. It also occurs in most valvular dysplasia and subvalvular and supravalvular lesions.

7.3.3.2 Effects and Issues of BPV

BPV has several direct effects on PV and RV [166–168]. First, the valve pressure gradient and right ventricular systolic pressure decreased significantly, while PA pressure almost did not increase. Second, the protuberance of PV decreases and its leaflets become more fluid. Finally, right ventricular function can be improved and tricuspid regurgitation reduced, even though cardiac output may increase only slightly. Recently, Mahfouz and his colleagues demonstrated that BPV improves right ventricular longitudinal motion and intraventricular and interventricular asynchrony in children [169]. Compared with the control group, the asynchrony of PS patients increased significantly and returned to near normal level after BPV. Acute complications of BPV are usually benign, with the main complication rate of 0.6% and mortality rate of 0.2% [170]. An important acute complication is transient, severe right ventricular outflow tract obstruction (also known as “suicidal right ventricle”), which can occur immediately after the remission of valve stenosis [171, 172]. This is similar to the pathophysiology of left ventricular obstruction in patients with hypertrophic obstructive cardiomyopathy. After treatment with β-blocker, it completely disappeared with the recovery of contractility and regression of hypertrophy [173]. In many cases, funnel obstruction may be implicated when residual gradients appear immediately after BPV and progressively decrease within a few months with the gradual reduction of RV hypertrophy [170].

The long-term efficacy of BPV has been studied in children, although the follow-up time is still shorter than that of valvulotomy. The two most common long-term complications after BPV are progressive PR and restenosis. Until recently, only one study in pediatric population reported results for more than 10 years [174]. In this study, all 48 patients developed advanced PR, but none had significant RV dysfunction or required PV replacement (PVR). Results of another study of 103 patients with more than 10 years of ultrasound follow-up showed that 60% of the patients had at least moderate PR, which is well tolerated in most patients. Only 5% had right ventricular dysfunction and only 3% needed PVR [135]. Generally speaking, PR after BPV tends to be progressive in pediatric population; however, severe RV expansion and dysfunction are rare [135, 174–177].

Concerning the effects of BPV in adults, an observation with 17-year follow-up of 85 adults after BPV (with an average of 10 years) found that the proportion of PR increased from mild (before BPV) to moderate (after BPV) was 2%, while the proportion of mild PR was 28%, with no PR progression or RV dilatation [178]. Other small clinical studies have suggested that PR progression and RV insufficiency are common after BPV [178, 179]. In a recent retrospective study of 41 adults, with an average follow-up of 11.3 years, 26.8% had mild PR at the last follow-up, and 7.3% had moderate PR [180]. It is worth noting that in many of the above studies, the ratio of balloon to annulus is between 1.2 and 1.4 [166, 173, 175], and higher ratio of balloon to annulus seems to be an important risk factor for PR [176, 181].

Another common long-term complication of BPV is restenosis. In a cohort study with 533 patients undergoing valvuloplasty and angioplasty for congenital heart defects enrolled, 23% of patients developed stenosis at a median follow-up period of more than 33 months, whereas 16% of the patients needed further intervention [34]. The risk factors for restenosis included dysplasia of valve morphology, small diameter of pulmonary artery annulus, high residual gradient immediately after BPV, and small balloon-to-annulus ratio (<1.2). In another study, 11% of 82 children and adolescents had restenosis (gradient) ≥50 mmHg after 2 years. Similarly, high residual gradient and small balloon to ring ratio (<1.2) were predictors of restenosis [177]. After a thorough analysis of the available data, a balloon-to-annulus ratio of 1.2–1.25 is recommended [171].

7.3.4 Transcatheter Pulmonic Valve Replacement

Transcatheter pulmonic valve replacement (TPVR) emerged in 2000 as a less invasive treatment option for PR and PS [182]. It initially found use in patients with RV-to-PA conduit dysfunction, given the need for an adequate landing zone [183]. Conduits are implanted during surgical management of certain congenital heart diseases such as pulmonary atresia and truncus arteriosus and as part of the Rastelli repair for d-TGA. All RVOT conduits eventually become dysfunctional (either obstructive or regurgitant) and require further intervention. TPVR has additionally become utilized as the procedure of choice in patients with failed bioprostheses via valve-in-valve implantation [184, 185].

7.3.4.1 Indications and Contradictions for TPVR

Devices for TPVR

Melody Valve
The first transcatheter valve to be developed was the Melody® valve (Medtronic, Minneapolis, MN), made of bovine jugular vein. It offers diameters between 18 and 22 mm, limiting its use to delivery zones that are ≤24 mm in diameter [186, 187]. Short-term studies have shown excellent results, with decrease in RV size and ECG QRS duration following valve implantation [188, 189], and midterm studies up to 9 years demonstrate good hemodynamic and clinical outcomes [190–192]. Moreover, the 5-year freedom from reintervention after Melody implantation exceeded 90% in two of the main clinical trials [193, 194]. The Melody was FDA approved for dysfunctional RVOT conduit or bioprostheses (valve-in-valve) who have moderate or greater pulmonary regurgitation and/or a mean RVOT gradient greater than 35 mmHg. A major cause of Melody valve dysfunction appears to be restenosis due to stent fracture. The incidence of this problem has decreased dramatically after pre-stenting has become widely adopted [195].

Sapien Valve Series
In 2016, the Edwards Sapien XT valve (Edwards Lifesciences, Irvine, CA) received FDA approval for use in failing pulmonary conduits. It is a bovine pericardial tissue valve that offers diameters from 23 to 29 mm [196]. This is extremely helpful for select patients who have larger native RVOTs, transannular-patched RVOTs, larger RV-to-PA conduits, as well as larger bioprostheses [197]. Figure 7.2 illustrates transcatheter implantation of the newest Sapien valve, the S3, into a patient with failed RV-to-PA conduit. Studies using the Sapien valves in the pulmonary position are few to date, with little long-term data available. Wilson and colleagues recently followed 25 patients for a mean of 3.5 years after Sapien TPVR [198]. Only one patient required reintervention for severe PR; no other patient had clinically significant PR, and valve function was preserved during follow up.

7.3.5 Issues for TPVR

7.3.5.1 Coronary Compression and Aortic Root Compression

In patients attempting transcatheter replacement, two major intraoperative complications need to be recognized and prevented: coronary artery compression and aortic root compression. Coronary artery compression is a potential complication of PPVI that may occur in approximately 5% of patients, while aortic root compression can cause aortic apex distortion and aortic regurgitation, affecting 10% of patients [199–201]. This risk exists even with PPVI in bioprosthetic valves. This risk should always be assessed during the procedure, which is achieved by performing balloon angioplasty (using a balloon of the expected PPVI size) at the conduit while assessing the coronary arteries with ascending aortography or selective coronary angiography. The study showed that 34 patients (71%) had TOF and abnormal coronary anatomy, and 7 of them had coronary artery compression at the time of testing [200]. 3D modeling and printing of high-resolution preoperative CT or cardiac MRI imaging can be used in order to predict and prevent such complications.

7.3.5.2 Paravalvular Leakage

An asymmetric RVOT creates a poor landing zone resulting in incomplete juxtaposition or PVL. The RVOT in many patients with TPVR indications is too much dilated for the available prosthesis. New technologies are also being developed to help overcome the limitations of current TPVR. A new method, which installs the 29 mm Sapien XT valve on a larger, more flexible (nucleus) balloon platform, has been shown making the valve oversized and reducing the PVL. A hybrid approach of PVR, which combines vascular access and landing area surgery with modified minimally invasive transcatheter valve delivery, has been used in many institutions. The success rate of operation is comparable to standard PVR and may reduce the incidence rate and mortality of perioperative [202, 203]. Other technologies currently under development include RVOT reducers/fillers to form smaller landing areas [204], self-expandable valve platforms that can cover larger RVOT diameters [119, 205], and finally tissue-engineered self-expandable scaffold platforms filled with autologous cells, which can prevent long-term hardening and calcification that plague xenograft implants [206]. However, most of these advances are still experimental.

7.3.5.3 Infective Endocarditis

Infective endocarditis is a complication of TPVR treatment that has been highlighted in recent years. Reports of infective endocarditis (IE) in the Melody (Dublin, Ireland) transcatheter pulmonary valve replacement (TPVR) raised safety concerns about the TPVR approach [207, 208]. A recent study analyzed 309 patients who underwent transcatheter pulmonary valve replacement (TPVR) and were discharged with a valve in place. The median follow-up duration was 5.1 years, and total observation until study exit was 1660.3 patient-years. Endocarditis was diagnosed in 46 patients (median 3.1 years after TPVR), and a total of 35 patients were reported to have TPV-related endocarditis (34 at the initial diagnosis, 1 with a second episode). The annualized incidence rate of endocarditis was 3.1% per patient-year and of TPV-related endocarditis was 2.4% per patient-year. At 5 years post-TPVR, freedom from a diagnosis of endocarditis was 89%, and freedom from TPV-related endocarditis was 92%. By multivariable analysis, age≦12 years at implant and immediate post-implant peak gradient ≧15mmHg were associated with development of endocarditis and with development of TPV-related endocarditis [209].

7.3.6 Conclusion

Congenital pulmonary venous disease is far more than acquired cases, among which valvular pulmonary venous disease is the most common. PBV is recognized as the gold standard intervention with good short-term and long-term effects. Recent studies have shown that more conservative balloon-to-annulus ratio (1.2 and below) can reduce complications, and PR is the most common. Severe PR, initially considered benign, inevitably leads to RV expansion and failure. Although surgical PVR remains the preferred intervention for most patients, TPVR provides a minimally invasive approach with short- and medium-term outcomes comparable to surgery. At this point, TPVR is approved for failed catheters and bioprosthetics, and off-label use is still severely limited by anatomy and device design issues. There is no doubt that new technological advances will increase the number of patients who can be successfully treated with this minimally invasive method in the future.

7.4 Transcatheter Treatment of Tricuspid Regurgitation

7.4.1 Introduction

Tricuspid regurgitation (TR) is one of the common heart valve diseases, and the prevalence of moderate/severe TR is about 0.55% in the general population and increases with age, reaching 3.96% in people over 75 years of age [210]. Moderate/severe TR is a predictor of cardiovascular death and all-cause mortality and is independent of left heart function, pulmonary hypertension, and right heart systolic function, suggesting that TR is not only a marker of disease severity but also a contributor to poor clinical outcomes and that reducing TR may improve long-term prognosis [211–213]. However, moderate/severe TR is almost always treated clinically with medications, with surgical treatment rates of only about 0.5–2.6%, the vast majority of which (85%) are simultaneous interventions of the tricuspid valve during other cardiac procedures [210, 214, 215]. Tricuspid valve repair or replacement alone is rarely performed, has an in-hospital mortality rate of 8.8%, and does not appear to increase long-term patient survival compared to pharmacological treatment. This may be due to the relatively insidious onset of TR, with more severe right heart insufficiency and/or end-organ impairment already occurring by the time signs and symptoms appear, leading to an increased risk of perioperative surgical death [216, 217]. With the successful promotion of transcatheter aortic valve interventions and the flourishing of mitral valve interventions, investigators have begun to focus on transcatheter tricuspid valve interventions in an effort to explore the best TR treatment options and reduce the burden of disease.

The development of devices for the percutaneous treatment of TR is quite challenging due to the structural complexity of the tricuspid valve in close proximity to each other, including the atrioventricular node, coronary sinus, right ventricular coronary sinus, right coronary artery, and aortic valve; moreover, the most common pathological mechanism of secondary TR is poor valve docking due to annular dilatation. Based on its structural and pathological characteristics, several devices for percutaneous treatment of TR have been proposed in the last decade with the aim of improving valve docking by clamping the leaflets together or resolving annular dilatation. There are several devices that have been converted from percutaneous mitral valve interventions. Transcatheter tricuspid valve replacement may also be considered when the lesion is too advanced to be treated with percutaneous repair, but clinical experience in this area is still in its infancy.

7.4.2 *Anatomy and Pathophysiology of AS*

7.4.2.1 Anatomy of TV

The tricuspid valve (TV) is the largest of the four heart valves. The tricuspid annulus (TA) is a more flattened semilunar shape rather than a planar structure. Its posterior septal portion is primarily adjacent to the ventricle, and the anterior septal portion is more frequently adjacent to the atrium. The structure of the TA is not fixed but varies dynamically. Its shape and size change throughout the cardiac cycle as the cardiac load changes. The normal tricuspid annulus circumference and area in healthy subjects, as measured by 3D echocardiography, are 12 ± 1 cm and 11 ± 2 cm^2, respectively, with a typical diameter of 30–35 mm, and are related to torso size. Because the septal portion of the annulus is fixed, expansion of the annulus occurs primarily in the free wall. Therefore, the size of the annulus for tricuspid valve repair is based on the size of the base of the atrial septal leaflets [218, 219].

The tricuspid valve consists of three leaflets, the anterior, posterior, and septal leaflets. The valve leaflets are attached to the TA and are connected to the papillary muscle of the right ventricle (RV) via the chordae [219]. Unlike the mitral valve, each leaflet of the tricuspid valve is attached to only one papillary muscle. The size of the leaflets varies, with the anterior valve being the largest and the septal valve the smallest.

Immediately adjacent to the TV are three other important structures. First, the right coronary artery (RCA) travels posterior to the right atrioventricular sulcus with its proximal segment distant from the tricuspid annulus, but as it travels inferior to the tricuspid annulus, the distance between the two gradually shortens to <3 mm, which would increase the risk of RCA injury by transcatheter tricuspid annuloplasty instrumentation. Second, the atrioventricular node and His bundle cross 3–5 mm posterior to the anterior septal union. Their injury can lead to complete heart block. Third, the coronary-free sinus of the Valsalva sinus is adjacent to the anterior septal union. Transcatheter TV repair devices that require anchoring in this area run the risk of aortic perforation [220].

7.4.2.2 Etiology and Pathophysiology of TR

Based on the underlying mechanism of TR, it can be divided into primary and secondary TR. Primary TR is estimated to have an incidence of 8–10%, and secondary is the most common mechanism of TR.

Primary TR is the result of abnormalities in the tricuspid valve structures, such as the tricuspid leaflets, tendon cords, papillary muscles, or annulus. The abnormalities can be both congenital and acquired. The most common congenital TV anomaly is Ebstein's disease. It is characterized by apical displacement of the leaflets arising directly from the atrial wall and the absence of clearly identifiable tendon cords. Acquired causes of primary TR include tumors, medically induced injury, drug injury, systemic disease, radiation, rheumatic, endocarditis, and trauma. Of these, TR caused by a transcatheter pacemaker or ICD lead is one of the most common etiologies. In patients without TR before pacemaker implantation, up to 38% may develop significant TR within the next 1–2 years [218–221].

Secondary TR usually presents with a dilated tricuspid annulus and poor leaflet docking, mainly due to right ventricular dilatation and dysfunction. The most common causes of secondary TR include combined left heart valve disease or myocardial disease. Moderate-to-severe TR can occur in 30–50% of patients with severe mitral regurgitation and in more than 25% of patients with severe aortic stenosis. These conditions can lead to increased left atrial pressure, causing pulmonary hypertension and, in turn, increased afterload on the right ventricle. Even after surgical or transcatheter treatment of mitral regurgitation or aortic stenosis, some TR may progress and manifest as delayed TR. Similarly, even after optimal pharmacological treatment of acute or chronic heart failure, or even cardiac resynchronization, some patients with reduced left ventricular (LV) ejection fraction may still develop TR. In addition to LV disease, other diseases that may increase RV afterload can cause RV dilation and dysfunction, which in turn can cause TR, such as primary pulmonary hypertension, pulmonary embolism, chronic lung disease, and pulmonary valve stenosis. In addition, diseases that directly affect RV function in the absence of pulmonary hypertension, such as inferior wall myocardial ischemia and arrhythmogenic right ventricular cardiomyopathy, may cause papillary muscle displacement and tendon elongation [221, 222].

In addition, another type of TR occurs in patients without pulmonary hypertension or coexisting left ß-heart disease and is referred to as isolated TR. Isolated TR is mainly seen in elderly patients, especially when combined with atrial fibrillation. Atrial fibrillation may be the main cause of RA enlargement, dilated TV annulus, and poor TV leaflet alignment in such patients [223].

TR usually has an insidious onset and develops slowly. There is a strong association between the appearance of symptoms and the severity of TR and right ventricular remodeling. In patients with secondary TR, the early manifestation is pulmonary vascular hyperfilling; in primary TR, the early change is central venous congestion. In severe stasis of the body circulation, peripheral edema, gastrointestinal stasis, and even hepatic stasis and ascites can occur. Over time, the compensatory mechanisms of the right ventricle and its ability to ensure adequate systolic output gradually decline, and a process of chronic right ventricular failure develops, manifesting symptoms of severe low cardiac output [218, 221].

7.4.3 *Indications and Contraindications for Transcatheter TV Repair*

The treatment strategy for TR should be developed according to the different stages of the disease. Pharmacological treatment aims to reduce volume load by diuretic therapy to reduce stasis in both the physical and pulmonary circulations, in addition to a combination of neurohormonal antagonists to control the compensatory mechanisms initiated by excessive blood volume load. In TR secondary to pulmonary hypertension, aggressive treatment of the primary disease may alleviate regurgitation to some extent. Surgical treatment remains the treatment of choice for severe TR. Patients with severe TR are often clinically complex because of their advanced age, history of other valve procedures, combined severe arrhythmias, and combined right ventricular dysfunction. Moreover, many patients are considered to undergo surgery at a time when they have reached an advanced stage of the disease. When surgery is performed at this time, the risk of right ventricular dysfunction is high, and the mortality rate is significantly increased. However, without treatment, these patients have poor survival and quality of life. On the other hand, after correction of TR, specific hemodynamic changes may be induced, such as the elimination of intracyclic vortices in the right ventricle (typical of functional TR), which may eventually lead to reverse remodeling and functional improvement of the right ventricle. Therefore, percutaneous therapy may be an important option for the treatment of TR in patients who are inoperable or at high risk for surgery. The current clinical practice guideline recommendations for surgical tricuspid repair or replacement are based on low-level evidence (B-NR or C), and the indications and timing of interventions for transcatheter tricuspid valve interventions need to be supported by evidence [12].

7.4.4 *Current Evidence and Limitations of Transcatheter Tricuspid Valve Repair (TTVr)*

TTVr includes types of edge-to-edge leaflet repair, valvuloplasty, and regurgitant port filling. The results of current salvage clinical application studies or feasibility clinical trials have demonstrated the safety and varying degrees of effectiveness of these TTVr procedures. However, further studies have also found that many TR patients may not be able to complete TTVr due to limitations in valve anatomy or poor transesophageal echocardiographic quality. Exploring effective transcatheter minimally invasive treatment options that can be adapted to most tricuspid valve anatomy remains an area that needs to be actively explored.

7.4.4.1 Leaflet/Coaptation Devices

TriClipTM

The TriClipTM Transcatheter Tricuspid Valve Repair System (Abbott Vascular, Santa Clara, CA, USA) uses the same clip technology as the MitraClip but has a different delivery system that includes a new introducer catheter specifically designed for TV. The most commonly used TriClip treatment strategy is to place the clip between the anterior and septal leaflets, a technique known as mitralization, which is currently the most feasible and frequently used method to achieve TR reduction using this device. A less common treatment strategy is to place the clip centrally between the septal and posterior leaflets and between the septal and anterior leaflets, the so-called triple orifice technique [224]. It has recently received the CE mark and is now approved for use in Europe.

Starting with the off-label use of the MitraClip system in the TV position, transcatheter TV edge-to-edge repair is the most mainstream percutaneous treatment for isolated severe TR or severe TR in combination with severe mitral regurgitation. In several reports and multicenter registries, the TV clipping technique has been shown to be safe and has shown immediate postoperative improvement of at least one grade of TR severity in >90% of patients. The mortality rate at 1 month of follow-up was 2.8% and 37.5% at 12 months. In addition, significant improvements in New York Heart Association (NYHA) functional class, 6-min walk test, and quality of life at 1 year were observed in both patients with severe TR and moderate TR [225–227]. Finally, in some cases, transcatheter TV margin repair can cause reverse right ventricular remodeling [228]. The TRILUMINATE trial (Trial to Evaluate Treatment with Abbott Transcatheter Clip Repair System in Patients with Moderate or Greater Tricuspid Regurgitation, NCT03227757) is a prospective, multicenter, single-arm study in 21 European and US centers in Europe and the United States to evaluate the safety and efficacy of the TriClip Repair System in treating patients with symptomatic grade 2 or greater TR. Preliminary results at 6 months showed that the TriClip System reduced TR by at least one grade, with significant clinical improvement observed at 6 months of follow-up, but we still need more longer follow-up results. Currently, TriClipTM has received the CE mark and is now approved for use in Europe.

FORMA

The FORMA Repair System (Edwards Lifesciences, Irvine, CA, USA) is a spacer that is fixed to the right ventricular apex by a track. The spacer is placed in the middle part of the tricuspid valve orifice, allowing the native valve to fit onto the spacer during closure, creating a new binding surface. The purpose of the device is to reduce TR by strengthening the binding of the leaflets [229].

The 3-year clinical and echocardiographic results of the first clinical use program of the FORMA system showed that the procedure was successful in 89% of cases, with a mortality rate of 0 within 30 days of the procedure. At 3-year follow-up, four (24%) patients died, and three (18%) patients required readmission for heart failure [230]. A follow-up SPACER trial (NCT02471807) is in progress.

PASCAL

The PASCAL system (Edwards Lifesciences, Irvine, CA, USA) consists of a central spacer and two paddles and clips. The design principle combines the features of the TriClip and FORMA devices, and the PASCAL system may overcome some of the technical limitations observed in the presence of large union gaps [231].

Immediate results with the PASCAL device for mitral regurgitation are excellent, but experience regarding its use in the treatment of TR is still in the early stages. The Edwards PASCAL TrAnScatheter Valve RePair System in Tricuspid Regurgitation (CLASP TR NCT03745313) study is an ongoing prospective, multicenter, single-arm study enrolling symptomatic patients with severe functional or degenerative TR with the aim of assessing the incidence of device- or procedure-related adverse events within 30 days, as well as clinical and echocardiographic outcomes at 1 year.

7.4.4.2 Annuloplasty Devices

Cardioband

The Cardioband (Edwards Lifesciences, Irvine, CA, USA system) system is an adjustable surgical-like Dacron band to achieve an undersized annuloplasty procedure. The band is secured to the atrial side of the tricuspid annulus by 12–17 anchors. After the band is fully secured to the TV valve annulus, the tricuspid orifice is reduced by constricting the band to reduce TR.

The TRIcuspid Regurgitation RePAIr with CaRdioband Transcatheter System (TRI-REPAIR) study enrolled 30 patients with moderate or severe functional TR, with an operative success rate of 100% and a 30-day incidence of the primary composite endpoint, serious adverse events, of 13.3%. All-cause mortality at 6 months and 1 year was 10.5% and 17.3%, respectively. The mean vena contracta at 6-month follow-up was reduced by 28%. In April 2018, the Cardioband system received the CE mark based on the results of the TRI-REPAIR study [232, 233]. A European prospective, single-arm, multicenter post-marketing follow-up study (TriBAND study NCT03779490) is currently enrolling 150 patients who will be followed for 5 years to assess the safety and efficacy of the Edwards Cardioband TR System in symptomatic, chronic, moderate-to-severe (>2+) functional TR.

TriAlign

The TriAlign device (Mitralign, Inc., Tewksbury, MA, USA) is designed to mimic the Kay procedure by plicating the posterior tricuspid valve leaflet through a transcatheter suture to achieve annuloplasty. One or more pairs of sutures are implanted in the anterior and septal posterior commissure, and these sutures are tightened with a dedicated plication lock device to obtain bicuspidization of the TV.

The SCOUT (Transcatheter Tricuspid Valve Anuloplasty System for Symptomatic Chronic Functional Tricuspid Regurgitation) trial recruited symptomatic patients with moderate or high functional TR patients to evaluate the feasibility and safety of the TriAlign device (Mitralign, Inc., Boston, MA, USA) and the 30-day treatment success rate was 80%, as demonstrated by a significant reduction in tricuspid annulus size and improvement in NYHA functional class [234]. The SCOUT II trial (NCT03225612) to further evaluate the safety and efficacy of the TriAlign system is an ongoing prospective, single-arm, multicenter study that will enroll up to 60 patients from 15 sites in Europe and the United States.

TriCinch

The TriCinch system (4Tech Cardio Ltd., Galway, Ireland) is a percutaneous annuloplasty device that reduces septolateral dimensions by cinching the tricuspid annulus at the level of the anterior commissure to achieve improved leaflet coaptation. The purpose is to improve leaflet coaptation. The system consists of a corkscrew element with a self-expandable nickel-titanium alloy stent [235]. The corkscrew is implanted and secured in the anterior tricuspid annulus, and the dimension of the annulus is tightened and altered under echocardiographic guidance to improve leaflet coaptation. The stent is then deployed in the inferior vena cava to maintain the applied tension. There are several studies on TriCinch in progress (NCT02098200, NCT03632967, NCT03294200).

7.4.5 The Current Development of Transcatheter Tricuspid Valve Replacement (TTVR)

TTVR can be divided by design concept into orthotopic-position and heterotopic-position TTVR. Orthotopic-position TTVR is a prosthetic valve placed in the anatomic position of the autologous tricuspid valve, replacing its function. While eliminating systolic regurgitation and improving stasis in the body circulation, it can reduce right atrial pressure and right ventricular preload and reverse right ventricular remodeling. Orthotopic-position TTVR involves the implantation of a prosthetic valve into the superior and/or inferior vena cava to reduce venous regurgitation and improve stasis in the body circulation, but it may not contribute to reverse right ventricular remodeling and functional recovery and is more often used as a palliative treatment approach or a complementary option to in situ TTVR. Several TTVR devices are currently in preclinical or early clinical trials, and the implementation of TTVR may break through the limitations faced by TTVr.

7.4.5.1 Orthotopic-Position TTVR

Lux-Valve

Lux-Valve (Jianshi Biotechnology (Zhejiang) Co.) is a self-expandable prosthetic valve consisting of a nickel-titanium alloy stent, a bovine pericardial leaflet, an adaptive polyester leakproof ring, a septal anchor device, and an anterior valve clamping device and is currently available in four sizes (30–55 mm). A previous study demonstrated that radial force-independent LuX-Valve was safe and no significant paravalvular leakage was observed by echocardiography with up to 180 days follow-up [236].

The results of a recently published study of partial salvage clinical applications (six cases) showed that all were successfully performed, and the degree of immediate postoperative tricuspid regurgitation was greatly reduced. Although one patient with moderate paravalvular leak died of right heart failure at the 3-month follow-up, the five living patients had no significant paravalvular leak and showed significant improvements in mean transvalvular gradient, right heart rate, conventional atrial function index, and overall longitudinal atrial strain at the 12-month follow-up, as well as a decrease in New York Heart Association functional class, suggesting a better safety and efficacy [237]. The registration premarket confirmatory clinical trial (TRAVEL) is ongoing and is planned to include 150 patients with severe TR, with promising results.

EVOQUE

EVOQUE (Edwards Lifesciences, Inc., USA) is also a self-expandable prosthetic valve consisting of a nickel-titanium alloy stent, a bovine pericardial leaflet, a braided skirt, and a leaflet clamping device, available in 44 and 48 mm sizes; the valve is delivered via the femoral vein route using a 28-F multidirectional adjustable bend delivery system to the tricuspid valve position. Subsequently, it is released coaxially using radial support forces and leaflet clamping techniques to achieve secure anchorage.

A recently published multicenter study of salvage clinical applications of severe TR (25 cases) showed that EVOQUE had an immediate operative success rate of 92%, a reduction in regurgitation severity to moderate or less in 96% of patients at 30-day follow-up, an increase in NYHA cardiac functional class I or II from 4 to 76% preoperatively, a pacemaker implantation rate of 8%, and a mortality rate of 0, initially demonstrating its safety and efficacy [238]. The upcoming TRISCEND I Pivotal Trial (NCT04482062) is planned to include 675 patients with severe TR, randomly assigned to the EVOQUE or optimized drug treatment groups, to further validate its safety and efficacy.

The Gate

The Gate (Navigate, USA) is composed of a self-expandable nitinol stent, a horse pericardial leaflet, a braided skirt, and a leaflet capture device in four sizes (40–52 mm); the valve is delivered to the tricuspid position using the 48-F delivery system via the right atrial incision route in the chest wall during operation and then released, with an anchoring mechanism similar to EVOQUE.

A recently published multicenter study of salvage clinical applications of severe TR (30 cases) showed an 87% success rate for immediate operation, 5% for intermediate open-heart surgery, 10% for pacemaker placement, and 10% for in-hospital mortality. Ninety-five percent of patients with successful TTVR at 30-day follow-up had regurgitation severity reduced to moderate or less, and the percentage of NYHA cardiac functional class I or II increased from 14 to 77% preoperatively. Preliminary indications are that its safety and efficacy are acceptable [239].

Other TTVR Systems

In addition, Cardiovalve (Boston Scientific, USA) and Intrepid (Medtronic, USA) TTVR systems via the femoral route are in early feasibility clinical studies, while TRiCares (TRiCares, France) via the femoral route and TriSol (TriSol, Israel) via the jugular route are in the preclinical phase. TriSol TTVR systems are still in the preclinical phase and no published data are available.

7.4.5.2 Heterotopic Position TTVR Systems

Several heterotopic position TTVR systems have initially demonstrated varying degrees of safety and efficacy. The transcatheter aortic valve replacement system SAPIEN XT (Edwards Lifesciences, Inc., USA) demonstrated high operational success rates in early salvage applications for severe TR [240], but the subsequent TRICAVAL randomized controlled trial was terminated early due to safety issues caused by valve displacement (4/14), and preliminary results did not appear to be superior to drug [241]. The TricValve [Pepperl+Fuchs (P+F), Australia] is a self-expandable nickel-titanium alloy stent with a bovine pericardial valve mounted on the stent, designed for the anatomical characteristics of the superior and inferior vena cava. The stent is designed to address the anatomical characteristics of the superior and inferior vena cava. Salvage clinical application studies have demonstrated a good safety profile, reducing vena cava regurgitation and improving clinical symptoms [242, 243]. Two early feasibility clinical studies (NCT03723239, NCT04141137) are underway in the United States and Europe, respectively. TriCento (New Valve Technology, Germany) is an ectopic TTVR system based on individualized CT imaging for implantation in the vena cava, with an overlapping stent facing the lateral opening of the tricuspid valve. A porcine pericardial valve is loaded on both ends to eliminate venous regurgitation. The results of a salvage clinical application study showed that it appears to be safe and can improve symptoms to some extent [244, 245].

7.4.6 Issues with TTVR/TTVr Devices

Currently, TTVR is performed in very few cases, the extent of associated complications is unknown, and standard definitions of endpoint events are lacking. The major complications of concern to investigators include poor valve anchorage, conduction block, bleeding events, valve thrombosis, and postoperative deterioration of right heart function. Poor valve anchoring may be due to selection of a nonoptimal size TTVR office-based membrane or failure of positioning release during operation, which will result in adverse conditions such as periprosthetic leakage, office-based membrane displacement or embolization, and intermediate open-heart intervention, for which careful multimodal imaging assessment, case screening, preoperative planning, and intraoperative maneuvers are essential [246]. Conduction block occurs as a result of direct compression of the TTVR valve by the atrioventricular node located adjacent to the tricuspid septal annulus. The rates of pacemaker placement after TTVR using the EVOQUE or GATE systems range from 8% to 10%. However, the Lux-Valve, which does not rely on radial support, has not been associated with postoperative permanent pacemaker placement events at present. The incidence of TTVR-related bleeding is approximately 8.3–13.0%, mainly related to device access, and further reduction of the delivery system size could help solve the problem, but this also involves improving TTVR valve design. Valve thrombosis is another important concern for investigators, and the risk of valve thrombosis is significantly higher in the right heart system because flow rates and pressures are much lower than in the left heart system, and optimization of postoperative antithrombotic regimens will be necessary for the maturation of TTVR [247]. The deterioration of right heart function after TTVR may occur in patients with severely impaired preoperative right heart function and/or severe pulmonary hypertension [248]. When regurgitant flow is significantly reduced or even eliminated, the risk of valve thrombosis is higher. After significant reduction or even elimination of flow, the right ventricular afterload increases, and right heart failure or inverse remodeling may occur, and benefit may be limited because the right ventricle is more sensitive to pressure changes. A comprehensive multiparametric assessment of right ventricular systolic function as well as pulmonary circulation hemodynamics and rigorous screening of patients with potential benefit can help avoid deterioration of right heart function after TTVR.

7.4.7 Conclusion

The burden of TR-related disease is gaining attention, and transcatheter tricuspid valve interventions offer hope for patients for whom surgical open-heart surgery is contraindicated or at high risk, but the current device designs are not yet satisfactory. TTVr is more demanding in terms of anatomic conditions and difficult to reliably reduce regurgitant flow, while TTVR has good promise for a broader population and better reduction of regurgitant flow. Preliminary studies have shown that TTVR with first-generation devices has good safety and efficacy, but its development still faces many challenges. However, with the accelerated emergence of innovative devices and the birth of high-quality studies, there is reason to be cautiously optimistic about TTVR treatment.

References

1. Horstkotte D, Loogen F. The natural history of aortic valve stenosis. Eur Heart J. 1988;9(Suppl E):57–64.
2. Braunwald E. Aortic stenosis: then and now. Circulation. 2018;137(20):2099–100.
3. Mas-Peiro S, Fichtlscherer S, Walther C, Vasa-Nicotera M. Current issues in transcatheter aortic valve replacement. J Thorac Dis. 2020;12(4):1665–80.
4. Lindroos M, Kupari M, HeikkilaJ TR. Prevalence of aortic valve abnormalities in the elderly: an echocardiographic study of a random population sample. J Am Coll Cardiol. 1993;21: 1220–5.
5. Roberts WC, Ko JM. Frequency by decades of unicuspid, bicuspid, and tricuspid aortic valves in adults having isolated aortic valve replacement for aortic stenosis, with or without associated aortic regurgitation. Circulation. 2005;111:920–5.
6. Michelena HI, Khanna AD, Mahoney D, Margaryan E, Topilsky Y, Suri RM, et al. Incidence of aortic complications inpatients with bicuspid aortic valves. JAMA. 2011;306:1104–12.
7. Michelena H, Prakash SK, Della Corte A, Bissell MM, Anavekar N, Mathieu P, et al. Bicuspid aortic valve: identifying knowledge gaps and rising to the challenge from the international bicuspid aortic valve consortium (BAVCon). Circulation. 2014;129:2691–704.
8. Bekeredjian R, Grayburn PA. Valvular heart disease: aortic regurgitation. Circulation. 2005;112:125–34.
9. Maurer G. Aortic regurgitation. Heart. 2006;92:994–1000.
10. Kanwar A, Thaden JJ, Nkomo VT. Management of patients with aortic valve stenosis. Mayo Clin Proc. 2018;93(4):488–508.
11. Otto CM, Nishimura RA, Bonow RO, et al. 2020 ACC/AHA guideline for the management of patients with valvular heart disease. J Am Coll Cardiol. 2020;76(25):e159–240.
12. Quintana RA, Monlezun DJ, DaSilva-DeAbreu A. One-year mortality in patients undergoing transcatheter aortic valve replacement for stenotic bicuspid versus tricuspid aortic valves: a meta-analysis and meta-regression. J Interv Cardiol. 2019;2(2019):8947204.
13. Cribier A, Eltchaninoff H, Bash A, et al. Percutaneous transcatheter implantation of an aortic valve prosthesis for calcific aortic stenosis: first human case description. Circulation. 2002;106:3006–8.
14. Grube E, Laborde JC, Zickmann B, et al. Firstreporton a human percutaneous transluminal implantation ofa self- expanding valve prosthesis for interventional treatment of aortic valve stenosis. Catheter Cardiovasc Interv. 2005;66:465–9.
15. Laborde JC, Borenstein N, Behr L, Farah B, FajadetJ. Percutaneous implantation ofan aortic valve prosthesis. Catheter Cardiovasc Interv. 2005;65:171–4.
16. Osman M, Ghaffar YA, Saleem M, et al. Meta-analysis comparing transcatheter aortic valve implantation with balloon versus self-expandable valves. Am J Cardiol. 2019;124:1252–6.
17. Rheude T, Blumenstein J, Möllmann H, Husser O. Spotlight on the SAPIEN 3 transcatheter heart valve. Med Devices (Auckl). 2018;1(11):353–60.
18. Webb JG, Murdoch DJ, Alu MC, et al. 3-year outcomes after valve-in-valve transcatheter aortic valve replacement for degenerated bioprostheses: the PARTNER 2 registry. J Am Coll Cardiol. 2019;73(21):2647–55.

19. Bleiziffer S, Bosmans J, Brecker S, et al. Insights on mid-term TAVR performance: 3-year clinical and echocardiographic results from the CoreValve ADVANCE study. Clin Res Cardiol. 2017 Oct;106(10):784–95.
20. Van Mieghem NM, Popma JJ, Michael Deeb G, et al. Omplete 2-year results confirm Bayesian analysis of the SURTAVI trial. JACC Cardiovasc Interv. 2020;13(3):323–31.
21. Stathogiannis K, Synetos A, Toutouzas K, et al. Medtronic CoreValve: achieving optimal outcomes. Hell J Cardiol. 2015;56(Suppl A):4–8.
22. Meredith IT, Hood KL, Haratani N, Allocco DJ, Dawkins KD. Boston scientific lotus valve. EuroIntervention. 2012;8(Suppl Q):Q70–4.
23. Seigerman ME, Nathan A, Anwaruddin S. The lotus valve system: an in-depth review of the technology. Curr Cardiol Rep. 2019;21(12):157.
24. Feldman TE, Reardon MJ, Rajagopal V, et al. Effect of mechanically expanded vs selfexpanding transcatheter aortic valve replacement on mortality and major adverse clinical events in high-risk patients with aortic stenosis: the REPRISE III randomized clinical trial. JAMA. 2018;319(1):27–37.
25. Cheng J, Chen M, Zhu D, Zhang J, Jia H, Guo Y. Successful trans-apical aortic valve implantation for a high risk patient with aortic stenosis using a new second-generation TAVI device—J-Valve system. J Cardiothorac Surg. 2015;17(10):5.
26. Mingwui Tung X, Wang FL, et al. A versatile transapical device for aortic valvular disease: one-year outcomes of a multicenter study on the J-Valve system. J Cardiol. 2018;72(5): 377–84.
27. Siontis GCM, Jini P, Pilgrim T, et al. Predictors of permanent pacemaker implantation in patients with severe aortic stenosis undergoing TAVRJ. Am Coll Cardiol. 2014;64:129–40.
28. Erkapic D, De Rosa S, Kelava A, Lehmann R, Fichtlscherer S, Hohnloser SH. Risk for permanent pacemaker after transcatheter aortic valve implantation: a comprehensive analysis of the literature. J Cardiovasc Electrophysiol. 2012;23:391–7.
29. Khatri PJ, Webb JG, Rodes-Cabau J, et al. Adverse effects associated with transcatheter aortic valve implantation: a meta-analysis of contemporary studies. Ann Intern Med. 2013;158:35–46.
30. Abdel-Wahab M, MehilliJ FC, et al. Comparison of balloon-expandable vs self-expandable valves in patients undergoing transcatheter aortic valve replacement. JAMA. 2014;311:1503.
31. van Rosendael PJ, Delgado V, BaxJJ. Pacemaker implantation rate after transcatheter aortic valve implantation with early and new-generation devices: a systematic review. Eur Heart J. 2018;39:2003–13.
32. Sanchez CE, Yakubov SJ, Arshi A. Innovations in transcatheter valve technology: what the next five years hold. Interv Cardiol Clin. 2018;7:489–501.
33. Urena M, Webb JG, Tamburino C, et al. Permanent pacemaker implantation after tanscatheter aortic valve implantation. Circulation. 2014;129:1233–43.
34. Fadahunsi O, Olowoyeye A, Ukaigwe A, et al. Incidence, predictors, and outcomes of permanent pacemaker implantation following transcatheter aortic valve replacement. JACC Cardiovasc Interv. 2016;9:2189–99.
35. Toggweiler S, Kobza R. Pacemaker implantation after transcatheter aortic valve: why is this still happening? J Thorac Dis. 2018;10:S3614–9.
36. Kodali S, Thourani VH, White J, et al. Early clinical and echocardiographic outcomes after SAPIEN 3 transcatheter aortic valve replacement in inoperable, high-risk and intermediate-risk patients with aortic stenosis. Eur Heart J. 2016;37:2252–62.
37. Wendler O, Schymik G, Treede H, et al. SOURCE3 registry. Circulation. 2017;135:1123–32.
38. Wendler O, Schymik G, Treede H, et al. SOURCE 3:1-year outcomes post-transcatheter aortic valve implantation using the latest generation of the balloon- expandable transcatheter heart valve. Eur Heart J. 2017;38:2717–26.
39. Pilgrim T, Stortecky S, Nietlispach F, et al. Repositionable versus balloon-expandable devices for transcatheter aortic valve implantation in patients with aortic stenosis. J Am Heart Assoc. 2016;5:e004088.

40. Popma JJ, Reardon M, Khabbaz K, et al. Early clinical outcomes after transcatheter aortic valve replacement using a novel self-expanding bioprosthesis in patients with severe aortic stenosis who are suboptimal for surgery. JACC Cardiovasc Interv. 2017;10:268–75.
41. Forrest JK, Mangi AA, Popma J, et al. Early outcomes with the evolut PRO repositionable self-expanding transcatheter aortic valve with pericardial Wrap. JACC Cardiovasc Interv. 2018;11:160–8.
42. Reichenspurner H, Schaefer A, Schafer U, et al. Self-expanding transcatheter aortic valve system for symptomatic high-risk patients with severe aortic stenosis. J Am Coll Cardiol. 2017;70:3127–36.
43. Husser O, Kim WK, Pellegrini C, et al. Multicenter comparison of novel self-expanding versus balloon-expandable transcatheter heart valves. JACC Cardiovasc Interv. 2017;10:2078–87.
44. Auffret V, Puri R, Urena M, et al. Conduction disturbances after tianscatheter aortic valve replacement: current status and future perspectives. Circulation. 2017;136:1049–69.
45. Pereira E, Ferreira N, Caeiro D, et al. Transcatheter aortic valve implantation and requirements of pacing over time. Pacing Clin Electrophysiol. 2013;36:559–69.
46. Naveh S, Perlman GY, Elitsur Y, et al. Electrocardiographic predictors of long-term cardiac pacing dependency following transcatheter aortic valve implantation. J Cardiovasc Electrophysiol. 2017;28:216–23.
47. van der Boon RMA, Van Mieghem NM, Theuns DA, et al. Pacemaker dependency after transcatheter aortic valve implantation with the self-expanding Medtronic CoreValve System. Int J Cardiol. 2013;168:1269–73.
48. Alasti M, Rashid H, Rangasamy K, et al. Long-term pacemaker dependency and impact of pacing on mortality following transcatheter aortic valve replacement with the LOTUS valve. Catheter Cardiovasc Interv. 2018;92:777–82.
49. Rodes-Cabau J, Ellenbogen KA, Krahn AD, et al. Management of conduction disturbances associated with transcatheter aortic valve replacement: JACC scientific expert panel. J Am Coll Cardiol. 2019;74:1086–106.
50. Van Belle E, Juthier F, Susen S, et al. Postprocedural aortic regurgitation in balloon-expandable and self- expandable transcatheter aortic valve replacement procedures. Circulation. 2014;129:1415–27.
51. Kodali S, Pibarot P, Douglas PS, et al. Paravalvular regurgitation after transcatheter aortic valve replacement with the Edwards sapien valve in the PARTNER trial: characterizing patients and impact on outcomes. Eur Heart J. 2015;36:449–56.
52. Dworakowski R, Wendler O, Halliday B, et al. Device-dependent association between paravalvar aortic regurgitation and outcome after TAVI. Heart. 2014;100:1939–45.
53. Athappan G, Patvardhan E, Tuzcu EM, et al. Incidence, predictors, and outcomes of aortic regurgitation after transcatheter aortic valve replacement: meta-analysis and systematic review of literature. J Am Coll Cardiol. 2013;61:1585–95.
54. Mack MJ, Leon MB, Smith CR, et al. 5-year outcomes of transcatheter aortic valve replacement or surgicalaortic valve replacement for high surgical risk patients with aortic stenosis (PARTNER 1): arandomised controlled trial. Lancet. 2015;385:2477–84.
55. Ong G, Annabi MS, Clavel MA, et al. Paravalvular regurgitation after transcatheter aortic valve replacement: is the problem solved? Interv Cardiol Clin. 2018;7:445–58.
56. Bhatheja S, PanchalHB BN, et al. Valvular performance and aortic regurgitation following transcatheter aortic valve replacement using Edwards valve versus CoreValve for severe aortic stenosis: a meta-analysis. Cardiovasc Revasc Med. 2016;17:248–55.
57. Abdel-Wahab M, Mehilli J, Frerker C, et al. Comparison of balloon-expandable vs self-expandable valves in patients undergoing transcatheter aortic valve replacement. JAMA. 2014;311:1503.
58. Thourani VH, Kodali S, Makkar RR, et al. Transcatheter aortic valve replacement versus surgical valve replacement in intermediate-risk patients: a propensity score analysis. Lancet. 2016;387:2218–25.

59. Meredith Am IT, Walters DL, Dumonteil N, et al. Transcatheter aortic valve replacement for severe symptomatic aortic stenosis using a repositionable valve system: 30-day primary endpoint results from the REPRISE II study. J Am Coll Cardiol. 2014;64:1339–48.
60. Mack MJ, Leon MB, Thourani VH, et al. Transcatheter aortic-valve replacement with a balloon-expandable valve in low-risk patients. N Engl J Med. 2019;380:1695–705.
61. Popma J, DeebG M, Yakubov SJ, et al. Transcatheter aortic-valve replacement with a self-expanding valve in low-risk patients. N Engl J Med. 2019;380:1706–15.
62. Laramasso M, Denegri A, Kuwata S, et al. Feasibility and safety oftransfemoral sheathless portico aortic valve implantation: preliminary results in a single center experience. Catheter Cardiovasc Interv. 2018;91:533–9.
63. Nombela-Franco L, Rodes-Cabau J, De Larochelliere R, et al. Predictive factors, efficacy, and safety of balloon post- dilation after transcatheter aortic valve implantation with a balloon-expandable valve. JACC Cardiovasc Interv. 2012;5:499–512.
64. Sinning JM, Hammerstingl C, Vasa-Nicotera M, et al. Aortic regurgitation index defnes severity of peri- prosthetic regurgitation and predicts outcome in patients after transcatheter aortic valve implantation. J Am Coll Cardiol. 2012;59:1134–41.
65. Toggweiler S, Wood DA, Rodes-Cabau J, et al. Transcatheter valve-in-valve implantation for failed balloon-expandable transcatheter aortic valves. JACC Cardiovasc Interv. 2012;5:571–7.
66. Ussia GP, Barbanti M, Ramondo A, et al. The valve- in-valve technique for treatment of aortic bioprosthesis malposition an analysis of incidence and 1-year clinical outcomes from the Italian CoreValve registry. J Am Coll Cardiol. 2011;57:1062–8.
67. Pibarot P, Hahn RT, Weissman NJ, et al. Association of paravalvular regurgitation with 1-year outcomes after transcatheter aortic valve replacement with the SAPIEN 3 valve. JAMA Cardiol. 2017;2:1208.
68. LohJK LSH, Abdelmoneim SS, et al. Regression of paravalvular aortic regurgitation and remodeling of self expanding transcatheter aortic valve: an observation from the CoreValve U.S. pivotal trial. JACC Cardiovasc Imaging. 2015;8:1364–75.
69. Pagnesi M, Martino EA, Chiarito M, et al. Silent cerebral injury after transcatheter aortic valveimplantation and the preventive role of embolic protection devices: asystematic review and meta-analysis. Int J Cardiol. 2016;221:97–106.
70. Athappan G, Gajulaplli RD, Sengodan P, et al. Infuence of transcatheter aortic valve replacement strategy and valve design on stroke after transcatheter aortic valve replacement: a meta-analysis and systematic review of literature. J Am Coll Cardiol. 2014;63:2101–10.
71. Werner N, Zeymer U, Schneider S, et al. Incidence and clinical impact of stroke complicating transcatheter aortic valve implantation: results from the German TAVI Registrrv. Catheter Cardiovasc Interv. 2016;88:644–53.
72. Bagur R, Solo K, Alghofaili S, et al. Cerebral embolic protection devices during transcatheter aortic valve implantation: systematic review and meta-analysis. Stroke. 2017;48:1306–15.
73. Kapadia SR, Kodali S, Makkar R, et al. Protection against cerebral embolism during transcatheter aortic valve replacement. J Am Coll Cardiol. 2017;69:367–77.
74. Mercanti F, Rosseel L, Neylon A, et al. Chimney stenting for coronary occlusion during TAVR: insights from the chimney registry. J Am Coll Cardiol Intv. 2020;13:751–61.
75. Khan JM, Greenbaum AB, Babaliaros VC, et al. The BASILICA trial: prospective multicenter investigation of intentional leaflet laceration to prevent TAVR coronary obstruction. JACC Cardiovasc Interv. 2019;12(13):1240–52.
76. Goel SS, Ige M, Tuzcu EM, et al. Severe aortic stenosis and coronary artery disease—implications for management in the transcatheter aortic valve replacement era: a comprehensive review. J Am Coll Cardiol. 2013;62:1–10.
77. Sankaramangalam K, Banerjee K, Kandregula K, et al. Impact of coronary artery disease on 30 day and 1 year mortality in patients undergoing transcatheter aortic valve replacement: a meta analysis. J Am Heart Assoc. 2017;6:e006092.

78. Witberg G, Zusman O, Codner P, Assali A, Kornowski R. Impact of coronary artery revascularization completeness on outcomes of patients with coronary artery disease undergoing transcatheter aortic valve replacement. J Am Heart Assoc. 2017;6(10):e006092.
79. Yudi MB, Sharma SK, Tang GHL, Kini A. Coronary Angiography and Percutaneous Coronary Intervention After Transcatheter Aortic Valve Replacement. J Am Coll Cardiol. 2018;71: 1360–78.
80. Une D, Ruel M, David TE. Twenty-year durability of the aortic Hancock II bioprosthesis in young patients: is it durable enough? Eur J Cardiothorac Surg. 2014;46:825–30.
81. Durand EL, Sokoloff A, Urena-Alcazar M, et al. Assessment of long-term structural deterioration of transcatheter aortic bioprosthetic valves using the new European definition. Circ Cardiovasc Interv. 2019;12:e007597.
82. Lancelltti P, Martinez C, Radermecker M. The long quest for the holy grailin transcatheter aortic bioprosthesis: durability and long-term performance. J Am Coll Cardiol. 2019;73:554–8.
83. Martinez C, Gach O, Radermecker MA, et al. Transcatheter aortic valve replacement: from a concept to a medical revolution. Rev Med Liege. 2019;74:S64–72.
84. Deeb GM, Reardon MJ, Chetcuti S, et al. 3-year outcomes in high-risk patients who underwent surgical or transcatheter aortic valve replacement. J Am Coll Cardiol. 2016;67:2565–74.
85. Pibarot P, Ternacle J, Jaber WA, et al. Structural deterioration of transcatheter versus surgical aortic valve bioprostheses in the PARTNER-2 trial. J Am Coll Cardiol. 2020;76(16):1830–43.
86. Toggweiler S, Humphries KH, Lee M, et al. 5-year outcome after transcatheter aortic valve implantation. J Am Coll Cardiol. 2013;61:413–9.
87. Gerckens U, Tamburino C, Bleiziffer S, et al. Final 5-year clinical and echocardiographic results for treatment of severe aortic stenosis with a self-expanding bioprosthesis from the ADVANCE Study. Eur Heart J. 2017;38(36):2729–38.
88. Blackman DJ, Saraf S, MacCarthy PA, et al. Long-term durability of transcatheter aortic valve prostheses. J Am Coll Cardiol. 2019;73:537–45.
89. Capodanno D, Petronio AS, Prendergast B, et al. Standardized definitions of structural deterioration and valve failure in assessing long-term durability of transcatheter and surgical aortic bioprosthetic valves: a consensus statement from the European Association of Percutaneous Cardiovascular Interventions (EAPCD) endorsed by the European Society of Cardiology (ESC) and the European Association for Cardio-Thoracic Surgery (EACTS). Eur Heart J. 2017;38:3382–90.
90. Barbanti M, Costa G, Zappulla P, et al. Incidence of long term structural valve dysfunction and bioprosthetic valve failure after transcatheter aortic valve replacement. J Am Heart Assoc. 2018;7:e008440.
91. Sondergaard L, Ihlemann N, Capodanno D, et al. Durability of transcatheter and surgical bioprosthetic aortic valves in patients at lower surgical risk. J Am Coll Cardiol. 2019;73:546–53.
92. Olson L, Subramanian R, Ackermann D, Orszulak T, Edwards W. Surgical pathology of the mitral valve: a study of 712 cases spanning 21 years. Mayo Clin Proc. 1987;62:22–34.
93. Nkomo VT, Gardin JM, Skelton TN, et al. Burden of valvular heart diseases: a population-based study. Lancet. 2006;368(9540):1005–11.
94. Li J, Pan W, Yin Y, et al. Prevalence and correlates of mitral regurgitation in the current era: an echocardiography study of a Chinese patient population. Acta Cardiol. 2016;71(1):55–60.
95. Hu P, Liu XB, Liang J, et al. A hospital-based survey of patients with severe valvular heart disease in China. Int J Cardiol. 2017;231:244–7.
96. Zhang B, Xu H, Zhang H, et al. Prognostic value of N-terminal pro- B-type natriuretic peptide in elderly patients with valvular heart disease. J Am Coll Cardiol. 2020;75(14):1659–72.
97. Head SJ, van Leeuwen WJ, Van Mieghem NM, et al. Surgical or transcatheter mitral valve intervention: complex disease requires complex decisions. EuroIntervention. 2014;9(10): 1133–5.

98. Gregg W. Stone. Transcatheter mitral repair and replacement: state-of-the art 2016. Presented at TVT, 17 June 2016[A].
99. 聂静雨, 朱振辉, 王浩,等. 35岁及以上人群二尖瓣反流现状及其影响因素研究. 中国循环杂志. 2017;032(005):475–9.
100. Maisano F, Torracca L, Oppizzi M, et al. The edge-to-edge technique: a simplified method to correct mitral insufficiency. Eur J Cardiothorac Surg. 1998;13(3):240–6.
101. Maisano F, La Canna G, Colombo A, et al. The evolution from surgery to percutaneous mitral valve interventions: the role of the edge-to-edge technique. J Am Coll Cardiol. 2011;58(21): 2174–82.
102. Feldman T, Wasserman HS, Herrmann HC, et al. Percutaneous mitral valve repair using the edge-to-edge technique: six-month results of the EVEREST Phase I Clinical Trial. J Am Coll Cardiol. 2005;46(11):2134–40.
103. Feldman T, Foster E, Glower DD, et al. Percutaneous repair or surgery for mitral regurgitation. N Engl J Med. 2011;364(15):1395–406.
104. Obadia JF, Messika-Zeitoun D, Leurent G, et al. Percutaneous repair or medical treatment for secondary mitral regurgitation. N Engl J Med. 2018;379(24):2297–306.
105. Stone GW, Lindenfeld J, Abraham WT, et al. Transcatheter mitral-valve repair in patients with heart failure. N Engl J Med. 2018;379(24):2307–18.
106. Praz F, Spargias K, Chrissoheris M, et al. Compassionate use of the PASCAL transcatheter mitral valve repair system for patients with severe mitral regurgitation: a multicentre, prospective, observational, first-in-man study. Lancet. 2017;390(10096):773–80.
107. Lim DS, Kar S, Spargias K, et al. Transcatheter valve repair for patients with mitral regurgitation: 30-day results of the CLASP study. JACC Cardiovasc Interv. 2019;12(14):1369–78.
108. Besler C, Noack T, von Roeder M, et al. Transcatheter edge-to-edge mitral valve repair with the PASCAL system: early results from a real-world series. EuroIntervention. 2020;16(10): 824–32.
109. Pan W, Zhou D, Wu Y, et al. First-in-human results of a novel user-friendly transcatheter edge-to-edge mitral valve repair device. JACC Cardiovasc Interv. 2019;12(23):2441–3.
110. Siminiak T, Firek L, Jerzykowska O, et al. Percutaneous valve repair for mitral regurgitation using the Carillon Mitral Contour System. Description of the method and case report. Kardiol Pol. 2007;65(3):272–8. discussion 279
111. Schofer J, Siminiak T, Haude M, et al. Percutaneous mitral annuloplasty for functional mitral regurgitation: results of the CARILLON mitral annuloplasty device European union study. Circulation. 2009;120(4):326–33.
112. Siminiak T, Wu JC, Haude M, et al. Treatment of functional mitral regurgitation by percutaneous annuloplasty: results of the TITAN Trial. Eur J Heart Fail. 2012;14(8):931–8.
113. Witte KK, Lipiecki J, Siminiak T, et al. The REDUCE FMR Trial: a randomized sham-controlled study of percutaneous mitral annuloplasty in functional mitral regurgitation. JACC Heart Fail. 2019;7(11):945–55.
114. Maisano F, La Canna G, Latib A, et al. First-in-man transseptal implantation of a "surgical-like" mitral valve annuloplasty device for functional mitral regurgitation. J ACC Cardiovasc Interv. 2014;7(11):1326–8.
115. Messika-Zeitoun D, Nickenig G, Latib A, et al. Transcatheter mitral valve repair for functional mitral regurgitation using the Cardioband system: 1 year outcomes. Eur Heart J. 2019;40(5): 466–72.
116. Siminiak T, Dankowski R, Baszko A, et al. Percutaneous direct mitral annuloplasty using the Mitralign Bident system: description of the method and a case report. Kardiol Pol. 2013;71 (12):1287–92.
117. Nickenig G, Schueler R, Dager A, et al. Treatment of chronic functional mitral valve regurgitation with a percutaneous annuloplasty system. J Am Coll Cardiol. 2016;67(25): 2927–36.

118. Colli A, Bellu R, Pittarello D, et al. Transapical off-pump NeoChord implantation on bileaflet prolapse to treat severe mitral regurgitation. Interact Cardiovasc Thorac Surg. 2015;21(4):554–6.
119. Promphan W, Prachasilchai P, Siripornpitak S, Qureshi SA, Layangool T. Percutaneous pulmonary valve implantation with the Venus P-valve: clinical experience and early results. Cardiol Young. 2016;26(4):698–710. This study reports early experience with a new self-expanding transcatheter valve, the Venus P-valve. All 6 patients described had severe PR with large RVOTs (range 24–32 mm). The valve was successfully implanted in all patients, but unexpected mild proximal valve migration occurred in one patient and resulted in mild paravalvar leak and significant tricuspid regurgitation
120. Colli A, Manzan E, Aidietis A, et al. An early European experience with transapical off-pump mitral valve repair with NeoChord implantation. Eur J Cardiothorac Surg. 2018;54(3):460–6.
121. Gammie JS, Bartus K, Gackowski A, et al. Beating-heart mitral valve repair using a novel ePTFE cordal implantation device. J Am Coll Cardiol. 2018;71(1):25–36.
122. Wang S, Meng X, Luo Z, et al. Transapical beating-heart mitral valve repair using a novel artificial chordae implantation system. Ann Thorac Surg. 2018;106(5):e265–7.
123. Bapat V, Rajagopal V, Meduri C, et al. Early experience with new transcatheter mitral valve replacement. J Am Coll Cardiol. 2018;71(1):12–21.
124. Sorajja P, Moat N, Badhwar V, et al. Initial feasibility study of a new transcatheter mitral prosthesis: the first 100 patients. J Am Coll Cardiol. 2019;73(11):1250–60.
125. Cheung A, Stub D, Moss R, et al. Transcatheter mitral valve implantation with Tiara bioprosthesis. EuroIntervention. 2014;10 Suppl U:U115–9.
126. Werner N, Kilkowski C, Sutor D, et al. Transcatheter mitral valve implantation (TMVI) using Edwards SAPIEN 3 Prostheses in patients at very high or prohibitive surgical risk: a single-center experience. J Interv Cardiol. 2020;2020:9485247.
127. Webb JG, Murdoch DJ, Boone RH, et al. Percutaneous transcatheter mitral valve replacement: first-in-human experience with a new transseptal system. J Am Coll Cardiol. 2019;73(11):1239–46.
128. Luo Z, Zhang H, Xie Y, et al. Clinical application of a fully ultrasound-guided transapical transcatheter mitral valve replacement device. JACC Cardiovasc Interv. 2020;13(17):e161–2.
129. Nobuyoshi M, Arita T, Shirai S, et al. Percutaneous balloon mitral valvuloplasty: a review. Circulation. 2009;119:e211–9.
130. Silbiger JJ. Anatomy, mechanics, and pathophysiology of the mitral annulus. Am Heart J. 2012;164:163–76.
131. Palacios IF, Sanchez PL, Harrell LC, et al. Which patients benefit from percutaneous mitral balloon valvuloplasty? Prevalvuloplasty and postvalvuloplasty variables that predict long-term outcome. Circulation. 2002;105(12):1465–71.
132. Mark H, Brown Richard A, Sudish L, et al. Transcatheter mitral valve replacement: an update on current techniques, technologies, and future directions. JACC Cardiovasc Interv. 2021;14:489–500.
133. Maurice E-S, Akins CW, Alec V. Mitral regurgitation. Lancet. 2009;373:1382–94.
134. Lloyd TR, Marvin WJ, Mahoney LT, Lauer RM. Balloon dilation valvuloplasty of bioprosthetic valves in extracardiac conduits. Am Heart J. 1987;114(2):268–74.
135. Devanagondi R, Peck D, Sagi J, Donohue J, Yu S, Pasquali SK, et al. Long-term outcomes of balloon Valvuloplasty for isolated pulmonary valve stenosis. Pediatr Cardiol. 2017;38(2):247–54.
136. Waller BF, Howard J, Fess S. Pathology of pulmonic valve steno- sis and pure regurgitation. Clin Cardiol. 1995;18(1):45–50.
137. Shaw AC, Kalidas K, Crosby AH, Jeffery S, Patton MA. The natural history of Noonan syndrome: a long-term follow-up study. Arch Dis Child. 2007;92(2):128–32.
138. Lin AE, Basson CT, Goldmuntz E, Magoulas PL, McDermott DA, McDonald-McGinn DM, et al. Adults with genetic syndromes and cardiovascular abnormalities: clinical history and management. Genet Med. 2008;10(7):469–94.

139. Kumar A, Stalker HJ, Williams CA. Concurrence of supravalvular aortic stenosis and peripheral pulmonary stenosis in three generations of a family: a form of arterial dysplasia. Am J Med Genet. 1993;45(6):739–42.
140. Driscoll DJ, Michels VV, Gersony WM, Hayes CJ, Keane JF, Kidd L, et al. Occurrence risk for congenital heart defects in relatives of patients with aortic stenosis, pulmonary stenosis, or ventricular septal defect. Circulation. 1993;87(2 Suppl):I114–20.
141. Bashore TM. Adult congenital heart disease: right ventricular out- flow tract lesions. Circulation. 2007;115(14):1933–47.
142. Stamm C, Anderson RH, Ho SY. Clinical anatomy of the normal pulmonary root compared with that in isolated pulmonary valvular stenosis. J Am Coll Cardiol. 1998;31(6):1420–5.
143. Vedanthan R, Sanz J, Halperin J. Bicuspid pulmonic valve. J Am Coll Cardiol. 2009;54(8):e5.
144. Rao BN, Anderson RC, Edwards JE. Anatomic variations in the tetralogy of Fallot. Am Heart J. 1971;81(3):361–71.
145. Ascione L, Lengo R, Tuccillo B, D'Andrea A, De Michele M, Porto A, et al. Quadricuspid pulmonary valve diagnosed bycardiac magnetic resonance. J Cardiovasc Med (Hagerstown). 2009;10(12):944–5.
146. Shah R, Shriki J, Shinbane JS. Cardiovascular magnetic resonance depiction of quadricuspid pulmonary valve with associated pulmonary regurgitation and pulmonary artery aneurysm. Tex Heart Inst J. 2014;41(3):349–50.
147. Hayes CJ, Gersony WM, Driscoll DJ, Keane JF, Kidd L, O'Fallon WM, et al. Second natural history study of congenital heart defects. Results of treatment of patients with pulmonary valvar stenosis. Circulation. 1993;87(2 Suppl):I28–37.
148. Jonsson B, Lee SJ. Haemodynamic effects of exercise in isolated pulmonary stenosis before and after surgery. Br Heart J. 1968;30(1):60–6.
149. Nishimura RA, Otto CM, Bonow RO, Carabello BA, Erwin JP, Guyton RA, et al. 2014 AHA/ACC guideline for the management of patients with valvular heart disease: a report of the American College of Cardiology/American Heart Association task force on practice guidelines. J Am Coll Cardiol. 2014;63(22):e57–185.
150. Hamza N, Ortiz J, Bonomo RA. Isolated pulmonic valve infective endocarditis: a persistent challenge. Infection. 2004;32(3):170–5.
151. Pellikka PA, Tajik AJ, Khandheria BK, Seward JB, Callahan JA, Pitot HC, et al. Carcinoid heart disease. Clinical and echocardiographic spectrum in 74 patients. Circulation. 1993;87(4): 1188–96.
152. Shimazaki Y, Blackstone EH, Kirklin JW. The natural history of isolated congenital pulmonary valve incompetence: surgical im- plications. Thorac Cardiovasc Surg. 1984;32(4):257–9.
153. Saremi F, Gera A, Ho SY, Hijazi ZM, Sánchez-Quintana D. CT and MR imaging of the pulmonary valve. Radiographics. 2014;34(1):51–71.
154. Davlouros PA, Kilner PJ, Hornung TS, Li W, Francis JM, Moon JC, et al. Right ventricular function in adults with repaired tetralogy of Fallot assessed with cardiovascular magnetic resonance imaging: detrimental role of right ventricular outflow aneurysms or akinesia and adverse right-to-left ventricular interaction. J Am Coll Cardiol. 2002;40(11):2044–52.
155. Bigdelian H, Mardani D, Sedighi M. The effect of pulmonary valve replacement (PVR) surgery on hemodynamics of patients who underwent repair of tetralogy of Fallot (TOF). J Cardiovasc Thorac Res. 2015;7(3):122–5.
156. Frigiola A, Giardini A, Taylor A, Tsang V, Derrick G, Khambadkone S, et al. Echocardiographic assessment of diastolic biventricular properties in patients operated for severe pulmonary regurgitation and association with exercise capacity. Eur Heart J Cardiovasc Imaging. 2012;13(8):697–702.
157. Frigiola A, Redington AN, Cullen S, Vogel M. Pulmonary regurgitation is an important determinant of right ventricular contractile dysfunction in patients with surgically repaired tetralogy of Fallot. Circulation. 2004;110(11 Suppl 1):II153–7.
158. Bouzas B, Kilner PJ, Gatzoulis MA. Pulmonary regurgitation: not a benign lesion. Eur Heart J. 2005;26(5):433–9.

159. Gatzoulis MA, Balaji S, Webber SA, Siu SC, Hokanson JS, Poile C, et al. Risk factors for arrhythmia and sudden cardiac death late after repair of tetralogy of Fallot: a multicentre study. Lancet. 2000;356(9234):975–81.
160. Knauth AL, Gauvreau K, Powell AJ, Landzberg MJ, Walsh EP, Lock JE, et al. Ventricular size and function assessed by cardiac MRI predict major adverse clinical outcomes late after tetralogy of Fallot repair. Heart. 2008;94(2):211–6.
161. Redington AN. Determinants and assessment of pulmonary regurgitation in tetralogy of Fallot: practice and pitfalls. Cardiol Clin. 2006;24(4):631–9. vii
162. Joynt MR, Yu S, Dorfman AL, Ghadimi Mahani M, Agarwal PP, Lu JC. Differential impact of pulmonary regurgitation on patients with surgically repaired pulmonary stenosis versus tetralogy of Fallot. Am J Cardiol. 2016;117(2):289–94.
163. Kan JS, White RI, Mitchell SE, Gardner TJ. Percutaneous balloon valvuloplasty: a new method for treating congenital pulmonary- valve stenosis. N Engl J Med. 1982;307(9):540–2.
164. Wang SZ, Ou-Yang WB, Hu SS, Pang KJ, Liu Y, Zhang FW, et al. First-in-human percutaneous balloon pulmonary Valvuloplasty under echocardiographic guidance only. Congenit Heart Dis. 2016;11(6):716–20.
165. Warnes CA, Williams RG, Bashore TM, Child JS, Connolly HM, Dearani JA, et al. ACC/AHA 2008 guidelines for the Management of Adults with congenital heart disease: a report of the American College of Cardiology/American Heart Association task force on practice guidelines (writing committee to develop guidelines on the management of adults with congenital heart disease). Circulation. 2008;118(23):e714–833.
166. Fawzy ME, Hassan W, Fadel BM, Sergani H, El Shaer F, El Widaa H, et al. Long-term results (up to 17 years) of pulmonary balloon valvuloplasty in adults and its effects on concomitant severe infundibular stenosis and tricuspid regurgitation. Am Heart J. 2007;153(3):433–8.
167. McCrindle BW. Independent predictors of long-term results after balloon pulmonary valvuloplasty. Valvuloplasty and angioplasty of congenital anomalies (VACA) registry investigators. Circulation. 1994;89(4):1751–9.
168. McCrindle BW, Kan JS. Long-term results after balloon pulmonary valvuloplasty. Circulation. 1991;83(6):1915–22.
169. Mahfouz RA, Moustafa TM, Gouda M, Gad M. Longitudinal function and ventricular dyssynchrony are restored in children with pulmonary stenosis after percutaneous balloon pulmonary valvuloplasty. Int J Cardiovasc Imaging. 2017;33(4):533–8.
170. Stanger P, Cassidy SC, Girod DA, Kan JS, Lababidi Z, Shapiro SR. Balloon pulmonary valvuloplasty: results of the Valvuloplasty and angioplasty of congenital anomalies registry. Am J Cardiol. 1990;65(11):775–83.
171. Rao PS. Percutaneous balloon pulmonary valvuloplasty: state of the art. Catheter Cardiovasc Interv. 2007;69(5):747–63.
172. Ben-Shachar G, Cohen MH, Sivakoff MC, Portman MA, Riemenschneider TA, Van Heeckeren DW. Development of infundibular obstruction after percutaneous pulmonary balloon valvuloplasty. J Am Coll Cardiol. 1985;5(3):754–6.
173. Fawzy ME, Galal O, Dunn B, Shaikh A, Sriram R, Duran CM. Regression of infundibular pulmonary stenosis after successful balloon pulmonary valvuloplasty in adults. Catheter Cardiovasc Diagn. 1990;21(2):77–81.
174. Merino-Ingelmo R, Santos-de Soto J, Coserria-Sánchez F, Descalzo- Señoran A, Valverde-Pérez I. Long-term results of percutaneous balloon valvuloplasty in pulmonary valve stenosis in the pediatric population. Rev Esp Cardiol (Engl Ed). 2014;67(5):374–9.
175. Garty Y, Veldtman G, Lee K, Benson L. Late outcomes after pulmonary valve balloon dilatation in neonates, infants and children. J Invasive Cardiol. 2005;17(6):318–22.
176. Harrild DM, Powell AJ, Tran TX, Trang TX, Geva T, Lock JE, et al. Long-term pulmonary regurgitation following balloon valvuloplasty for pulmonary stenosis risk factors and relationship to exercise capacity and ventricular volume and function. J Am Coll Cardiol. 2010;55 (10):1041–7.

177. Rao PS, Galal O, Patnana M, Buck SH, Wilson AD. Results of three to 10 year follow up of balloon dilatation of the pulmonary valve. Heart. 1998;80(6):591–5.
178. Sadr-Ameli MA, Sheikholeslami F, Firoozi I, Azarnik H. Late results of balloon pulmonary valvuloplasty in adults. Am J Cardiol. 1998;82(3):398–400.
179. Taggart NW, Cetta F, Cabalka AK, Hagler DJ. Outcomes for balloon pulmonary valvuloplasty in adults: comparison with a concur- rent pediatric cohort. Catheter Cardiovasc Interv. 2013;82 (5):811–5.
180. Qian X, Qian Y, Zhou Y, Yang X. Percutaneous pulmonary balloon Valvuloplasty provides good long-term outcomes in adults with pulmonary valve stenosis. J Invasive Cardiol. 2015;27 (12):E291–6.
181. Berman W, Fripp RR, Raisher BD, Yabek SM. Significant pulmonary valve incompetence following oversize balloon pulmonary valveplasty in small infants: a long-term follow-up study. Catheter Cardiovasc Interv. 1999;48(1):61–5.
182. Bonhoeffer P, Boudjemline Y, Saliba Z, Merckx J, Aggoun Y, Bonnet D, et al. Percutaneous replacement of pulmonary valve in a right-ventricle to pulmonary-artery prosthetic conduit with valve dysfunction. Lancet. 2000;356(9239):1403–5.
183. Feltes TF, Bacha E, Beekman RH, Cheatham JP, Feinstein JA, Gomes AS, et al. Indications for cardiac catheterization and intervention in pediatric cardiac disease: a scientific statement from the American Heart Association. Circulation. 2011;123(22):2607–52.
184. Gillespie MJ, Rome JJ, Levi DS, Williams RJ, Rhodes JF, Cheatham JP, et al. Melody valve implant within failed bioprosthetic valves in the pulmonary position: a multicenter experience. Circ Cardiovasc Interv. 2012;5(6):862–70.
185. Webb JG, Wood DA, Ye J, Gurvitch R, Masson JB, Rodés-Cabau J, et al. Transcatheter valve-in-valve implantation for failed bioprosthetic heart valves. Circulation. 2010;121(16): 1848–57.
186. Cheatham SL, Holzer RJ, Chisolm JL, Cheatham JP. The Medtronic melody® transcatheter pulmonary valve implanted at 24-mm diameter—it works. Catheter Cardiovasc Interv. 2013;82(5):816–23.
187. Meadows JJ, Moore PM, Berman DP, Cheatham JP, Cheatham SL, Porras D, et al. Use and performance of the melody Transcatheter pulmonary valve in native and postsurgical, nonconduit right ventricular outflow tracts. Circ Cardiovasc Interv. 2014;7(3):374–80.
188. Coats L, Khambadkone S, Derrick G, Sridharan S, Schievano S, Mist B, et al. Physiological and clinical consequences of relief of right ventricular outflow tract obstruction late after repair of con- genital heart defects. Circulation. 2006;113(17):2037–44.
189. Plymen CM, Bolger AP, Lurz P, Nordmeyer J, Lee TY, Kabir A, et al. Electrical remodeling following percutaneous pulmonary valve implantation. Am J Cardiol. 2011;107(2):309–14.
190. Khambadkone S, Coats L, Taylor A, Boudjemline Y, Derrick G, Tsang V, et al. Percutaneous pulmonary valve implantation in humans: results in 59 consecutive patients. Circulation. 2005;112(8):1189–97.
191. Morray BH, McElhinney DB, Boudjemline Y, Gewillig M, Kim DW, Grant EK, et al. Multicenter experience evaluating transcatheter pulmonary valve replacement in bovine jugular vein (contegra) right ventricle to pulmonary artery conduits. Circ Cardiovasc Interv. 2017;10(6):e004914.
192. Lurz P, Nordmeyer J, Giardini A, Khambadkone S, Muthurangu V, Schievano S, et al. Early versus late functional outcome after successful percutaneous pulmonary valve implantation: are the acute effects of altered right ventricular loading all we can expect? J Am Coll Cardiol. 2011;57(6):724–31.
193. Cheatham JP, Hellenbrand WE, Zahn EM, Jones TK, Berman DP, Vincent JA, et al. Clinical and hemodynamic outcomes up to 7 years after transcatheter pulmonary valve replacement in the US melody valve investigational device exemption trial. Circulation. 2015;131(22): 1960–70.
194. McElhinney DB, Cheatham JP, Jones TK, Lock JE, Vincent JA, Zahn EM, et al. Stent fracture, valve dysfunction, and right ventricular outflow tract reintervention after transcatheter

pulmonary valve implantation: patient-related and procedural risk factors in the US melody valve trial. Circ Cardiovasc Interv. 2011;4(6):602–14.
195. McElhinney DB, Hellenbrand WE, Zahn EM, Jones TK, Cheatham JP, Lock JE, et al. Short- and medium-term outcomes after trans- catheter pulmonary valve placement in the expanded multicenter US melody valve trial. Circulation. 2010;122(5):507–16.
196. Ewert P, Horlick E, Berger F. First implantation of the CE-marked transcatheter Sapien pulmonic valve in Europe. Clin Res Cardiol. 2011;100(1):85–7.
197. Haas NA, Moysich A, Neudorf U, Mortezaeian H, Abdel-Wahab M, Schneider H, et al. Percutaneous implantation of the Edwards SAPIEN (TM) pulmonic valve: initial results in the first 22 patients. Clin Res Cardiol. 2013;102(2):119–28.
198. Wilson WM, Benson LN, Osten MD, Shah A, Horlick EM. Transcatheter pulmonary valve replacement with the Edwards Sapien system: the Toronto experience. JACC Cardiovasc Interv. 2015;8(14):1819–27.
199. Fraisse A, Assaidi A, Mauri L, Malekzadeh-Milani S, Thambo JB, Bonnet D, et al. Coronary artery compression during intention to treat right ventricle outflow with percutaneous pulmonary valve implantation: incidence, diagnosis, and outcome. Catheter Cardiovasc Interv. 2014;83(7):E260–8.
200. Morray BH, McElhinney DB, Cheatham JP, Zahn EM, Berman DP, Sullivan PM, et al. Risk of coronary artery compression among patients referred for transcatheter pulmonary valve implantation: a multicenter experience. Circ Cardiovasc Interv. 2013;6(5):535–42.
201. Lindsay I, Aboulhosn J, Salem M, Levi D. Aortic root compression during transcatheter pulmonary valve replacement. Catheter Cardiovasc Interv. 2016;88(5):814–21. A retrospective study of 174 patients who underwent transcatheter pulmonic valve replacement. Intra-procedural aortic root compression occurred in 9% of the patients, and 31% of those had concomitant coronary compression. Most of cases of aortic root com- pression occurred in TOF with native or transannular- patched RVOT
202. Porras D, Gurvitz M, Marshall AC, Emani SM. Hybrid approach for off-pump pulmonary valve replacement in patients with a dilated right ventricular outflow tract. Ann Thorac Surg. 2015;100(5):e99–101.
203. Chen Q, Turner M, Caputo M, Stoica S, Marianeschi S, Parry A. Pulmonary valve implantation using self-expanding tissue valve without cardiopulmonary bypass reduces operation time and blood product use. J Thorac Cardiovasc Surg. 2013;145(4):1040–5.
204. Sizarov A, Boudjemline Y. Novel materials and devices in the transcatheter management of congenital heart diseases-the future comes slowly (part 3). Arch Cardiovasc Dis. 2016;109(5): 348–58.
205. Schoonbeek RC, Takebayashi S, Aoki C, Shimaoka T, Harris MA, Fu GL, et al. Implantation of the Medtronic harmony transcatheter pulmonary valve improves right ventricular size and function in an ovine model of postoperative chronic pulmonary insufficiency. Circ Cardiovasc Interv. 2016;9(10):e003920.
206. Schlegel F, Salameh A, Oelmann K, Halling M, Dhein S, Mohr FW, et al. Injectable tissue engineered pulmonary heart valve implantation into the pig model: a feasibility study. Med Sci Monit Basic Res. 2015;21:135–40.
207. McElhinney DB, Benson LN, Eicken A, Kreutzer J, Padera RF, Zahn EM. Infective endocarditis after transcatheter pulmonary valve replacement using the Melody valve: combined results of 3 prospective North American and European studies. Circ Cardiovasc Interv. 2013;6:292–300.
208. Amat-Santos IJ, Ribeiro HB, Urena M, et al. Prosthetic valve endocarditis after transcatheter valve replacement: a systematic review. J Am Coll Cardiol. 2015;8:334–46.
209. McElhinney DB, Sondergaard L, Armstrong AK, et al. Endocarditis after transcatheter pulmonary valve replacement. J Am Coll Cardiol. 2018;72:2717–28.
210. Topilsky Y, Maltais S, Medina Inojosa J, et al. Burden of tricuspid regurgitation in patients diagnosed in the community setting. JACC Cardiovasc Imaging. 2019;12(3):433–42.

211. Zhan Y, Debs D, Khan MA, et al. Natural history of functional tricuspid regurgitation quantified by cardiovascular magnetic resonancel. J Am Coll Cardiol. 2020;76(11):1291–301.
212. Nath J, Foster E, Heidenreich PA. Impact of tricuspid regurgitation on long-term survival. J Am Coll Cardiol. 2004;43(3):405–9.
213. Messika-Zeitoun D, Verta P, Gregson J, et al. Impact of tricuspid regurgitation on survival in patients with heart failure: a large electronic health record patient-level database analysis. Eur J Heart Fail. 2020;22(10):1803–13.
214. Stuge O, Liddicoat J. Emerging opportunities for cardiac surgeons within structural heart disease. J Thorac Cardiovasc Surg. 2006;132(6):1258–61.
215. Alqahtani F, Berzingi CO, Aljohani S, et al. Contemporary trends in the use and outcomes of surgical treatment of tricuspid regurgitation. J Am Heart Assoc. 2017;6(12):e007597.
216. Zack CJ, Fender EA, Chandrashekar P, et al. National trends and outcomes in isolated tricuspid valve surgery. J Am Coll Cardiol. 2017;70(24):2953–60.
217. Axtell AL, Bhambhani V, Moonsamy P, et al. Surgery does not improve survival in patients with isolated severe tricuspid regurgitation. J Am Coll Cardiol. 2019;74(6):715–25.
218. Arsalan M, Walther T, Smith RL, Grayburn PA. Tricuspid regurgitation diagnosis and treatment. Eur Heart J. 2017;38(9):634–8.
219. Prihadi EA, Delgado V, Leon MB, Enriquez-Sarano M, Topilsky Y, Bax JJ. Morphologic Types of Tricuspid Regurgitation: characteristics and Prognostic Implications. JACC Cardiovasc Imaging. 2019 Mar;12(3):491–9.
220. Ascione G, Del Forno B, Carino D, et al. Treatment of isolated tricuspid regurgitation in 2020: an update. Fac Rev. 2020;21(9):26.
221. Hahn RT, Chandrashekhar Y. Tricuspid regurgitation: a voyage of discovery. JACC Cardiovasc Imaging. 2019 Mar;12(3):572–5.
222. Bartko PE, Arfsten H, Frey MK, et al. Natural history of functional tricuspid regurgitation: implications of quantitative doppler assessment. JACC Cardiovasc Imaging. 2019;12(3): 389–97.
223. Topilsky Y, Nkomo VT, Vatury O, et al. Clinical outcome of isolated tricuspid regurgitation. JACC Cardiovasc Imaging. 2014 Dec;7(12):1185–94.
224. Braun D, Orban M, Orban M, et al. Transcatheter edge-to-edge repair for severe tricuspid regurgitation using the triple-orifice technique versus the bicuspidalization technique. JACC Cardiovasc Interv. 2018;11(17):1790–2.
225. Nickenig G, Kowalski M, Hausleiter J, et al. Transcatheter treatment of severe tricuspid regurgitation with the edge-to-edge MitraClip technique. Circulation. 2017;135(19):1802–14.
226. Sorban M, Besler C, Braun D, et al. Six-month outcome after transcatheter edge-to-edge repair of severe tricuspid regurgitation in patients with heart failure. Eur J Heart Fail. 2018;20(6): 1055–62.
227. Braun D, Nabauer M, Orban M, et al. One-year results of transcatheter treatment of severe tricuspid regurgitation using the edge-to-edge repair technique. Eurolntervention. 2018;14(4): e413–5.
228. Orban M, Braun D, Deseive S, et al. Transcatheter edge-to-edge repair for tricuspid regurgitation is associated with right ventricular reverse remodeling in patients with right-sided heart failure. JACC Cardiovasc lmaging. 2019;12(3):559–60.
229. Puri R, Rodes-Cabau J. The FORMA repair system. Interv Cardiol Clin. 2018;7(1):47–55.
230. Asmarats L, Perlman G, Praz F, et al. Long-term outcomes of the forma transcatheter tricuspid valve repair system for the treatment of severe tricuspid regurgitation: insights from the first-in-human experience. JACC Cardiovasc Interv. 2019;12(15):1438–47.
231. Fam NP, Ho EC, Zahrani M, et al. Transcatheter tricuspid valve repair with the PASCAL system. JACC Cardiovasc lnterv. 2018;11(4):407–8.
232. Nickenig G, Weber M, Schueler R, et al. 6-month outcomes of tricuspid valve reconstruction for patients with severe tricuspid regurgitation. J Am Coll Cardiol. 2019;73(15):1905–15.

233. Anderson A, Hausleiter J, von Bardeleben RS, et al. One-year outcomes of the TRI-REPAIR study assessing cardioband tricuspid valve reconstruction system for patients with functional tricuspid regurgitation. J Card Fail. 2019;25:S11.
234. Hahn RT, Meduri CU, Davidson CJ, et al. Early feasibility study of a transcatheter tricuspid valve annuloplasty: SCOUT trial 30-day results. J Am Coll Cardiol. 2017;69(14):1795–806.
235. Mangieri A, Lim S, Rogers JH, et al. Percutaneous tricuspid annuloplasty. Interv Cardiol Clin. 2018;7:31–6.
236. Lu F, Qiao F, Lv Y, et al. A radial force-independent bioprosthesis for transcatheter tricuspid valve implantation in a preclinical model. Int J Cardiol. 2020;319:120–6.
237. Sun Z, Zhang Z, Li H, Xie Y, et al. Twelve-month outcomes of the LuX-valve for transcatheter treatment of severe tricuspid regurgitation. EuroIntervention. 2021;17:818–26.
238. Fam NP, von Bardeleben RS, Hensey M, et al. Transfemoral transcatheter tricuspid valve replacement with the EVOQUE system: a multicenter, observational, first-in-human experience. JACC Cardiovasc Interv. 2021;14(5):501–11.
239. Hahn RT, Kodali S, Fam N, et al. Early multinational experience of transcatheter tricuspid valve replacement for treating severe tricuspid regurgitation. JACC Cardiovasc Interv. 2020;13(21):2482–93.
240. Laule M, Stangl V, Sanad W, et al. Percutaneous transfemoral management of severe secondary tricuspid regurgitation with Edwards Sapien XT bioprosthesis: first-in-man experience. J Am Coll Cardiol. 2013;61(18):1929–31.
241. Dreger H, Mattig I, Hewing B, et al. Treatment of severe tricuspid regurgitation in patients with advanced heart failure with CAval vein implantation of the Edwards Sapien XT VALve (TRICAVAL): a randomised controlled trail. EuroIntervention. 2020;15(17):1506–13.
242. Figulla HR, Kiss K, Lauten A. Transcatheter interventions for tricuspid regurgitation-heterotopic technology: TricValve. EuroIntervention. 2016;12:Y116–8.
243. Lauten A, Ferrari M, Hekmat K, et al. Heterotopic transcatheter tricuspid valve implantation: first-in-man application of a novel approach to tricuspid regurgitation. Eur Heart J. 2011;32(10):1207–13.
244. Toggweiler S, De Boeck B, Brinkert M, et al. First-in-man implantation of the Tricento transcatheter heart valve for the treatment of severe tricuspid regurgitation. EuroIntervention. 2018;14(7):758–61.
245. Cruz-Gonzalez I, Gonzalez-Ferreiro R, Amat-Santos IJ, et al. TRICENTO transcatheter heart valve for severe tricuspid regurgitation. Initial experience and mid-term follow-up. Rev Esp Cardiol (Engl Ed). 2021;74(4):351–4.
246. Asmarats L, Puri R, Latib A, et al. Transcatheter tricuspid valve interventions: landscape, challenges, and future directions. J Am Coll Cardiol. 2018;71(25):2935–56.
247. Hwang HY, Kim KH, Kim KB, et al. Propensity score matching analysis of mechanical versus bioprosthetic tricuspid valve replacements. Ann Thorac Surg. 2014;97(4):1294–9.
248. Lurz P, Orban M, Besler C, et al. Clinical characteristics, diagnosis, and risk stratification of pulmonary hypertension in severe tricuspid regurgitation and implications for transcatheter tricuspid valve repair. Eur Heart J. 2020;41(29):2785–95.

Chapter 8
Occlusion Devices for Cardiovascular Disease

Yong Zeng, Yang Zhang, and Chenchen Tu

8.1 Transcatheter Closure of Atrial Septal Defect

8.1.1 Introduction

Atrial septal defect (ASD) is one of the most common congenital heart diseases. The prevalence of ASD is estimated to be 1.441 per 1000 live births worldwide (95% CI: 1.215–1.687) [1]. With the development of diagnosis and treatment technology, leading to an increased life expectancy of patients with ASD. Although transcatheter closure of ASD has been demonstrated to significantly reduce mortality, ASD patients with or without occlusion therapy still have a higher mortality rate than a general population cohort matched on birth year and gender [2].

8.1.2 Anatomy and Pathophysiology of ASD

8.1.2.1 Embryology of ASD

In the fourth week of gestation, the appearance of primary septum at the top of the atrium marks the beginning of atrial septum formation. During development, the free edge of crescent shaped primary septum is covered by mesenchymal cells, derived from embryonic endocardium. The primary septum continued to grow downward and eventually adhered to and fused with the endocardial cushion to close the primary foramen and separate the original atrium. Subsequently, programmed cell death occurred in the more cranial part, resulting in fenestrations that become the ostium secundum. Around 33 days of gestation, remodeling occurs in the atrial roof

Y. Zeng (✉) · Y. Zhang · C. Tu
Beijing Anzhen Hospital, Capital Medical University, Beijing, China

Y.-D. Tang et al. (eds.), *Application of Biomaterials in the Treatment of Cardiovascular Diseases*, https://doi.org/10.1007/978-981-19-7712-1_8

between the septum primum and the superior caval vein, and the septum secundum forms as an infolding of the atrial wall. With the development of the secondary septum, a foramen ovale adjacent to the primary septum but not fused with it was formed. Between the foramen ovale and the ostium secundum, a loop-like structure allowing blood to flow from the right to the left was formed, so that the arterial blood from the placenta could enter the fetal circulation through this channel [3]. After the newborn begins to breathe, the pulmonary pressure and right atrial pressure decrease, which makes the primary septum cover the foramen ovale, close the flap valve structure, and close the traffic between the two rooms.

8.1.2.2 Types of ASD

Because of the similar hemodynamic state, atrial septal communication (ASD) not only occurs on the atrial septum but also includes other defects that can cause bilateral atrial communication. According to the different anatomical characteristics, it can be divided into the following categories:

(a) *Secundum atrial septal defect*

Secundum atrial septal defect is the most common type of ASD, accounting for 75–80% of all ASDs. Most of the defects occurred in the oval fossa, mainly due to deficiency in septum primum or, rarely, due to deficiency in septum secundum. The secundum ASD type, size, and shape can vary greatly. Although the vast majority of atrial septal defects are sporadic with no identifiable cause, secundum defects are often encountered in genetic syndromes such as Holt Oram, Noonan, bud Chiari, Ellis-van Creveld, and Down [4–8].

(b) *Primum atrial septal defects*

This type of ASD accounts for about 15% of all ASDs. It is usually caused by incomplete fusion of the primary septum and endocardial cushion. The primary ASD often does not exist in isolation and is often accompanied by atrioventricular canal defects, including atrioventricular valve abnormalities and ventricular septal defects.

(c) *Sinus venosus defect*

This type of ASD accounts for 5–10% of all ASDs, and this defect is located outside of the true septum. The superior sinus venosus defect allows the vena cava to straddle the atrial septum, resulting in communication between left and right atrium. Because the orifice of the superior vein cava is above the defect, there is no complete muscular border. Usually, there is associated partial anomalous connection of the right superior pulmonary vein to the superior vein cava, leading to communication between the left atrium and the orifice of the superior vena cava.

Inferior sinus venosus defect: this type of ASD is far less common than the superior sinus venosus defect, accounting for less than 1% of all ASD patients. However, the two types of ASD have similar characteristics such as the orifice of the inferior vena cava overriding the atrial septum and the lack of a complete muscular border. Furthermore, the right inferior pulmonary vein can often be connected anomalously to the inferior vena cava.

(d) *Coronary sinus defects*

This type of ASD is also less than 1% of all ASDs and is not true defect in the atrial septum. Rather there is an interatrial communication that exists resulting from partial or complete unroofing of the tissue separating the coronary sinus from the left atrium, allowing a shunt through the defect and the orifice of coronary sinus. In addition, this type of ASD is often accompanied by persistent left superior vena cava, and in the completely unroofed defect, the left superior vena cava attaches to the roof of the left atrium.

8.1.2.3 Pathophysiology of ASD

The direction and magnitude of blood flow through an atrial communication mainly depend on the size of the defect, the relative relationship between the compliance of the two ventricles, and the state of the atrioventricular valve. Both the size of the defect and the compliances of the ventricles can change over time [9]. The right ventricular compliance is low at birth, and the left-to-right shunt is negligible. Within a few weeks after birth, with the decrease of pulmonary vascular resistance, the right ventricular compliance increased, so the left-to-right shunt also increased gradually. When left ventricular compliance is impaired (such as mitral regurgitation, hypertension, etc.), the degree of left-to-right shunt may be further increased.

At the early stages of the disease, the increase of pulmonary artery blood flow only dilated the pulmonary vascular bed and did not cause significant changes in pulmonary vascular resistance. The long-term high blood flow state will lead to endothelial cell damage, depletion of the vasodilator reserve, and the remodeling of pulmonary vascular bed, eventually leading to severe pulmonary hypertension. The left-to-right shunt between atrium level results in overcirculation of the right atrium, right ventricle, and pulmonary vascular. Chronic volume overload to the lungs from an ASD may lead to progressive dyspnea, recurrent respiratory infections, right ventricular dilatation, and myocardial cell hypertrophy and fibrosis. The increase of volume load and pressure load eventually leads to right ventricular dysfunction.

The dilated right ventricle compresses the left ventricle, and the systolic function of the left ventricle is affected by the abnormal movement of the interventricular septum. Furthermore, adverse interventricular interaction results in decreased left ventricular compliance. These changes result in decreased left ventricular diastolic filling, increased pulmonary-to-systemic flow ratio through the defect, and diminished systemic output [10]. The pathological changes of the structure and function of heart and pulmonary vascular often cause a series of secondary diseases, such as repeated pneumonia, pulmonary hypertension, atrial arrhythmia, etc.

8.1.3 Indications and Contraindications for ASD Closure

Percutaneous or surgical closure of the atrial septal defect is indicated in cases of impaired function, right atrial and/or atrial enlargement, and a net left-to-right shunt large enough to cause physiological sequelae [11]. (Class of recommendation: I, Level of evidence: B) In addition, closure of an ASD is reasonable in the presence of paradoxical embolism (Class of recommendation: IIa, Level of evidence: C) or documented orthodeoxia-platypnea irrespective of shunt size (Class of recommendation: IIa, Level of evidence: B) [12].

Closure can be contraindicated in adults with PA systolic pressure greater than two thirds systemic, pulmonary vascular resistance greater than two thirds systemic, and/or a net right-to-left shunt.

Due to the limitation of equipment, operation, and imaging technology, interventional closure therapy has strict requirements on the patient's age, weight, pathological type, concomitant diseases, combined malformations, and so on. However, with the development of technology and clinical exploration, the indications of transcatheter closure are also being relaxed.

8.1.4 Devices for ASD Closure

With the continuous innovation of materials and design, the success rate of transcatheter closure is improved, while the risk of complications is reduced. However, different types of occlusion devices have their own advantages and disadvantages. Therefore, it is critical to accurately understand the anatomy of atrial septum and clinical characteristics of patients and select the most appropriate device for optimal occlusion treatment.

8.1.4.1 Amplatzer Septal Occluder (Fig. 8.1a)

Amplatzer Septal Occluder (AGA Medical Corporation, Plymouth, Minnesota) is one of the most popular devices for occlusion of ASDs. It is a self-centering, self-expanding device that consists of two discs made from 0.004–0.005 inch Nitinol wire mesh and linked together by a short connecting waist. The polyester fiber patch is sewn into both of discs and waist to promote complete occlusion and endothelialization. The thickness of the waist is 4 mm, which connects the left and right discs and also plays the role of positioning and fixing the device. The diameter of the left disc is 12–14 mm larger than in diameter than the central waist, and the right disc is 10–12 mm larger than the waist. The left atrial disc is larger than the right atrial disc because of the higher left atrial pressure. The two discs, respectively, extend radially 6–7 mm and 4–5 mm around the connecting waist to play a fixed role. The device size is corresponding to the diameter of the waist; there are 27 device

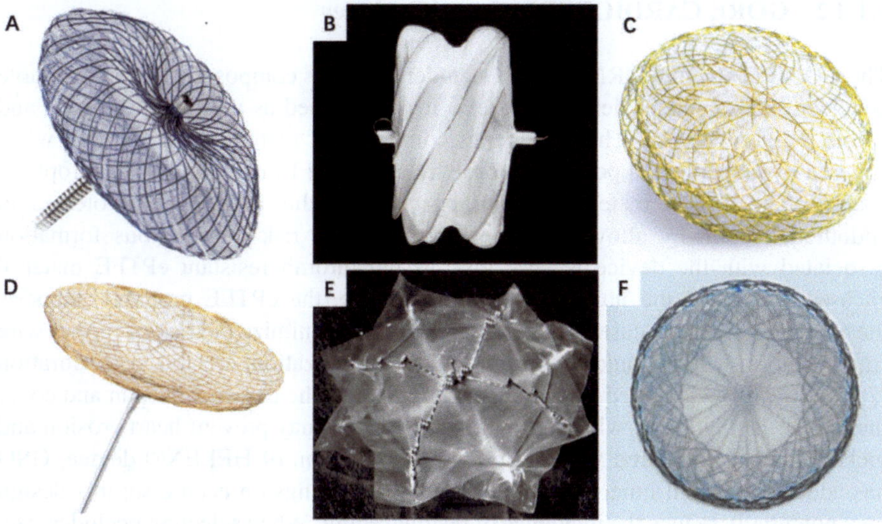

Fig. 8.1 Devices for atrial septal defect closure. (**a**) Amplatzer Septal Occluder; (**b**) Gore CARDIOFORM ASD Occluder; (**c**) Figulla Flex II ASD Occluder; (**d**) CeraFlex ASD Occluder; (**e**) BioSTAR; (**f**) Nit-Occlud ASD-R

sizes from 4 to 40 mm (in 1 mm increments between 4 to 20 mm and 2 mm increments between 20 and 40 mm). The corresponding delivery sheath sizes range from 6 to 12 French. The self-centering, repositionable, and recapturable characteristics of this device provide an ideal convenience of deployment. The Amplatzer Cribriform Multifenestrated Septal Occluder is designed to use in multiple atrial septal defects. This device has a narrower waist and atrial discs with equal diameter on both sides to place it through one of the central defects and effectively cover the surrounding defects.

As one of the most commonly used devices for occlusion of ASDs, Amplatzer Septal Occluder (ASO) has accumulated a lot of experience. Early study showed that success rate of ASO (95.7%) is comparable with surgical closure and major and total adverse effect is 1.6% and 7.2%, respectively [13]. The IMPACT registry showed similar success and adverse event rate [14]. Of the 1000 patients with ASDs for closure studied by Daniel R et al. [15], procedural closure occurred in 97.9%, with 1-month and 2-year closure of 98.5% and 97.9%, respectively. The total adverse event rate associated with the ASO or delivery system at 2-year follow-up was 6.56% (61/930). It is worth noting that oversized device, aortic or superior vena cava rim deficiencies, smaller weight, device size ratio, septal malalignment, and dynamic ASD are considered as risk factors of heart erosion [16–18], although the exact mechanism is not yet clear. In addition, a comprehensive evaluation of the rims that surround the ASD is paramount to successful decision-making and good outcomes [19].

8.1.4.2 GORE CARDIOFORM Septal Occluder

The GORE CARDIOFORM Septal Occluder (GSO) is composed of an implantable occluder and a catheter delivery system. It is preloaded as a kit of the device and whole delivery system. The occluder is composed of platinum-filled nitinol wires covered by an expanded polytetrafluoroethylene (ePTFE) material with hydrophilic coating. The ePTFE material is microporous, and this property promotes rapid endothelialization by allowing tissue ingrowth. The risk of thrombus formation associated with the device is very low, as the thromboresistant ePTFE material encased almost all the nitinol wires. Furthermore, the ePTFE probably supports the embedded wires in case of any fracture, so as to minimize the risk of further wire fragmentation, which could trigger migration, embolization, erosion, or perforation [20]. When fully deployed, two independent discs of the device will span and cover the defect. Its flexibility with minimal metal content may prevent heart erosion and nickel allergy. Compared with the previous generation of HELEX® device, GSO has added five small inner petal-shaped supporting rings on each disc; this design may potentially reduce the incidence of residual shunt. When selecting occluder size, it is generally recommended that the device size should be more than twice the defect size, and the device diameter should be no more than 90% of the measured septal length. There are four device sizes from 15 to 30 mm with 5 mm increments. Stable placement is not likely in defects >18 mm owing to its non-self-centering feature. The GORE CARDIOFORM ASD Occluder (GCO) (Fig. 8.1b) has an adaptable central large inter-disc region that occupies the septal defect in contrast to the GSO. The self-centering feature allows GCO to be used to treat larger ASDs. The available sizes are 27, 32, 37, 44, and 48 mm to treat a range of defects from 8 to 35 mm.

Results from preliminary clinical trials have verified efficacy and safety of this device in children and adolescents with ASDs [21, 22]. In the paper reported by Søndergaard et al. [23], during the short-term follow-up of 11 patients with PFO or ASD for closure, there is no residual shunt detected in any of them. Grohmann et al. studied 206 patients with secundum ASDs [22]. In their series, 131 patients (76%) were classified as single defects, 42 patients (24%) were classified as multi-fenestrated defects, and 33 patients (19%) were classified as deficient retro-aortic rims. Midterm follow-up showed excellent overall closure rate (95.4%) and low complication rate (5.2%). Recently, the initial clinical results also suggested that GCO implantation is feasible and effective in closing ASDs [24, 25].

8.1.4.3 Occlutech Figulla Flex II ASD Occluder (Fig. 8.1c)

Figulla Flex II ASD Occluder (FSO) is a double-disc, self-expanding, and self-centering device for ASD closure, similar to the Amplatzer devices. FSO is already the third generation of the series. This device is constructed from 0.082 to 0.186 mm Nitinol wire coated with titanium oxide. A unique braiding technology of this device minimized the amount of material in the left disc that results in a lower and greater flexible profile. Polyethylene fabric is sewn within both discs and the waist to facilitate thrombogenicity. In contrast to the ASO, FSO has no hub on the left atrial side, which may provide less impingement feature and promote rapid endothelialization as well as reduce the risk for thrombosis. More importantly, due to this design, the left atrial disc develops a ball-like shape during deployment, which may mitigate against prolapse of the left-sided disc during the implantation [26]. Another advantage of FSO is the unique connection mechanism between the device and the delivery cable. The device connects to the delivery cable through a bioptome-like system, which enables full circular movement. The delivery cable is also shapeable. The above characteristics of FSO can be improved flexibility for optimal septum alignment without tension on the delivery cable. There are 20 available device sizes, ranging from 4 to 40 mm (1–1.5 mm increments between 4 and 21 mm, 3 mm increments between 21 and 39 mm, and 40 mm).

Nikolaus et al. reviewed 1315 patients who received closure of ASD by using the Occlutech Occluder Devices [27]. In their retrospective series, successful implantation was achieved in $n = 1291/1315$ (98.2%) patients. Only eight episodes of significant complications occurred, during a mean follow-up of 2.7 years. Of note, no erosion or death was reported. The drawbacks of FSO are its relatively limited available sizes, and delivery sheath for the FSO is larger than ASO which may increase the risk of vascular damage [28]. Various head-to-head studies, nevertheless, have shown that FSO is not inferior to ASO in terms of complications and effectiveness [29–33].

8.1.4.4 CeraFlex ASD Occluder (Fig. 8.1d)

CeraFlex ASD Occluder (LifeTech Scientific Co., Shenzhen, China) is the fourth generation device manufactured by LifeTech Scientific for interventional ASD closure. The self-expandable, double-disk device is weaved of nitinol wire mesh coated with titanium nitride. Ceraflex is preloaded as a package of the device and has some similar features to FSO such as LA disk without hub and the flexible connection system. The design of no hub on the LA disk may reduce the damage to the atrial wall, and the flexible rotation feature proved helpful with correct alignment to the septum. The size of the device available ranges from 6 to 42 mm at 2 mm increments. This device also has a couple of shared disadvantages with FSO such as the limited size options and the requirement of a slightly larger delivery sheath than with ASO devices.

In a study by Astarcioglu et al., it was found that Ceraflex has a compatible efficacy and safety comparing to ASO [34]. Another investigator had similar experience [35]. Another large study also showed excellent success rate of 94.3%; moreover, there were no deaths or embolizations [36]. It is worth noting that during closure of ASD with a Ceraflex device, the left atrial disk of the occluder did not unfold correctly assuming a conal formation. However, the device was retracted immediately, and the patient underwent uneventful closure with another larger occluder. The development of such malformations may be associated with softer, hubless left disk occluders.

8.1.4.5 Bioabsorbable Devices

During the past decades, various types of occlusion devices based on nondegradable materials have been widely applied in clinic due to the excellent success rate. However, complications such as cardiac erosion, device fracture, endothelialization insufficiency, thrombus formation, and metal allergy also raised more and more concerns. In addition, the traditional occluder renders the prosthetic redundant after being implanted. The redundant not only causes persistence of a local inflammatory response but also will obstruct transseptal access to the left atrium. By comparison, the bioabsorbable device may reduce immunological response and preserve transseptal access due to its absorbable characteristic. Most of the above complications may be alleviated by applying the device based on bioabsorbable materials. An ideal material allows absorption in a controlled and predictable fashion to ensure the overall structural integrity of the device before endothelialization is completed while minimizing the inflammatory response.

The first partially bioabsorbable device for ASD closure, BioSTAR (Fig. 8.1e), was introduced by NMT Medical, in 2003. The design of the BioSTAR device is predicated on STARFlex (NMT products) framework. A nitinol wire is attached to the ends of the MP35N "double-umbrella" framework. The device retains the self-centering mechanism of the STARFlex device, which improves its use in larger ASDs. BioSTAR uses an acellular porcine intestinal collagen layer (ICL) matrix mounted on the framework, and this matrix is completely resorbed rapidly after implantation. A heparin benzalkonium chloride complex (HBAC) is coated on the patch to minimize the risk of thrombus formation after deployment. The available size of BioSTAR is 23, 28, and 33 mm, and 1.5–2 times the balloon-stretched diameter is considered to be the appropriate device size. The BioSTAR Evaluation STudy (BEST) has demonstrated the efficacy and safety of BioSTAR for the treatment of ASD. This study showed remarkable procedural success rate of 98% that may be superior to those reported previously with conventional devices. Five patients presented transient atrial arrhythmia after implantation, but with no major adverse events at implant or during follow-up [37]. In this multicenter trial, the advanced age and the presence of an atrial septal aneurysm may be important risk factors for the generation of arrhythmias. A comparative study showed a comparable closure rate of BioSTAR to ASO in small-to-moderate defects after 6 months of follow-up [38]. In the paper reported by Christoph et al., it was found that a failure rate of implantation of the BioSTAR device is 12%. In addition, 9% and 12% of the subjects had short-term and long-term complications, respectively. They concluded that the deficient aortic rims seem to be related to the failure of occlusion, and the long-term complications seen with the BioSTAR might be attributable to unique immunological features of the bioabsorbable device [20].

However, the BioSTAR occluders are no longer produced, because of the closure of manufacturer. Many novel designs based on bioabsorbable materials have been brought into existence, such as the Carag Bioresorbable Septal Occluder [39], Improved Amplatzer Occluder [40], and fully degradable PLLA occluders [41]. The safety and efficacy of the abovementioned devices still need to be verified in further clinical application.

8.1.4.6 Nit-Occlud ASD-R (Fig. 8.1f)

Like other transcatheter occlusion systems developed for the closure of ASDs, the Nit-Occlud ASD-R (NOASD-R) (pfm Medical, Cologne, Germany) is a double-disc, self-expandable, self-centering device. However, there are still many differences between this device and other nitrol-mesh devices in terms of design. It consists of two equal-size retaining discs jointed by a connecting waist. The device is knitted from a single nitinol wire without any soldering or protruding screwing or clamp hubs in either side of it, resulting in decreasing amount of metal on the left atrial disk. A polyester membrane is facing the left atrium by suturing to the nitinol rims of the left disc. The distinct reverse configuration of the left atrial disc provides an anchoring mechanism that enables the device to be attached to the interatrial septum firmly. The "Snare-like" release mechanism of this device is composed of a central locking wire that travels across the device and a pusher with a distal wire noose. By pulling the locking and retaining wires, the device is released with minimal malposition because no rotatory movements are required for the release of the device. In conclusion, the unique design of NOASD-R may be helpful to overcome some drawbacks of conventional devices for ASD closure such as occurrence of erosion, thrombus formation, device embolization, and delayed endothelialization. The NOASD-R is available in 12 various sizes, ranging from 8 to 30 mm with 2 mm increments.

An initial study showed standout success rate of 98.6% (73/74 patients), only one case of complete heart block required removal of the device. Of note, albeit the safety and efficacy of NOASD-R demonstrated in this study, the no-hub design of the left disc increases the difficulty of device retrieval when necessary [42]. In a study by Mustafa et al., NOASD-R also achieved optimal success rate. However, a device-related erosion was observed in one patient with aortic rim deficiency. The author concludes that the relative decrease of right atrial diameter after occlusion and the equal-diameter design of NOASD-R may be the main causes of erosion [43]. In one most recent report by Alexandra et al., the efficacy and safety of the NOASD-R has also been described [44]. What was noteworthy was that there are two cases of erosion including one lethal case [43, 45]. The cases mentioned above illustrated that NOASD-R is not free of cardiac erosion, and the exact mechanism of erosion remains unclarified.

8.1.5 Issues with ASD Occlusion Devices

With the significant refinement in device technology, the complication incidence of ASD occluders has been kept in a very low degree. In the meantime, the growing experience of closure devices brought some rare and delayed complications to light [46, 47]. Due to the different design, the incidence and types of complications among various devices are obviously different. It is worth noting that compared with long-term complications, immediate complications are more prone to be detected and can be addressed by immediate device retrieval and replacement or even surgical closure. For long-term complications, it is not only lack of effective detection methods, but also resulting in fatal consequences or long-term serious impact on patients.

8.1.5.1 Thromboembolic Events

As a foreign material, occluder increases the risk of thrombus development in the left or right atrium after implantation [48, 49]. A meta-analysis of 203 case series showed that the pooled estimate rate of device thrombosis after ASD closure was 1% (95% CI, 0.8–1.0) [50]. The device-related thrombosis is mostly related to the healing response of ASD occluded prosthesis. Due to the differences in histopathological characteristics and biocompatibility of healing reactions, there may be different thrombogenic risks among various occluders. A review of complications associated with transcatheter atrial septal occluder implantation found that thromboembolism was more frequently encountered on the NMD devices than on the ASO [51]. In an early study by Ulrike et al. among the 407 ASD patients, the incidence of thrombus was 1.2% [52]. The difference in incidence of thrombus formation among devices was significant, 7.1–22% in the CardioSEAL device, 5.7% in the StarFLEX device, 0.8% in the HELEX® device, and 0% in the Amplatzer device. Furthermore, postprocedure atrial fibrillation and persistent atrial septal aneurysm have been seen as significant risk factors for thrombus formation.

In the 2020 ESC guidelines, it is still recommended that antiplatelet therapy is required for at least 6 months (aspirin 75 mg) after device implantation [53]. However, literature reported that device-related thrombus has been detected more than 1 year after device implantation [54–57]. For this reason, the follow-up and reevaluation after implantation are necessary. Cardiologists should reserve a measure of suspicion for late-thrombus development in patients with high risk of thrombosis, and it is also reasonable to extend the antiplatelet therapy in this patient population.

8.1.5.2 Cardiac Erosion

No cases of cardiac erosion (CE) or perforation have been reported in the early pivotal studies of ASO. However, with the accumulation of experience of this device in clinical applications, this rare but potentially fatal complication has gradually come to light and arose more and more attention.

In 2004, a series reporting on the issue of device erosion was published by a panel of experts selected by AGA Medical. In the light of the number of implants and erosion cases in the United States, they concluded the estimated incidence of cardiac erosion is 0.1%. Of the 28 cases, 12 involved perforations occurring at the dome of the atria and the aorta, while no atrial perforation was found in 3 cases. The data from this review showed that the time to erosion varied. Nineteen of the 28 patients (68%) developed symptoms within the first 72 h of the procedure and 8 cases (29%) within 5 days to 8 months after the procedure, and only 1 patient was diagnosed at 3 years after the implantation [58].

At the Pediatric and Adult Interventional Cardiac Symposium (PICS–AICS) 2012 meeting, the advisory panel reviewed the data from literature review, field events reported to the company, post-approval study (PAS) data, post-approval study investigator query, and the Manufacturer and User Facility Device Experience (MAUDE) database. There are 97 cases reported worldwide (48 in the United States and 49 outside) leading to an estimated incidence of erosion of 0.07–0.11% in the United States and 0.04–0.17% worldwide. Based on the data available, they estimated that the mortality rate in the United States is 0.008% or 0.016%, compared with 0.004–0.015% worldwide (according to number of devices sold or implanted, respectively). Similar to the conclusion of the 2004, data summarized at PICS 2012 also revealed that most of the CE occurred at the roof of the atrium. As regard to the time to erosion, most of the erosions still occurred within the first year after implant (87.6%). In addition, more than half of children (57% of cases) presented CE within 72 h after the procedure [59]. A case-control study involving 125 cardiac erosion cases found that the median duration from implant to erosion diagnosis was 14 days, and more than 30% of the patients ($n = 40$) were diagnosed on or within 1 day, >1 year in 16%, and >5 years in 6% [16].

As mentioned above, CE commonly occurs within hours or even days after the successful deployment of the occluder; thus, it seems unreasonable to attribute the occurrence of CE entirely to intraoperative complications. The occurrence of this injury is more likely due to the interaction between the newly implanted device and the atrial structure, but there is still no clear mechanism to explain the occurrence of this interaction. The changes of atrial size and structure following closure of the ASD may affect the spatial relationship between the device and the atrial wall. With the shrinkage of the right atrium by eliminating the left-to-right atrial shunt, the device could come into contact with the atrial wall. The seesaw-like movement of the edge of the device near the superior rim with each cardiac cycle may result in a stretching of the free atrial wall and the erosion of the cardiac structure. The aorta adjacent to the anterosuperior rim of the defect is more vulnerable to erosion once the atrial roof has been eroded [59]. In view of the rarity of the complications and the heterogeneity of the finding of multiple case reports, reviews, and analyses of the MAUDE database, no absolute risk factor has been identified as a contraindication for transcatheter closure of ASD [51, 60]. There might be, however, device and patient-related factors that increase the risk of cardiac erosion in some patients relative to the general population of patients receiving ASD closure. In the 2004 review [58], the panel of experts postulated that patients with at least one of the following conditions may have higher risk of cardiac erosion who patients may having high risk of cardiac erosion who with at least one of the following conditions: (1) patients who require significantly larger ASO (>1.5 times) than the native diameter of the ASD, (2) patients with development of small pericardial effusion at 24-h follow-up, (3) patients with deformation of the ASO at the aortic root (significant splaying of the device edges by the aorta), and (4) patients with high defects (minimal aortic and superior rims). A study to explore echocardiographic predictors of cardiac erosion also confirmed the increased risk of aortic rim absence [17]. In addition, the study also found that poor posterior rim consistency, septal malalignment, and the presence of dynamic ASD (defined as at least 50% decrease in size of the ASD during atrial systole and diastole) may be the factors that significantly increase the risk of cardiac erosion. A case-control study by Varna et al. found that deficiency of any rim (including the aortic rim), device >5 mm larger than ASD diameter, and weight/device size ratio were significantly related to erosion [16]. However, there is tendency of operators in general to choose the larger device to avoid device embolization in patients with deficient aortic rim. In addition, some interventionalists believed that device oversizing may stabilize the device by allowing aortic straddling to minimize the risk of erosion [61]. Therefore, there is still no consensus in the relation between the device size and the occurrence of CE. It was somewhat of a surprise that no case of cardiac erosion caused by HELEX device has been reported. This might be partially explained by its relative less rigid design. On the other hand, it could be attributed to the fact that HELEX is only recommended for defects less than 18 mm.

It is evident that there is neither consensus about mechanisms of cardiac erosion nor guidelines on how to prevent erosion. The following recommendations are still noteworthy:

1. Follow instructions for use when performing balloon sizing.
2. Avoid overstretching the balloon when balloon sizing the defect.
3. Use stop-flow technique for maximum inflation of balloon sizing.
4. Be gentle with to and fro of the device (Minnesota wiggle) while the device is attached to the delivery cable.
5. Identify patients who may be at higher risk and will require closer follow-up.
6. Mandatory 24-h follow-up in all patients.
7. Educate patients about the risk and need for echocardiography with symptoms [58].

Given that the events are more frequent within the first year, it is reasonable to closely follow up the first 12 months (e.g., serial echo at 1 week, 1 month, 6 months) and clinically follow up yearly thereafter. Furthermore, mandatory device tracking was strongly recommended for both devices [59].

The current analysis or review regarding this rare but potentially fatal complication is all based on retrospective data, which may hinder the exploration of the real etiology of this complication. To extend understanding of erosion and further reduce the risk of this complication of transcatheter ASD closure, an adequately potent prospective research will be critical.

8.1.5.3 Nickel Hypersensitivity

Nitinol is extensively used in implantable medical devices due to its excellent radiopacity, elasticity, resistance to fatigue and corrosion, and memory shape properties. The material is an alloy composed of 55% nickel and 45% titanium. The use of device which is constructed of this alloy can predispose the patient to revealing nickel hypersensitivity. Allergy symptoms can develop immediately or months after the insertion and can manifest as atypical chest pain and increased migraine frequency. In addition, dyspnea, palpitations, rash/urticaria, or pericardial tamponade have also been described as possible symptoms associated with hypersensitivity of nickel [62–65]. Although delayed type 4 reaction is still considered to be the main contributing mechanisms, there are potentially other mechanisms such as cytotoxic reaction or innate immunity [66]. The pathogenesis may include the formation of platelet adhesions and inflammatory mediators at the site of the device, which then circulate systemically [67].

A study conducted by Breet et al. included 190 patients who received a PFO or an ASD closure device [68]. There were seven patients (5%) who developed new-onset migraine headache with or without aura (MHA) or increased frequency and severity of MHA or aura postprocedure. Of the seven patients, six received the test for nickel allergy, and four of these patients (67%) showed a positive nickel skin test result. Reddy et al. studied 95 patients who underwent percutaneous secundum ASD or PFO closure [65]. Among these, patch testing was performed on eight patients who reported to have a history of nickel allergy, and six patients were found to be positive. Six patients with a documented nickel allergy were recommended to undergoing percutaneous defect closure using the less nickel-eluting devices; none developed allergic reaction to the implanted device. Therefore, inquiring about a history of nickel allergy is very important for patients as it can potentially avoid allergic complications. In the paper reported by Kim et al., 4 of 5 nickel-allergic patients and 10 of the 33 non-nickel-allergic patients had post-closure complications that proved related to nickel allergy [69]. Since positive reactions are not invariably relevant to complications and negative reactions did not rule out adverse events, it was concluded that a patch test for nickel allergy does not appear to be necessary as a routine test. A survey aimed to investigate nickel allergy management strategies in intracardiac occluders was performed by the American Contact Dermatitis Society. A total of 70 individuals responded to the survey, including 56 dermatologists and 14 allergists/immunologists. There was no consensus on the ability of patch testing to accurately determine allergic reactions within cardiac tissue. However, most would inquire about the history of nickel allergy before implantation. With a positive patch test, nickel-based intracardiac devices should be avoided, or the decision should be left to the discretion of the cardiologist [70]. Commonly, symptoms of device syndrome can require treatment with oral corticosteroids and antiplatelet therapy [68]. In some rare cases, however, surgical removal of the device is considered reasonable when severe allergic symptoms cannot be alleviated by medical management alone [71].

8.1.5.4 Atrioventricular Block

The occurrence of conduction abnormalities, including complete AV block (AVB), is considered to be a complication of ASD closure. The atrioventricular node located in the triangle of Koch is very close to the rims of ASD, so it is easy to be damaged after device placement. However, the exact incidence rate of this complication is still unclear. A retrospective study showed a low risk of heart block occurring in 0.3% of patients [72]. In a study by Astarcioglu et al. among 706 patients, 6 had experienced the development of AVB, giving an incidence of 0.85% [73]. The risk factors included younger age, larger size of the ASD, and larger size of the device.

In most cases, AVB is transient and recovers spontaneously or after corticosteroid therapy, with a good prognosis [74–76]. Conversely, there still are cases of late complete AVB [77, 78]. These observations underline the fact that careful cardiac rhythm monitoring is critical during long-term follow-up after percutaneous ASD closure, even in patients without early postprocedural cardiac conduction abnormalities.

8.1.5.5 Atrial Tachyarrhythmias

The natural history of patients with ASD and intra-atrial shunt leads to increased right atrial volume load and leads to myocardial fiber expansion and stretching. In the long run, there will be myocardial fibrosis and electrical remodeling, eventually leading to conduction delay at the crista terminalis and sinus node dysfunction. Conduction delay at the crista terminalis persists beyond ASD closure and may cause the long-term atrial arrhythmia [79]. Atrial fibrillation is extremely rare in ASD patients younger than 40 years old without transcatheter closure. However, the incidence of atrial fibrillation is more than 50% in patients over 60 years old [80, 81]. A meta-analysis by Sami et al. showed that the incidence of atrial arrhythmia and atrial fibrillation after ASD transcatheter closure was 6.5% (95% CI 3.5–11.7%) and 4.9% (95% CI 1.9–11.7%), respectively [82]. The retrospective study of Kyoung min et al. enrolled 427 patients with secundum ASD who underwent transcatheter closure. It was found that most new-onset atrial tachyarrhythmias occurred within 6 months (median: 2.6 [IQR: 1.2–4.1] months) after closure. Multivariate analysis confirmed that the risk for new-onset atrial tachyarrhythmia was significant in patients with transient atrial fibrillation or flutter during closure (hazard ratio [HR]: 9.90, 95% confidence interval [CI]: 2.86–34.20; $P < 0.001$), deficient posteroinferior rim (HR: 5.48, 95% CI: 1.15–25.72; $P = 0.04$), and age of closure over 48 years (HR: 3.30, 95% CI: 1.30–8.38; $P = 0.01$).

The current mainstream view is that ASD closure can reduce the right atrial volume load to prevent the progress of electrical remodeling, thereby reducing the risk of atrial tachyarrhythmia. In a meta-analysis by Vecht et al., atrial septal defect closure can reduce the prevalence of atrial tachyarrhythmias in patients who had preexisting atrial tachyarrhythmias (OR = 0.66 [95% CI 0.57–0.77]) [83]. In addition, the prevailing view is that patients who underwent ASD closure at young age show a lower incidence of supraventricular arrhythmias [84, 85]. It is worth mentioning that patients with atrial septal defect still had a higher risk of first time AF than the comparison cohort, even after transcatheter closure (adjusted HR 8.2; 95% CI 6.6–10.2), with no difference between surgical group and transcatheter group (HR 1.5; 95% CI 0.6–3.5). Furthermore, compared to the comparison cohort, a higher risk of stroke is associated with patients with ASD regardless of with or without closure [86]. As regards to the treatment, atrial arrhythmias should be treated to restore and maintain sinus rhythm if possible. If atrial fibrillation occurs, both antiarrhythmic therapy and anticoagulation should be recommended [87]. Although transseptal access through the device after deployment of occluders is difficult, there are still reports proving that catheter ablation is safe and feasible in refractory cases [88].

8.1.5.6 Infective Endocarditis

Device-related infective endocarditis (IE) is a rare complication after transcatheter closure. Even under strict aseptic operating conditions and prophylactic use of antibiotics, there are still related cases reported [89–94]. In these retrospective cases, the vast majority of infective endocarditis involved ASO devices. According to the blood culture results, it is suggested that *Staphylococcus aureus* may be responsible for the most endocarditis cases. What was noteworthy was that one patient had multiple untreated dental caries. Surgical removal of the device was performed uneventfully in all cases.

At present, the mechanism of device-related endocarditis is not clear. It is generally believed that the device with incomplete endothelialization increases the risk of IE. At present, it is still unable to accurately evaluate the degree of endothelialization in the long-term follow-up after the intervention. The experience of complete endothelialization of the device is mostly from the results of animal experiments [95]. Currently, the length of antibiotic prophylaxis of endocarditis has been recommended for 6 months after device implantation [96]. In one case, however, the occluder was not completely endothelialized at autopsy 5 months after ASO implantation, although previous echocardiography showed that ASD was completely closed [97]. Therefore, it is reasonable to prolong the duration of endocarditis prophylaxis after transcatheter ASD closure in some selective patients.

Moreover, postponing the catheter procedure in patients with recent infectious events and keeping oral health are also critical to prevent infective endocarditis after ASO implantation.

8.1.5.7 Device Embolization

Potential risk factors for device embolism include inadequate or floppy rim, undersized device, operator-related technical reasons such as malposition during the "push-pull" maneuver, or excessive tension on delivery cable during device deployment [98]. A survey from AGA proctors reported that the incidence of device embolism was 21 out of 3824 patients (0.55%) [99]. Another analysis based on the US FDA Manufacturer and User Facility Device Experience (MAUDE) database shows that there were 114 (0.62%) device embolizations out of 18,333 Amplatzer septal occluder implants [51]. Other meta-analyses of transcatheter closure of ASD/PFO revealed the rate of device embolization was 0.2–0.43% [15, 50]. Most of device embolizations occurred within 24 h after occlusion, and the most frequent sites of embolization were the left side of the heart. The device embolized into the atrium may induce arrhythmias such as ventricular premature contractions [99].

In different reports, the proportion of embolic devices that need surgical removal is different [99, 100]. At present, the strategies of device removal are mainly based on the experience of the operator and the embolic site of the device. Although transcatheter removal has many advantages such as shorter operation time and shorter hospital stay, there was still case of valve damage caused by repeated violent operation being reported [101]. To sum up, accurate imaging assessment can not only better guide the choice of interventional occlusion strategy but also help to find the device embolism timely. Once the device embolization is defined, the first objective is simply to get the device into a position in which it will not cause harm. The device may then be stabilized and moved or removed from the body.

8.1.6 Summary

Because the success rate of percutaneous transcatheter closure is not inferior to surgical closure and has a smaller risk of complications, it has gradually become the preferred treatment for more and more patients with secundum atrial septal defect [13, 102]. Extensive clinical application allows clinicians to accumulate a lot of experience of periprocedural evaluation and manipulation and also have a comprehensive understanding of various potential risks.

The improvement of treatment strategy and device design continuously expands the indications of interventional occlusion, which enables many patients who were previously considered to have lost the opportunity of occlusion to achieve successful occlusion and improve prognosis through this "simpler" treatment. For example, drug therapy combined with custom-made opening device can effectively delay the disease progression of PAH-ASD patients by reducing the volume load of heart.

However, the continuous innovation of technology and equipment not only reduces the risk of complications and improves the convenience of operation but also faces new challenges. For example, bioabsorbable materials, which can effectively reduce the risk of thrombosis and promote complete endothelialization, are faced with the risk of premature degradation in clinical application; the no-hub design of the left disc, which can significantly reduce the metal content, increases the difficulty of device retrieval when the device is embolized. In addition, there is still no consensus on the exact mechanism of cardiac erosion and effective measures to avoid it. Various complications caused by incomplete endothelialization are not uncommon. Currently, antiplatelet therapy and prophylaxis of endocarditis are recommended for 6 months after device implantation; however, these guidelines are based on the results of animal experiments. Many confusions still need to be answered by further clinical exploring. In the face of the complexity and diversity of ASD lesions, we should realize that accurate identification and evaluation of the disease and appropriate strategy choice are essential for successful occlusion.

8.2 Transcatheter Closure of Patent Foramen Ovale

8.2.1 Introduction

Patent foramen ovale (PFO) is a frequent congenital heart abnormality of fetal origin in healthy population. The prevalence of PFO in the general population is 15–35% in autopsy and 25% on echocardiography studies, respectively [103–106]. The prevalence of this heart abnormality is much higher than that of bicuspid aortic valve, which is regarded as the most common adult congenital heart disease [107]. Furthermore, autopsy-derived study has shown that the prevalence of PFOs decreases with each decade of life, while the average size of PFO increases with each decade of life [106]. Despite the lack of evidence from longitudinal studies, the spontaneous closure of smaller defects over time may be responsible for this finding. However, PFO plays a critical role in the fetal blood circulation. The persistent opening of patent foramen ovale after birth has been linked to various pathological conditions, including but not limited to stroke and migraine headache with aura [108]. With the continuous innovation of treatment technology, several clinical trials have explored the efficacy of PFO occlusion technology in the prevention of stroke.

8.2.2 Anatomy and Pathophysiology of PFO

The embryonic development of the atrial septum and foramen ovale has been elucidated in detail in above content of ASD. Before birth, the fetus relies on the maternal circulation for oxygenation, and the foramen ovale allows the oxygenated blood from the maternal circulation to enter the systemic circulation. After birth, with the neonate starting to breathe autonomously, the right atrial and pulmonary artery pressure began to decrease and the left atrial pressure began to increase gradually. The inverted pressure gradient between the left and right atrium makes the septum primum closure against the fossa ovalis rim [109]. The foramen ovale is obliterated in 75% of individuals within a few years of birth [110]. However, the reason why the PFOs fail to close in one quarter of the population is still unclear.

8.2.2.1 Anatomy of PFO

The size of a PFO in adults, defined as the maximum opening of the interatrial communication, ranges from 1 to 19 mm (mean 4.9 mm) [106]. The autopsy-derived study demonstrated that the average size of PFO increases with each decade of life. The mean diameter in the first decade is 3.4 mm and in the tenth decade is 5.8 mm [106]. It is noteworthy that greater sized PFOs seem to have higher risk of cryptogenic stroke and paradoxical embolism [111, 112], but another study holds the opposite view [113].

The typical PFO is a long and narrow tunnel. However, the morphology of PFOs is highly heterogeneous among different individuals. In practice, PFO is usually associated with various types of cardiac malformations, including atrial septal aneurysm and Chiari network and Eustachian valve, which may increase the risk of stroke.

Atrial septal aneurysm (ASA) is a localized deformity of interatrial septum (IAS), usually at the levels of fossa ovalis which protrudes into the right or left atrium or even swings left and right with the cardiac cycle [114]. The prevalence of ASA in different studies was close to 2–3% [114, 115]. Although previous study contradicted this [116], it has been widely accepted that the presence of ASA is associated with a higher risk of stroke in patients with PFO [113, 117–119]. It should be noted that the size of PFO coexisted with ASA is generally larger than that of PFO alone [120]. This kind of PFO has a higher probability of opening up, which seems to explain that the coexistence of the two malformations is always accompanied by an increased risk of paradoxical embolism.

In addition, there are other anatomical variations such as Eustachian valve and Chiari network, which are frequently associated with PFO. The Chiari network is a weblike network of threads and fibers connecting the Eustachian valve to other atrial structures [121]. Eustachian valve is an embryological remnant of the right valve of the sinus venosus. Although it has been confirmed that these abnormal structures are also associated with stroke risk in patients with PFO, the exact mechanism has not been well elucidated [117, 118]. In conclusion, the evaluation of PFO in size and morphology is a crucial part of guiding PFO occlusion therapy.

8.2.2.2 Pathophysiology of PFO

As the main reason why PFO has attracted clinical attention, paradoxical embolism is related to the opening of PFO and the formation of right heart derived thrombus.

In normal physiological state, PFO is mostly closed due to the higher left atrial pressure. However, in some special conditions, elevated right atrial pressure causes or increases the degree of right-to-left shunting through the PFO and therefore the likelihood of paradoxical embolization. For example, pulmonary embolism, right heart infarction, or severe tricuspid regurgitation may cause the rise of right atrial pressure and has been associated with increased blood flow through a PFO [122].

Deep vein thrombosis (DVT) is a potential source of thromboembolism through PFO; therefore, some clinical features that induce DVT may increase the risk of stroke [123]. Paradoxical emboli originating from the pelvis might be another important source of cryptogenic stroke in patients with PFO. In addition, recent surgery, trauma, and use of oral contraceptives are related with blood hypercoagulability and consequently with an increased risk of paradoxical embolization [124].

8.2.3 Indications for PFO Closure

Although several early RCTs have failed to demonstrate that PFO device closure is superior to medical therapy alone in reducing stroke recurrence [125–127], the promising results of recent RCTs provide evidence for various guidelines for PFO treatment [128–130].

Statement from the American Academy of Neurology stated that PFO closure with Amplatzer Occluder device could be available in case with recurrent strokes despite adequate medical therapy with no other mechanism identified (evidence level C) [131]. Canadian guideline for stroke recommended PFO device closure plus long-term antiplatelet therapy to carefully selected patients [(a) age, 18–60 years; (b) the diagnosis of the index stroke event is confirmed by imaging as a nonlacunar embolic ischemic stroke or a TIA with positive neuroimaging or cortical symptoms; (c) the patient has been evaluated by a neurologist or clinician with stroke expertise, and the PFO is felt to be the most likely cause for the index stroke event following a thorough etiological evaluation to exclude alternate etiologies] with a recent ischemic stroke or TIA attributed to a PFO [132]. In the European Society of Cardiology position statement of 2019, PFO closure is recommended to carefully select patients aged from 18 to 65 years with a confirmed cryptogenic stroke, TIA, or systemic embolism and an estimated high probability of a causal role of the PFO as assessed by clinical, anatomical, and imaging features [133].

Of note, two meta-analyses showed that PFO closure prevents the recurrence of stroke while increasing the risk of new-onset atrial fibrillation. The recurrence rate of stroke was decreased by 2.9–3.3%, but the incidence of atrial fibrillation was increased by 3.3% in the PFO closure group compared with the drug treatment group [134, 135].

8.2.4 Devices for PFO Closure

Anatomic characteristics of PFO and conformation of the closure device for PFO are important determinants for the success of PFO closure. Unlike atrial septal defects that have an oval-shaped hole, the PFO often has a horseshoe shape with a narrow tunnel resulting from the failed fusion between the septum primum and septum secundum. To promote adequate endothelialization, the PFO closure device has to approximate the ends of both septal and mechanically hold them together [136]. In addition, the connecting waist of the PFO device should be very narrow and flexible.

8.2.4.1 Amplatzer Patent Foramen Ovale Occluder (Fig. 8.2a)

Similar to the Amplatzer ASD devices, the Amplatzer PFO Occluder is also a self-expanding double-disk device which is composed of nitinol wire and fabric patch

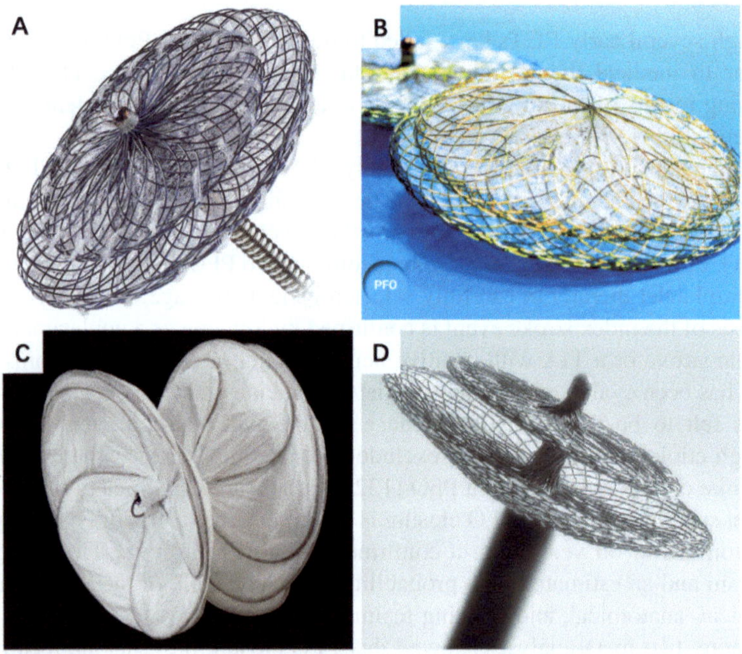

Fig. 8.2 Devices for patent foramen ovale closure. (**a**) Amplatzer Patent Foramen Ovale Occluder. (**b**) The Flex II PFO Occluder. (**c**) GORE CARDIOFORM Septal Occluder. (**d**) Cardi-O-fix PFO Occluder

sewn into the two disks. However, the connecting waist of this device is relatively narrow. The waist is 3 mm long, and attached slightly eccentrically to the two discs. It allows a flexible suitable of the disks to the septum even in a long tunnel PFO, and avoiding the device centering within the PFO thereby minimizing the risk of erosion. Because of the unidirectional valve characteristic of PFO and the higher right atrial pressure in this kind of malformation, the right atrial disc is slightly larger than the left atrial disc of this device. The available device sizes are 18, 25, 30, or 35 mm, which correspond to the diameter of the right atrial disks.

The results from the CLOSE trial showed that percutaneous PFO closure with the Amplatzer PFO Occluder (AMP) has a lower recurrence risk of cryptogenic stroke as compared with medical therapy alone. In addition, there was no significant difference in the risk of serious adverse events between the PFO closure group and medical therapy group. Seven periprocedural events of atrial fibrillation that occurred in PFO closure group were all resolved before the patients' discharge [128]. In the same period, two other RCTs which also involved the Amplatzer PFO Occluder concluded similar results [130, 137]. The conclusion of the above trials not only strongly proves the safety of the Amplatzer PFO Occluder but also provides solid evidence support for the use of this device in secondary prevention of cryptogenic stroke in PFO patients.

In a head-to-head study involving 406 patients, a high procedural success was observed in both the Amplatzer PFO Occluder group and the Occlutech figure device group. However, the Occlutech figure appears to have a lower risk of supraventricular tachycardia (9% vs. 17%, $P = 0.02$) [138]. According to the conclusion of a meta-analysis, AMP showed a lower risk of recurrent TIA or ischemic stroke compared to the STARFlex Septal Occluder (RR 0.35, 95% CI 0.14–0.91). In the clustered ranking plot, however, on the risk of recurrent neurologic events against new-onset atrial fibrillation, AMP appeared to be superior to the GORE HELEX/CARDIOFORM Septal Occluders [139].

8.2.4.2 The Flex II PFO Occluder (Fig. 8.2b)

The Figulla Flex II PFO device is composed by a nitinol wire mesh forming two flexible retention discs. The unique braiding technique applied to the discs allows for a single layer left atrial disc which is covered by an ultrathin polyethylene terephthalate patch. The right atrial disc of the device is slightly 2 mm larger than the left, and the two discs are connected by a flexible 3 mm waist in the center. The available device sizes are 16/18, 23/25, 27/30, and 31/35 mm, which correspond to the diameter of the two discs. The delivery system has a flexible ball and socket joint which allows angulation of the occluder during placement. The characteristic of this device can improve flexibility for optimal septum alignment without tension on the delivery cable.

In a head-to-head observational study, 406 patients received percutaneous closure of PFO with either the Amplatzer ($n = 179$) or the Figulla ($n = 227$) device after a stroke or a transient ischemic attack scrambled to the PFO [138]. However, a trend of higher incidence of acute residual shunt after device deployment is observed in Figulla occluder group. There is no significant difference between the two devices in terms of procedural and fluoroscopy times. In addition, Figulla occluder group seems to have a lower incidence of supraventricular tachycardia. Another single-arm prospective study (OPPOSE) also showed excellent efficacy and safety of Figulla Flex II PFO occluder [140]. The primary outcome of grade 0 or 1 shunt at 6 months was seen in 78% of patients. There were two access site procedural complications, but no new onset of atrial fibrillation was observed.

8.2.4.3 GORE CARDIOFORM Septal Occluder (Fig. 8.2c)

Because of the homogeneity of ASD and PFO in structure and the design characteristics of the CARDIOFORM Septal Occluder, this device is indicated for ASD closure for defects up to 17 mm and patent foramen ovale (PFO) closure to prevent recurrent ischemic stroke. The detailed description of this device has been introduced in the segment of ASD.

The REDUCE trial revealed benefit of PFO closure to prevent recurrent stroke. The CARDIOFORM Septal Occluder was used as one of closure devices. In this trial, the incidence of new brain infarctions was significantly lower in the PFO closure group than in the antiplatelet-only group [129]. Another multicenter study by Stefan et al. also demonstrated that CARDIOFORM Septal Occluder has good efficacy and low risk of new-onset paroxysmal atrial fibrillation [141].

8.2.4.4 Cardi-O-Fix PFO Occluder (Fig. 8.2d)

A single-center prospective study by Zhao et al. enrolled 246 patients with a PFO [142]. All patients were treated by PFO interventional closure, with the Cardi-O-fix PFO Occluder or Amplatzer PFO Occluder. There was no statistically significant difference in the complete closure rate and effective occlusion rate ($P > 0.05$) between the devices. A retrospective study by Han et al. showed that Cardi-O-fix PFO Occluder has promising efficacy in terms of the interventional closure of PFO [143].

Other catheter-based cures that have been developed or are in development include device-based closure with only minimal residuals within the heart, such as small "in-tunnel" devices that focus solely on closing the PFO tunnel [144].

8.2.5 Issues with PFO Occlusion Devices

Because the anatomical location of the defect is somewhat similar to the ASD, the device-related complications are similar to complications related to the ASD. Some of the complications are inherent to the device and remains the same as described earlier, in the segment of ASD closure.

According to meta-analysis regarding this procedure, the most frequent late complication is device thrombosis that is seen in 1.0–2.0%. Device embolism is a serious complication and occurs at a rate of 0.9–1.3% [50]. The protocol for anticoagulation therapy may need to be revisited. These patients may need lifelong antiplatelet therapy or for a longer time period than 6 months. A battery of tests to rule out inherent clotting disorders is also recommended prior to device closure. These will help in sorting out patients who may be candidates for antiplatelet and/or anticoagulation therapy.

In an ASD patient who had significant left-to-right shunt before device procedure and had small residual shunt after device placement, the primary objective of improving the hemodynamic compromise has been achieved as opposed to a patient with PFO. The PFO patients usually have a small shunt to begin with, and if the very small shunt persists, the procedure should not be labeled as a clinical success.

On the other hand, the PFO closure group showed a higher rate of new-onset atrial fibrillation, and a higher flutter was observed than in the drug treatment group. The incidence of new-onset atrial fibrillation was 3.2% in the device closure group compared to 0.47% in the control group ($P < 0.001$) [145].

8.2.6 Summary

Transcatheter PFO closure is a straightforward, safe procedure that in randomized clinical trials reduces the risk of recurrent cryptogenic stroke. Fastidious technique, combined with intracardiac imaging under conscious sedation, can minimize procedural complications and ensure procedural success.

8.3 Transcatheter Closure of Ventricular Septal Defect

8.3.1 Introduction

Ventricular septal defect (VSD) is a frequent congenital heart abnormality in children. Abnormal communication and shunt formation between the left and right ventricles are the major mechanisms of hemodynamic damage in patients with VSD. Most VSD close spontaneously; however, if not close, these defects may result in harmful complications, including arrhythmia, ventricular dysfunction, and pulmonary hypertension [146, 147]. VSD occurs in almost 40% of children with congenital heart lesions. The incidence of isolated VSD accounts for 0.3% in newborns. Since up to 90% of VSD may ultimately shut down, its incidence rate is obviously decreased in adults. There was no gender preference for VSD [148].

8.3.2 Anatomy and Pathophysiology of VSD

8.3.2.1 Embryology of VSD

VSD occurs when there is abnormal development or disruption of ventricular septum during complex embryonic cardiac morphogenesis. VSD is often isolated, but it may also be related to other congenital heart defects (e.g., pulmonary artery stenosis, patent ductus arteriosus (PDA), right aortic arch, and atrial septal defect). They are also related to subaortic stenosis aortic coarctation and are common components of congenital heart diseases (e.g., transposition of the great arteries and tetralogy of Fallot). Some genetic factors, such as single gene, chromosome, and polygene inheritance, have been shown to be associated with VSD. Recently, a TBX5 mutation was found to cause VSD in Holt-Oram syndrome patients. Nongenetic risk factors are also associated with the occurrence of VSD, including phenylketonuria, maternal diabetes, and maternal infections (e.g., febrile diseases, influenza, and rubella). Exposure to cocaine, marijuana, alcohol, and other toxins as well as drugs (e.g., ibuprofen and metronidazole) is also associated with VSD [149, 150].

8.3.2.2 Types of Interventricular Communications

Anatomically, the ventricular septum consists of a muscular and membranous septum [151]. The septum is found at the bottom of the heart below the right and noncoronary tips of the aortic valve. The septal leaflet of the tricuspid valve can be classified into two parts. The right side of the muscular septum is composed of outlet, apical trabecular, and inlet [152].

There are many anatomical classification systems of VSD, but a generally accepted standard is still lacking. Hence, the European Association for Cardiothoracic Surgery and Society for Thoracic Surgery-Congenital Heart Surgery Database Committee jointly proposed a VSD unified reporting system. The classification system divides VSDs into four different anatomical types according to the edge and location of VSDs [153].

(a) Type 1 defect

Type 1 defect is a defect located at the outlet of the muscular septum, which is called cone, subpulmonary, funnel, or supracristal defect. Also in this category, there is a parallel artery VSDs located directly below the semilunar valve. The top of type 1 defect is generated by the fibrous continuity of the pulmonary valve and aortic valve leaflets.

(b) Type 2 defect

Type 2 defect confluence with diaphragm. This defect usually extends to one of the three parts of the muscular septum. Type 2 defect can be defined as the term peri membrane [151, 154].

(c) Type 3 defect

Type 3 defect is in the entrance part of the muscular septum, below the atrioventricular valve, also known as entrance or atrioventricular tube/septal defect.

(d) Type 4 defect

The type 4 defect is in the trabecular part of the muscle septum surrounded by muscles [155]. Trabecular muscular septal defect (muscular VSD) can be in the middle, top, rear, front, or multiple parts of the muscle. Multiple muscle VSDs (Swiss cheese VSD) are not clearly seen from the right during operation, as it is covered with rough trabeculae.

8.3.2.3 Pathophysiological Mechanisms

The main pathophysiological mechanisms of VSD are shunt between the left and right ventricles. The volume and direction of shunt blood can be used to estimate the hemodynamic changes in VSDs. These factors can be affected by the location and size of VSDs and pulmonary vascular resistance (VR).

For the long-term, enlarged, left-to-right shunts, irreversible changes occur in pulmonary vascular endothelium, resulting in persistent polycyclic aromatic hydrocarbons. When the pressure in the pulmonary circulation is higher than that in the systemic circulation, the shunt direction is reversed and turns into right-to-left shunts. This is called Eisenmenger syndrome, which accounts for 10–15% among VSD patients.

8.3.3 Indications and Contraindications for VSD Closure

Depending on the European Society of Cardiology, the following are recommendations for intervention in VSD [53]:

- In cases with left ventricular volume overload and no pulmonary arterial hypertension (no noninvasive signs of pulmonary artery pressure elevation or invasive confirmation of pulmonary VR <3 WU [Wood units] in case of such signs), it is recommended to close VSD (class of recommendation: I; level of evidence: C).
- In cases with a history of repeated episodes of infective endocarditis but no obvious left-to-right shunts, it is required to close VSDs (class of recommendation: IIa; level of evidence: C).
- In cases with VSD-associated aortic valve cusp prolapse and subsequent progressive aortic regurgitation, surgery should be performed (class of recommendation: IIa; level of evidence: C).
- In patients with pulmonary arterial hypertension (pulmonary VR 3-5 WU), it is necessary to close VSDs when there remains obvious left-to-right shunt (Qp:Qs >1.5) (pulmonary-to-systemic flow ratio) (class of recommendation: IIa; level of evidence: C).

- In cases with pulmonary arterial hypertension (pulmonary VR ≥5 WU), VSDs should be shut down when there remains obvious left-to-right shunt (Qp:Qs >1.5), but a decision from experts is needed (class of recommendation: IIb; level of evidence: C).
- In cases with Eisenmenger syndrome and those with severe pulmonary arterial hypertension (pulmonary VR ≥5 WU) presenting with exercise-induced desaturation, surgery to close VSDs is not recommended (class of recommendation: III; level of evidence: C).

8.3.4 Devices for VSD Closure

In the past, a wide variety of devices (e.g., Rashkind double-umbrella device, buckle device, and vascular coil) have been proposed for transcatheter closure of VSDs. These devices are used to turn off various VSD types in an off-label manner.

8.3.4.1 Amplatzer Membranous VSD Occluder (Fig. 8.3a)

To treat VSDs without distorting the aortic valve, Amplatzer Membranous Occluders have been developed.

At the beginning of the twenty-first century, Amplatzer Membranous and Muscle VSD Occluders began to be used. The phase I trial of American membranous

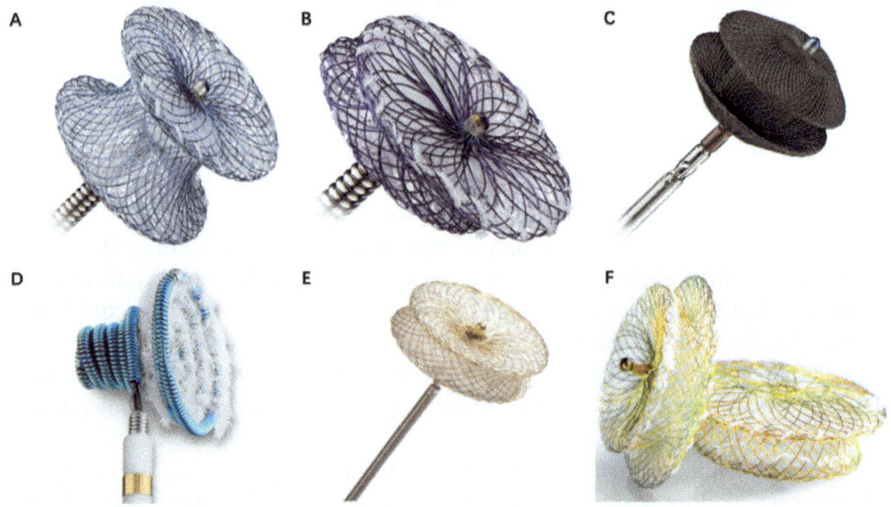

Fig. 8.3 Devices for VSD closure. (**a**) Amplatzer Membranous Occluder; (**b**) Amplatzer Muscular Occluder; (**c**) Amplatzer Duct Occluder II; (**d**) Nit-Occlud coil; (**e**) Cera VSD Occluder; (**f**) Occlutech muscular VSD occluder

devices recorded that the success rate of full VCD closure in patients was 91% after 6 months of follow-up and only one case of complete heart block [156]. The findings are also reflected in other small studies, which show that rare patients with complete heart block have a high surgical success rate and a closure rate of more than 95% [157, 158]. After these preliminary reports, more studies began to show that the incidence rate of complete heart block was higher (2–22%) [159–162]. Complete heart block patients who required pacemaker implantation are reported several months after the procedure, and usually there is no change in electrocardiogram or previous symptoms. The risk factors for the development of complete heart block are not easy to clarify. In a report describing the high incidence of complete heart block, the number of patients was significantly reduced compared with previous studies [162]. In view of concerns about conduction disorders, especially the long distance from implantation, the clinical approval of Amplatzer Membranous Occluder has not been granted. Amplatzer Membranous Occluder II is developed to avoid conduction defects by promoting equipment stability and decreasing radial force. It has been reported that 95% of patients ($n = 19$) successfully closed the heart defect, no increase in tricuspid or aortic regurgitation was observed during 1-year follow-up, and there were no complete heart block patients who required pacemaker implantation [163].

8.3.4.2 Amplatzer Muscular VSD Occluder (Fig. 8.3b)

The Amplatzer Muscular Occluder is a self-expanding device made of nickel-titanium alloy wire. It is composed of two flat discs, which are connected through the central connecting waist, and its diameter can estimate the size of the device. To promote thrombosis, polyester fabric is placed in the two disks.

The Amplatzer Muscular Occluder was first tested in 1999 in animal models [164]. Following that, a number of clinical studies documented a high success rate of VSD closure with this occluder [165, 166]. In 2004, the results of 75 patients registered as part of the multicenter register trying to close were published [167]. The operation was successfully performed on the majority of patients (85%). Most devices use a right ventricular anterograde approach. Severe adverse events were reported in a minority of patients (10%), including two surgery-related deaths. Conduction defects or arrhythmias occurred in 20% of patients, but no complete heart block patients who required pacemaker implantation were found. After operation, the defect was completely closed in 47% of patients, which could increase to 92% during the follow-up of 12 months. Low body weight (<50 kg) was significantly related to adverse events and postoperative residual shunt. In 2007, the use of muscle VSD occluder was approved by the Food and Drug Administration (USA). Muscular VSD Occluders are also applied for the closure of membranous and post-myocardial infarction VSD [168–170].

8.3.4.3 Amplatzer Duct Occluders I and II (Fig. 8.3c)

Amplatzer Duct Occluders I and II have been applied for the closure of muscular, membranous and postsurgical residual VSDs. Amplatzer Duct Occluder I is specifically designed to close membranous VSDs with "winebox" aneurysm formation on the right ventricular side or excess tricuspid valve tissue, with a success rate of >90%, a low residual shunt rate (>95% full closure rate), and no pacemaker required of complete heart block cases [171, 172]. During the closure, an arteriovenous loop was created, and the delivery sheath was advanced anteriorly. The retained disc of the device is pulled into the wide portion of the aneurysm, and the remainder of the device is deployed in the opening of the aneurysm into the RV, creating a slight waist on the device.

Recently, Amplatzer Duct Occluder II has been applied for transcatheter closure of small-to-moderate muscular and membranous VSDs. The retrograde arterial deployment is induced by a low volume delivery sheath. A recently reported series describes a success rate of over 90% for the procedure, with a full closure rate of 95% and only a very small number of patients having persistent conduction defects [173–176]. The size of the device is usually 1–2 mm larger than the defect size. The lumbar diameter of an Amplatzer Duct Occluder II is 6 mm, which is larger than the defect size (<6 mm). Additionally, the diameter of the retained disc is 6 mm greater than that of the waist. Thus, an adequate distance from the defect edge to the aortic valve is needed (usually >23 mm).

8.3.4.4 Nit-Occlud Coil (Fig. 8.3d)

The PFM Nit-Occlud coil is designated for use with both muscular and membranous VSD closures and was awarded the CE mark in 2010. A polyester fiber is attached to this device, which displays an asymmetric pattern with a smaller right-hand cone and a larger left-hand cone. It is available in a variety of sizes and can be used for VSDs with/without restrictive aneurysmal tissue. In a European multicenter registry, Nit-Occlud coil has been successfully implanted in 92% of patients [177]. The remaining 8% failure patients are attributed to improper device positioning, large residual shunts, aortic valve impingement, transient heart blockage, and defect mismatch. Hemolysis was reported in four cases, with surgical closure of the VSD in one case, implantation of a second device in one case, and spontaneous resolution in two cases. Device embolism is found in one case, and the device is surgically removed. One patient developed complete heart block following device placement, which recovered after a course of steroid therapy. However, there are no cases of conduction defects that require pacemaker implantation or surgical removal of the device. At recent follow-up, 99% of patients attained a full closure rate.

8.3.4.5 Other Technologies

The Cera series of VSD occluders (Lifetech, China) (Fig. 8.3e) can be applied to close post-infarct, muscular, and membranous VSDs. The Cera device includes one asymmetrical and two symmetrical designs and has a self-expanding nitinol stent. In a multicenter study ($n = 55$), the success rate of the procedure was 91%. One device was embolized and retrieved percutaneously, one case has complete heart block and requires pacemaker implantation, and one patient developed aortic regurgitation and the device was surgically removed [178]. At recent follow-up, 91% of cases achieved a full closure rate.

Occlutech myocardial and membranous VSD occluders (Helsingborg, Sweden) (Fig. 8.3f) are self-expanding, nickel-titanium alloy devices. However, there is a lack of studies reporting on this device.

8.3.5 *Issues with VSD Occlusion Devices*

8.3.5.1 Conduction Defects

Conduction disturbance remains a major problem, especially after membranous VSD closure, due to the proximity of VSD to the conduction systems. The suggested mechanisms include the device exerting radial forces that result in damage to the surrounding tissues. Previous studies have shown that the Amplatzer Membranous Occluder contributed to a complete heart block rate of 2–22%; thus, it was not clinically approved [159–162]. As described previously [159, 179], complete heart block may be developed several months after implantation, or even years. At present, it remains difficult to identify the risk factors for complete heart block development after device placement. Univariate analysis revealed that the location of VSDs and application of Amplatzer Membranous Occluder may be risk factors for complete heart block, in a multicenter European transcatheter VSD occlusion registry, but these factors were not significantly observed in multivariate analysis [160]. Other potential risk factors may include small sample size and oversized devices [159]. Implantation with Lifetech symmetric Amplatzer Membranous VSD Occluders in 228 subjects indicated that a shorter distance of VSDs from the aortic valves to the tricuspid valves conferred higher risk of complete heart block, demonstrating a role for VSD location and its association with the conduction systems in the development of complete heart block [179].

It has been recently reported that the incidence of complete heart block is low (<2%) in patients implanted with Nit-Occlud coils or Amplatzer Duct Occluders I and II [171–177]. Conduction defects may occur during muscular VSD closure, but this side effect is uncommon. Steroid therapy has been used to treat device-related complete heart block in several case reports [179, 180], but no larger studies have been conducted. The prophylactic use of steroids in these patients needs to be validated by larger placebo-controlled studies [181].

8.3.5.2 Residual Shunts

Residual shunts can be classified as large (>4 mm), moderate (2–4 mm), small (1–2 mm), and trivial (<1 mm), depending on the width of the color Doppler jet. Large residual shunts have been reported during implantation, often resulting in the replacement of a new device. A full closure rate of >90% can be achieved at the last follow-up [159, 171, 173]. Surgical resection for serious residual shunts has been reported rarely [160, 172].

8.3.5.3 Aortic Valve Distortion

Aortic valve distortion associated with the membranous VSD device closure has been reported in 9% of patients implanted with Amplatzer Membranous VSD sealer, which is usually detected by careful echocardiographic examination during the procedure [161]. Aortic regurgitation after device implantation is usually mild; thus, the removal of the device is not necessary. The distance between the aortic valves and VSDs should be kept at least 2 mm, especially the Amplatzer muscle-type occluder or Amplatzer Duct Occluder II. Implanting the device in the aneurysmal tissues on the right ventricular side can increase the distance between the aortic valves and the device, thereby decreasing the risk of aortic regurgitation [171, 172].

8.3.5.4 Hemolysis

In most large studies, hemolysis occurred in 1–2% among different devices. This is usually mild and resolves spontaneously within 2–3 days of implantation. Cases that required surgical removal of the device or blood transfusion are rarely found [160, 171]. It remains hard to assess the risk factors for hemolysis following device implantation.

8.3.5.5 Endocarditis

Several studies have documented that endocarditis infections are associated with the devices in patients with/without any residual VSDs [182–184]. It is a severe adverse event, which may result in the need of antibiotic therapy and even surgical removal of the device. Due to the rarity of this adverse event, the risk factors may be difficult to identify.

8.3.6 *Summary*

Transcatheter VSD closure has developed considerably in the past two decades with improved techniques and tools. Larger delivery systems and the risks of conductive obstruction and residual shunts are more challenging. The off-label use of other devices and development of low volume delivery systems have made the procedures more effective and safer. The incidence of adverse events was low in this era. Despite these drawbacks, the risk of complications, especially severe conduction defects, remains and therefore requires patient selection and careful techniques. Therefore, it is necessary to develop novel VSD devices for a better success rate.

8.4 Transcatheter Closure of PDA

8.4.1 *Introduction*

The ductus arteriosus links the left pulmonary artery to the aorta descending behind the left subclavian artery, which is an important fetal structure that closes spontaneously within 2 days of birth in approximately 90% of full-term infants. A prolonged PDA is regarded as uncommon and is seen in premature babies or in newborns with circulatory or respiratory abnormalities. PDA is a form of congenital heart defect or known as persistent patency in term infants in the third month of life [185]; it may be related to a variety of congenital heart defects. PDA is often an isolated lesion in adults.

The incidence of PDA in full-term infants is approximately 1 in 2000 live births and accounts for 5–10% of congenital heart diseases [186]. When "silent" PDA is added, the incidence increases dramatically to 1 in 500 live births [187]. PDA is a frequent issue in premature newborns and very low birth weight infants, with 65% of newborns with a birth weight ≤1000 g, and is related to a variety of neonatal disorders [188]. In most reports, the ratio of females to males is 2:1 [185].

8.4.2 *Anatomy and Pathophysiology of PDA*

8.4.2.1 Embryology of PDA

During the fetal period, the pulmonary vascular system is not functioning, and the oxygen pressure is low. The ducts are kept open due to the circulating prostaglandins levels in the placenta. After removing the placenta at birth, the levels of prostaglandins are reduced, causing the reversal of blood flow into the lung, and the remaining prostaglandins are diminished. Besides, after the baby takes its first breath, the oxygen pressure will be increased, thus leading to a decrease in pulmonary VR.

Under normal conditions, the ducts may close in the first 24 h after birth due to the muscle contraction of the ducts. These muscles are sensitive to endothelin, bradykinin, acetylcholine, and oxygen. Although functional closure appears in the first few hours after birth, anatomical closure may occur within several weeks. At the later stages, the duct can become fibrous, so that no blood flow passes through it. In most infants, the duct closes in 2–3 weeks and should be closed if it has not closed after 8 weeks. Spontaneous closure of the catheter is very rare.

8.4.2.2 Types of PDA

PDA are usually isolated lesions in adult patients. The size and shape of PDA vary considerably and have implications for both pathophysiological properties and occluder type during catheter intervention [189]. PDA in adults can be graded as follows [190]:

(a) *Silent*
 Only a small PDA is examined by echocardiography; no cardiac murmur is audible.
(b) *Small*
 A continuous or ejection systolic murmur may be heard and may radiate posteriorly. Minor hemodynamic changes, absence of pulmonary hypertension, normal size of left atrium and left ventricle, and normal peripheral pulses.
(c) *Moderate*
 Dynamic changes in arterial pressure pulses (e.g., marked aortic regurgitation) with a persistent murmur. The left ventricle and the left atrium are enlarged and a reversible pulmonary hypertension.
(d) *Large*
 Adults present with the physiology of Eisenmenger syndrome. Obvious pulmonary hypertension is observed. The murmur is high pitched, and the continuous murmur is usually not present. Differential cyanosis (higher oxygen saturation in the upper body than that in the lower body) and toe deformities are related to this.

8.4.2.3 Pathophysiology

Prostaglandin E2 promotes patency of the ducts. If the arterial ducts remain patent after birth, they are related to pulmonary hemorrhage and pulmonary edema, necrotizing small bowel colitis, bronchopulmonary dysplasia, kidney failure, congestive heart diseases, and intracerebral hemorrhage. PDA causes the blood to flow from the descending aorta through the unclosed arterial ducts into the pulmonary circulation, which leads to pulmonary edema. Diastolic “theft” from the aorta needs to be compensated for by increased cardiac output. Extremely preterm babies have a limited ability to increase the volume per beat, so the heart rate is increased to enhance the cardiac output. Reduced blood flow in the lower half of the body leads to an increased risk of necrotizing small bowel colitis and renal failure [191–193].

The size of the right-to-left shunts may depend on various factors, including the ratios of pulmonary VR to systemic VR. If the systemic VR is high and the pulmonary VR is low, the blood flow to the lung through the shunts may be high. This may cause the reversal of blood to the left ventricle and left atrium, subsequently leading to hypertrophy of the left ventricle and left atrium.

8.4.3 Indications and Contraindications for PDA Closure

Depending on the European Society of Cardiology, recommendations for intervention in PDA are as follows [53]:

- In cases with left ventricular volume overload but no pulmonary arterial hypertension (no increase in pulmonary artery pressure, pulmonary VR <3 WU), PDA closure should be performed (class of recommendation: I; level of evidence: C).
- The most appropriate device should be chosen for the closure (class of recommendation: I; level of evidence: C).
- In cases with pulmonary arterial hypertension (pulmonary VR 3–5 WU), PDA closure is needed if the left-to-right shunt is >1.5 (Qp:Qs) (class of recommendation: IIa; level of evidence: C).
- In cases with pulmonary arterial hypertension (pulmonary VR ≥5 WU), PDA closure should be performed if the left-to-right shunt is >1.5 (Qp:Qs), but an expert decision is highly recommended (class of recommendation: IIb; level of evidence: C).
- In cases of Eisenmenger syndrome and those presenting with exercise-induced desaturation, PDA closure should not be performed (class of recommendation: III; level of evidence: C).

8.4.4 Devices for PDA Closure

In most centers today, device closure is a suitable approach for most patients with PDA. The presence of catheter calcification increases the risk of the procedure and may favor device closure. This can be performed simultaneously as a "diagnostic" cardiac catheterization. Typically, vascular closure by a transcatheter device is a standard procedure. Pre-intervention aortography allows the configuration, size, and association of the catheter with nearby anatomical structures (tracheal shadow, aorta, and pulmonary artery) to be determined, which has a significant implication for catheter closure. The combination of angiographic and echocardiographic guidance of transvenous access can be applied to reduce the risk of complications following arterial puncture, thus enabling same-day patient discharge.

By using a variety of devices, successful closure is achieved in most attempts. Approximately 95% of catheters are completely closed within 12 months of device placement. At present, the Amplatzer Duct Occluder (standard/modified) and the Gianturco coil are the most appropriate devices for the closures of medium-to-large (>2 mm) and small (≤2 mm) PDA, respectively [194, 195]. The design of the Amplatzer Muscular VSD Occluder, with a large connecting waist and two retained discs, appears to be more appropriate for large PDA closure with high pressure [196].

8.4.4.1 Amplatzer Duct Occluder

For the closures of large and moderate PDA, Amplatzer Duct Occluder I is the most appropriate device (Fig. 8.4a) that contains a mesh of nickel-titanium wire woven (0.004 inch thick) with a mushroomlike structure and self-expandable design, and the polyester fibers are sewn into the device [197, 198]. Recessed screws are fitted to the end of the pulmonary and attached to the delivery line. A sheath tube of 6–8 F is used for the delivery. In a previous trial, Amplatzer Duct Occluder I was successfully implanted in 99% patients (n = 435/439) with an immediate full closure rate of 76% (n = 384/435), increasing to 99.7% complete closure at 1 year (n = 359/360). Amplatzer Duct Occluder II has two holding surfaces (Fig. 8.4b), forming six potential closure planes to conform to the PDA's anatomical structure. Amplatzer

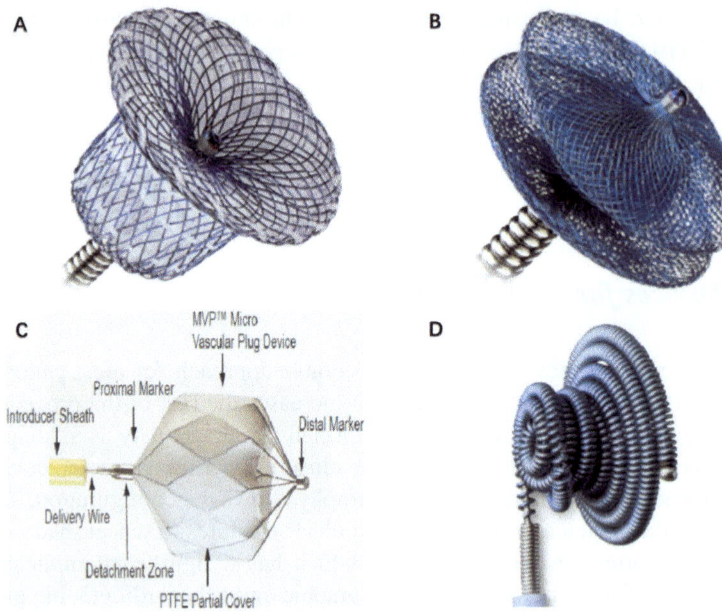

Fig. 8.4 Devices for PDA closure. (**a**) Amplatzer Ductal Occluder I; (**b**) Amplatzer Ductal Occluder II; (**c**) Micro Vascular Plug System; (**d**) Nit-Occlud Coils

Duct Occluder II is made of two layers of nickel-titanium alloy wire woven in place of the polyester fibers of ADO, thus allowing for low entry angles and delivery via 4-F and 5-F catheters. It is particularly valuable for younger children with a large catheter diameter and different forms of PDA [199, 200].

8.4.4.2 Microvascular Plug System (Fig. 8.4c)

The flexible, laser cutting, microvascular plug system consists of a cylinder with two extremities by tapering down. The deployment in the tubular body portion is necessary for vascular closure. PTFE membranes are used to cover the tubular part starting from the proximal part. A radiopaque marker is observed at each end of the distal tapering. A delivery cable with a plastic sleeve (4 cm) is coupled with the device to improve the effect of loading. A rotational detachable cable was used to replace the electrolytic detachment as consistent with Amplatzer Duct Occluder. There are two sizes for this device: MVP-3 and MVP-5, with diameters of 5.3 and 6.5 mm, respectively. The unconstrained length is 12 mm for both sizes. MVP-3 and MVP-5 are recommended for vascular closure of 3 and 3–5 mm, respectively. Microcatheters (0.021 or 0.027 inch) are placed in all devices [201].

8.4.4.3 Nit-Occlud Coils (Fig. 8.4d)

The Nit-Occlud device is specifically designed for PDA closure with large size (6 mm) and various shapes, which can be delivered through 4-F or 5-F catheters. This device shows a progressive stiffness from the aortal to the proximal windings to optimally adapt to the catheter anatomy, and the pull-through can be avoided by strong winding. The compact and tight winding of the middle section ensures fast and effective occlusion. It can be repositioned before release with flexible delivery and a nontraumatic design that enables the coil to be fully retracted until reaching an optimal position. The delivery system contains an opaque guide sheath with a distal marking ring and is pre-mounted on a disposable handle, thus simplifying the implantation procedure. The Nit-Occlud coil system is available in a variety of designs, including Nit-Occlud Double Disk (large, short aortopulmonary windows), Stiff (a reinforced coil for medium-to-large PDA), Medium (a reinforced coil for medium-sized PDA), and Nit-Occlud Flex (a flexible coil for closing small PDA). When using the Nit-Occlud Flex and Medium, a transport sheath is required so that they can be easily transferred into the guiding sheath. The radiopaque coil is compatible with MRI [202].

8.4.5 Problems with PDA Occluders

The side effects of PDA occluders are rare but may require surgical removal. The main adverse advents are coil-related embolism and Amplatzer Duct Occluder-related stenosis or obstruction.

If the coils are not deployed in the appropriate positions or the size of catheter is not fixed, coil embolism may occur, and the coils are easy to retrieve by interventional means. However, when multiple coils are migrated, coil embolization often causes a major problem. The incidence of embolism is much lower with Amplatzer Duct Occluder compared to coils. However, embolization associated with Amplatzer Duct Occluder may occur later, leading to the need for surgical removal [203].

Device-induced aortic coarctation or pulmonary stenosis occurs rarely, especially in low-weight PDA patients who required a large device. Therefore, routine echocardiographic and clinical follow-ups are needed. In certain cases, as the patients grow, Doppler flow acceleration may resolve spontaneously. In other cases, the obstruction may become worse, thus requiring surgical removal of the device or transcatheter. To prevent this possible adverse event, aortography should be performed after device deployment, before and after device release. Even after device release, if necessary, the device can be retrieved percutaneously, and the catheter blocked with a new device [204].

8.4.6 Summary

Transcatheter PDA closure is the standard of practice in many patients, and PDA closure is required in patients with catheter-induced left ventricular volume overload. For those with left-to-right PDA and severe PAH, closure can be carried out according to the individual patient's condition. Management of the very minimal or clinically silent PDA is still extremely controversial. Technology is evolving, and the transcatheter closure of PDA has become more feasible and safe using the latest equipment. Globally, Amplatzer Duct Occluders and coils are most often used for PDA closure with higher closure rates and lesser complications. Transcatheter PDA occlusion in preterm or low birth weight infants may be a great challenge, and further development of the device and catheter design is required before transcatheter occlusion can become the most ideal therapeutic approach in this vulnerable population. The evolution of transcatheter PDA occlusion from an 18-F sheath 40 years ago to a 3-F sheath today is impressive, and further improvement may enhance the efficacy and safety of transcatheter PDA occluders.

8.5 Transcatheter Closure of Left Atrial Appendage

8.5.1 Introduction

Atrial fibrillation is significantly related to congestive heart failure, cerebral infarction, and overall mortality [205, 206]. In atrial fibrillation patients with cerebral infarction, the left atrial appendage (LAA) is the most common source of thrombus [207, 208]. The specific LAA shape, accompanying trabeculae, pectus carinatum morphology, hypercoagulable state, atrial remodeling, and atrial fibrillation-related inflammation all contribute to thrombus formation [209]. At present, oral anticoagulation with direct oral anticoagulants or warfarin is the treatment of choice to prevent cerebral infarction and systemic thromboembolism in atrial fibrillation patients with a CHA2DS2-VASc score ≥2; and the survival rate can be improved by oral anticoagulants [210, 211]. Despite the proven efficacy of OAC, a large proportion of patients at risk of cerebral infarction may have relative or absolute contraindications to oral anticoagulant due to patient nonadherence, drug interactions, high bleeding risk, and cost [212, 213]. In addition, there are extensive data indicating that 45% of patients discontinue oral anticoagulants (including direct oral anticoagulants) within 2 years [214, 215]. In recent years, different LAA blocking strategies have been developed as effective and safe alternatives for the prevention of systemic thromboembolism and cerebral infarction in atrial fibrillation patients who were recommended for oral anticoagulants [216–218]. Apart from the prevention of cerebral infarction, several studies have suggested that epicardial LAA blockade reduces the burden of arrhythmias and exhibits a remarkable impact on electrolyte and fluid balance, systemic homeostasis (e.g., blood pressure), and atrial remodeling [218–221].

8.5.2 Anatomy and Stroke Risk

The LAA is a trabecular, long tubular structure with a narrow junction with the venous portion of the left atrium. It varies greatly in size and shape, with a curved or spiral axis in 70% of patients [222]. Anatomically, the LAA is well-divided into an orifice, neck, and lobe [223]. In patients with chronic atrial fibrillation, the LAA remodeling results in dilatation, stretching, and a decrease in the volume of the pectinate muscle [224].

The LAA is a source of thrombus in approximately 57% of valvular atrial fibrillation and 90% of non-valvular atrial fibrillation [207, 208]. This high thrombotic burden is due to its shape, neck skeleton, and range of trabeculae [222]. Di Biase and coworkers demonstrated a close relationship of the LAA morphology and shape with stroke risk by computed tomography (CT) imaging [225]. There are four types of LAA shapes: (1) cauliflower, (2) cactus, (3) windsock, and (4) chicken wing. It was found that 79% of patients with a chicken wing shaped LAA did not have a history of stroke or transient ischemic attack (odds ratio [OR]: 0.21; 95% confidence interval [CI]: 0.05–0.91; $P = 0.036$). In addition, patients with other LAA morphologies tended to have a stroke or transient ischemic attack compared with the chicken wing morphology: the OR for the cactus morphology was 4.08 (95% CI: 1.04–17. 27; $P = 0.046$), the OR for the pinwheel morphology was 4.8 (95% CI: 1.89–22.50; $P = 0.038$), and the OR for the flowerlike LAA morphology was 8.02 (95% CI: 0.92–27.86; $P = 0.056$) [225].

Other studies showed that the extent of LAA trabeculae and smaller mouths were related to an epidemiological history of systemic thromboembolism and cerebral infarction by identifying low LAA flow rates [226, 227].

8.5.3 Indications for the LAA Closure

Indication for LAA occluder placement will be discussed for the following five clinical conditions [228]:

1. Non-valvular atrial fibrillation cases who are eligible for chronic long-term oral anticoagulant
2. Cases with a contraindication for oral anticoagulant
3. Cases with a high bleeding risk who have received chronic oral anticoagulant
4. Noncompliant cases (unable or unwilling to take oral anticoagulant)
5. Specific subgroups
   (a) Inefficient oral anticoagulant (“stroke on oral anticoagulant”)
   (b) Electrically isolated LAA after catheter ablation
   (c) Combination of LAA occluder placement and atrial fibrillation ablation
   (d) LAA closure for “primary” prevention

8.5.4 Devices for the LAA Closure

Current catheter-based LAA occlusion devices are designed according to three principles:

(a) *The plug*

The intravascular delivery device leaflet or umbrella obstructs the LAA neck (WATCHMAN), thus blocking blood flow into the LAA. The exclusion of the LAA is dependent on the sealing/endocardialization of the device leaflet/umbrella [229, 230].

(b) *Pacifier principle*

A device with a leaflet or umbrella is delivered intravascularly, along with an additional disc that seals the mouth of the LAA from the left atrial side. Exclusion of the LAA is dependent on sealing/endocardialization of the device leaflet/umbrella (Amplatzer cardiac plug, Amulet) and/or sealing disc (LAmbre) [231, 232].

(c) *Ligation*

The LARIAT uses endocardial and epicardial approaches to snare and ligate the body of the LAA. Exclusion of the LAA is dependent on complete ligation of the LAA neck [216].

The endovascular device is based on a nickel-titanium alloy and is deployed via a dedicated sheath that can be retracted by pushing the device out or retracting the sheath or a staged combination of both. There are few contraindications to anatomy, and they are technically feasible in over 95% of patients. These procedures are usually conducted under general anesthesia or deep sedation with transesophageal echocardiogram (TEE) guidance. Experienced operators can use fluoroscopy or intracardiac echocardiography alone, both of which allow for local anesthesia and an awake patient [233, 234].

8.5.4.1 WATCHMAN and WATCHMAN FLX

The WATCHMAN device contains a self-expanding nickel-titanium alloy ten-branch frame with a 160 μm permeable polyethylene terephthalate membrane fabric cap facing the left atrium. The open distal end is held in one row by ten active fixation anchors. This device is available in five different sizes: 33, 30, 27, 24, and 21 mm. The transabdominal access sheath has an external diameter of 14 F and is available in three different prefabricated curve shapes: single curve, double curve (applied for more than 90% of procedures), and anterior curve. The three proximal radiolucent markers correspond to the approximate deployment levels of devices 33, 27, and 21, respectively.

The second-generation WATCHMAN FLX device (Fig 8.5a) is completely covered by a polyethylene terephthalate membrane cap to reduce leakage around the device [235]. This device is 10–20% smaller than the initial version and is available in five different sizes (35, 31, 27, 24, and 20 mm) for the LAA bores of 15–32 mm. The higher number of brackets and anchors, in combination with the greater radial strength, can provide better tissue retention. While initially allowing for greater pressures (10–27%), it has been observed that oversizing increases the risk of device embolism.

8.5.4.2 Amplatzer Cardiac Plug and Amulet

The Amplatzer cardiac plug (Fig. 8.5b) is a self-expanding nickel-titanium alloy mesh with a proximal polyester disc and a distal leaflet connected by an articulated waist [236]. The length of Amplatzer cardiac plug is shorter than its diameter so that it can accommodate the shorter LAA. The valve has six pairs of stabilizing wires designed to be implanted 10 mm inside the LAA orifice, while the disc is designed to seal the LAA orifice on the left atrial side. Eight different sizes are available, ranging

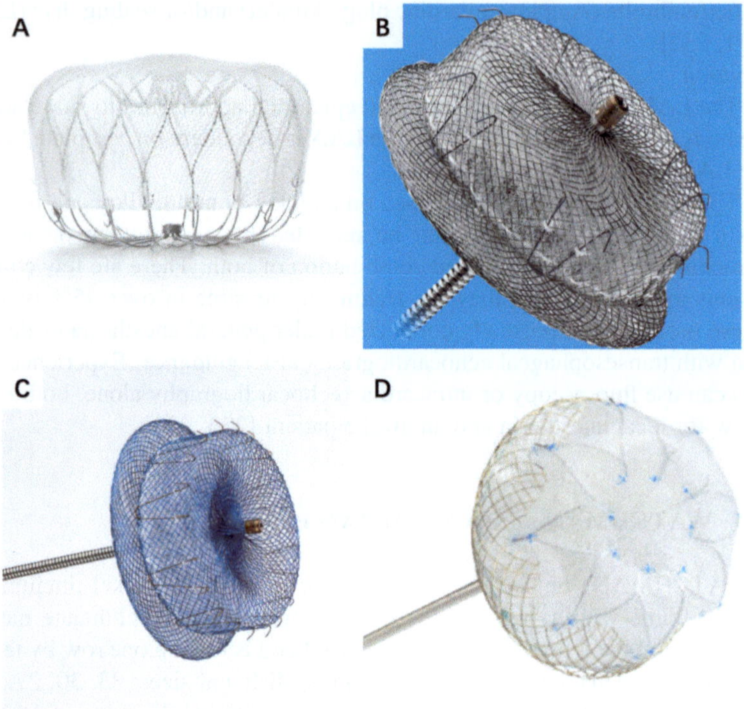

Fig. 8.5 Devices for closure of the LAA. (**a**) WATCHMAN FLX; (**b**) Amplatzer cardiac plug; (**c**) Amplatzer Amulet; (**d**) LAmbre LAA closure system

from 16 to 30 mm, depending on the size of the lobe dimensions (the landing zone size range = 12.6–28.5 mm).

The Amulet device (Fig. 8.5c) is the most recent device that has the same design to the Amplatzer cardiac plug device but modified to help implantation and decrease postoperative adverse effects [236]. The device is preloaded with eight different sizes (range = 16–34 mm) to fit the LAA sizes (landing zone measurements) of 11–31 mm, with a minimum LAA depth of 12 mm. Compared with the Amplatzer cardiac plug device, the distal lobe is 2–3 mm longer, and the proximal disc is 6–7 mm larger in diameter than the distal lobe and has a longer waist and more stabilizing wires (six to ten pairs), giving totally greater stability. In addition, the proximal screws are recessed to decrease the risk of thrombosis. Like the WATCHMAN FLX unit, fewer oversized devices [237] are required given its more stable design.

8.5.4.3 LAmbre LAA Closure System (Fig. 8.5d)

The LAmbre LAA closure system is a nickel-titanium alloy device consisting of a distal self-expanding umbrella and a left atrial cover, covered by a double polyethylene terephthalate membrane and connected by a central articulated waist [238]. The device is held in place by a dual stabilization mechanism with eight proximal U-shaped anchors and eight distal hooks. The distal umbrella deploys first, the hooks are recessed until they deploy, and then the proximal cover is released. Two types of LAmbre device are available, the standard and the special, for single or double leaf anatomy, respectively, the latter for multi-leaf or small LAA. The main advantages of the LAmbre device are its low profile (8–10 F) and the fact that it is fully repositionable and recapturable.

8.5.5 *Issues with LAA Occlusion Devices*

There are several problems associated with LAA closure. These problems can be categorized as equipment-related thrombosis, equipment embolism, and leakage around equipment.

8.5.5.1 Device-Related Thrombus

Device-related thrombosis is recognized as the Achilles' heel of LAA closure because of its unclarified risk factors, unpredictable timing, and close relationship with ischemic events. In a meta-analysis of >10,000 patients who underwent LAA closure, device-related thrombosis occurred in 3.8% of patients who had a fivefold increase in systemic embolism or cerebral infarction [239]. The timing of device-related thrombosis is unpredictable, with 42%, 57%, and 1% of patients diagnosed at less than 90 days, 90 to 365 days, and more than 365 days, respectively. Although most patients achieved thrombolysis with short-term oral anticoagulant, device-related thrombosis recurred in 50% of patients who discontinued oral anticoagulant [240]. However, findings on the predictors of device-related thrombosis are heterogeneous and hampered by the small number of events [241, 242]. It remains unclear whether device-related thrombosis is primarily associated with patient factors, surgical factors (e.g., peri-device leakage or deep implantation), type and timing of adjuvant antithrombotic therapy, or a combination of these factors. More research is needed to develop a device-related thrombosis scoring model to identify patients with high device-related thrombosis risk and to identify its modifiable predictors. Optimizing the LAA closure device is also needed to further decrease the risk of device-related thrombosis. The WATCHMAN FLX device is designed to minimize exposure to metallic materials to reduce the risk of device-related thrombosis. Early experience with the device has shown a device-related thrombosis rate of 1.8%

(7 out of 395 patients) at 1 year, which is lower than that reported in previous studies. In addition, some in vivo studies of next-generation LAA closure devices suggest that the new antithrombotic coating has the potential to further reduce the odds of device-related thrombosis after LAA closure.

8.5.5.2 Device Embolization

Device embolism associated with endocardial LAA devices, including nitrocellulose cage percutaneous LAA closure devices and transcatheter self-expanding nitrocellulose plugs, is rare but potentially challenging, with an incidence rate of <4% [243]. Most of the reported devices-associated embolisms are acute, occurring during the postprocedure hospital stay (65%), while approximately 30% are late embolisms identified during routine TEE follow-up [244]. Although the majority of devices-associated embolisms were removed percutaneously in published reports [244], in some complex cases, a surgical approach is needed. It is recommended that the inexperienced operator minimize this complication by adequately imaging the anatomy of the LAA with CT and TEE scans to optimize the size of the devices. In addition, 45-day, 6-month, and 12-month TEE follow-up periods are recommended.

8.5.5.3 Postprocedural Leaks

Postoperative leakage may occur following percutaneous LAA occlusion and has been observed with both endocardial and percutaneous endocardial occlusion, but not with LAA exclusion system occlusion. Depending on the type of LAA device and how often and how it is monitored, the reported incidence of leakage ranges from 0% to 63% [245, 246]. The consequences of leaks around such devices remain unclear, and studies have reported inconsistent results regarding stroke risk [247–249].

The conflicting nature of previous findings exposes the drawbacks of imaging modalities and the significant role of appropriate device sizing. The prevention of cerebral infarction in patients with atrial fibrillation has received increasing attention. Thus, it is important to gain insight into the management of leakage after deployment, while continuous TEE monitoring and anticoagulant therapy are also recommended.

8.5.5.4 Catheter Ablation with LAA Closure

Patients undergoing LAA closure are often older patients with high-risk atrial fibrillation who require additional radiofrequency catheter ablation procedures targeting the pulmonary veins and other non-pulmonary vein sites to treat symptomatic AF. Endocardial device-related ablation, particularly around the LAA, is challenging because of the LAA skeleton remodeling, which may increase the risk of new peri-device leakage [250]. As only limited data are available, this is speculative.

8.5.6 Summary

LAA closure has emerged as a promising alternative to oral anticoagulant in patients with atrial fibrillation. While randomized clinical data in this area are limited to WATCHMAN devices in patients eligible for oral anticoagulant, real-world registries show an obvious shift toward higher-risk populations with contraindications to oral anticoagulant or considered to be at risk of severe bleeding as the long-term efficacy and safety of the procedure continues to improve. However, in this challenging situation, the lack of consistent evidence from randomized trials may have hindered stronger support from the International Heart Association and the wider dissemination of this technology. Ongoing randomized trials focusing on unresolved issues, such as LAA closure in patients not eligible for oral anticoagulant, head-to-head comparisons with the gold standard direct oral anticoagulant, emerging devices, or the safety of combined procedures, may facilitate the inevitable growth of LAA closure procedures in the near future.

Acknowledgments The authors would like to express their gratitude to EditSprings (https://www.editsprings.com/) for the expert linguistic services provided.

References

1. Liu Y, Chen S, Zühlke L, Black GC, Choy MK, Li N, et al. Global birth prevalence of congenital heart defects 1970–2017: updated systematic review and meta-analysis of 260 studies. Int J Epidemiol. 2019;48(2):455–63.
2. Nyboe C, Karunanithi Z, Nielsen-Kudsk JE, Hjortdal VE. Long-term mortality in patients with atrial septal defect: a nationwide cohort-study. Eur Heart J. 2018;39(12):993–8.
3. Kloesel B, DiNardo JA, Body SC. Cardiac embryology and molecular mechanisms of congenital heart disease: a primer for anesthesiologists. Anesth Analg. 2016;123(3):551–69.
4. Sinha R, Nema C. Rare cardiac defect in Holt-Oram syndrome. Cardiovasc J Afr. 2012;23(2):e3–4.
5. Aypar E, Atalay S, Tutar E, Demir F. Unusual cardiac phenotype in a newborn with noonan syndrome. Congenit Heart Dis. 2010;5(2):178–81.
6. Diegeler A, Van Son JA, Mohr FW. Budd-Chiari syndrome as late complication of secundum atrial septal defect closure. Eur J Cardiothorac Surg. 1997;12(3):501–3.
7. Baujat G, Le Merrer M. Ellis-van Creveld syndrome. Orphanet J Rare Dis. 2007;2:27.
8. Stoll C, Dott B, Alembik Y, Roth MP. Associated congenital anomalies among cases with Down syndrome. Eur J Med Genet. 2015;58(12):674–80.
9. Fuse S, Tomita H, Hatakeyama K, Kubo N, Abe N. Effect of size of a secundum atrial septal defect on shunt volume. Am J Cardiol. 2001;88(12):1447–50, a9.
10. Geva T, Martins JD, Wald RM. Atrial septal defects. Lancet. 2014;383(9932):1921–32.

11. Stout KK, Daniels CJ, Aboulhosn JA, Bozkurt B, Broberg CS, Colman JM, et al. 2018 AHA/ACC Guideline for the Management of adults with congenital heart Disease: a report of the American College of Cardiology/American Heart Association Task Force on Clinical Practice Guidelines. J Am Coll Cardiol. 2019;73(12):e81–e192.
12. Galiè N, Hoeper MM, Humbert M, Torbicki A, Vachiery JL, Barbera JA, et al. Guidelines for the diagnosis and treatment of pulmonary hypertension: the Task Force for the Diagnosis and Treatment of Pulmonary Hypertension of the European Society of Cardiology (ESC) and the European Respiratory Society (ERS), endorsed by the International Society of Heart and Lung Transplantation (ISHLT). Eur Heart J. 2009;30(20):2493–537.
13. Du ZD, Hijazi ZM, Kleinman CS, Silverman NH, Larntz K. Comparison between transcatheter and surgical closure of secundum atrial septal defect in children and adults: results of a multicenter nonrandomized trial. J Am Coll Cardiol. 2002;39(11):1836–44.
14. Moore JW, Vincent RN, Beekman RH 3rd, Benson L, Bergersen L, Holzer R, et al. Procedural results and safety of common interventional procedures in congenital heart disease: initial report from the National Cardiovascular Data Registry. J Am Coll Cardiol. 2014;64(23): 2439–51.
15. Turner DR, Owada CY, Sang CJ Jr, Khan M, Lim DS. Closure of secundum atrial septal defects with the AMPLATZER septal occluder: a prospective, multicenter, post-approval study. Circ Cardiovasc Interv. 2017;10(8):e004212.
16. McElhinney DB, Quartermain MD, Kenny D, Alboliras E, Amin Z. Relative risk factors for cardiac erosion following transcatheter closure of atrial septal defects: a case-control study. Circulation. 2016;133(18):1738–46.
17. Amin Z. Echocardiographic predictors of cardiac erosion after Amplatzer septal occluder placement. Catheter Cardiovasc Interv. 2014;83(1):84–92.
18. Roberts WT, Parmar J, Rajathurai T. Very late erosion of Amplatzer septal occluder device presenting as pericardial pain and effusion 8 years after placement. Catheter Cardiovasc Interv. 2013;82(4):E592–E4.
19. Mallula K, Amin Z. Recent changes in instructions for use for the Amplatzer atrial septal defect occluder: how to incorporate these changes while using transesophageal echocardiography or intracardiac echocardiography? Pediatr Cardiol. 2012;33(7):995–1000.
20. Happel CM, Laser KT, Sigler M, Kececioglu D, Sandica E, Haas NA. Single center experience: implantation failures, early, and late complications after implantation of a partially biodegradable ASD/PFO-device (BioStar®). Catheter Cardiovasc Interv. 2015;85(6):990–7.
21. Grohmann J, Höhn R, Fleck T, Schmoor C, Stiller B. Transcatheter closure of atrial septal defects in children and adolescents: single-center experience with the GORE® septal occluder. Catheter Cardiovasc Interv. 2014;84(6):E51–7.
22. Lombardi M, Tagliente MR, Pirolo T, Massari E, Milella L, Vairo U. Feasibility and safety of a new generation of gore septal occluder device in children. J Cardiovasc Med (Hagerstown). 2016;17(Suppl 2):e249–e51.
23. Søndergaard L, Loh PH, Franzen O, Ihlemann N, Vejlstrup N. The first clinical experience with the new GORE® septal occluder (GSO). EuroIntervention. 2013;9(8):959–63.
24. de Hemptinne Q, Horlick EM, Osten MD, Millán X, Tadros VX, Pighi M, et al. Initial clinical experience with the GORE(®) CARDIOFORM ASD occluder for transcatheter atrial septal defect closure. Catheter Cardiovasc Interv. 2017;90(3):495–503.
25. Sommer RJ, Love BA, Paolillo JA, Gray RG, Goldstein BH, Morgan GJ, et al. ASSURED clinical study: new GORE® CARDIOFORM ASD occluder for transcatheter closure of atrial septal defect. Catheter Cardiovasc Interv. 2020;95(7):1285–95.
26. Haas NA, Happel CM, Soetemann DB, Hanslik A, Moysich A, Kececioglu D, et al. Optimal septum alignment of the Figulla Flex occluder to the atrial septum in patients with secundum atrial septal defects. EuroIntervention. 2016;11(10):1153–60.
27. Haas NA, Soetemann DB, Ates I, Baspinar O, Ditkivskyy I, Duke C, et al. Closure of secundum atrial septal defects by using the occlutech occluder devices in more than 1300

patients: the IRFACODE project: a retrospective case series. Catheter Cardiovasc Interv. 2016;88(4):571–81.

28. Cansel M, Pekdemir H, Yağmur J, Tasolar H, Ermis N, Kurtoglu E, et al. Early single clinical experience with the new Figulla ASD Occluder for transcatheter closure of atrial septal defect in adults. Arch Cardiovasc Dis. 2011;104(3):155–60.
29. Godart F, Houeijeh A, Recher M, Francart C, Polge AS, Richardson M, et al. Transcatheter closure of atrial septal defect with the Figulla(®) ASD Occluder: a comparative study with the Amplatzer(®) Septal Occluder. Arch Cardiovasc Dis. 2015;108(1):57–63.
30. Kenny D, Eicken A, Dähnert I, Boudjemline Y, Sievert H, Schneider MB, et al. A randomized, controlled, multi-center trial of the efficacy and safety of the Occlutech Figulla Flex-II Occluder compared to the Amplatzer septal occluder for transcatheter closure of secundum atrial septal defects. Catheter Cardiovasc Interv. 2019;93(2):316–21.
31. Bhattacharjya S, Pillai LS, Doraiswamy V, Satyanarayana RM, Chandrasekaran R, Pavithran S, et al. Prospective concurrent head-to head comparison of three different types of nitinol occluder device for transcatheter closure of secundum atrial septal defects. EuroIntervention. 2019;15(4):e321–e8.
32. Kim AY, Jung SY, Chang JY, Jung JW, Choi JY. Early to mid-term follow-up outcomes of percutaneous closure of atrial septal defects using recent Generation devices: a single-center experience. Korean Circ J. 2019;49(4):326–35.
33. Snijder RJR, Renes LE, Bosshardt D, Suttorp MJ, Ten Berg JM, Post MC. Percutaneous atrial septal defect closure using the occlutech figulla device in adults: more than 800 patient-years of follow-up. J Interv Cardiol. 2020;2020:7136802.
34. Astarcioglu MA, Kalcik M, Sen T, Aykan AC, Gokdeniz T, Gursoy OM, et al. Ceraflex versus Amplatzer occluder for secundum atrial septal defect closure. Multicenter clinical experience. Herz. 2015;40(Suppl 2):146–50.
35. Aydın Şahin D, Başpınar O, Sülü A, Karslıgil T, Kul S. A comparison of the in vivo neoendothelialization and wound healing processes of three atrial septal defect occluders used during childhood in a nonrandomized prospective trial. Anatol J Cardiol. 2017;18(3): 229–34.
36. Apostolopoulou SC, Tsoutsinos A, Laskari C, Kiaffas M, Rammos S. Large single centre experience with the Cera™ and CeraFlex™ occluders for closure of interatrial communications: usefulness of the flexible rotation feature. Cardiovasc Interv Ther. 2018;33(1):70–6.
37. Mullen MJ, Hildick-Smith D, De Giovanni JV, Duke C, Hillis WS, Morrison WL, et al. BioSTAR Evaluation STudy (BEST): a prospective, multicenter, phase I clinical trial to evaluate the feasibility, efficacy, and safety of the BioSTAR bioabsorbable septal repair implant for the closure of atrial-level shunts. Circulation. 2006;114(18):1962–7.
38. Morgan G, Lee KJ, Chaturvedi R, Benson L. A biodegradable device (BioSTAR) for atrial septal defect closure in children. Catheter Cardiovasc Interv. 2010;76(2):241–5.
39. Sigler M, Söderberg B, Schmitt B, Mellmann A, Bernhard J. Carag bioresorbable septal occluder (CBSO): histopathology of experimental implants. EuroIntervention. 2018;13(14): 1655–61.
40. Huang XM, Zhu YF, Cao J, Hu JQ, Bai Y, Jiang HB, et al. Development and preclinical evaluation of a biodegradable ventricular septal defect occluder. Catheter Cardiovasc Interv. 2013;81(2):324–30.
41. Li BN, Xie YM, Xie ZF, Chen XM, Zhang G, Zhang DY, et al. Study of biodegradable occluder of atrial septal defect in a porcine model. Catheter Cardiovasc Interv. 2019;93(1): E38–e45.
42. Peirone A, Contreras A, Ferrero A, da Costa RN, Pedra SF, Pedra CA. Immediate and short-term outcomes after percutaneous atrial septal defect closure using the new nit-occlud ASD-R device. Catheter Cardiovasc Interv. 2014;84(3):464–70.
43. Bulut MO, Yucel IK, Kucuk M, Balli S, Basar EZ, Celebi A. Initial experience with the Nit-Occlud ASD-R: short-term results. Pediatr Cardiol. 2016;37(7):1258–65.

44. Heath A, Alvensleben IV, Navarro J, Echazú G, Kozlik-Feldmann R, Freudenthal F. Developing high medical technology, a challenge for developing countries: the percutaneous closure of atrial septal defects using Nit-Occlud ASD-R: early and mid-term results. World J Pediatr Congenit Heart Surg. 2019;10(4):433–9.
45. Kister T, Dähnert I, Lurz P. Fatal erosion atrial septal defect device. Catheter Cardiovasc Interv. 2016;87(5):951–4.
46. Jalal Z, Hascoet S, Baruteau AE, Iriart X, Kreitmann B, Boudjemline Y, et al. Long-term complications after transcatheter atrial septal defect closure: a review of the medical literature. Can J Cardiol. 2016;32(11):1315.e11–8.
47. Tanghöj G, Odermarsky M, Naumburg E, Liuba P. Early complications After percutaneous closure of atrial septal defect in infants with procedural weight less than 15 kg. Pediatr Cardiol. 2017;38(2):255–63.
48. Sherman JM, Hagler DJ, Cetta F. Thrombosis after septal closure device placement: a review of the current literature. Catheter Cardiovasc Interv. 2004;63(4):486–9.
49. Kalaycı B, Kalaycı S. Right atrial ball thrombus associated with atrial septal occluder device: a late complication of transcatheter atrial septal defect closure. Anatol J Cardiol. 2017;18(5):E9.
50. Abaci A, Unlu S, Alsancak Y, Kaya U, Sezenoz B. Short and long term complications of device closure of atrial septal defect and patent foramen ovale: meta-analysis of 28,142 patients from 203 studies. Catheter Cardiovasc Interv. 2013;82(7):1123–38.
51. Delaney JW, Li JS, Rhodes JF. Major complications associated with transcatheter atrial septal occluder implantation: a review of the medical literature and the manufacturer and user facility device experience (MAUDE) database. Congenit Heart Dis. 2007;2(4):256–64.
52. Krumsdorf U, Ostermayer S, Billinger K, Trepels T, Zadan E, Horvath K, et al. Incidence and clinical course of thrombus formation on atrial septal defect and patient foramen ovale closure devices in 1,000 consecutive patients. J Am Coll Cardiol. 2004;43(2):302–9.
53. Baumgartner H, De Backer J, Babu-Narayan SV, Budts W, Chessa M, Diller GP, et al. 2020 ESC Guidelines for the management of adult congenital heart disease. Eur Heart J. 2020;42(6):563–645.
54. Olabiyi OO, Morales DL, Franklin WJ. De novo thrombus on an atrial septal defect device 3 years after its implantation. Pediatr Cardiol. 2013;34(5):1269–71.
55. Uysal F, Bostan ÖM, Şenkaya Sığnak I, Güneş M, Çil E. Huge thrombus formation 1 year after percutaneous closure of an atrial septal defect with an Amplatzer septal occluder. Anatol J Cardiol. 2016;16(1):63–4.
56. Snijder RJ, Suttorp MJ, Berg JM, Post MC. Percutaneous closure of secundum type atrial septal defects: more than 5-year follow-up. World J Cardiol. 2015;7(3):150–6.
57. Furuta A, Nagashima M, Sugiyama H, Sakamoto T, Yamazaki K. Thrombus formation 4 years after percutaneous transcatheter closure of an atrial septal defect. J Card Surg. 2017;32(8):492–3.
58. Amin Z, Hijazi ZM, Bass JL, Cheatham JP, Hellenbrand WE, Kleinman CS. Erosion of Amplatzer septal occluder device after closure of secundum atrial septal defects: review of registry of complications and recommendations to minimize future risk. Catheter Cardiovasc Interv. 2004;63(4):496–502.
59. Diab K, Kenny D, Hijazi ZM. Erosions, erosions, and erosions! Device closure of atrial septal defects: how safe is safe? Catheter Cardiovasc Interv. 2012;80(2):168–74.
60. Butera G, Romagnoli E, Carminati M, Chessa M, Piazza L, Negura D, et al. Treatment of isolated secundum atrial septal defects: impact of age and defect morphology in 1,013 consecutive patients. Am Heart J. 2008;156(4):706–12.
61. El-Said HG, Moore JW. Erosion by the Amplatzer septal occluder: experienced operator opinions at odds with manufacturer recommendations? Catheter Cardiovasc Interv. 2009;73(7):925–30.
62. Dickison P, Harris V, Smith SD. Nickel hypersensitivity following closure of atrial septal defect: a case report and review of the literature. Australas J Dermatol. 2018;59(3):220–2.

63. Lai DW, Saver JL, Araujo JA, Reidl M, Tobis J. Pericarditis associated with nickel hypersensitivity to the Amplatzer occluder device: a case report. Catheter Cardiovasc Interv. 2005;66 (3):424–6.
64. Belohlavek J, Belohlavkova S, Hlubocky J, Mrazek V, Linhart A, Podzimek S. Severe allergic dermatitis after closure of foramen ovale with Amplatzer occluder. Ann Thorac Surg. 2013;96 (3):e57–9.
65. Reddy BT, Patel JB, Powell DL, Michaels AD. Interatrial shunt closure devices in patients with nickel allergy. Catheter Cardiovasc Interv. 2009;74(4):647–51.
66. Schalock PC, Crawford G, Nedorost S, Scheinman PL, Atwater AR, Mowad C, et al. Patch testing for evaluation of hypersensitivity to implanted metal devices: a perspective from the American Contact Dermatitis Society. Dermatitis. 2016;27(5):241–7.
67. Rigatelli G, Cardaioli P, Giordan M, Aggio S, Chinaglia M, Braggion G, et al. Nickel allergy in interatrial shunt device-based closure patients. Congenit Heart Dis. 2007;2(6):416–20.
68. Wertman B, Azarbal B, Riedl M, Tobis J. Adverse events associated with nickel allergy in patients undergoing percutaneous atrial septal defect or patent foramen ovale closure. J Am Coll Cardiol. 2006;47(6):1226–7.
69. Kim HJ, Shin JU, Lee J, Lee H, Jin S, Kim SH, et al. Positive reactions to nickel on a patch test do not predict clinical outcome of nickel alloy-based atrial septal defect occluder implantation. Dermatology. 2015;230(2):184–8.
70. Bahrani B, Moghaddam N, DeKoven J. Cross-sectional survey of nickel allergy management in the context of intracardiac device implantation. Dermatitis. 2019;30(3):213–21.
71. Fernandes P, Sharma SR, Magee A, Michielon G, Fraisse A. Severe migraine associated with nickel allergy requiring surgical removal of atrial septal device. Ann Thorac Surg. 2019;108 (3):e183–e4.
72. Johnson JN, Marquardt ML, Ackerman MJ, Asirvatham SJ, Reeder GS, Cabalka AK, et al. Electrocardiographic changes and arrhythmias following percutaneous atrial septal defect and patent foramen ovale device closure. Catheter Cardiovasc Interv. 2011;78(2):254–61.
73. Wang Y, Hua Y, Li L, Wang X, Qiao L, Shi X, et al. Risk factors and prognosis of atrioventricular block after atrial septum defect closure using the Amplatzer device. Pediatr Cardiol. 2014;35(3):550–5.
74. Suda K, Raboisson MJ, Piette E, Dahdah NS, Miró J. Reversible atrioventricular block associated with closure of atrial septal defects using the Amplatzer device. J Am Coll Cardiol. 2004;43(9):1677–82.
75. Lin SM, Hwang HK, Chen MR. Amplatzer septal occluder-induced transient complete atrioventricular block. J Formos Med Assoc. 2007;106(12):1052–6.
76. Celiker A, Ozkutlu S, Karakurt C, Karagöz T. Cardiac dysrhythmias after transcatheter closure of ASD with Amplatzer device. Turk J Pediatr. 2005;47(4):323–6.
77. Amoozgar H, Ahmadipoor M, Amirghofran AA. Complete heart block following transcatheter closure of atrial septal defect due to growth of inflammatory tissue. Pediatr Cardiol. 2014;35 (7):1301–3.
78. Dittrich S, Sigler M, Priessmann H. Late complete atrioventricular block after closure of an atrial septal defect with a gore septal occluder (GSO™). Catheter Cardiovasc Interv. 2016;87 (5):945–50.
79. Morton JB, Sanders P, Vohra JK, Sparks PB, Morgan JG, Spence SJ, et al. Effect of chronic right atrial stretch on atrial electrical remodeling in patients with an atrial septal defect. Circulation. 2003;107(13):1775–82.
80. Gatzoulis MA, Freeman MA, Siu SC, Webb GD, Harris L. Atrial arrhythmia after surgical closure of atrial septal defects in adults. N Engl J Med. 1999;340(11):839–46.
81. John Sutton MG, Tajik AJ, McGoon DC. Atrial septal defect in patients ages 60 years or older: operative results and long-term postoperative follow-up. Circulation. 1981;64(2):402–9.
82. Alnasser S, Lee D, Austin PC, Labos C, Osten M, Lightfoot DT, et al. Long term outcomes among adults post transcatheter atrial septal defect closure: systematic review and meta-analysis. Int J Cardiol. 2018;270:126–32.

83. Vecht JA, Saso S, Rao C, Dimopoulos K, Grapsa J, Terracciano CM, et al. Atrial septal defect closure is associated with a reduced prevalence of atrial tachyarrhythmia in the short to medium term: a systematic review and meta-analysis. Heart. 2010;96(22):1789–97.
84. Roos-Hesselink JW, Meijboom FJ, Spitaels SE, van Domburg R, van Rijen EH, Utens EM, et al. Excellent survival and low incidence of arrhythmias, stroke and heart failure long-term after surgical ASD closure at young age. A prospective follow-up study of 21–33 years. Eur Heart J. 2003;24(2):190–7.
85. Webb G, Gatzoulis MA. Atrial septal defects in the adult: recent progress and overview. Circulation. 2006;114(15):1645–53.
86. Nyboe C, Olsen MS, Nielsen-Kudsk JE, Hjortdal VE. Atrial fibrillation and stroke in adult patients with atrial septal defect and the long-term effect of closure. Heart. 2015;101(9): 706–11.
87. Warnes CA, Williams RG, Bashore TM, Child JS, Connolly HM, Dearani JA, et al. ACC/AHA 2008 guidelines for the management of adults with congenital heart disease: a report of the American College of Cardiology/American Heart Association Task Force on Practice Guidelines (Writing Committee to Develop Guidelines on the Management of Adults With Congenital Heart Disease). Developed in Collaboration With the American Society of Echocardiography, Heart Rhythm Society, International Society for Adult Congenital Heart Disease, Society for Cardiovascular Angiography and Interventions, and Society of Thoracic Surgeons. J Am Coll Cardiol. 2008;52(23):e143–263.
88. Li X, Wissner E, Kamioka M, Makimoto H, Rausch P, Metzner A, et al. Safety and feasibility of transseptal puncture for atrial fibrillation ablation in patients with atrial septal defect closure devices. Heart Rhythm. 2014;11(2):330–5.
89. Amedro P, Soulatges C, Fraisse A. Infective endocarditis after device closure of atrial septal defects: case report and review of the literature. Catheter Cardiovasc Interv. 2017;89(2): 324–34.
90. Jha NK, Kiraly L, Murala JS, Tamas C, Talo H, El Badaoui H, et al. Late endocarditis of Amplatzer atrial septal occluder device in a child. World J Cardiol. 2015;7(10):703–6.
91. Kim DJ, Shim CY, You SC, Lee SH, Hong GR. Late bacterial endocarditis and abscess formation after implantation of an Amplatzer septal occluder device. Circulation. 2015;131 (25):e536–8.
92. Thibodeau-Jarry N, Ibrahim R, Ducharme A, Sia YT. Late infection of an atrial septal defect closure device: a possible complication. Can J Cardiol. 2015;31(9):1204.e9–11.
93. Białkowski J, Pawlak S, Banaszak P. Incomplete endothelialisation of an Amplatzer septal occluder device followed by meningitis and late acute bacterial endocarditis. Cardiol Young. 2016;26(4):808–10.
94. Ng J, Downton T, Davidson N, Marangou J. Corynebacterium diphtheriae-infective endocarditis in a patient with an atrial septal defect closure device. BMJ Case Rep. 2019;12(5): e229478.
95. Lock JE, Rome JJ, Davis R, Van Praagh S, Perry SB, Van Praagh R, et al. Transcatheter closure of atrial septal defects. Experimental studies. Circulation. 1989;79(5):1091–9.
96. Wilson W, Taubert KA, Gewitz M, Lockhart PB, Baddour LM, Levison M, et al. Prevention of infective endocarditis: guidelines from the American Heart Association: a guideline from the American Heart Association Rheumatic Fever, Endocarditis, and Kawasaki Disease Committee, Council on Cardiovascular Disease in the Young, and the Council on Clinical Cardiology, Council on Cardiovascular Surgery and Anesthesia, and the Quality of Care and Outcomes Research Interdisciplinary Working Group. Circulation. 2007;116(15):1736–54.
97. Kawamura A, Kigasawa H, Kamma H. Autopsy findings of Amplatzer septal occluder at 5 months after closure of atrial septal defect: how long does it take to be endothelialized? J Invasive Cardiol. 2013;25(8):E167–8.
98. Moore J, Hegde S, El-Said H, Beekman R 3rd, Benson L, Bergersen L, et al. Transcatheter device closure of atrial septal defects: a safety review. JACC Cardiovasc Interv. 2013;6(5): 433–42.

99. Levi DS, Moore JW. Embolization and retrieval of the Amplatzer septal occluder. Catheter Cardiovasc Interv. 2004;61(4):543–7.
100. DiBardino DJ, McElhinney DB, Kaza AK, Mayer JE Jr. Analysis of the US Food and Drug Administration Manufacturer and User Facility Device Experience database for adverse events involving Amplatzer septal occluder devices and comparison with the Society of Thoracic Surgery congenital cardiac surgery database. J Thorac Cardiovasc Surg. 2009;137(6):1334–41.
101. Koçyildirim E, Kanani M, Bonhoeffer P, Elliott MJ. Amplatzer device embolization: hazards of multiple attempts at catheter retrieval. Anadolu Kardiyol Derg. 2007;7(3):329–30.
102. Kodaira M, Kawamura A, Okamoto K, Kanazawa H, Minakata Y, Murata M, et al. Comparison of clinical outcomes after transcatheter vs. minimally invasive cardiac surgery closure for atrial septal defect. Circ J. 2017;81(4):543–51.
103. Meissner I, Khandheria BK, Heit JA, Petty GW, Sheps SG, Schwartz GL, et al. Patent foramen ovale: innocent or guilty? Evidence from a prospective population-based study. J Am Coll Cardiol. 2006;47(2):440–5.
104. Di Tullio MR. Patent foramen ovale: echocardiographic detection and clinical relevance in stroke. J Am Soc Echocardiogr. 2010;23(2):144–55. quiz 220
105. Penther P. Patent foramen ovale: an anatomical study. Apropos of 500 consecutive autopsies. Arch Mal Coeur Vaiss. 1994;87(1):15–21.
106. Hagen PT, Scholz DG, Edwards WD. Incidence and size of patent foramen ovale during the first 10 decades of life: an autopsy study of 965 normal hearts. Mayo Clin Proc. 1984;59(1):17–20.
107. Siu SC, Silversides CK. Bicuspid aortic valve disease. J Am Coll Cardiol. 2010;55(25):2789–800.
108. Rundek T, Elkind MS, Di Tullio MR, Carrera E, Jin Z, Sacco RL, et al. Patent foramen ovale and migraine: a cross-sectional study from the Northern Manhattan Study (NOMAS). Circulation. 2008;118(14):1419–24.
109. van Vonderen JJ, Roest AA, Siew ML, Walther FJ, Hooper SB, te Pas AB. Measuring physiological changes during the transition to life after birth. Neonatology. 2014;105(3):230–42.
110. Clark EB. Pathogenetic mechanisms of congenital cardiovascular malformations revisited. Semin Perinatol. 1996;20(6):465–72.
111. Homma S, Di Tullio MR, Sacco RL, Mihalatos D, Li Mandri G, Mohr JP. Characteristics of patent foramen ovale associated with cryptogenic stroke. A biplane transesophageal echocardiographic study. Stroke. 1994;25(3):582–6.
112. Hausmann D, Mügge A, Daniel WG. Identification of patent foramen ovale permitting paradoxic embolism. J Am Coll Cardiol. 1995;26(4):1030–8.
113. Turc G, Lee JY, Brochet E, Kim JS, Song JK, Mas JL. Atrial septal aneurysm, shunt size, and recurrent stroke risk in patients with patent foramen ovale. J Am Coll Cardiol. 2020;75(18):2312–20.
114. Yetkin E, Atalay H, Ileri M. Atrial septal aneurysm: prevalence and covariates in adults. Int J Cardiol. 2016;223:656–9.
115. Olivares-Reyes A, Chan S, Lazar EJ, Bandlamudi K, Narla V, Ong K. Atrial septal aneurysm: a new classification in two hundred five adults. J Am Soc Echocardiogr. 1997;10(6):644–56.
116. Di Tullio MR, Sacco RL, Sciacca RR, Jin Z, Homma S. Patent foramen ovale and the risk of ischemic stroke in a multiethnic population. J Am Coll Cardiol. 2007;49(7):797–802.
117. Windecker S, Stortecky S, Meier B. Paradoxical embolism. J Am Coll Cardiol. 2014;64(4):403–15.
118. Nakayama R, Takaya Y, Akagi T, Watanabe N, Ikeda M, Nakagawa K, et al. Identification of high-risk patent foramen ovale associated with cryptogenic stroke: development of a scoring system. J Am Soc Echocardiogr. 2019;32(7):811–6.

119. Mas JL, Arquizan C, Lamy C, Zuber M, Cabanes L, Derumeaux G, et al. Recurrent cerebrovascular events associated with patent foramen ovale, atrial septal aneurysm, or both. N Engl J Med. 2001;345(24):1740–6.
120. Homma S, Sacco RL, Di Tullio MR, Sciacca RR, Mohr JP. Atrial anatomy in non-cardioembolic stroke patients: effect of medical therapy. J Am Coll Cardiol. 2003;42 (6):1066–72.
121. Schneider B, Hofmann T, Justen MH, Meinertz T. Chiari's network: normal anatomic variant or risk factor for arterial embolic events? J Am Coll Cardiol. 1995;26(1):203–10.
122. Cramer SC, Rordorf G, Kaufman JA, Buonanno F, Koroshetz WJ, Schwamm L. Clinically occult pelvic-vein thrombosis in cryptogenic stroke. Lancet. 1998;351(9120):1927–8.
123. Miranda B, Fonseca AC, Ferro JM. Patent foramen ovale and stroke. J Neurol. 2018;265(8): 1943–9.
124. Meacham RR 3rd, Headley AS, Bronze MS, Lewis JB, Rester MM. Impending paradoxical embolism. Arch Intern Med. 1998;158(5):438–48.
125. Furlan AJ, Reisman M, Massaro J, Mauri L, Adams H, Albers GW, et al. Closure or medical therapy for cryptogenic stroke with patent foramen ovale. N Engl J Med. 2012;366(11):991–9.
126. Meier B, Kalesan B, Mattle HP, Khattab AA, Hildick-Smith D, Dudek D, et al. Percutaneous closure of patent foramen ovale in cryptogenic embolism. N Engl J Med. 2013;368(12): 1083–91.
127. Carroll JD, Saver JL, Thaler DE, Smalling RW, Berry S, MacDonald LA, et al. Closure of patent foramen ovale versus medical therapy after cryptogenic stroke. N Engl J Med. 2013;368 (12):1092–100.
128. Saver JL, Carroll JD, Thaler DE, Smalling RW, MacDonald LA, Marks DS, et al. Long-term outcomes of patent foramen ovale closure or medical therapy after stroke. N Engl J Med. 2017;377(11):1022–32.
129. Søndergaard L, Kasner SE, Rhodes JF, Andersen G, Iversen HK, Nielsen-Kudsk JE, et al. Patent foramen ovale closure or antiplatelet therapy for cryptogenic stroke. N Engl J Med. 2017;377(11):1033–42.
130. Mas JL, Derumeaux G, Guillon B, Massardier E, Hosseini H, Mechtouff L, et al. Patent foramen ovale closure or anticoagulation vs. antiplatelets after stroke. N Engl J Med. 2017;377 (11):1011–21.
131. Messé SR, Gronseth G, Kent DM, Kizer JR, Homma S, Rosterman L, et al. Practice advisory: recurrent stroke with patent foramen ovale (update of practice parameter): report of the guideline development, dissemination, and implementation subcommittee of the American Academy of Neurology. Neurology. 2016;87(8):815–21.
132. Wein T, Lindsay MP, Côté R, Foley N, Berlingieri J, Bhogal S, et al. Canadian stroke best practice recommendations: secondary prevention of stroke, sixth edition practice guidelines, update 2017. Int J Stroke. 2018;13(4):420–43.
133. Pristipino C, Sievert H, D'Ascenzo F, Louis Mas J, Meier B, Scacciatella P, et al. European position paper on the management of patients with patent foramen ovale. General approach and left circulation thromboembolism. Eur Heart J. 2019;40(38):3182–95.
134. De Rosa S, Sievert H, Sabatino J, Polimeni A, Sorrentino S, Indolfi C. Percutaneous closure versus medical treatment in stroke patients with patent foramen ovale: a systematic review and meta-analysis. Ann Intern Med. 2018;168(5):343–50.
135. Correction: device closure versus medical therapy alone for patent foramen ovale. Ann Intern Med. 2018;169(6):428.
136. Gevorgyan Fleming R, Kumar P, West B, Noureddin N, Rusheen J, Aboulhosn J, et al. Comparison of residual shunt rate and complications across 6 different closure devices for patent foramen ovale. Catheter Cardiovasc Interv. 2020;95(3):365–72.
137. Lee PH, Song JK, Kim JS, Heo R, Lee S, Kim DH, et al. Cryptogenic stroke and high-risk patent foramen ovale: the DEFENSE-PFO trial. J Am Coll Cardiol. 2018;71(20):2335–42.

138. Trabattoni D, Gaspardone A, Sgueglia GA, Fabbiocchi F, Gioffrè G, Montorsi P, et al. AMPLATZER versus Figulla occluder for transcatheter patent foramen ovale closure. EuroIntervention. 2017;12(17):2092–9.
139. Tsivgoulis G, Katsanos AH, Mavridis D, Frogoudaki A, Vrettou AR, Ikonomidis I, et al. Percutaneous patent foramen ovale closure for secondary stroke prevention: network meta-analysis. Neurology. 2018;91(1):e8–e18.
140. Hildick-Smith D, Williams T, MacCarthy P, Melikian N, Monaghan M, Spence M, et al. Occlutech percutaneous patent foramen ovale closure: safety and efficacy registry (OPPOSE). Int J Cardiol. 2017;245:99–104.
141. Hardt SE, Eicken A, Berger F, Schubert S, Carminati M, Butera G, et al. Closure of patent foramen ovale defects using GORE® CARDIOFORM septal occluder: results from a prospective European multicenter study. Catheter Cardiovasc Interv. 2017;90(5):824–9.
142. Zhao E, Liu W, Zhang Y, Cheng G, Du Y, He L, et al. Safety and efficacy of Cardi-O-fix occluder for percutaneous closure of a patent foramen ovale: a single-center prospective study. Medicine (Baltimore). 2017;96(14):e6572.
143. Han Y, Zhang X, Zhang F. Patent foramen ovale closure by using transesophageal echocardiography for cryptogenic stroke: single center experience in 132 consecutive patients. J Cardiothorac Surg. 2020;15(1):11.
144. Zimmermann WJ, Heinisch C, Majunke N, Staubach S, Russell S, Wunderlich N, et al. Patent foramen ovale closure with the SeptRx device initial experience with the first "In-Tunnel" device. JACC Cardiovasc Interv. 2010;3(9):963–7.
145. Elgendy AY, Elgendy IY, Mojadidi MK, Mahmoud AN, Barry JS, Jneid H, et al. New-onset atrial fibrillation following percutaneous patent foramen ovale closure: a systematic review and meta-analysis of randomised trials. EuroIntervention. 2019;14(17):1788–90.
146. Hopkins MK, Goldstein SA, Ward CC, Kuller JA. Evaluation and management of maternal congenital heart disease: a review. Obstet Gynecol Surv. 2018;73(2):116–24.
147. Kenny D. Interventional cardiology for congenital heart disease. Korean Circ J. 2018;48(5):350–64.
148. Pinto NM, Waitzman N, Nelson R, Minich LL, Krikov S, Botto LD. Early childhood inpatient costs of critical congenital heart disease. J Pediatr. 2018;203:371–9.e7.
149. Muthialu N, Balakrishnan S, Sundar R. Single patch closure of multiple VSDs through right atrial approach. Indian Heart J. 2018;70(4):578–9.
150. Durden RE, Turek JW, Reinking BE, Bansal M. Acquired VSD due to infective endocarditis. Ann Pediatr Cardiol. 2018;11(1):100–2.
151. Baker EJ, Leung MP, Anderson RH, Fischer DR, Zuberbuhler JR. The cross sectional anatomy of VSD: a reappraisal. Br Heart J. 1988;59(3):339–51.
152. Ho S, McCarthy KP, Josen M, Rigby ML. Anatomic-echocardiographic correlates: an introduction to normal and congenitally malformed hearts. Heart. 2001;86(Suppl 2):Ii3–11.
153. Jacobs JP, Burke RP, Quintessenza JA, Mavroudis C. Congenital Heart Surgery Nomenclature and Database Project: ventricular septal defect. Ann Thorac Surg. 2000;69(4 Suppl):S25–35.
154. Van Praagh R, Geva T, Kreutzer J. Ventricular septal defects: how shall we describe, name and classify them? J Am Coll Cardiol. 1989;14(5):1298–9.
155. Anderson RH, Wilcox BR. The surgical anatomy of ventricular septal defect. J Card Surg. 1992;7(1):17–35.
156. Fu YC, Bass J, Amin Z, Radtke W, Cheatham JP, Hellenbrand WE, et al. Transcatheter closure of perimembranous ventricular septal defects using the new Amplatzer membranous VSD occluder: results of the U.S. phase I trial. J Am Coll Cardiol. 2006;47(2):319–25.
157. Hijazi ZM, Hakim F, Haweleh AA, Madani A, Tarawna W, Hiari A, et al. Catheter closure of perimembranous ventricular septal defects using the new Amplatzer membranous VSD occluder: initial clinical experience. Catheter Cardiovasc Interv. 2002;56(4):508–15.
158. Bass JL, Kalra GS, Arora R, Masura J, Gavora P, Thanopoulos BD, et al. Initial human experience with the Amplatzer perimembranous ventricular septal occluder device. Catheter Cardiovasc Interv. 2003;58(2):238–45.

159. Butera G, Carminati M, Chessa M, Piazza L, Micheletti A, Negura DG, et al. Transcatheter closure of perimembranous ventricular septal defects: early and long-term results. J Am Coll Cardiol. 2007;50(12):1189–95.
160. Carminati M, Butera G, Chessa M, De Giovanni J, Fisher G, Gewillig M, et al. Transcatheter closure of congenital ventricular septal defects: results of the European Registry. Eur Heart J. 2007;28(19):2361–8.
161. Holzer R, de Giovanni J, Walsh KP, Tometzki A, Goh T, Hakim F, et al. Transcatheter closure of perimembranous ventricular septal defects using the amplatzer membranous VSD occluder: immediate and midterm results of an international registry. Catheter Cardiovasc Interv. 2006;68(4):620–8.
162. Predescu D, Chaturvedi RR, Friedberg MK, Benson LN, Ozawa A, Lee KJ. Complete heart block associated with device closure of perimembranous ventricular septal defects. J Thorac Cardiovasc Surg. 2008;136(5):1223–8.
163. Tzikas A, Ibrahim R, Velasco-Sanchez D, Freixa X, Alburquenque M, Khairy P, et al. Transcatheter closure of perimembranous ventricular septal defect with the Amplatzer(®) membranous VSD occluder 2: initial world experience and one-year follow-up. Catheter Cardiovasc Interv. 2014;83(4):571–80.
164. Amin Z, Gu X, Berry JM, Bass JL, Titus JL, Urness M, et al. New device for closure of muscular ventricular septal defects in a canine model. Circulation. 1999;100(3):320–8.
165. Thanopoulos BD, Tsaousis GS, Konstadopoulou GN, Zarayelyan AG. Transcatheter closure of muscular ventricular septal defects with the amplatzer ventricular septal defect occluder: initial clinical applications in children. J Am Coll Cardiol. 1999;33(5):1395–9.
166. Hijazi ZM, Hakim F, Al-Fadley F, Abdelhamid J, Cao QL. Transcatheter closure of single muscular ventricular septal defects using the amplatzer muscular VSD occluder: initial results and technical considerations. Catheter Cardiovasc Interv. 2000;49(2):167–72.
167. Holzer R, Balzer D, Cao QL, Lock K, Hijazi ZM. Device closure of muscular ventricular septal defects using the Amplatzer muscular ventricular septal defect occluder: immediate and mid-term results of a U.S. registry. J Am Coll Cardiol. 2004;43(7):1257–63.
168. Muthusamy K. Retrograde closure of perimembranous ventricular septal defect using muscular ventricular septal occluder: a single-center experience of a novel technique. Pediatr Cardiol. 2015;36(1):106–10.
169. Holzer R, Balzer D, Amin Z, Ruiz CE, Feinstein J, Bass J, et al. Transcatheter closure of postinfarction ventricular septal defects using the new Amplatzer muscular VSD occluder: results of a U.S. Registry. Catheter Cardiovasc Interv. 2004;61(2):196–201.
170. Egbe AC, Poterucha JT, Rihal CS, Taggart NW, Cetta F, Cabalka AK, et al. Transcatheter closure of postmyocardial infarction, iatrogenic, and postoperative ventricular septal defects: the Mayo Clinic experience. Catheter Cardiovasc Interv. 2015;86(7):1264–70.
171. El Said HG, Bratincsak A, Gordon BM, Moore JW. Closure of perimembranous ventricular septal defects with aneurysmal tissue using the Amplazter Duct Occluder I: lessons learned and medium term follow up. Catheter Cardiovasc Interv. 2012;80(6):895–903.
172. Landman G, Kipps A, Moore P, Teitel D, Meadows J. Outcomes of a modified approach to transcatheter closure of perimembranous ventricular septal defects. Catheter Cardiovasc Interv. 2013;82(1):143–9.
173. Koneti NR, Sreeram N, Penumatsa RR, Arramraj SK, Karunakar V, Trieschmann U. Transcatheter retrograde closure of perimembranous ventricular septal defects in children with the Amplatzer duct occluder II device. J Am Coll Cardiol. 2012;60(23):2421–2.
174. El-Sisi A, Sobhy R, Jaccoub V, Hamza H. Perimembranous ventricular septal defect device closure: choosing between Amplatzer duct occluder I and II. Pediatr Cardiol. 2017;38(3): 596–602.
175. Ghosh S, Sridhar A, Sivaprakasam M. Complete heart block following transcatheter closure of perimembranous VSD using amplatzer duct occluder II. Catheter Cardiovasc Interv. 2018;92 (5):921–4.

176. Narin N, Pamukcu O, Tuncay A, Baykan A, Sunkak S, Tasci O, et al. Percutaneous ventricular septal defect closure in patients under 1 year of age. Pediatr Cardiol. 2018;39(5):1009–15.
177. Haas NA, Kock L, Bertram H, Boekenkamp R, De Wolf D, Ditkivskyy I, et al. Interventional VSD-closure with the Nit-Occlud(®) Lê VSD-coil in 110 patients: early and midterm results of the EUREVECO-registry. Pediatr Cardiol. 2017;38(2):215–27.
178. Esteves CA, Solarewicz LA, Cassar R, Neves JR, Esteves V, Arrieta R. Occlusion of the perimembranous ventricular septal defect using CERA® devices. Catheter Cardiovasc Interv. 2012;80(2):182–7.
179. Yang R, Kong XQ, Sheng YH, Zhou L, Xu D, Yong YH, et al. Risk factors and outcomes of post-procedure heart blocks after transcatheter device closure of perimembranous ventricular septal defect. JACC Cardiovasc Interv. 2012;5(4):422–7.
180. Yip WC, Zimmerman F, Hijazi ZM. Heart block and empirical therapy after transcatheter closure of perimembranous ventricular septal defect. Catheter Cardiovasc Interv. 2005;66(3): 436–41.
181. Thakkar B, Patel N, Bohora S, Bhalodiya D, Singh T, Madan T, et al. Transcatheter device closure of perimembranous ventricular septal defect in children treated with prophylactic oral steroids: acute and mid-term results of a single-centre, prospective, observational study. Cardiol Young. 2016;26(4):669–76.
182. El-Sisi AM, Menaissy YM, Bekheet SA. Infective endocarditis following coil occlusion of perimembranous ventricular septal defect with the Nit-Occlud((®)) Le device. Ann Pediatr Cardiol. 2016;9(1):59–61.
183. Kassis I, Shachor-Meyouhas Y, Khatib I, Khoury A, Le TP, Lorber A. Kingella endocarditis after closure of ventricular septal defect with a transcatheter device. Pediatr Infect Dis J. 2012;31(1):105–6.
184. Scheuerman O, Bruckheimer E, Marcus N, Hoffer V, Garty BZ. Endocarditis after closure of ventricular septal defect by transcatheter device. Pediatrics. 2006;117(6):e1256–8.
185. Forsey JT, Elmasry OA, Martin RP. Patent arterial duct. Orphanet J Rare Dis. 2009;4:17.
186. Mitchell SC, Korones SB, Berendes HW. Congenital heart disease in 56,109 births. Incidence and natural history. Circulation. 1971;43(3):323–32.
187. Lloyd TR, Beekman RH 3rd. Clinically silent patent ductus arteriosus. Am Heart J. 1994;127 (6):1664–5.
188. Koch J, Hensley G, Roy L, Brown S, Ramaciotti C, Rosenfeld CR. Prevalence of spontaneous closure of the ductus arteriosus in neonates at a birth weight of 1000 grams or less. Pediatrics. 2006;117(4):1113–21.
189. Baruteau AE, Hascoët S, Baruteau J, Boudjemline Y, Lambert V, Angel CY, et al. Transcatheter closure of patent ductus arteriosus: past, present and future. Arch Cardiovasc Dis. 2014;107(2):122–32.
190. Schneider DJ, Moore JW. Patent ductus arteriosus. Circulation. 2006;114(17):1873–82.
191. Sudhakar P, Jose J, George OK. Contemporary outcomes of percutaneous closure of patent ductus arteriosus in adolescents and adults. Indian Heart J. 2018;70(2):308–15.
192. Weisz DE, Giesinger RE. Surgical management of a patent ductus arteriosus: is this still an option? Semin Fetal Neonatal Med. 2018;23(4):255–66.
193. de Boode WP, Kluckow M, McNamara PJ, Gupta S. Role of neonatologist-performed echocardiography in the assessment and management of patent ductus arteriosus physiology in the newborn. Semin Fetal Neonatal Med. 2018;23(4):292–7.
194. Thanopoulos BV, Eleftherakis N, Tzannos K, Stefanadis C, Giannopoulos A. Further experience with catheter closure of patent ductus arteriosus using the new Amplatzer duct occluder in children. Am J Cardiol. 2010;105(7):1005–9.
195. Thanopoulos BD, Hakim FA, Hiari A, Tsaousis GS, Paphitis C, Hijazi ZM. Patent ductus arteriosus equipment and technique. Amplatzer duct occluder: intermediate-term follow-up and technical considerations. J Interv Cardiol. 2001;14(2):247–54.
196. Yan C, Zhao S, Jiang S, Xu Z, Huang L, Zheng H, et al. Transcatheter closure of patent ductus arteriosus with severe pulmonary arterial hypertension in adults. Heart. 2007;93(4):514–8.

197. Masura J, Walsh KP, Thanopoulous B, Chan C, Bass J, Goussous Y, et al. Catheter closure of moderate- to large-sized patent ductus arteriosus using the new Amplatzer duct occluder: immediate and short-term results. J Am Coll Cardiol. 1998;31(4):878–82.
198. Sharafuddin MJ, Gu X, Titus JL, Sakinis AK, Pozza CH, Coleman CC, et al. Experimental evaluation of a new self-expanding patent ductus arteriosus occluder in a canine model. J Vasc Interv Radiol. 1996;7(6):877–87.
199. Dua J, Chessa M, Piazza L, Negura D, Micheletti A, Bussadori C, et al. Initial experience with the new Amplatzer Duct Occluder II. J Invasive Cardiol. 2009;21(8):401–5.
200. Forsey J, Kenny D, Morgan G, Hayes A, Turner M, Tometzki A, et al. Early clinical experience with the new Amplatzer Ductal Occluder II for closure of the persistent arterial duct. Catheter Cardiovasc Interv. 2009;74(4):615–23.
201. Boudjemline Y. Covidien micro vascular plug in congenital heart diseases and vascular anomalies: a new kid on the block for premature babies and older patients. Catheter Cardiovasc Interv. 2017;89(1):114–9.
202. Celiker A, Aypar E, Karagöz T, Dilber E, Ceviz N. Transcatheter closure of patent ductus arteriosus with Nit-Occlud coils. Catheter Cardiovasc Interv. 2005;65(4):569–76.
203. Huang Y, Kong JF, Venkatraman SS. Biomaterials and design in occlusion devices for cardiac defects: a review. Acta Biomater. 2014;10(3):1088–101.
204. Agrawal H, Waller BR 3rd, Surendan S, Sathanandam S. New patent ductus arteriosus closure devices and techniques. Interv Cardiol Clin. 2019;8(1):23–32.
205. Schnabel RB, Yin X, Gona P, Larson MG, Beiser AS, McManus DD, et al. 50 year trends in atrial fibrillation prevalence, incidence, risk factors, and mortality in the Framingham Heart Study: a cohort study. Lancet. 2015;386(9989):154–62.
206. Andersson T, Magnuson A, Bryngelsson IL, Frøbert O, Henriksson KM, Edvardsson N, et al. All-cause mortality in 272,186 patients hospitalized with incident atrial fibrillation 1995-2008: a Swedish nationwide long-term case-control study. Eur Heart J. 2013;34(14):1061–7.
207. Blackshear JL, Odell JA. Appendage obliteration to reduce stroke in cardiac surgical patients with atrial fibrillation. Ann Thorac Surg. 1996;61(2):755–9.
208. Manning WJ, Silverman DI, Keighley CS, Oettgen P, Douglas PS. Transesophageal echocardiographically facilitated early cardioversion from atrial fibrillation using short-term anticoagulation: final results of a prospective 4.5-year study. J Am Coll Cardiol. 1995;25(6): 1354–61.
209. Mitusch R, Siemens HJ, Garbe M, Wagner T, Sheikhzadeh A, Diederich KW. Detection of a hypercoagulable state in nonvalvular atrial fibrillation and the effect of anticoagulant therapy. Thromb Haemost. 1996;75(2):219–23.
210. Hart RG, Pearce LA, Aguilar MI. Meta-analysis: antithrombotic therapy to prevent stroke in patients who have nonvalvular atrial fibrillation. Ann Intern Med. 2007;146(12):857–67.
211. Ruff CT, Giugliano RP, Braunwald E, Hoffman EB, Deenadayalu N, Ezekowitz MD, et al. Comparison of the efficacy and safety of new oral anticoagulants with warfarin in patients with atrial fibrillation: a meta-analysis of randomised trials. Lancet. 2014;383(9921):955–62.
212. Hsu JC, Maddox TM, Kennedy KF, Katz DF, Marzec LN, Lubitz SA, et al. Oral anticoagulant therapy prescription in patients with atrial fibrillation across the spectrum of stroke risk: insights from the NCDR PINNACLE registry. JAMA Cardiol. 2016;1(1):55–62.
213. Katz DF, Maddox TM, Turakhia M, Gehi A, O'Brien EC, Lubitz SA, et al. Contemporary trends in oral anticoagulant prescription in atrial fibrillation patients at low to moderate risk of stroke after guideline-recommended change in use of the CHADS(2) to the CHA(2)DS(2)-VASc score for thromboembolic risk assessment: analysis from the national cardiovascular data registry's outpatient practice innovation and clinical excellence atrial fibrillation registry. Circ Cardiovasc Qual Outcomes. 2017;10(5):e003476.
214. Kachroo S, Hamilton M, Liu X, Pan X, Brixner D, Marrouche N, et al. Oral anticoagulant discontinuation in patients with nonvalvular atrial fibrillation. Am J Manag Care. 2016;22(1): e1–8.

215. Yao X, Abraham NS, Alexander GC, Crown W, Montori VM, Sangaralingham LR, et al. Effect of adherence to oral anticoagulants on risk of stroke and major bleeding among patients with atrial fibrillation. J Am Heart Assoc. 2016;5(2):e003074.
216. Bartus K, Han FT, Bednarek J, Myc J, Kapelak B, Sadowski J, et al. Percutaneous left atrial appendage suture ligation using the LARIAT device in patients with atrial fibrillation: initial clinical experience. J Am Coll Cardiol. 2013;62(2):108–18.
217. Holmes DR, Reddy VY, Turi ZG, Doshi SK, Sievert H, Buchbinder M, et al. Percutaneous closure of the left atrial appendage versus warfarin therapy for prevention of stroke in patients with atrial fibrillation: a randomised non-inferiority trial. Lancet. 2009;374(9689):534–42.
218. Maybrook R, Pillarisetti J, Yarlagadda V, Gunda S, Sridhar AR, Deibert B, et al. Electrolyte and hemodynamic changes following percutaneous left atrial appendage ligation with the LARIAT device. J Interv Card Electrophysiol. 2015;43(3):245–51.
219. Di Biase L, Burkhardt JD, Mohanty P, Sanchez J, Mohanty S, Horton R, et al. Left atrial appendage: an underrecognized trigger site of atrial fibrillation. Circulation. 2010;122(2): 109–18.
220. Yamada T, McElderry HT, Allison JS, Kay GN. Focal atrial tachycardia originating from the epicardial left atrial appendage. Heart Rhythm. 2008;5(5):766–7.
221. Dar T, Afzal MR, Yarlagadda B, Kutty S, Shang Q, Gunda S, et al. Mechanical function of the left atrium is improved with epicardial ligation of the left atrial appendage: insights from the LAFIT-LARIAT Registry. Heart Rhythm. 2018;15(7):955–9.
222. Al-Saady NM, Obel OA, Camm AJ. Left atrial appendage: structure, function, and role in thromboembolism. Heart. 1999;82(5):547–54.
223. Wunderlich NC, Beigel R, Swaans MJ, Ho SY, Siegel RJ. Percutaneous interventions for left atrial appendage exclusion: options, assessment, and imaging using 2D and 3D echocardiography. JACC Cardiovasc Imaging. 2015;8(4):472–88.
224. Shirani J, Alaeddini J. Structural remodeling of the left atrial appendage in patients with chronic non-valvular atrial fibrillation: implications for thrombus formation, systemic embolism, and assessment by transesophageal echocardiography. Cardiovasc Pathol. 2000;9(2): 95–101.
225. Di Biase L, Santangeli P, Anselmino M, Mohanty P, Salvetti I, Gili S, et al. Does the left atrial appendage morphology correlate with the risk of stroke in patients with atrial fibrillation? Results from a multicenter study. J Am Coll Cardiol. 2012;60(6):531–8.
226. Khurram IM, Dewire J, Mager M, Maqbool F, Zimmerman SL, Zipunnikov V, et al. Relationship between left atrial appendage morphology and stroke in patients with atrial fibrillation. Heart Rhythm. 2013;10(12):1843–9.
227. Fukushima K, Fukushima N, Kato K, Ejima K, Sato H, Fukushima K, et al. Correlation between left atrial appendage morphology and flow velocity in patients with paroxysmal atrial fibrillation. Eur Heart J Cardiovasc Imaging. 2016;17(1):59–66.
228. Glikson M, Wolff R, Hindricks G, Mandrola J, Camm AJ, Lip GYH, et al. EHRA/EAPCI expert consensus statement on catheter-based left atrial appendage occlusion—an update. Europace. 2019;22(2):184.
229. Price MJ. The WATCHMAN left atrial appendage closure device: technical considerations and procedural approach. Interv Cardiol Clin. 2018;7(2):201–12.
230. Whisenant B, Weiss P. Left atrial appendage closure with transcatheter-delivered devices. Interv Cardiol Clin. 2014;3(2):209–18.
231. Gloekler S, Shakir S, Doblies J, Khattab AA, Praz F, Guerios Ê, et al. Early results of first versus second generation Amplatzer occluders for left atrial appendage closure in patients with atrial fibrillation. Clin Res Cardiol. 2015;104(8):656–65.
232. Huang H, Liu Y, Xu Y, Wang Z, Li Y, Cao K, et al. Percutaneous left atrial appendage closure with the LAmbre device for stroke prevention in atrial fibrillation: a prospective, multicenter clinical study. JACC Cardiovasc Interv. 2017;10(21):2188–94.

233. Berti S, Paradossi U, Meucci F, Trianni G, Tzikas A, Rezzaghi M, et al. Periprocedural intracardiac echocardiography for left atrial appendage closure: a dual-center experience. JACC Cardiovasc Interv. 2014;7(9):1036–44.
234. Nietlispach F, Gloekler S, Krause R, Shakir S, Schmid M, Khattab AA, et al. Amplatzer left atrial appendage occlusion: single center 10-year experience. Catheter Cardiovasc Interv. 2013;82(2):283–9.
235. Seeger J, Birkemeyer R, Rottbauer W, Wöhrle J. First experience with the Watchman FLX occluder for percutaneous left atrial appendage closure. Cardiovasc Revasc Med. 2017;18(7): 512–6.
236. Swaans MJ, Wintgens LI, Alipour A, Rensing BJ, Boersma LV. Percutaneous left atrial appendage closure devices: safety, efficacy, and clinical utility. Med Devices (Auckl). 2016;9:309–16.
237. Tzikas A, Gafoor S, Meerkin D, Freixa X, Cruz-Gonzalez I, Lewalter T, et al. Left atrial appendage occlusion with the AMPLATZER Amulet device: an expert consensus step-by-step approach. EuroIntervention. 2016;11(13):1512–21.
238. Cruz-González I, Freixa X, Fernández-Díaz JA, Moreno-Samos JC, Martín-Yuste V, Goicolea J. Left atrial appendage occlusion with the LAmbre device: initial experience. Rev Esp Cardiol (Engl Ed). 2018;71(9):755–6.
239. Alkhouli M, Busu T, Shah K, Osman M, Alqahtani F, Raybuck B. Incidence and clinical impact of device-related thrombus following percutaneous left atrial appendage occlusion: a meta-analysis. JACC Clin Electrophysiol. 2018;4(12):1629–37.
240. Asmarats L, Cruz-González I, Nombela-Franco L, Arzamendi D, Peral V, Nietlispach F, et al. Recurrence of device-related thrombus after percutaneous left atrial appendage closure. Circulation. 2019;140(17):1441–3.
241. Fauchier L, Cinaud A, Brigadeau F, Lepillier A, Pierre B, Abbey S, et al. Device-related thrombosis after percutaneous left atrial appendage occlusion for atrial fibrillation. J Am Coll Cardiol. 2018;71(14):1528–36.
242. Pracon R, Bangalore S, Dzielinska Z, Konka M, Kepka C, Kruk M, et al. Device thrombosis after percutaneous left atrial appendage occlusion is related to patient and procedural characteristics but not to duration of postimplantation dual antiplatelet therapy. Circ Cardiovasc Interv. 2018;11(3):e005997.
243. Bajaj NS, Parashar A, Agarwal S, Sodhi N, Poddar KL, Garg A, et al. Percutaneous left atrial appendage occlusion for stroke prophylaxis in nonvalvular atrial fibrillation: a systematic review and analysis of observational studies. JACC Cardiovasc Interv. 2014;7(3):296–304.
244. Aminian A, Lalmand J, Tzikas A, Budts W, Benit E, Kefer J. Embolization of left atrial appendage closure devices: a systematic review of cases reported with the watchman device and the amplatzer cardiac plug. Catheter Cardiovasc Interv. 2015;86(1):128–35.
245. Saw J, Fahmy P, DeJong P, Lempereur M, Spencer R, Tsang M, et al. Cardiac CT angiography for device surveillance after endovascular left atrial appendage closure. Eur Heart J Cardiovasc Imaging. 2015;16(11):1198–206.
246. Freixa X, Abualsaud A, Chan J, Nosair M, Tzikas A, Garceau P, et al. Left atrial appendage occlusion: initial experience with the Amplatzer™ Amulet™. Int J Cardiol. 2014;174(3): 492–6.
247. Pillarisetti J, Reddy YM, Gunda S, Swarup V, Lee R, Rasekh A, et al. Endocardial (Watchman) vs epicardial (Lariat) left atrial appendage exclusion devices: understanding the differences in the location and type of leaks and their clinical implications. Heart Rhythm. 2015;12 (7):1501–7.
248. Gianni C, Di Biase L, Trivedi C, Mohanty S, Gökoğlan Y, Güneş MF, et al. Clinical implications of leaks following left atrial appendage ligation with the LARIAT device. JACC Cardiovasc Interv. 2016;9(10):1051–7.
249. Viles-Gonzalez JF, Kar S, Douglas P, Dukkipati S, Feldman T, Horton R, et al. The clinical impact of incomplete left atrial appendage closure with the Watchman Device in patients with atrial fibrillation: a PROTECT AF (Percutaneous Closure of the Left Atrial Appendage Versus

Warfarin Therapy for Prevention of Stroke in Patients With Atrial Fibrillation) substudy. J Am Coll Cardiol. 2012;59(10):923–9.
250. Turagam MK, Lavu M, Afzal MR, Vuddanda V, Jazayeri MA, Parikh V, et al. Catheter ablation for atrial fibrillation in patients with watchman left atrial appendage occlusion device: results from a multicenter registry. J Cardiovasc Electrophysiol. 2017;28(2):139–46.

Chapter 9
Prosthetic Vascular Grafts: Past, Present, and Future

Yongjun Li and Zhiyuan Wu

9.1 A Brief History of Prosthetic Vascular Grafts

Prosthetic vascular grafts (PVGs) have made a striking progress during the past more than 100 years. In 1906, Carrel and Guthrie first reported the use of jugular vein repair on the carotid artery in dogs, laying the foundation for the field of vascular surgery [1]. In the same year, Goyanes applied popliteal vein to patch the defect of the popliteal artery for the first time in clinical practice [2]. Since then, the reports of vascular repair using artery or vein grafts were increasing, i.e., Gross reported that artery grafts can be used in human patients with satisfied effect [3]. However, it was not started until the first application of PVGs—Vinyon-N—in dog's vessel described by Voorhees and Blakemore in 1952 [4, 5], creating the time of PVGs in modern vascular surgery.

After then, many kinds of materials have been explored to make PVGs including nylon, Teflon, Orlon, Dacron, polyurethane (PU), etc. Nylon was tried by Edwards et al. [6] but finally failed with the degradation which can result in aneurysm, while Teflon, Orlon, and Dacron didn't have such issues. A study reported that Dacron may be the most promising candidates among these materials in respect of mechanical and physical characteristics [7]. And the history also proves this view. After a series of modifications, PVGs made of Dacron have been widely applied in clinical practice since 1957 [8, 9].

Another extensively used PVG is polymer polytetrafluorethylene (PTFE). However, one of the disadvantages is its poor flexibility, confining its applications in curve area. To solve this shortcoming, modifications were made to transform PTFE

Y. Li (✉) · Z. Wu
Department of Vascular Surgery, Beijing Hospital, National Center of Gerontology, Beijing, China
e-mail: liyongjun4679@bjhmoh.cn; wuzhiyuan5025@bjhmoh.cn

Y.-D. Tang et al. (eds.), *Application of Biomaterials in the Treatment of Cardiovascular Diseases*, https://doi.org/10.1007/978-981-19-7712-1_9

into expanded PTFE (ePTFE) which has been successfully adopted in clinical practice, as reported by Soyer [10] and Campbell [11] in the early 1970s.

PVGs made of these materials play an important role in the artery reconstruction in vascular surgery. However, their applications are still limited with several factors such as graft infection, children development, and small-caliber vessels (≤6 mm). Vascular surgeons and bioengineering scholars have been exploring the products suitable for small-caliber vessels. Currently, the main ideas are about the tissue-engineered vascular grafts (TEVGs) and in situ regenerated tissue-engineered vessels, which are promising for the commercial small-caliber PVGs in the future.

It's worthwhile to note that other materials such as metal, glass, and silk have also been tested. Even though these materials finally failed, they have helped the surgeons and researchers realize the importance of porosity and antithrombosis. Besides, other allogeneic grafts, xenografts, and human umbilical vein grafts have also been widely tested for the exploration of PVGs. In the 1990s, the success of endovascular repair has largely encouraged the studies of another graft, vascular stents. These grafts and stents have tremendously contributed to the development of vascular surgery, shifted the vascular surgeons' concept, and also improved patient's outcomes.

9.2 Two Major Materials in PVGs [12, 13]

9.2.1 Dacron

Dacron is polyethylene terephthalate (PET), which can trace back to 1939. The PVGs made of Dacron were reported of clinical application by Julian in 1957 and by DeBakey in 1958 [8, 9], mainly for the reconstruction of thoracic or abdominal aorta or aortic branches.

Dacron used clinically is mainly made by weaving or knitting technology. Woven polyester is developed by the form of stranded fibers stacked up and down, so it owns the characteristics of low porosity and high strength. Knitted type mainly uses textile technology to make it have typical porosity and radial expansion; therefore, before use, it is often necessary to use gelatin, collagen, or albumin to seal the holes of knitted type Dacron to further prevent blood overflow. Dacron has good stability and will not degrade significantly more than 10 years after implantation. But in comparison, knitted Dacron is easier to expand after being used as an arterial substitute, which is mainly caused by the knitting technology, not PET. Many grafts made of Dacron material are very durable. A single-center study with an average follow-up of 12 years showed that the graft failure rate was only 0.2% [14].

After being implanted in the body, the blood and tissue response to the graft begins when the blood flow is recanalized. The first is the Vroman effect, followed by platelet adsorption, inflammatory cell infiltration, and endothelial cell and vascular smooth muscle cell migration. In the first few hours to a few days, a thrombus plane composed of fibrin, platelets, and blood cells can be formed. Despite this phenomenon, it does not affect its wide use as a vascular graft. In addition, there are a variety of modified Dacron artificial blood vessels in clinical practice, either with heparin coating or silver coating.

9.2.2 ePTFE

PTFE is another material developed by the DuPont Company. It was firstly introduced as Teflon in 1937. The inertness, hydrophobicity, and durability of the material itself make it useful as an insulator material. The PTFE molecule itself is biologically stable, and its vascular graft is difficult to degrade in the body. Its surface is negatively charged, which can reduce its reaction with components in blood vessels.

Its hydrophobic nature prevents blood from leaking out of the graft. In 1969, Gore obtained a patent for expanded polytetrafluoroethylene (expanded PTFE, ePTFE). At present, ePTFE is one of the main forms of artificial blood vessel grafts. In the production process of ePTFE, micropores are generated mainly through operations such as heating, stretching, and extrusion. From the naked eye, its surface is smoother, softer, and more flexible than Dacron. Its porous morphology can be seen under a scanning microscope. This porous morphology occupies about 20% of the ePTFE structure, more than Dacron's. In addition, the size of the holes on this ePTFE material is smaller than that of woven or knitted materials.

There are two types of ePTFE grafts: standard and thin-walled. The standard ePTFE has a wall thickness of 400–600 μm, and the thin-walled ePTFE has a thickness of about 200–300 μm. Compared with standard ePTFE, thin-walled compliance and maneuverability are more advantageous, but the corresponding price is that some thin-walled grafts will lose structural endurance and are not suitable for dialysis access. A major disadvantage of ePTFE is its relative stiffness, which may lead to a mismatch in compliance between the graft and the target artery. The distortion resistance of ePTFE can be achieved by adding an outer plastic ring.

Like Dacron, ePTFE is as effective as a vascular graft, with a 5-year primary patency rate of 95%. Other similarities include the initial host response to ePTFE graft implantation, such as fibrin agglutinates and platelet aggregation. The improvement of ePTFE is mainly manifested in increasing the permeability of the tissue (the inward growth of the tissue is related to the porosity of the graft), changing the inner surface of the graft cavity, and coating the inner surface of the graft with anticoagulant or antithrombotic drugs.

9.3 Applications of PVGs in Vascular Surgery

Even though vascular surgery has advanced into the endovascular era, open surgery still plays essential roles as one of the treatment [15, 16]; thus, PVGs still serve as critical equipment in vascular surgical practices. Comparing to the stents for endovascular management, the applications of PVGs have never shrunk; instead, PVGs can now even be applied for rescuing the challenging complications of vascular stents or serve as candidates for some complex lesions which vascular stents only can't manage.

9.3.1 Reconstruction of the Thoracic, Abdominal Aorta

As mentioned before, one of PVGs' roles is to serve as aortic substitutions, maintaining the stability and integrity of blood circulation. With the advancement of endovascular techniques, such as parallel/sandwich/chimney stents [17–20], fenestrated techniques [21], and stent assisted coil embolism [22], it is now widely suggested to adopt endovascular aortic repair (EVAR) to deal with lesions like aortic aneurysm and aortic dissection. However, for some challenging surgical anatomies or complex lesions, EVAR can't manage efficiently or feasibly, and then open surgery still holds the its advantages. A recent meta-analysis revealed that open surgery with PVGs carried lower re-intervention rates (OR 0.38, 95% CI 0.24–0.64) [23] compared to EVAR with stents during the over 5-year follow-up.

Besides, for some lesions like aortic sarcoma, endovascular technique may help with pathological diagnosis by the minimally invasive technique. But open surgery still holds the advantages where PVGs are essential [24]. In other acute vascular diseases like aortic intimal intussusception, open repair with PVGs can save patient's life [25, 26].

9.3.2 Applications of PVGs in the Reconstruction of Lower Limb Arteries

Like the management of thoracic-abdominal diseases, the first treatment of the reconstruction of lower extremity arteries is also endovascular management. Some complex lesions can't be managed by endovascular technique; instead, open bypass surgeries can help to recanalize these arteries. During these processes, autologous venous is usually the first option. While still some patients lack proper autologous venous, PVGs are then the candidate options for vascular surgeons.

In the bypass surgeries, ePTFE coated with heparins showed higher patency rates than the normal PTFEs. Besides, ePTFE coated with heparins can also be used to rescue the arteries which were treated with stents but finally failed. This application was reported of a 6-month primary patency rate up to 83.3% and a 2-year primary patency of 49.1% [27]. In 2010, Puli et al. reported a result of multiple center registered study involving 425 ePTFE PVGs and showing that the primary patency rate, secondary patency rate, and limb salvage rate were 75%, 82%, and 88%, respectively, at the first postoperative year and were 66%, 75%, and 85% at the second postoperative year.

Usually, the above-knee bypass grafts hold higher patency rate and limb salvage rate. It may be due to the intimal hyperplasia, stenosis, or thrombosis in the anastomotic area of the below-knee grafts. For the anastomotic stoma in the distal below-knee artery, a combination of PVGs with autologous venous can help to improve the bypass graft patency rates [27].

9.3.3 Other Extra-Anatomy Bypass Reconstructions

Other bypass ways can also be seen in axillary-axillary artery bypass, axillary-femoral artery bypass, and femoral-femoral artery bypass.

In addition, in the surgical treatment of aortic arteritis, due to the characteristics of lesions involving blood vessels, the first choice for short-segment stenotic lesions is mainly endovascular treatment [28], while the treatment of lesions in other parts requires various forms of extra-anatomy vascular reconstruction such as carotid-carotid, axillary-carotid, aorta-carotid, aorta-aorta, aorta-renal recopnstructions, etc. [29–31]. A recent retrospective study by our team on 116 patients with arteritis (145 vascular surgeries in total) [30] showed that the primary patency rate of open surgery was 97.3%, 86.2%, 70.5%, and 48.8% in the first 1, 3, 5, and 10 years, respectively, and correspondingly it was 93.3%, 73.1%, 57.5%, and 31.8% in the endovascular management group. In terms of the primary assisted patency rate, it was 98.2%, 90.3%, 73.2%, and 47.2%, respectively, in open surgery; the corresponding rate was 95%, 81%, 64.4%, and 35% in the endovascular management group. In terms of the secondary patency rate, open surgery was 98.2%, 96.1%, 83.5%, and 56.2%; endovascular management were 98.7%, 86%, 71.2%, and 44.9%.

In addition to the abovementioned applications, PVGs can also be used to reconstruct some veins such as portal vein and vena cava.

9.4 PVG Related Clinical Complications

9.4.1 Thrombosis

Thrombosis is the most common cause of vascular graft failure, especially vascular grafts involving arteries of the lower extremities. Postoperative vascular graft thrombosis can be caused by multiple factors, such as the patient's age, gender, risk factors, concomitant diseases, and the angle of suture during arterial reconstruction. Early graft thrombosis is more common in anastomotic thrombosis, mostly due to technical reasons. In the early postoperative and long-term follow-up process, regular monitoring and timely correction of appeal factors are required. Postoperative monitoring and evaluation methods can be ultrasound, arteriography, CT, etc. Once the thrombosis is determined, the surgeon needs to conduct comprehensive diagnosis and treatment based on the cause of the patient's thrombosis. At present, there are many ways to choose: drug therapy (anticoagulation, antiplatelet), thrombectomy, thrombolytic therapy, revascularization, or even amputation if necessary.

9.4.2 Infection

Infection of vascular graft is one of its most serious complications. The most common pathogenic bacteria are *Staphylococcus aureus*, followed by Gram-negative bacteria and fungi. The main reasons are preoperative contamination, bacterial planting of vascular grafts, and infection around vascular grafts. The ways of preoperative contamination are direct planting (such as bacterial manipulation, contact of exuded intestinal juice with the graft, surgical incision, etc.). Bacterial planting in vascular grafts usually occurs when bacteria spread through the bloodstream and are eventually planted in the vascular graft; if the patient's immunity declines in the later stage, it may also lead to graft infection. Infections around vascular grafts are mainly caused by contamination of the skin, digestive tract, and genitourinary tract. The infection of vascular grafts needs to be prevented from the three links of preoperative, intraoperative, and postoperative, and antibiotics should be applied. When an infection occurs, the infected graft should be removed, and the circulation should be reestablished.

9.5 Preclinical, Translational Studies

9.5.1 Research on Vascular Grafts in Large Animal Models [32]

Before vascular grafts are used in humans, their safety and effectiveness must be verified by animal models. Among them, in the selection of animal models, we need to consider availability, price, immunogenicity, implantability, reproducibility, and the most critical similarity with normal human physiological mechanisms. Although rats and rabbits are widely used in experiments and are cheap, they lack the similarity to human physiology, especially in anatomy and hemodynamics. In contrast, large animal models such as dogs, pigs, and sheep, although they are more expensive in daily maintenance, can better simulate human physiological functions.

9.5.2 *Research on Biomedical Engineering of Small-Caliber Blood Vessels [33, 34]*

PVGs made of materials such as Dacron or ePTFE have been successfully used in the reconstruction of human large-caliber blood vessels, but the development of small-caliber ($\leq$6 mm) blood vessels has not been successfully used in clinical practice. In the research and development of small-caliber artificial blood vessels, the current main research ideas are tissue-engineered vascular grafts (TEVGs) and in situ regenerative tissue-engineered blood vessels. Among them, TEVGs are more extensive. The diameters of small arteries, small veins, and capillaries are all in the millimeter range, so the method of constructing large arteries is not suitable for constructing these small blood vessels. The ideal small-caliber blood vessel needs the following conditions: (1) thrombosis resistance; (2) mechanical properties, minimum burst pressure of 1700 mmHg; (3) fatigue strength; and (4) biocompatibility. In addition, the cellular components on TEVGs also need to have the potential for remodeling and some biochemical signaling functions.

TEVGs can be divided into scaffold-based vascular grafts and self-assembled vascular grafts. The reason for the popularity of the latter is that its physical support ensures that cells follow a specific pathway during the clonal proliferation process. The main materials used are synthetic materials, natural materials, or both. The advantage of artificial materials here is that the appropriate materials and construction technology can be selected according to the clinical needs. However, the ensuing technical requirements and long processing time limit its clinical application. In addition, due to the lack of cell binding sites of these artificial materials, the antithrombotic properties are greatly weakened. Although there are many preclinical studies of this type of vascular graft, the research to evaluate its clinical efficacy is extremely limited. Among them, a polyglycolic acid vascular graft implanted with vascular smooth muscle has a half-year patency rate of 88% in large animals. However, subsequent clinical trials showed that the half-year patency rate was 63%, but the 1-year patency rate was only 28%.

Natural materials can be divided into extracellular matrix-based materials and decellularized natural materials. Among the materials based on extracellular matrix, collagen, gelatin, elastin, fibrin, etc. are most used. The protein derived from the extracellular matrix can maintain the binding site for cell adsorption, improve the biomimetic property and biocompatibility of the vascular graft, and at the same time stimulate the further proliferation of the recruited cells. However, such materials have insufficient mechanical stress and are easily degraded, that is, if poorly controlled, they can lead to aneurysm formation or even rupture. The natural sources of decellularized materials are mostly allogeneic or heterogeneous materials, and the process of decellularization helps to avoid immune responses from the recipient. This type of material has been widely used in clinical research, but its high price, low patency rate, and immune response due to decellularization efficiency still limit its development.

Although a series of advances have been made in the research of vascular grafts based on "scaffold," some scholars believe that the "scaffold" method will cause abnormal cell proliferation, so they put forward the concept of self-assembled blood vessel (self-assembled) graft. That is, if the cells in the 3D environment are properly stimulated, the cells can spontaneously form new vascular grafts. To achieve this goal, there are currently three main strategies: cell collection, micro-tissue aggregation, and cell printing technology. However, the main drawback of this type of technology is the limited shape of the graft.

In summary, vascular surgery has undergone a century of development and has experienced the pre-prosthetic graft stage, the prosthetic graft stage, and the endovascular treatment stage. Although vascular surgery has entered the endovascular era, artificial blood vessel grafts still play a pivotal role. The application of prosthetic grafts in large blood vessels is mature, and vascular surgeons still need to beware of complications such as thrombosis and graft infection. In the research of small-caliber prosthetic grafts, various preclinical studies have made progress, but the clinical trial results are not good, and we still lack effective cell sources, materials, methods, and products. Continuous innovation using vascular engineering, biomedical engineering, and stem cell engineering will still bring hope to the commercialization of small-caliber vascular grafts and ultimately meet the needs of patients.

References

1. Carrel A, Guthrie CC. Uniterminal and biterminal venous transplantations 2; 1906. p. 266.
2. Hiektonn T. Arterial homografts: an experimental study in dogs. Acta Orthop Scand. 1952;23 (sup10):1–114.
3. Gross RE, Hurwitt ES, et al. Preliminary observations on the use of human arterial grafts in the treatment of certain cardiovascular defects. N Engl J Med. 1948;239(16):578.
4. Blakemore AH, Voorhees AB Jr. The use of tubes constructed from vinyon N cloth in bridging arterial defects; experimental and clinical. Ann Surg. 1954;140(3):324–34.
5. Voorhees AB Jr, Jaretzki A 3rd, Blakemore AH. The use of tubes constructed from vinyon "N" cloth in bridging arterial defects. Ann Surg. 1952;135(3):332–6.
6. Edwards WS. Arterial grafts: past, present, and future. Arch Surg. 1978;113(11):1225–33.
7. Hess F. History of (MICRO) vascular surgery and the development of small-caliber blood vessel prostheses (with some notes on patency rates and re-endothelialization). Microsurgery. 1985;6(2):59–69.
8. Belio ML, et al. Dacron tube and bifurcation arterial prostheses produced to specification; experimental and clinical use. Surgery. 1957;41(1):50–61.
9. De Bakey ME, et al. Clinical application of a new flexible knitted Dacron arterial substitute. Am Surg. 1958;24(12):862–9.
10. Soyer T, et al. A new venous prosthesis. Surgery. 1972;72(6):864–72.

11. Campbell CD, et al. Expanded polytetrafluoro-ethylene as a small artery substitute. Trans Am Soc Artif Intern Organs. 1974;20a:86–90.
12. Johnston JCKW. Rutherford's vascular surgery. Saunders; 2014.
13. Xue L, Greisler HP. Biomaterials in the development and future of vascular grafts. J Vasc Surg. 2003;37(2):472–80.
14. Van Damme H, et al. Intrinsic structural failure of polyester (Dacron) vascular grafts. A general review. Acta Chir Belg. 2005;105(3):249–55.
15. Chen Z, Yang Y. The foundation of the family: open surgery in vascular surgery. Chin J Vasc Surg. 2020;05(01):1–5.
16. Li Y, et al. The status and significance of traditional open surgery for abdominal aortic aneurysms in the era of endovascular treatment. Chin J Vasc Surg. 2020;05(01):10–3.
17. Wu ZY, et al. Outcomes of chimney and/or periscope techniques in the endovascular management of complex aortic pathologies. Chin Med J. 2017;130(17):2095–100.
18. Zhi-yuan W, et al. Endovascular repair of complex aortoiliac aneurysm with the sandwich technique in sixteen patients. Ann Vasc Surg. 2018:S0890509618305004.
19. Zhi-Yuan W, Chen Z-G, Diao Y-P, Yan S, Li Y-J, Chen Y-X. Total endovascular repair by sandwich technique thoracoabdominal aortic dissecting aneurysm with aneurysm related Disseminated Intravascular Coagulation (DIC). J Vasc Endovasc Surg. 2016;1(3):15.
20. Zhi-Yuan W, Li Y-j. Prevention of branch occlusion after endovascular repair of aortic lesion. Chin J Cardiovasc Med. 2019;2(24):195–7.
21. Shuting Y, et al. One case report of endovascular repair of aortic dissection with guidewire-assisted technique and in situ arch fenestration. J Vasc Endovasc Surg. 2017;003(001):641–3.
22. Li Z, et al. Two cases of renal artery branch aneurysm treated with stent-assisted coil embolization. Chin J Vasc Surg. 2017;009(4):293–5.
23. Chen Z-G, et al. The long-term outcomes of open and endovascular repair for abdominal aortic aneurysm: a meta-analysis. Asian J Surg. 2019;42:899.
24. Wu Z-Y, et al. Primary aortic sarcoma in arch and descending aorta: a case report and literature review. J Thorac Dis. 2018;10(4):E289–95.
25. Wu ZY, et al. Aortic intimo-intimal intussusception: a pooled analysis of published reports. Ann Vasc Surg. 2021;75:471.
26. Wu ZY, et al. Aortic intimal intussusception during acute type B aortic dissection endovascular repair. Ann Transl Med. 2019;7(22):700.
27. Ma L, et al. Application of heparin-coated vascular prosthesis in revascularization of severe lower extremity ischemia after failure of endovascular therapy. Chin J Vasc Surg. 2020;05(01): 17–22.
28. Diao Y, et al. Surgical operation and endovascular treatment effect and safety analysis of 116 cases of Takayasu arteritis. Natl Med J China. 2016;096(006):447–50.
29. Chen Z, et al. Research progress on surgical classification of Takayasu arteritis. Natl Med J China. 2016;96:2280.
30. Diao Y, et al. Surgery and endovascular management in patients with Takayasu's arteritis: a ten-year retrospective study. Ann Vasc Surg. 2020;63:34–44.
31. Chen ZG, et al. Simultaneous multi-supra-aortic artery bypass successfully implemented in 17 patients with type I Takayasu arteritis. Eur J Vasc Endovasc Surg. 2018;56:903.
32. Liu RH, et al. Review of vascular graft studies in large animal models. Tissue Eng Part B Rev. 2018;24(2):133–43.
33. Carrabba M, Madeddu P. Current strategies for the manufacture of small size tissue engineering vascular grafts. Front Bioeng Biotechnol. 2018;6:41.
34. Song HG, et al. Vascular tissue engineering: progress, challenges, and clinical promise. Cell Stem Cell. 2018;22(3):340–54.

Chapter 10
Cardiac Patch

Zhihong Wang, Weiwei Wang, Jing Liu, and Jing Yang

The WHO accounts on that cardiovascular diseases (CVDs) will continue to be the major cause of death around the world during the next 10–15 years, which will lead to more than 23,600,000 million deaths in 2030 [1]. Therein, MI is the main cause of death [2], taking up to 23% each year. In common, MI can be complicated by arrhythmia, heart failure, and even heart shock, which will eventually lead to the inability to draw enough blood into the body [3]. Pathologically, MI usually leads to important changes in ventricle geometry, architecture, and function in damaged tissue (Fig. 10.1). This change, also known as "ventricular remodeling," is characterized by ventricular dilatation, ventricular hypertrophy, loss of cardiac function, interstitial fibrosis, and increased sphericity index [4]. Therefore, MI can be confirmed by some quantifiable methods [e.g., electrocardiogram (ECG) monitoring] and some imageology methods [e.g., magnetic resonance imaging (MRI)].

The treatments of MI include medication, mini-invasive or surgical interventions, heart transplant devices, and heart assistance devices, which can ease patients' pain, reduce cardiac burden, and prevent cardiac complications. Therein, modern medicine has developed many treatments, but the clinical effects are not satisfactory [5]. The gold standard treatment of acute MI is reperfusion therapy with coronary stent or bypass surgery by opening the blocked arteries to restore the blood flow to the heart damaged area. For the end-stage patients, cardiac replacement therapy is required [6]. However, the availability of donor organs is very limited, causing many patients to die while waiting for transplantation [7]. Besides, this method is not suitable for many patients especially the old because of the need for antirejection therapy. With the development of biomedical science and technology, new alternative approaches based on TREM have been emerging in the last decade. Some have been investigated in phase I and II clinical trials, with promising results in restoring

Z. Wang · W. Wang · J. Liu · J. Yang (✉)
Institute of Biomedical Engineering, Chinese Academy of Medical Science and Peking Union Medical College, Tianjin, China
e-mail: liujing@bme.pumc.edu.cn; yangjing@bme.cams.cn

Y.-D. Tang et al. (eds.), *Application of Biomaterials in the Treatment of Cardiovascular Diseases*, https://doi.org/10.1007/978-981-19-7712-1_10

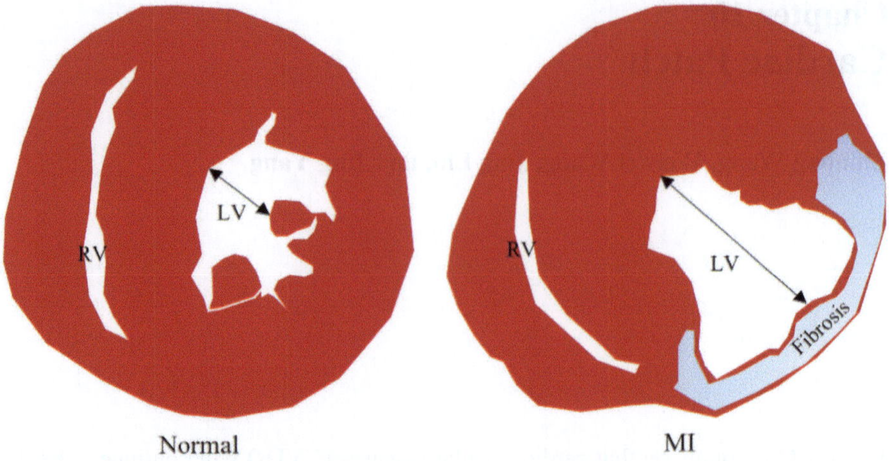

Fig. 10.1 Changes in ventricle architecture, geometry, and function after an MI. *RV* right ventricle, *LV* left ventricle. Arrow indicates ventricle dilation

cardiac function [8]. As is shown in Fig. 10.2, three types of TERM approaches were provided, including stem cell injection therapy, cellularized cardiac patch, or cell-free cardiac patch implants [9].

In recent years, the field of biomedical materials has continuously developed a variety of technologies to reconstruct myocardial function after infarction, such as injectable biomaterials and cardiac patch, which have become a research hotspot in myocardial tissue engineering [10]. The combination of biomedical materials and myocardial tissue engineering, as well as the requirements of performance design, has developed many types of functional heart patches. A cardiac patch could be defined as a piece of heart tissue cultured in a laboratory that can be used to replace or even repair injury area in a patient. In general, an artificial cardiac path was designed and located at the impaired heart tissue to physically stabilize the scar tissue, thereby inhibiting dilatation of the myocardium and alleviating wall stress and paradoxical systolic bulging. Additionally, it is also expected to create a temporary supportive microenvironment for angiogenesis in the infarct zone.

10.1 Basic Requirement for Cardiac Patch

In the last decade, cardiac patch designed to be attached onto the damaged heart surface becomes one of the most prospective therapies to induce healing and repair to the injured myocardium area. Several studies have shown that cardiac patch is feasible in reducing left ventricular remodeling and improving cardiac function [11–13]. Grafted cells loaded in myocardial patches can secrete many cytokines to help repair myocardial function [12–15]. The design and constituent materials of the

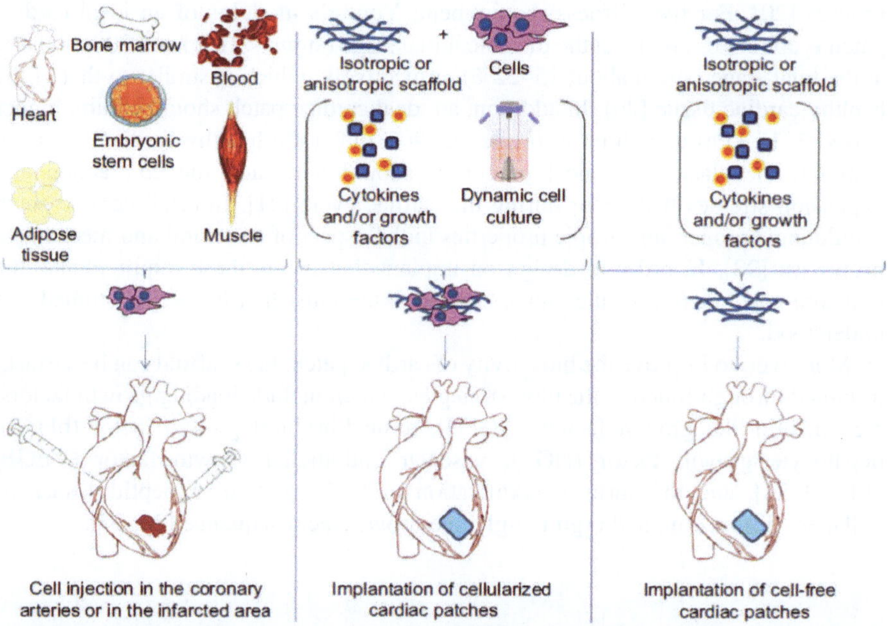

Fig. 10.2 TERM approaches in the treatment of infarcted cardiac area. Three types are provided: (1) cell injection therapies, in which cells directly are injected in the damaged tissue or the coronary arteries; (2) cellularized cardiac patch, with cells pre-seeded in a scaffold to form a native assemble cardiac tissue and then implanted in the infarcted area; and (3) cell-free cardiac patch, constructed only by scaffolds and bioactive molecules. Reused with permission from M. Boffito, S. Sartori, C. Mattu, G. Ciardelli, Polyurethanes for cardiac applications, in: S.L. Cooper, J. Guan (Eds.), Advances in Polyurethane Biomaterials, Elsevier Ltd., 2016

scaffold are critical for MI treatment. In order to be an ideal functional biomimetic tissue substitute, TERM-based cardiac patch must be biocompatible and biodegradable and have proper microstructure, porosity, and mechanical properties, resembling the native cardiac extracellular matrix (ECM) as soon as possible. Besides, the rate of its degradation should match with that of tissue repair, without toxic degradation products [16, 17].

In general, an ideal TERM-based patch scaffold should be an interconnected porous microstructure and possess high surface-to-volume ratio, in order for cell growth and migration through scaffold, vascularization, as well as oxygen, nutrient, and metabolic waste exchange. To be specific, (1) the porosity should be >80% to satisfy nutrient/waste exchange. (2) Pore size should range between 8 and 12 μm to allow vascularization. (3) The capillary density should range between 2400 and 3300 capillaries/mm^2, and the intercapillary distance should be about 20 μm, to mimic the organization of healthy myocardial tissues and make sure the proper blood supply [18, 19]. For the mechanical property, the optimal cardiac patch should finely approach the mechanical properties of the native myocardial tissue and provide a proper mechanical support for newly formed tissues during the tissue regeneration process [20]. For the stiffness requirement, Young's modulus of an ideal cardiac patch is about 10–20 kPa at the diastole initial stage (strain <10%) and 200–300 kPa at the final stage (strain about 15–22%), respectively, which is similar to that of the healthy cardiac tissue [20]. In addition, an ideal cardiac patch should exhibit higher stress (3–15 kPa) and strain at break (22–90%) than the healthy cardiac tissue. In addition, the elastomechanical properties should withstand the contraction and expansion stresses that occur during the cardiac cycle [21]. In brief, cardiac patch should have proper anisotropic properties in the aspect of structural and mechanical properties [22]. In order to design an implant that meets these requirements, the structure and function of the native cardiac tissue must first be deeply studied and understood.

Moreover, to improve the bioactivity of cardiac patch, the scaffold can be surface-modified through functionalization strategies, which include loading growth factors, e.g., insulin-like growth factor-I (IGF-I); basic fibroblast growth factor (bFGF); hepatocyte growth factor (HGF); vascular endothelial growth factor (VEGF) [14, 23–25], and the surface modification of ECM protein or peptides such as collagen, fibronectin, and arginine-glycine-aspartic acid sequence [26, 27].

10.2 Cellularized Cardiac Patch

Various approaches have been developed to construct cardiac patch (Fig. 10.3) [28]. Cardiac patch is fabricated to encapsulate stem cells and provide stimulus signal that promote stem cells to differentiate into active cardiac cells. In addition, adult stem/progenitor cells are regulated by the systemic effect of cytokines, such as granulocyte colony-stimulating growth factor which can enhance cell circulation and cardiac tissue repair efficiency [29]. To improve cell retention in the cardiac patch, many advanced cell delivery techniques have been investigated [30]. These mainly include (a) cells grown onto degradable scaffolds [31], (b) cell-containing hydrogels [32], or (c) cellular sheets without any scaffold [33]. Scientists have considered that ideal cardiac patch should be (a) contractile and flexible, (b) mechanically stable/robust, (c) easily vascularized, (d) electrophysiologically stable, and (e) nonimmunogenic [32].

The creation of cell sheet preparation begun from the mid-1970s by Green and his assistants [34]. After extensive research on the suitable culture media for human keratinocyte proliferation to fabricate a cell sheet, a successful method was designed to enzymatically detach the human epithelium sheet. Soon after, this sheet was successfully applied to treat the seriously burned patients clinically [35], which is the first cell-loaded cardiac patch treatment. Afterward, temperature-sensitive polymer poly(*N*-isopropylacrylamide) coated cell culture plates were developed by Okano et al., which can form thin cell sheet composed of epithelial cells or mesenchymal cells [33]. The fully intact cell sheets could be obtained by lowering the ambient temperature. However, there are still several limitations of the cell sheet strategy. First, cultured sheets of mesenchymal stem cells or cardiomyocytes are

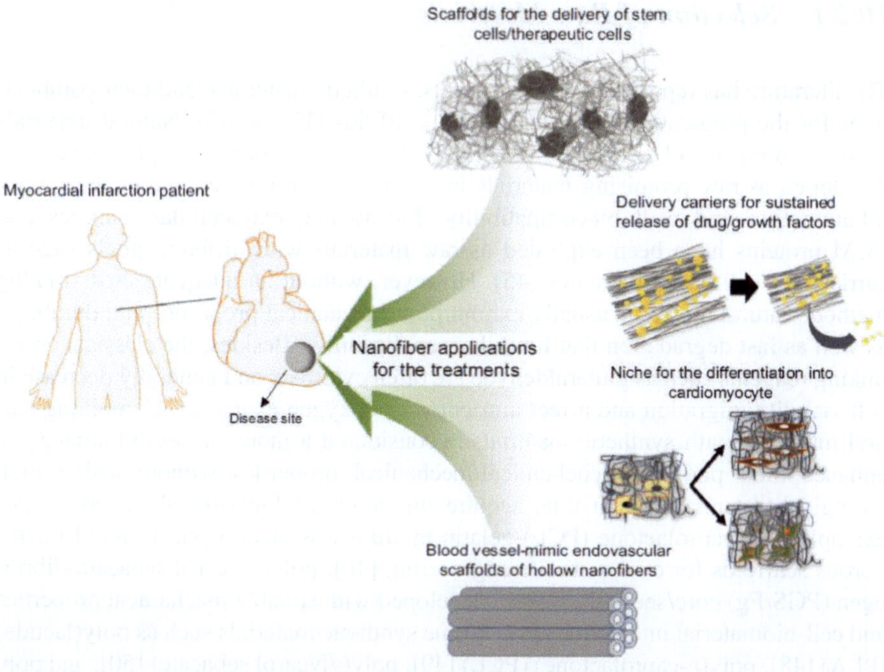

Fig. 10.3 Schematic overview of some applicable methods of multifunctional cardiac patch for cardiac tissue recovery. Reused with permission from P.-H. Kim, J.-Y. Cho, Myocardial tissue engineering using electrospun nanofiber composites, BMB Rep. (2016);49(1):26–36

notoriously fragile due to a lack of ECM in these constructs [36]. Thus, cells were further cultured with ascorbate compounds to promote more collagen production. A recent study showed that cell sheets were created by this method of the cultures of fibroblasts and smooth muscle cell (SMC) sheets [37]. Moreover, longer culture periods could realize more complex and mature ECM structures. In addition, introducing a synthetic scaffold could enhance the mechanical properties, and in vitro results indicated a better tissue regeneration and functionality [12, 38]. Second, the thickness of the construct tissues is limited due to the diffusion technology. This is very essential for clinically relevant cardiac patch because the ventricle wall thickness of the healthy human heart is more than 1 cm. To solve this limitation, endothelial cells were included into mesenchymal cells using a cell sheet stacking manipulation technology by Okano's group [39]. In vitro results showed that the endothelial cells could assemble into a network similar to the native microvasculature and can also connect with the host circulation to avoid hypoxia of cells within the center part of the scaffolds. Another strategy is to build serial sheets that allow each layer to be vascularized in vivo by the host body and then implanted the next layer [40]. This concept has been confirmed in the rat model, and this methodology has become a popular trend in recent years [41, 42].

10.2.1 Selection of Raw Materials

The literature has reported natural materials, synthetic materials, and their combinations for the preparation of suitable patch scaffolds [12, 14, 15]. Natural materials such as collagen [43], chitosan and silk [12], and hyaluronic acid [44] have been developed as raw promising materials in tissue engineering because of their weak inflammation and well biocompatibility. Meanwhile, extracellular matrices and ECM proteins have been exploited as raw materials with different goals such as carriers for cell-based therapies [45]. However, without an adequate cross-linking method, natural materials usually exhibit poor mechanical properties and durability as well as fast degradation that limit their applicability. Besides, the classical cross-linking reagents such as glutaraldehyde are often cytotoxic and generally decrease in cell viability/migration and affect nutrient/waste/oxygen exchange. Combining natural materials with synthetic materials is considered a more successful strategy to enhance their poor physicochemical/mechanical properties without losing high biological potential, which thus acquire the so-called bioartificial materials. For example, polycaprolactone (PCL)-gelatin mixture was electrospun to build microporous scaffolds for cardiac tissue engineering [46]; poly(glycerol sebacate)/fibrinogen (PGS/Fg) core/shell fibers were developed with suitable mechanical properties and cell-biomaterial interactions [47]. Some synthetic materials such as poly(lactide) (PLA) [48], poly(ε-caprolactone) (PCL) [49], poly(glycerol sebacate) [50], and poly (lactide-co-glycolide) (PLGA) [51] have also been extensively studied as raw materials in TERM, as they show many merits like its biocompatibility, biodegradability, processability, and controllable mechanical properties that can be adjusted to suit the needs of the final application. In comparison with natural materials, synthetic materials usually cause a higher level of inflammatory reaction while avoiding pathogen transmission risk caused by some biological-origin materials.

10.2.2 Fabrication Techniques

A wide variety of preparation techniques have been reported in the literature on manufacturing suitable patches or matrices in repairing the infarcted myocardial tissue using natural, synthetic, or bioartificial materials. These techniques can be divided into traditional and advanced technologies. The traditional technologies include phase separation, gas foaming, salt leaching, and porous structure formation, which are regulated by some external stimuli like thermodynamic instabilities and pore-forming agents. As a consequence, the obtained patches usually end up with poor repeatability and reproducibility caused by the unrestricted fabrication parameters, which will further lead to the inhomogeneous gas cavities and porosity and the difficulty in replicating native ECM. Contrarily, advanced fabrication technologies such as 3D-printing [13] or rapid prototyping (RP) technique have provided a whole new approach to designing and fabricating customizable constructs with an accurately defined geometry with the computer's aid, both of which can optimize the internal construction and external shape of patch scaffolds to better satisfy the needs in specific applications.

In this aspect, electrospinning stands between traditional and advanced fabrication technique. Although electrospinning cannot manufacture high-thickness biomedical products with the accurate predefined geometry, it can better simulate the geometric structure of native ECM with high repeatability and versatility, which applies to almost all natural, synthetic, and bioartificial polymers [52]. Furthermore, the simplicity and inexpensive character of the electrospinning devices shall avail the large-scale production and further industrial manufacture.

10.2.3 Challenges of Cardiac Patch Constructs

Tissue-engineered myocardial mesh has many advantages and shows a considerable future, but it still faces several challenges before its application [32]. These limitations include (a) the possibility of inflammatory response, (b) potential toxic degradation products [33], (c) different compliance or mechanical compatibility between polymers and heart tissue [32], and (d) the lack of interconnected relationship between cells. While it is possible to avoid all the above limitations by using completely biological cell sheets, they may still be limited due to the poor mechanical properties.

10.3 Acellular Biomaterials-Based Cardiac Patches

10.3.1 Engineered Polymer Cardiac Patch

Wagner's research group prepared a double-layer polyurethane/extracellular matrix (ECM) composite heart patch (Fig. 10.4) [53]. The purpose was to improve the load on the ischemic ventricular wall caused by myocardial infarction through the unification of mechanical support and biological activity. This 6-mm-diameter heart patch was prepared by a two-stream electrospinning process. The double-layer design concept can effectively improve the heart function after being sutured to the heart surface. The patch results in a lower ESA (end systolic) and EDA (end diastole) than the infarct group and higher FAC (area change percentage). At the same time, the patch group has the highest level of host cell infiltration. From the quantitative analysis of newly formed tissue and scar area, this double-layer patch leads to a significant recovery of myocardial tissue function.

A Cryogel initiated by APS and TEMED with three cross-linkers including MA-G/Ppy, PEGDA, and DOPA was manufactured as cardiac patch [54]. The design concept is to simultaneously introduce conductivity and cardiomyocytes into the heart patch. Polypyrrole nanoparticles were used to improve the electrical

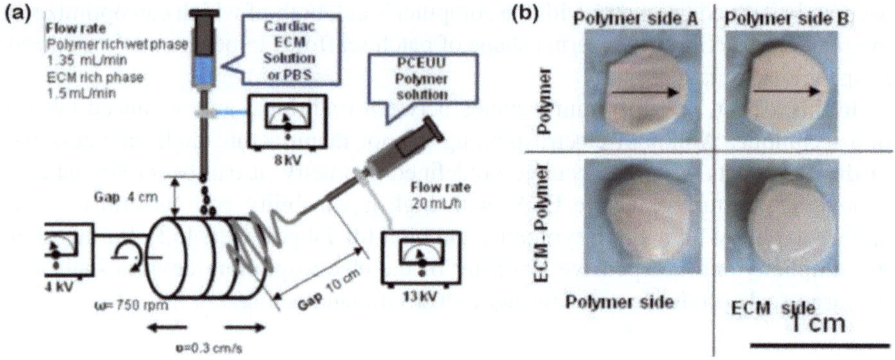

Fig. 10.4 (**a**) Fabrication overview for bilayer biohybrid scaffold. (**b**) Cardiac patch samples. Top: polymer only. Bottom: bilayer patch. Reused with permission from Sci. Transl. Med. 2014;6: 218ra6

signal, thereby stimulating the electrical contraction coupling behavior of the myocardium, promoting the expression of the heart-specific proteins and enhancing the conduction of calcium ion signals. After culturing cardiomyocytes in the gel patch for 8 days, contraction activity of the patch was obtained in vitro. The implantation of conductive patch in infarcted myocardium can significantly improve the heart function. Masson staining of repaired heart obviously showed the growth of new myocardium, with a reduction in the infarct area by 42.6%. Although the current research and design of patch materials have significantly improved the function of the infarcted heart in terms of performance, fixing the patch by suture is always a challenge in practical clinical applications.

10.3.2 Light-Triggered Adhesive Cardiac Patch

To relieve the tissue damage caused by patch sutures, a UV-cross-linked hydrophobic adhesive that can maintain the adhesion in the dynamic and humid environment of internal pressure and blood flow was prepared by Lang and colleagues (Fig. 10.5) [55]. Sebacic acid and glycerol that were naturally produced and easily metabolized in the body were polymerized for preparing the adhesive. The hydrophobic light-activated adhesive (HLAA) prepolymer after copolymerization of glycerol and sebacic acid undergoes C=C cross-linking under ultraviolet light to form a 3D polymer network, and finally a flexible polymer membrane with a thickness of 200 μm was obtained. The compressive modulus of this membrane patch was as high as 3.8 MPa, and the adhesion strength can reach 20 kPa. This HLAA-coated patch could efficiently attach to the myocardium defect in interventricular septum and remained stable in the dynamic body fluid environment. However, the modulus of the heart patch was too high relative to the myocardial tissue, which will cause potential pressure on the synchronized contraction of the heart.

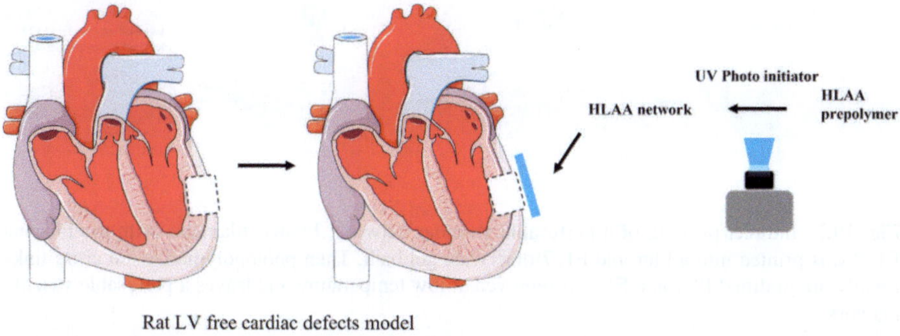

Fig. 10.5 The schematic diagram of light-triggered adhesive cardiac patch

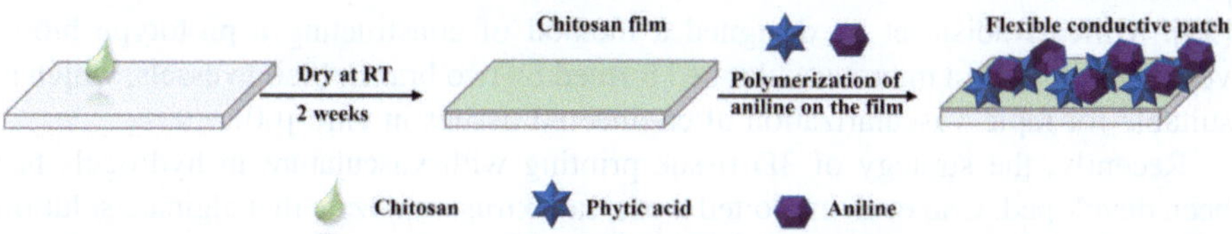

Fig. 10.6 Chemical components and fabrication steps of the PANI patch

Based on the reduction of electrical signal conduction level in the infarct area after myocardial infarction [56], and the principle of photo-adhesion to address the shortcomings of the short duration of clinical electrical activity of conductive polymers, chitosan film with surface doped polyaniline was designed by Mawad et al. The strong binding effect between phytic acid and chitosan was used for fixing the dopant in the conductive support layer to inhibit the deterioration of electrical activity of resulting patch (Fig. 10.6) [57]. In addition, rose bengal was incorporated into the patch to produce a film that can be activated by light, and then a green laser was used to irradiate the patch, avoiding new wounds added to the tissue caused by suture method. The conductive patch can stick to the surface of tissue intactly for 2 weeks without significant arrhythmia, which proved the biocompatibility of polymer patch. Although the light-adhesive patch avoids the fixation of heart patch by suture, the application of light still has potential secondary damage to the tissue, and it is also easy to cause immune response or multilevel oxidative stress.

10.3.3 3D-Printing Cardiac Patch

Although many current studies have focused on mixing cells with hydrogels, the cells are effectively differentiated into cardiomyocytes to create a patient-specific immune compatible cardiac patch [58]. However, these cardiac patches haven't a well blood vessel network that can match the patient's vascular structure. It has

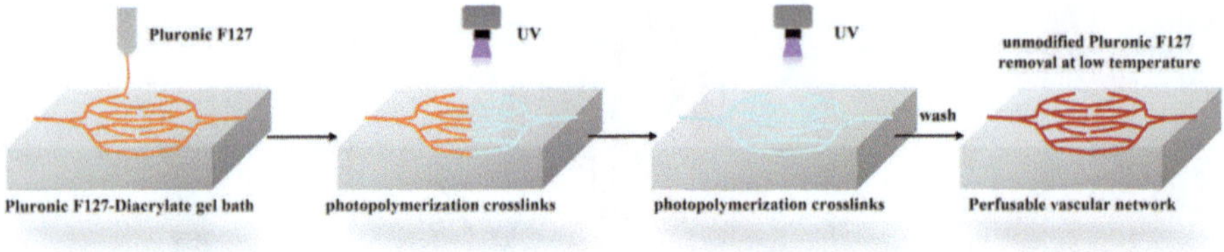

Fig. 10.7 Indirect printing of a perfusable vascular network. Firstly, filaments with the Pluronic F127 was printed into a Pluronic F127-diacrylate gel bath. Then photopolymerization cross-links. Finally, unmodified Pluronic F127 is removed at low temperature and leaves a perfusable vascular network

previously been demonstrated that the predesigned blood vessel network within the parenchymal tissue is critical for the function of the plaque after transplantation [59]. Milica Radisic et al. designed a method of constructing a prototype blood vessel network by a microvascular bed formed by two branch blood vessels, which is suitable for rapid vascularization of engineered tissues in vitro [60].

Recently, the strategy of 3D tissue printing with vasculature in hydrogels has been developed. Gao et al. exploited a coaxial extrusion nozzle that alginate solution flows externally and calcium solution flows internally to form endogenous and perfusable microchannel structures [61]. These hollow microchannels are printed on a platform, which is gradually lowered into the calcium bath solution for secondary cross-linking. Besides, Hinton et al. developed an extrusion method using hydrogels to help direct structure print in a sacrificial gelatin particle bath to strengthen the cross-linking [62]. Although alginate provides a direct and biocompatible extrusion 3D bioprinting method, its poor self-supporting property makes it difficult to construct a complex bioengineering scaffold. Wu et al. adopted an indirect method to extrude Pluronic F127 filaments into a Pluronic F127-diacrylate gel container that can provide external support for printing 3D perfusion vascular trees (Fig. 10.7) [63]. However, in most investigations, blood vessels were printed using only ECs, without parenchymal tissue, which was subsequently cast on top of the blood vessel.

Current studies have accomplished the co-printing of ECs and surrounding thin tissue together [64]. However, ECs could not open blood vessels, and the obtained tissue was thin. Tal Dvir et al. reported a new 3D-printing technology using personalized hydrogels derived from patients' omentum tissue as bio-inks (Fig. 10.8). In this strategy, the engineered patch did not cause immune responses after transplantation, thereby reducing the need for immunosuppressive therapy. When the hydrogel was mixed with the patient's own cells, it can perfectly match the patient's immune, biochemical, and anatomical properties [65].

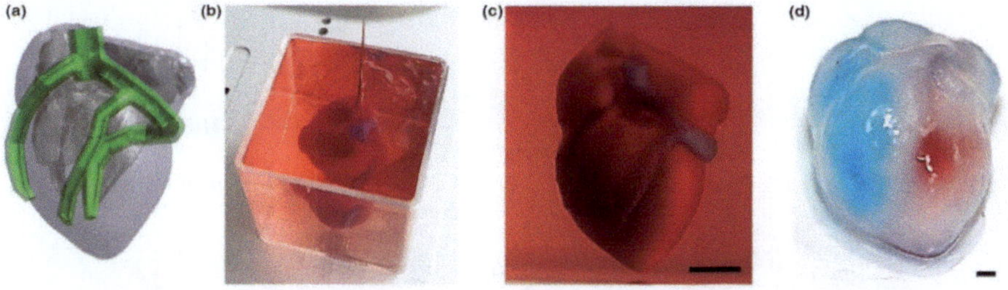

Fig. 10.8 (**a**) CAD model designed from human heart. (**b**, **c**) Support the printed heart in the bath. (**d**) After extraction, red and blue dyes are injected into both ventricles, respectively, to show the hollow chambers as well as the septum between them. Reused with permission from Advanced Science 2019;6:10

10.4 Bio-Patches Loaded with Cytokines

In addition to heart patches based on biomaterials alone, treatment cells including mesenchymal stromal cells (MSCs), bone marrow-derived progenitor cells (BMPCs), induced pluripotent stem cells (iPSCs) and adult or embryonic stem cells (ASCs or ESCs), or cytokines could also be directly delivered by bio-patch to the site of damaged heart. Cell therapy in cardiac repair has several limitations in clinical application. For example, cells are easily broken and needed to be carefully preserved before transplantation to maintain high viability and function [66]. Considering labor and material costs, cell therapy is extremely time-consuming and expensive. Due to the use of living cells, some unmatured cells may show irregular cell growth or tumorigenicity after transplantation [67]. However, the advantage of this approach is that cells delivered into the damaged myocardium can secrete cytokines, thereby improving myocardial perfusion and ventricular function. Recently, various scaffold materials encapsulating cytokines have been designed for use as heart patches for myocardial regeneration [68].

10.4.1 Paracrine Factors Delivered in Patch for Myocardial Infarction (MI)

The focus of cell therapy is to use different cells to improve myocardial tissue recovery by influencing the heart remodeling and reducing the size of the scar caused by MI [69]. Paracrine mechanisms determine tissue repair mediated by adult stem cells. For example, the regeneration factors from adult stem cells can perform genetic repair of damaged myocardium at close range [59, 70]. Artificial heart patch, which can potentially solve the problems with respect to the use of live cells, can provide effective cell therapy for the injured area by delivering paracrine factors. As shown in Fig. 10.9, Cheng et al. constructed a scaffold matrix from the heart tissue of decellularized myocardial extracellular matrix to deliver myocardial

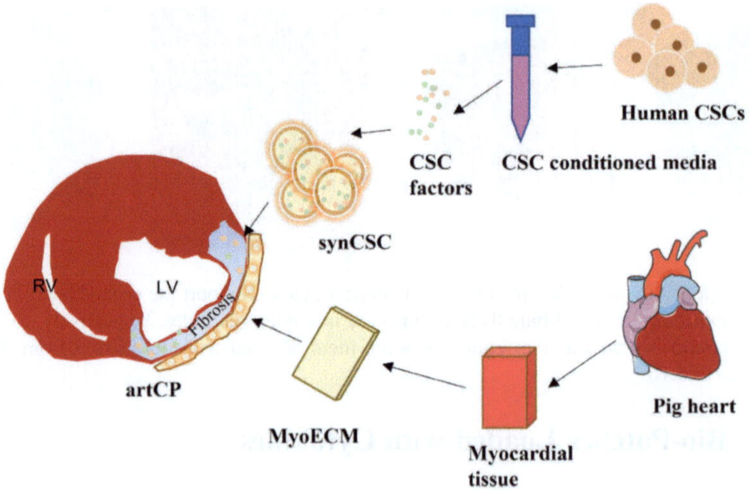

Fig. 10.9 Schematic diagram of fabrication paracrine factors delivered in artificial heart patch

stromal cell-derived repair factors encapsulated in synthetic biodegradable polymers. The resulting patch contains all therapeutic factors secreted by cells and does not trigger the patient's immune response to live cells.

After 3 weeks of artificial heart patch treatment in the rat heart model, cardiac function improved by about 50%, and scars at the injury site were reduced by about 30%, compared with untreated rats. A preliminary 7-day study of a pig's heart disease model revealed that compared with untreated controls, the scars in parts of pig's heart were reduced by about 30%, and the heart function was stable. Additional experiments have shown that this patch can be frozen and safely stored for at least 30 days and frozen ones are as effective as newly created patches. Because there are no living cells involved, it will not trigger the patient's immune system, which is the first step toward a truly ready-made heart patch treatment solution.

In spite of the paracrine mechanism caused by cytokine being very suitable for slowing down pathological tissue remodeling and disease development, repairing important ventricular scar bulges may need the growth of new myocardium. Therefore, a strategy that combines the advantage of paracrine factors with the generation and integration of engineered heart tissue will be very attractive for reducing left ventricular dilation and left ventricular remodeling after infarction [71]. To increase the rate of cell implantation, bioengineering methods including injectable biomaterials, heart patches, and magnetic/molecular targeting can greatly improve cell retention [72, 73]. However, the incorporation of transplanted biomaterial construct with the host myocardium remains a huge challenge. Herein, Gu Zhen team invented a polymer microneedle (MN) patch loaded with cardiac stromal cells aiming at improving cardiac regeneration, as shown in Fig. 10.10 [74].

Hydrogel attached to the surface of the patch contains cardiac stromal cells, which can secrete proteins, RNA, and small molecules to promote the growth of

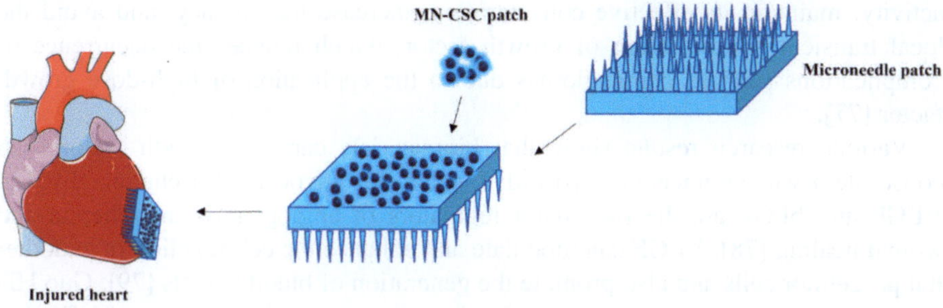

Fig. 10.10 Schematic diagram of the design and the therapeutic effect of MN-CSCs

cardiomyocytes. Human cardiomyocytes do not have the ability to regenerate. However, ESCs as well as iPS can differentiate into a great diversity of body cells under different inducible factors. The medical community is trying to use artificially induced cardiomyocytes as "Band-Aid" to replace dead cardiomyocytes and repair damaged hearts. At present, this patch has only been successful in animal experiments. Before using these microneedle patches for human treatment, the research team also plans to replace the polymers in the patch with substances that can be naturally dissolved in the body while also seeking a less invasive implantation method.

10.4.2 Growth Factors Delivered in Patch for Myocardial Infarction

Generally speaking, the exogenous therapeutic molecules used in myocardial infarction mainly can accelerate the regeneration of myocardial cells and even blood vessels or reduce the apoptosis of cardiac cells. Therefore, growth factors and cytokines are usually the best choice. Growth factors promote tissue regeneration at different stages inducing neovascularization [71]. Angiogenic cytokines can be used to induce blood vessel formation in heart plaque [75]. Thus, it could be used to improve the repair and regeneration of injury site; the life span of growth factors in the body is very short, which is disadvantageous for maintaining the biological activity. Therefore, it is very important to regulate the dose of growth factors to enhance the efficacy in vivo [76]. The progress of biomaterials and tissue engineering provides help in solving this problem. Biomaterials influence the stability, activity, release, and distribution of factors. The combination of growth factors and biological materials can be accomplished through physical affinity or chemical cross-linking. Growth factors can be compounded with these carrier materials and released by free diffusion or material degradation. At the same time, biomaterials can protect the growth factors from inactivation by temperature, proteolytic enzymes, and acids. Slow release of growth factors in the body can preserve its biological activity, maintain its effective concentration, increase the efficacy, and avoid the local transient concentration of growth factor, which reduces the occurrence of complications such as hemangiomas due to the application of high-dose growth factor [77].

Various research results show that biomaterials carrying growth factors can constitute tissue-engineered myocardial patch for myocardial ischemic disease. VEGF and bFGF are the two major regulators of angiogenesis in ischemia and wound healing [78]. VEGF can modulate and enhance the cell function of endothelial progenitor cells and also promote the generation of blood vessels [79]. Guo HD and colleagues used short peptide scaffolds to ensure sustained release of VEGF in infarcted myocardium for at least 1 month. Four weeks after transplantation, the preservation of cells in the infarcted heart was enhanced, and angiogenesis was promoted to improve cardiac function, which also indicates that sustained delivery of VEGF is an effective treatment for MI [80]. Tang Y et al. designed an experiment on the treatment of MI by targeting mesenchymal stem cells (MSCs) combined with overexpressed P-selectin and VEGF-encapsulated immunoliposomes targeting the infarcted myocardium [81]. Besides, fibroblast growth factor 2 (FGF2) has been shown to be a cardioprotective agent independent of stimulating new blood vessel formation. Exogenously administered FGF2 can reduce the degree of contractile dysfunction or myocardial infarction [82].

Insulin-like growth factor 1 (SDF-1) is a chemoattractant aimed at BMSCs and endothelial progenitor cells, which plays a role as a cardioprotectant. IGF-1 is a cardioprotective and angiogenic agent. A stent composed of alginate/alginate sulfate was created to improve the angiogenesis and viability of cardiac plaque by binding a blending of survival and angiogenic factors including IGF-1, VEGF, and SDF. It is constructed by inoculating a mixture of neonatal heart cells with survival factors and angiogenic factors into an alginate scaffold [83]. This patch can be used as an effective bioreactor to produce a heart patch with a functional vascular network. After successful transplantation into the infarct, it can alleviate left ventricular remodeling and dysfunction in myocardial damaged rats. The functional vessel network that develops in cardiac plaques can maintain the structure of transplanted and resident viable myocardium. After transplantation into scar tissue, the heart plaque will electrically and structurally fasten with the host myocardium, thereby benefiting the contraction and relaxation function of the left ventricular. It has been verified that IGF-1 can stimulate hypertrophy and postpone the death of cardiomyocyte after myocardial infarction, thus reducing ventricular dilation [84].

Another growth factor that works in protecting and repairing role in the heart muscle is hepatocyte growth factor (HGF) [85]. Exogenous administration of HGF can promote endothelial cell regeneration, increase capillary density, and improve blood supply to ischemic myocardium [86]. HGF can induce angiogenesis through PI3K/Akt and MARK/ERK signaling pathways [87], without increasing vascular permeability or inflammatory response, unlike VEGF and b-FGF [88]. Co-delivery of IGF and HGF cytokines in combination with Sca-1+/CD31− cells could improve cardiac function by changing the hostile microenvironment after MI. HGF and IGF-1 can significantly increase the survival of cardiac stem cells as well as promote

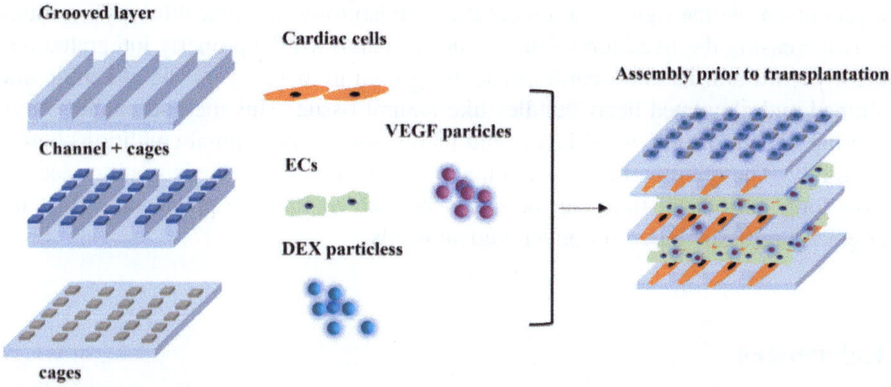

Fig. 10.11 Steps to assemble a functional cardiac patch in the bottom-up method

myocardial regeneration and angiogenesis [89]. Neuroregulin-1 (NRG) is another important growth factor that can promote cell survival, myocardial cell proliferation, and blood vessel formation. NRG-encapsulated microparticles composed of poly (lactic acid-glycolic acid) copolymer as a patch adhering with ADSCs (ADSC-NRG-MPs) efficiently converted macrophages from pro-inflammatory to healing type after 1 week of treatment for MI. After 3 months, the survival rate of transplanted cells and the proliferation of cardiomyocytes increased, the formation of arterioles and capillaries increased significantly, and the infarct size decreased, indicating that ADSC-NRG-MPs can facilitate the repair of infarcted cardiac tissue [90].

The heart muscle possesses a highly 3D structure, consisting of a complex fiber network of cellular components and structural extracellular matrix (ECM) [91]. To meet the metabolic demands of contracted tissue, the condensed network of blood vessels between cell layers can provide enough oxygen and nutrients into the surrounding cells. Since the function of myocardium largely depends on its unique structural organization, thus reproducing anisotropic structure in the ventricular wall and internal vasculature is necessary for MI treatment [92]. To conquer these challenges, some advanced lithography techniques have enabled anisotropic 3D heart tissue and blood vessel-embedded septa through layer-by-layer assembly [93]. In a bottom-up design shown in Fig. 10.11, laser patterning on the electrospun fiber layer created a unique predefined structure for heart cells and endothelial cells. This type of scaffold has three different structures: (1) microgrooves used to facilitate the aligned heart tissue assembly, (2) microtunnels with side cages to allow endothelial cells and microparticles to release VEGF slowly, and (3) a structure similar to microcage used to capture dextran particles that can control anti-inflammatory drugs releasing to the external microenvironment.

This technology provides a methodology of assembling a modular, thick heart patch composed of layers including heart tissue, blood vessels, and anti-inflammatory and pro-angiogenic factors. The 3D structure of fibrous ECM is summarized by micro-patterned electrospun layer assembly, which mimics the changes in the arrangement of collagens throughout the left ventricular wall. These layers resemble the rigidity and mechanical anisotropy of the healthy myocardium, thus increasing the likelihood that the cardiac patch will be properly integrated into the myocardium. Besides, cardiomyocytes grown in these layers can assemble into aligned and elongated heart bundles like natural tissue. This method enables engineering design of individual layers and incubation under optimal conditions before assembly. The microparticle system that releases growth factors or small molecules over long periods of time can be easily assembled into the patch and affect the engineering tissue based on physiological needs.

References

1. World Health Organization. Noncommunicable diseases country profiles 2018. Geneva: World Health Organization; 2018.
2. Du X, Patel A, Anderson CS, Dong J, Ma C. Epidemiology of cardiovascular disease in China and opportunities for improvement: JACC International. J Am Coll Cardiol. 2019;73:3135–47.
3. Plotkin M, Vaibavi SR, Rufaihah AJ, Nithya V, Wang J, Shachaf Y, Kofidis T, Seliktar D. The effect of matrix stiffness of injectable hydrogels on the preservation of cardiac function after a heart attack. Biomaterials. 2014;35:1429–38.
4. Zamilpa R, Lindsey ML. Extracellular matrix turnover and signaling during cardiac remodeling following MI: causes and consequences. J Mol Cell Cardiol. 2010;48:558–63.
5. Huang K, Ozpinar EW, Su T, Tang J, Shen D, Qiao L, Hu S, Li Z, Liang H, Mathews K, Scharf V, Freytes DO, Cheng K. An off-the-shelf artificial cardiac patch improves cardiac repair after myocardial infarction in rats and pigs. Sci Transl Med. 2020;12:eaat9683.
6. Ramakrishna H, Pajaro O. Heart transplantation in the era of continuous flow ventricular assist devices and the total artificial heart: will new technologies surpass the gold standard? Ann Card Anaesth. 2011;14:174–5.
7. Christman KL, Vardanian AJ, Fang Q, Sievers RE, Fok HH, Lee RJ. Injectable fibrin scaffold improves cell transplant survival, reduces infarct expansion, and induces neovasculature formation in ischemic myocardium. J Am Coll Cardiol. 2004;44:654–60.
8. Chachques JC, Trainini JC, Lago N, Cortes-Morichetti M, Schussler O, Carpentier A. Myocardial assistance by grafting a new bioartificial upgraded myocardium (MAGNUM Trial): clinical feasibility study. Ann Thorac Surg. 2008;85:901–8.
9. Boffito M, Sartori S, Mattu C, Ciardelli G. Polyurethanes for cardiac applications. In: Advances in polyurethane biomaterials. Duxford: Woodhead; 2016.
10. Hosoyama K, Ahumada M, McTiernan CD, Davis DR, Variola F, Ruel M, Liang W, Suuronen EJ, Alarcon EI. Nanoengineered electroconductive collagen-based cardiac patch for infarcted myocardium repair. ACS Appl Mater Interfaces. 2018;10:44668–77.
11. Sarig U, Sarig H, De-Berardinis E, Chaw SY, Nguyen EBV, Ramanujam VS, Thang VD, Al-Haddawi M, Liao S, Seliktar D. Natural myocardial ECM patch drives cardiac progenitor based restoration even after scarring. Acta Biomater. 2016;44:209–20.
12. Chen J, Zhan Y, Wang Y, Han D, Tao B, Luo Z, Ma S, Wang Q, Li X, Fan L, Li C, Deng H, Cao F. Chitosan/silk fibroin modified nanofibrous patches with mesenchymal stem cells prevent heart remodeling post-myocardial infarction in rats. Acta Biomater. 2018;80:154–68.
13. Liu J, Miller K, Ma X, Dewan S, Lawrence N, Whang G, Chung P, McCulloch AD, Chen S. Direct 3D bioprinting of cardiac micro-tissues mimicking native myocardium. Biomaterials. 2020;256:120204.
14. Boccardo S, Gaudiello E, Melly L, Cerino G, Ricci D, Martin I, Eckstein F, Banfi A, Marsano A. Engineered mesenchymal cell-based patches as controlled VEGF delivery systems to induce extrinsic angiogenesis. Acta Biomater. 2016;42:127–35.

15. Wang Y, Zhang J, Qin Z, Fan Z, Cheng C. Preparation of high bioactivity multilayered bone-marrow mesenchymal stem cell sheets for myocardial infarction using a 3D-dynamic system. Acta Biomater. 2018;72:182–95.
16. Boffito M, Sartori S, Ciardelli G. Polymeric scaffolds for cardiac tissue engineering: requirements and fabrication technologies. Polym Int. 2014;63:2–11.
17. Ma PX. Biomimetic materials for tissue engineering. Adv Drug Deliv Rev. 2008;60:184–98.
18. Davis ME, Hsieh PCH, Grodzinsky AJ, Lee RT. Custom design of the cardiac microenvironment with biomaterials. Circ Res. 2005;97:8–15.
19. Eschenhagen T. Engineering myocardial tissue. Circ Res. 2005;97:1220–31.
20. Bouten CVC, Dankers PYW, Driessen-Mol A, Pedron S, Brizard AMA, Baaijens FPT. Substrates for cardiovascular tissue engineering. Adv Drug Deliv Rev. 2011;63:221–41.
21. Hidalgo-Bastida LA, Barry JJA, Everitt NM, Rose FRAJ, Buttery LD, Hall IP, Claycomb WC, Shakesheff KM. Cell adhesion and mechanical properties of a flexible scaffold for cardiac tissue engineering. Acta Biomater. 2007;3:457–62.
22. Courtney T, Sacks MS, Stankus J, Guan J, Wagner WR. Design and analysis of tissue engineering scaffolds that mimic soft tissue mechanical anisotropy. Biomaterials. 2006;27:3631–8.
23. Zhang YH, Zhang GW, Gu TX, Li-Ling J, Wen T, Zhao Y, Wang C, Fang Q, Yu L, Liu B. Exogenous basic fibroblast growth factor promotes cardiac stem cell-mediated myocardial regeneration after miniswine acute myocardial infarction. Coron Artery Dis. 2011;22:279.
24. Munarin F, Kant RJ, Rupert CE, Khoo A, Coulombe KLK. Engineered human myocardium with local release of angiogenic proteins improves vascularization and cardiac function in injured rat hearts. Biomaterials. 2020;251:120033.
25. Nelson DM, Baraniak PR, Ma Z, Guan J, Mason NS, Wagner WR. Controlled release of IGF-1 and HGF from a biodegradable polyurethane scaffold. Pharm Res. 2011;28:1282–93.
26. Schussler O, Coirault C, Louis-Tisserand M, Al-Chare W, Oliviero P, Menard C, Michelot RJ, Bochet P, Salomon DR, Chachques JC, Carpentier A, Lecarpentier Y. Use of arginine–glycine–aspartic acid adhesion peptides coupled with a new collagen scaffold to engineer a myocardium-like tissue graft. Nat Rev Cardiol. 2009;6:240–9.
27. Kuraitis D, Hosoyama K, Blackburn NJR, Deng C, Zhong Z, Suuronen EJ. Functionalization of soft materials for cardiac repair and regeneration. Crit Rev Biotechnol. 2019;39:451–68.
28. Kim PH, Cho JY. Myocardial tissue engineering using electrospun nanofiber composites. BMB Rep. 2016;49(1):26–36.
29. Tongers J, Losordo DW, Landmesser U. Stem and progenitor cell-based therapy in ischaemic heart disease: promise, uncertainties, and challenges. Eur Heart J. 2011;32:1197–206.
30. Dvir T, Kedem A, Ruvinov E, Levy O, Freeman I, Landa N, Holbova R, Feinberg MS, Dror S, Etzion Y, Leor J, Cohen S. Prevascularization of cardiac patch on the omentum improves its therapeutic outcome. Proc Natl Acad Sci. 2009;106:14990–5.
31. Kellar RS, Landeen LK, Shepherd BR, Naughton GK, Ratcliffe A, Williams SK. Scaffold-based three-dimensional human fibroblast culture provides a structural matrix that supports angiogenesis in infarcted heart tissue. Circulation. 2001;104:2063–8.
32. Zimmermann W-H, Melnychenko I, Wasmeier G, Didié M, Naito H, Nixdorff U, Hess A, Budinsky L, Brune K, Michaelis B, Dhein S, Schwoerer A, Ehmke H, Eschenhagen T. Engineered heart tissue grafts improve systolic and diastolic function in infarcted rat hearts. Nat Med. 2006;12:452–8.
33. Shimizu T, Yamato M, Kikuchi A, Okano T. Cell sheet engineering for myocardial tissue reconstruction. Biomaterials. 2003;24:2309–16.
34. Green H, Kehinde O, Thomas J. Growth of cultured human epidermal cells into multiple epithelia suitable for grafting. Proc Natl Acad Sci. 1979;76:5665–8.
35. O'Connor N, Mulliken J, Banks-Schlegel S, Kehinde O, Green H. Grafting of burns with cultured epithelium prepared from autologous epidermal cells. Lancet. 1981;317:75–8.

36. Miyahara Y, Nagaya N, Kataoka M, Yanagawa B, Tanaka K, Hao H, Ishino K, Ishida H, Shimizu T, Kangawa K, Sano S, Okano T, Kitamura S, Mori H. Monolayered mesenchymal stem cells repair scarred myocardium after myocardial infarction. Nat Med. 2006;12:459–65.
37. Hobo K, Shimizu T, Sekine H, Shin'oka T, Okano T, Kurosawa H. Therapeutic angiogenesis using tissue engineered human smooth muscle cell sheets. Arterioscler Thromb Vasc Biol. 2008;28:637–43.
38. Kai D, Wang QL, Wang HJ, Prabhakaran MP, Zhang Y, Tan YZ, Ramakrishna S. Stem cell-loaded nanofibrous patch promotes the regeneration of infarcted myocardium with functional improvement in rat model. Acta Biomater. 2014;10:2727–38.
39. Sasagawa T, Shimizu T, Sekiya S, Haraguchi Y, Yamato M, Sawa Y, Okano T. Design of prevascularized three-dimensional cell-dense tissues using a cell sheet stacking manipulation technology. Biomaterials. 2010;31:1646–54.
40. Shimizu T, Sekine H, Yang J, Isoi Y, Yamato M, Kikuchi A, Kobayashi E, Okano T. Polysurgery of cell sheet grafts overcomes diffusion limits to produce thick, vascularized myocardial tissues. FASEB J. 2006;20:708–10.
41. Campbell GR, Turnbull G, Xiang L, Haines M, Armstrong S, Rolfe BE, Campbell JH. The peritoneal cavity as a bioreactor for tissue engineering visceral organs: bladder, uterus and vas deferens. J Tissue Eng Regen Med. 2008;2:50–60.
42. Zhu M, Li W, Dong X, Yuan X, Midgley AC, Chang H, Wang Y, Wang H, Wang K, Ma PX, Wang H, Kong D. In vivo engineered extracellular matrix scaffolds with instructive niches for oriented tissue regeneration. Nat Commun. 2019;10:4620.
43. Allison S, Ahumada M, Andronic C, Mcneill B, Variola F, Griffith M, Ruel M, Hamel V, Liang W, Suuronen EJ. Electroconductive nanoengineered biomimetic hybrid fibers for cardiac tissue engineering. J Mater Chem B. 2017;5:2402–6.
44. Fiumana E, Pasquinelli G, Foroni L, Carboni M, Bonafé F, Orrico C, Nardo B, Tsivian M, Neri F, Arpesella G. Localization of mesenchymal stem cells grafted with a hyaluronan-based scaffold in the infarcted heart. J Surg Res. 2013;179:e21–9.
45. Smith MJ, McClure MJ, Sell SA, Barnes CP, Walpoth BH, Simpson DG, Bowlin GL. Suture-reinforced electrospun polydioxanone–elastin small-diameter tubes for use in vascular tissue engineering: a feasibility study. Acta Biomater. 2008;4:58–66.
46. Shevach M, Maoz BM, Feiner R, Shapira A, Dvir T. Nanoengineering gold particle composite fibers for cardiac tissue engineering. J Mater Chem B. 2013;1:5210–7.
47. Ravichandran R, Venugopal JR, Sundarrajan S, Mukherjee S, Sridhar R, Ramakrishna S. Expression of cardiac proteins in neonatal cardiomyocytes on PGS/fibrinogen core/shell substrate for Cardiac tissue engineering. Int J Cardiol. 2013;167:1461–8.
48. Liu Q, Tian S, Zhao C, Chen X, Lei I, Wang Z, Ma PX. Porous nanofibrous poly(l-lactic acid) scaffolds supporting cardiovascular progenitor cells for cardiac tissue engineering. Acta Biomater. 2015;26:105–14.
49. Shin M, Ishii O, Sueda T, Vacanti JP. Contractile cardiac grafts using a novel nanofibrous mesh. Biomaterials. 2004;25:3717–23.
50. Ahmadkhani L, Mostafavi E, Ghasemali S, Baghban R, Pazoki-Toroudi H, Davaran S, Malakootikhah J, Asadi N, Mammadova L, Saghfi S, Webster TJ, Akbarzadeh A. Development and characterization of a novel conductive polyaniline-g-polystyrene/Fe3O4 nanocomposite for the treatment of cancer. Artif Cells Nanomed Biotechnol. 2019;47:873–81.
51. Senel Ayaz HG, Perets A, Ayaz H, Gilroy KD, Govindaraj M, Brookstein D, Lelkes PI. Textile-templated electrospun anisotropic scaffolds for regenerative cardiac tissue engineering. Biomaterials. 2014;35:8540–52.
52. Bhardwaj N, Kundu SC. Electrospinning: a fascinating fiber fabrication technique. Biotechnol Adv. 2010;28:325–47.
53. D'Amore A, Yoshizumi T, Luketich SK, Wolf MT, Gu XZ, Cammarata M, Hoff R, Badylak SF, Wagner WR. Bi-layered polyurethane—extracellular matrix cardiac patch improves ischemic ventricular wall remodeling in a rat model. Biomaterials. 2016;107:1–14.

54. Wang LY, Jiang JZ, Hua WX, Darabi A, Song XP, Song C, Zhong W, Xing MMQ, Qiu XZ. Mussel-inspired conductive cryogel as cardiac tissue patch to repair myocardial infarction by migration of conductive nanoparticles. Adv Funct Mater. 2016;26:4293–305.
55. Lang N, Pereira MJ, Lee Y, Friehs I, Vasilyev NV, Feins EN, Ablasser K, O'Cearbhaill ED, Xu CJ, Fabozzo A, Padera R, Wasserman S, Freudenthal F, Ferreira LS, Langer R, Karp JM, del Nido PJ. A blood-resistant surgical glue for minimally invasive repair of vessels and heart defects. Sci Transl Med. 2014;6:218ra6.
56. Mawad D, Mansfield C, Lauto A, Perbellini F, Nelson GW, Tonkin J, Bello SO, Carrad DJ, Micolich AP, Mahat MM, Furman J, Payne DJ, Lyon AR, Gooding JJ, Harding SE, Terracciano CM, Stevens MM. A conducting polymer with enhanced electronic stability applied in cardiac models. Sci Adv. 2016;2:e1601007.
57. Mawad D, Mansfield C, Lauto A, Perbellini F. A conducting polymer with enhanced electronic stability applied in cardiac models. Sci Adv. 2016;2(11):e1601007.
58. Edri R, Gal I, Noor N, Harel T, Fleischer S, Adadi N, Green O, Shabat D, Heller L, Shapira A, Gat-Viks I, Peer D, Dvir T. Personalized hydrogels for engineering diverse fully autologous tissue implants. Adv Mater. 2019;31:9.
59. Hodgkinson CP, Bareja A, Gomez JA, Dzau VJ. Emerging concepts in paracrine mechanisms in regenerative cardiovascular medicine and biology. Circ Res. 2016;118:95–107.
60. Chiu LLY, Montgomery M, Liang Y, Liu HJ, Radisic M. Perfusable branching microvessel bed for vascularization of engineered tissues. Proc Natl Acad Sci U S A. 2012;109:E3414–23.
61. Gao Q, He Y, Fu JZ, Liu A, Ma L. Coaxial nozzle-assisted 3D bioprinting with built-in microchannels for nutrients delivery. Biomaterials. 2015;61:203–15.
62. Hinton TJ, Jallerat Q, Palchesko RN, Park JH, Grodzicki MS, Shue HJ, Ramadan MH, Hudson AR, Feinberg AW. Three-dimensional printing of complex biological structures by freeform reversible embedding of suspended hydrogels. Sci Adv. 2015;1:10.
63. Wu W, DeConinck A, Lewis JA. Omnidirectional printing of 3D microvascular networks. Adv Mater. 2011;23:H178–83.
64. Zhang YS, Arneri A, Bersini S, Shin SR, Zhu K, Goli-Malekabadi Z, Aleman J, Colosi C, Busignani F, Dell'Erba V, Bishop C, Shupe T, Demarchi D, Moretti M, Rasponi M, Dokmeci MR, Atala A, Khademhosseini A. Bioprinting 3D microfibrous scaffolds for engineering endothelialized myocardium and heart-on-a-chip. Biomaterials. 2016;110:45–59.
65. Noor N, Shapira A, Edri R, Gal I, Wertheim L, Dvir T. 3D printing of personalized thick and perfusable cardiac patches and hearts. Adv Sci. 2019;6:10.
66. Strobel J, Brenner L, Zimmermann R, Weiss D, Zingsem J, Eckstein R, Weisbach V. Influence of duration and temperature of transport and storage prior to processing on cell quality of cord blood units—a German experience. Clin Lab. 2015;61:1453–61.
67. Tang JN, Cores J, Huang K, Cui XL, Luo L, Zhang JY, Li TS, Qian L, Cheng K. Concise review: is cardiac cell therapy dead? Embarrassing trial outcomes and new directions for the future. Stem Cells Transl Med. 2018;7:354–9.
68. Jackman CP, Ganapathi AM, Asfour H, Qian Y, Allen BW, Li Y, Bursac N. Engineered cardiac tissue patch maintains structural and electrical properties after epicardial implantation. Biomaterials. 2018;159:48–58.
69. Tompkins BA, Balkan W, Winkler J, Gyongyosi M, Goliasch G, Fernandez-Aviles F, Hare JM. Preclinical studies of stem cell therapy for heart disease. Circ Res. 2018;122:1006–20.
70. Huang K, Ozpinar EW, Su T, Tang J, Shen D, Qiao L. An off-the-shelf artificial cardiac patch improves cardiac repair after myocardial infarction in rats and pigs. Sci Transl Med. 2020;12 (538):eaat9683.
71. Ye L, Zimmermann W-H, Garry DJ, Zhang J. Patching the heart cardiac repair from within and outside. Circ Res. 2013;113:922–32.
72. Li M, Izpisua Belmonte JC. Mending a faltering heart. Circ Res. 2016;118:344–51.
73. Madonna R, Van Laake L, Davidson SM, Engel FB, Hausenloy DJ, Lecour S, Leor J, Perrino C, Schulz R, Ytrehus K, Landmesser U, Mummery CL, Janssens S, Willerson J, Eschenhagen T, Ferdinandy P, Sluijter JPG. Position paper of the European Society of Cardiology Working

Group Cellular Biology of the Heart: cell-based therapies for myocardial repair and regeneration in ischemic heart disease and heart failure. Eur Heart J. 2016;37:1789–1798D.
74. Tang J, Wang J, Huang K, Ye Y, Su T. Cardiac cell-integrated microneedle patch for treating myocardial infarction. Sci Adv. 2018;4(11):eaat9365. https://doi.org/10.1126/sciadv.aat9365.
75. Takahashi H, Yokota T, Uchimura E, Miyagawa S, Ota T, Torikai K, Saito A, Hirakawa K, Kitabayashi K, Okada K, Sawa Y, Okita Y. Newly developed tissue-engineered material for reconstruction of vascular wall without cell seeding. Ann Thorac Surg. 2009;88:1269–76.
76. Tabata Y, Ikada Y. Vascularization effect of basic fibroblast growth factor released from gelatin hydrogels with different biodegradabilities. Biomaterials. 1999;20:2169–75.
77. Vajanto I, Rissanen TT, Rutanen J, Hiltunen MO, Tuomisto TT, Arve K, Narvanen O, Manninen H, Rasanen H, Hippelainen M, Alhava E, Yla-Herttuala S. Evaluation of angiogenesis and side effects in ischemic rabbit hindlimbs after intramuscular injection of adenoviral vectors encoding VEGF and LacZ. J Gene Med. 2002;4:371–80.
78. Ferrara N, Alitalo K. Clinical applications of angiogenic growth factors and their inhibitors. Nat Med. 1999;5:1359–64.
79. Moore MAS, Hattori K, Heissig B, Shieh JH, Dias S, Crystal RG, Rafii S. Mobilization of endothelial and hematopoietic stem and progenitor cells by adenovector-mediated elevation of serum levels of SDF-1, VEGF, and angiopoietin-1. Ann NY Acad Sci. 2001;938:36–47.
80. Guo HD, Cui GH, Yang JJ, Wang C, Zhu J, Zhang LS, Jiang J, Shao SJ. Sustained delivery of VEGF from designer self-assembling peptides improves cardiac function after myocardial infarction. Biochem Biophys Res Commun. 2012;424:105–11.
81. Owens GK, Kumar MS, Wamhoff BR. Molecular regulation of vascular smooth muscle cell differentiation in development and disease. Physiol Rev. 2004;84:767–801.
82. Detillieux KA, Sheikh F, Kardami E, Cattini PA. Biological activities of fibroblast growth factor-2 in the adult myocardium. Cardiovasc Res. 2003;57:8–19.
83. Dvir T, Kedem A, Ruvinov E, Levy O, Freeman I, Landa N, Holbova R, Feinberg MS, Dror S, Etzion Y, Leor J, Cohen S. Prevascularization of cardiac patch on the omentum improves its therapeutic outcome. Proc Natl Acad Sci U S A. 2009;106:14990–5.
84. Padin-Iruegas ME, Misao Y, Davis ME, Segers VFM, Esposito G, Tokunou T, Urbanek K, Hosoda T, Rota M, Anversa P, Leri A, Lee RT, Kajstura J. Cardiac progenitor cells and biotinylated insulin-like growth factor-1 nanofibers improve endogenous and exogenous myocardial regeneration after infarction. Circulation. 2009;120:876–87.
85. Jin HK, Wyss JM, Yang RH, Schwall R. The therapeutic potential of hepatocyte growth factor for myocardial infarction and heart failure. Curr Pharm Des. 2004;10:2525–33.
86. Sala V, Crepaldi T. Novel therapy for myocardial infarction: can HGF/Met be beneficial? Cell Mol Life Sci. 2011;68:1703–17.
87. Sengupta S, Gherardi E, Sellers LA, Wood JM, Sasisekharan R, Fan TPD. Hepatocyte growth factor/scatter factor can induce angiogenesis independently of vascular endothelial growth factor. Arterioscler Thromb Vasc Biol. 2003;23:69–75.
88. Kaga T, Kawano H, Sakaguchi M, Nakazawa T, Taniyama Y, Morishita R. Hepatocyte growth factor stimulated angiogenesis without inflammation: differential actions between hepatocyte growth factor, vascular endothelial growth factor and basic fibroblast growth factor. Vasc Pharmacol. 2012;57:3–9.
89. Wang X, Li Q, Hu Q, Suntharalingam P, From AHL, Zhang J. Intra-myocardial injection of both growth factors and heart derived Sca-1(+)/CD31(-) cells attenuates post-MI LV remodeling more than does cell transplantation alone: neither intervention enhances functionally significant cardiomyocyte regeneration. PLoS One. 2014;9:e95247.

90. Diaz-Herraez P, Saludas L, Pascual-Gil S, Simon-Yarza T, Abizanda G, Prosper F, Garbayo E, Jose Blanco-Prieto M. Transplantation of adipose-derived stem cells combined with neuregulin-microparticles promotes efficient cardiac repair in a rat myocardial infarction model. J Control Release. 2017;249:23–31.
91. Pope AJ, Sands GB, Smaill BH, LeGrice IJ. Three-dimensional transmural organization of perimysial collagen in the heart. Am J Physiol Heart Circ Physiol. 2008;295:H1243–52.
92. Engelmayr GC, Cheng MY, Bettinger CJ, Borenstein JT, Langer R, Freed LE. Accordion-like honeycombs for tissue engineering of cardiac anisotropy. Nat Mater. 2008;7:1003–10.
93. Zhang BY, Montgomery M, Chamberlain MD, Ogawa S, Korolj A, Pahnke A, Wells LA, Masse S, Kim J, Reis L, Momen A, Nunes SS, Wheeler AR, Nanthakumar K, Keller G, Sefton MV, Radisic M. Biodegradable scaffold with built-in vasculature for organ-on-a-chip engineering and direct surgical anastomosis. Nat Mater. 2016;15:669–78.

Chapter 11
Total Artificial Heart

Yongjun Li and Zhengxi Xu

11.1 Introduction

Cardiovascular disease (CVD) is the leading cause of morbidity and mortality in the United States and a significant public health problem in most industrialized nations. Since 1900, it has been the leading cause of death in the United States every year except 1918 [1]. In 2005, CVD afflicted 80 million Americans and caused 864,500 deaths. Meanwhile, the number of people with CVD, and especially its advanced forms, has been increasing. There are two main reasons for this. First, although there is still no cure for CVD, palliative therapy has improved to the point that more people are surviving past their initial episodes of CVD to live on with some form of the disease. Second, the average age of the US population is rising as the "baby boom" generation ages.

An increasingly prevalent form of advanced CVD is heart failure (HF). Almost 5.7 million Americans (approximately 3.2 million men and 2.5 million women) are living with HF [1]. Its etiology can be ischemic, idiopathic, or viral. More than $37 [2] billion is spent each year on the care of HF patients, and many therapeutic advances have been made. In 2005, HF directly caused 53,000 deaths and indirectly caused another 250,000. As large as the problem is now, its magnitude is expected to worsen as more cardiac patients are able to survive and live longer with their disease and thus increase their chances of developing end-stage HF.

At present, treatment of advanced HF takes three forms: medical therapy, surgical therapy, and cardiac support or replacement [2, 3]. Medical therapy (e.g., intravenous inotropes and vasodilators) relieves symptoms by reducing cardiac work and increasing myocardial contractility. However, although advances in medical therapy

Y. Li (✉) · Z. Xu
Department of Vascular Surgery, Beijing Hospital, National Center of Gerontology, Beijing, China
e-mail: liyongjun4679@bjhmoh.cn; xuzhengxi4772@bjhmoh.cn

Y.-D. Tang et al. (eds.), *Application of Biomaterials in the Treatment of Cardiovascular Diseases*, https://doi.org/10.1007/978-981-19-7712-1_11

have helped improve quality of life for those with heart failure, mortality remains unaffected. Surgical therapy (e.g., revascularization and valve replacement or repair) relieves symptoms of ischemia and valvular dysfunction but in most cases does not stop the underlying disease process from progressing. When conventional medical and surgical therapies for HF are exhausted, cardiac support or replacement (e.g., heart transplantation or implantation of an artificial heart or ventricular assist device) may in some cases become the only therapeutic alternative.

Heart transplantation has evolved into a suitable treatment for advanced HF. However, it has severe limitations related to patient selection, organ procurement and distribution, and cost-effectiveness. Slightly more than 2000 patients with end-stage heart failure receive heart transplants each year in the United States. However, about 3000 patients are on the active heart transplant waiting list at any given time, and as many as 40,000 more are potential candidates for heart transplantation [4]. Heart transplantation for the relatively young (<40 years old) is not very promising because the life expectancy of a donor heart recipient is about 10 years on average and 20 years at most. In 2004, 460 patients on the active waiting list died while awaiting a donor heart. Heart transplantation is also associated with continuous, lifelong, expensive medical therapy.

To help overcome these limitations, engineers and physicians have continued efforts begun over four decades ago to develop systems for providing either temporary or permanent mechanical circulatory support (MCS). Originally, such systems were intended to support patients indefinitely because other forms of heart replacement did not appear to be feasible. Temporary MCS has been shown to be a suitable option for some HF patients who are awaiting heart transplants [5, 6] and for others who are not transplant candidates but need support for indefinite periods [7]. In recent clinical studies, myocardial function improved sufficiently in some cases to allow removal of the MCS device and avoid heart transplantation [8, 9]. Nevertheless, in light of the shortcomings of medical therapy, surgical therapy, and heart transplantation, efforts have continued to develop a total artificial heart (TAH) that would not only save the lives of critically ill HF patients but also allow them to resume relatively normal lifestyles. Here, we review the historical development and current status of TAH technology.

11.2 Early Development and Experience

In 1812, LeGallois first proposed the idea of supporting a failing heart with either a permanent or temporary device [10]. In the 1920s, Lindbergh and Carrel discussed and planned an artificial heart [11]. Throughout the 1940s, researchers including Dennis and Gibbon were developing a machine that would bypass the circulation of the heart and lungs to allow open-heart surgery. The modern era of MCS began in 1951 when Dennis first used a heart-lung machine to sustain the circulation while the heart was opened to repair an atrial septal defect [12]. Two years later, Gibbon repeated this procedure [13]. However, high mortality in the first few cases led both

Dennis and Gibbon to abandon the use of their heart-lung machines. In 1954, Lillehei began to use cross-circulation (human-to-human perfusion) as a means to support heart and lung function during congenital heart defect [4]. However, because of the controversy created by Lillehei's procedure, researchers continued efforts to develop a machine that would allow open-heart surgery.

By 1955, Kirklin and associates at the Mayo Clinic had refined the Mayo-Gibbon machine and the techniques that allowed open-heart surgery [14]. Likewise, Lillehei and associates had developed their machine that also allowed for safe open-heart operations [15]. By 1960, Kirklin and Lillehei in Minneapolis and DeBakey and Cooley in Houston had perfected their machines and techniques to the point where heart surgery was becoming routine in Minnesota and Texas. The early developmental work on the use of mechanical circulatory systems by Dennis, Lillehei, De Wall, Gibbon, and Kirklin allowed for many new cardiac operations, including coronary artery bypass, heart transplantation, valve repair, and implantation of a TAH. After refinements of the heart-lung machines, DeBakey and Cooley began to develop many of the surgical techniques that eventually made open-heart surgery routine around the world. In 1957, Akutsu and Kolff became the first to implant a TAH in vivo [16]. Inserted into the chest of a dog, the pump adequately maintained the circulation for approximately 90 min. However, Akutsu and Kolff never applied their TAH technology clinically. In 1964, the National Heart Institute established the Artificial Heart Program to promote the development of the TAH and other cardiac assist devices. In the early 1960s, DeBakey and researchers at Baylor College of Medicine in Houston began developing a TAH. In 1963, DeBakey implanted the first clinical left ventricular assist device (LVAD) into a 42-year-old patient [17]. The pump functioned well, but the patient died of pulmonary complications after 4 days of support. In 1967, DeBakey implanted an LVAD into a 37-year-old who presented with symptoms of HF, including easy fatigability and severe dyspnea on slight exertion. This patient also had history of rheumatic heart disease since age 18 and closed mitral valvulotomy at age 25. The intention was to use the LVAD until sufficient myocardial recovery could be gained. The LVAD supported the patient's circulation for 10 days and was then electively removed. The patient was discharged from the hospital on postoperative day 29 and later resumed normal activity. On follow-up at 18 months after LVAD removal, the patient remained free of HF symptoms, and a chest X-ray showed a significant reduction in cardiac size.

11.3 First Total Artificial Heart

The first implantation of a TAH into a human was done by Cooley on April 4, 1969, in a 47-year-old man who could not be weaned from cardiopulmonary bypass (CPB) following left ventricular aneurysmectomy [2]. The intent was to support the patient until a donor heart could be found. The TAH designed by Liotta was a pneumatically powered, double-chambered pump with Dacron-lined right and left inflow cuffs and outflow grafts. Wada-Cutter hingeless valves controlled the direction of blood flow through the pump. The TAH itself was connected to a large external power unit, which unfortunately severely restricted patient mobility. The TAH performed adequately for 64 h until transplantation. The donor heart also functioned well, but the patient died of pseudomonal pneumonia 32 h after transplantation. Though the Liotta device performed as designed, it was never used clinically again. Nevertheless, this case clearly demonstrated that a TAH could be used safely and effectively in a human as a bridge to transplantation.

11.4 Akutsu-III Total Artificial Heart

The second implantation of a TAH in a human was also done by Cooley. On July 23, 1981, Cooley implanted the Akutsu-III TAH into a critically ill 26-year-old man who suffered heart failure after undergoing coronary artery bypass surgery for severe arteriosclerosis. Unable to be weaned from CPB after surgery, the patient was fitted with the TAH in a final effort to sustain his life. The Akutsu-III TAH consisted of two pneumatically powered, double-chambered pumps featuring reciprocating hemispherical diaphragms [18]. The TAH provided excellent hemodynamics and supported the patient in stable condition for a total of 55 h until a suitable donor heart was found. The patient finally received a transplant but died of infectious, renal, and pulmonary complications 10 days later. Despite the fatal outcome, this case demonstrated that a TAH could adequately sustain a patient for several days, with no evidence of hemolysis or thromboembolism, until heart transplantation.

11.5 Jarvik-7/CardioWest Total Artificial Heart

In the late 1970s, Kolff and his team at the University of Utah developed the Jarvik-7 TAH. In 1982, DeVries became the first to permanently implant a TAH when he implanted a Jarvik-7 into a dying patient [19]. The Jarvik-7 TAH was a pneumatically powered, biventricular pulsatile device that replaced the heart [3, 20]. The pumps were connected to their respective native atria by synthetic cuffs and connectors. Each pump had chambers for air and blood separated by a smooth, flexible polyurethane diaphragm. The inflow and outflow conduits contained Medtronic-Hall tilting disk valves. The filling of the pumps was aided by vacuum. Pneumatic drivelines, brought out through the chest wall to connect with an external console, shuttled air to the pumps during systole, causing collapse of the pump sac and blood ejection. Pump rate, drive pressure, and systolic duration were monitored and optimized from the external console. The Jarvik-7 had a stroke volume of 70 mL and a normal cardiac output of 6–8 L/min (maximum, 15 L/min).

In the initial clinical experience with the Jarvik-7 TAH, a total of five patients were permanently supported for periods ranging from 10 to 620 days. The TAH was able to adequately support circulation, but its large drive console and frequent medical complications limited patient activity. Four patients were able to make brief trips out of the hospital to see family and friends. Long-term outcomes, however, were poor. Patients supported by the Jarvik-7 for longer periods suffered several complications, including thromboembolism, stroke, infection, and multiorgan failure.

Despite these mixed results, in 1985 the Jarvik-7 (renamed the Symbion) TAH entered clinical trials as a bridge to transplantation. In 1986, Copeland reported the first successful use of this device for this indication [21]. Between 1985 and 1991, approximately 170 patients were supported with the Symbion TAH as a bridge to transplantation [22]. Sixty-six percent underwent successful heart transplantation, a rate similar to those in bridge-to-transplantation studies of left ventricular assist devices. Sepsis and multiple organ failure were the primary causes of death during TAH support.

Although the bridge-to-transplantation study demonstrated that the Jarvik-7 (Symbion) TAH was clinically safe and effective, the US Food and Drug Administration (FDA) withdrew the device's investigational device exemption (IDE) for clinical use in January 1991 because of inadequate compliance with FDA regulations [23]. In January 1993, the IDE was restored to the CardioWest TAH, which differed little from the original Jarvik-7. In 2001, Syncardia, Inc. was formed, and the CardioWest TAH-t has since been used successfully in the United States, Canada, and France [24]. In the US clinical trials, survival to transplantation and overall 1-year survival rates were significantly higher for CardioWest TAH-t recipients ($n = 81$) than for controls ($n = 35$) (79% vs. 46% and 70% vs. 31%, respectively). These results led to market approval of the CardioWest TAH-t as a bridge-to-transplantation device in 2004. By 2009, more than 800 implants have been performed at more than 40 centers worldwide.

11.6 AbioCor Implantable Replacement Heart

On July 2, 2001, as part of an FDA-sponsored phase I clinical trial, surgeons at Jewish Hospital in Louisville, Kentucky, performed the first implantation of the AbioCor TAH in a 59-year-old man with end-stage HF [25]. The AbioCor Implantable Replacement Heart is a self-contained, electrohydraulic TAH that has been developed and tested by ABIOMED, Inc. (Danvers, MA) and the Texas Heart Institute, with the support of the National Heart, Lung, and Blood Institute (NHLBI) [26, 27]. It is designed to sustain the circulation and extend the lives of patients with end-stage HF who have suffered irreversible left and right ventricular failure, for whom surgery or medical therapy is inadequate, and who would otherwise soon die. The AbioCor is the first TAH to be used clinically that is fully implantable and communicates to external hardware without penetrating the skin. The device utilizes a transcutaneous energy transfer system (TET) and a radiofrequency communication system that allows it to be powered and controlled by signals transmitted across intact skin. A unique feature of the AbioCor is a right-left flow-balancing mechanism that eliminates the need for an external vent or internal compliance chamber [28].

The internal components of the AbioCor system consist of a thoracic unit, internal TET coil, controller, and battery [29]. The thoracic unit (pump) weighs about 2 lb and consists of two artificial ventricles, four valves, and an innovative motor-driven hydraulic pumping system. The pump's motor rotates at 6000–8000 rpm, which allows sufficient hydraulic fluid pressure to compress the diaphragm around the blood chamber and eject blood. A miniaturized electronics package implanted in the patient's abdomen monitors and controls the pump rate, right-left balance, and motor speed. An internal rechargeable battery, also implanted within the abdomen, provides emergency or backup power. The internal battery is continually recharged via the TET system and can provide up to 30 min of untethered operation. The AbioCor's external components include a computer console, an external TET coil, and external battery packs.

The external computer communicates via the RF Comm system with the abdominally implanted controller, which controls the pump. The external TET coil provides the pump with power from the console or from the external battery packs. The external battery packs can power the AbioCor TAH for 2–4 h.

In the phase I feasibility trial, 14 patients received the AbioCor as destination therapy. Thromboembolism was a significant adverse event in this group of patients, and there were two pump failures. However, compared with the known rates of infection for implantable systems with percutaneous drivelines, device-related infections in this trial were relatively rare. The longest surviving patient was supported for 512 days, most of which lived at home with a good quality of life. In 2006, the FDA approved the AbioCor for commercial use under a Humanitarian Device Exemption.

11.7 Complications

Use of a TAH is associated with serious complications. The most frequent complications are infection, severe postoperative bleeding, and thromboembolism [30–33]. Potentially serious but less frequent complications are renal, hepatic, pulmonary, and neurologic dysfunction and complications caused by technical problems [30, 34]. Complicating factors include patient selection, device size, implantation timing and location, the need for extensive surgery at implantation, and the reliability of support equipment.

Life-threatening infections have been the most important complication for patients being supported permanently by a TAH [35]. In the Jarvik-7 experience, all patients supported for many months developed serious infections that eventually contributed to their deaths [9, 36]. Patients supported by a TAH for shorter periods while awaiting heart transplantation had infection rates of 30–40% [31–35]. During the 1980s, driveline and mediastinal infections in TAH-supported patients were frequent and severe, regardless of the duration of support. However, in the more recent bridge-to-transplantation experience with the CardioWest TAH-t, serious infections contributing to death or delayed transplantation occurred in only 21% of patients [37, 38].

Patients supported with a TAH, regardless of the intended use, are very susceptible to infection. Predisposing factors include the tissue trauma associated with surgical implantation; contamination of the implanted device; depression of the body's immune defenses; the large amount of foreign material presented on the surface of the device; and use of the drivelines, tubes, catheters, and other devices that are necessary for the care of these patients. Infections can occur at any point during TAH support. Once an internal component of the TAH system is infectiously colonized, treatment is difficult and often ineffective. Infections are more likely to occur in the early postoperative period, especially in the most critically ill patients, as a result of device contamination during the course of care and postoperative bleeding caused by intensive care procedures and exposure during reoperation. Meticulous care and numerous infection prevention measures are viral in all TAH patients.

Postoperative bleeding is a frequent and serious complication of TAH implantation. It generally occurs in 40–50% of TAH or ventricular assist device recipients [27]. In the more recent CardioWest TAH-t experience, the rate was approximately 25%. Contributing factors include severe HF and associated hepatic dysfunction, the extensive surgery and lengthy CPB time required for implantation, and the necessity for postoperative anticoagulation therapy. Severe HF often leads to hepatic dysfunction, long operation time and extracorporeal circulation time, postoperative anticoagulation treatment, etc. Severe heart failure leads to hepatic dysfunction and impairs the patient's blood clotting function. Patients with severe heart failure receive anticoagulation and antiplatelet therapy before surgery, and the anticoagulation effect is difficult to reverse when TAH is implanted. The prolonged operation time and extracorporeal circulation time will lead to the loss of the clotting factor. Postoperative anticoagulation therapy needs to establish a balance between coagulation and fibrinolysis and better control of the coagulation system.

Thrombosis in TAH is of special concern. In the first six Jarvik-7 TAH implants, five patients had thromboembolic events. However, currently, the incidence of such thrombotic events is only 10–25% [30, 31, 37]. Measures taken include close monitoring of the patient's coagulation and fibrinolysis system and adequate anticoagulation and antiplatelet therapy. Generally speaking, heparin and warfarin are used as anticoagulant measures to maintain the PT time, and the value of ACT or INR is two to three times the normal value. At the same time, aspirin or (and) dipyridamole can also be used.

The relationship between infection, bleeding, and thromboembolism is very complicated. The formation of blood clots can lead to the formation of infections, and infections can also lead to the formation of blood clots. Bacterial culture of thrombus in the device often yields positive results [38]. Anti-infection and white blood cells can embed bacteria in the thrombus. Bacteria, endotoxin, and inflammatory cells can affect platelet aggregation and accelerate the formation of thrombus [38]. Bacterial endotoxin can induce platelet aggregation and endothelial damage and increase the activity of endothelial cell thromboplastin. Excessive bleeding leads to reopening of the chest to stop the bleeding, which increases the risk of contamination again. Both blood transfusion and intravascular monitoring can expose patients to the external environment. Infection, bleeding, and thromboembolism can lead to multiple organ failure, which is an important cause of death after TAH implantation.

In addition, complications related to device dysfunction after TAH implantation also need to be paid attention to. The mismatch between TAH and the patient, social and belief issues, activity, and nutrition issues are also key factors determining the outcome of treatment. Complications with catastrophic consequences due to device obstructions are relatively rare. The mismatch between TAH and patients is still a problem, and all TAHs currently only apply to patients with a body surface area greater than 1.7 m^2. Because TAH is expensive, it obviously increases the social and economic burden. The TAH volume is usually large, which limits the patient's activities and rehabilitation exercises. Further design can consider appropriately increasing the patient's postoperative space for activities.

11.8 Total Artificial Heart

11.8.1 Research and Development

It has been shown that HF can be treated with volume-displacement TAHs, such as the pneumatic CardioWest TAH-t or the AbioCor TAH. The CardioWest TAH-t has successfully supported patients awaiting transplantation [39]. Although these TAHs have been used successfully in some patients, their large size, mechanical complexity, and poor durability have also limited their application.

Recently, new design applications using continuous-flow rotary blood pumps have emerged to create a smaller and more compact alternative to the larger volume-displacement TAHs. Unlike the pulsatile, volume-displacement TAHs, these rotary pumps have no flexible membranes, pusher-plates, or prosthetic valves that can wear out. They are actuated by rapidly spinning impellers, which reduce their mechanical complexity. Rotary pumps are more durable, making them less expensive to produce. In addition, these pumps are smaller, quieter, and more energy efficient than are volume-displacement pumps. Lastly, rotary pumps are inflow pressure sensitive—they closely imitate the native heart by autonomously increasing pump flow in response to an increasing preload. Surprisingly, this technology has never been used clinically in a TAH configuration, despite the significant need for a cardiac replacement device that is smaller, simpler, and less expensive. Therefore, constant-flow technology may be ideally suited for integration into the next generation of permanently implantable CRDs.

Recently, at the Texas Heart Institute, we have begun extensive testing of rotary blood pumps to assess the feasibility of using two continuous-flow pumps for totally implantable biventricular replacement. Axial-flow and centrifugal-type blood pumps are being developed for use as CRDs. Initially, the constant-flow pumps were designed and used clinically as left ventricular assist devices (LVADs). However, it was noted, particularly in pumps placed outside the ventricle, that the ventricle itself acted as a reservoir and that the aortic valve did not open. Therefore, technically, there was no pulse. Despite this, patients had normal physiologic function, even years after the device was implanted. This fact supports our belief that constant-flow pumps could be a feasible option for a CRD. The use of two independent constant-flow pumps modifies and may even eliminate the need for a separate control mechanism for atrial balance. Each pump's flow will be regulated by the pressure differential between the inflow and the outflow, even at a constant rpm. Testing of that concept in our laboratory has also proved promising, with animal survival times of 7 weeks after total cardiac replacement with two rotary pumps. This was achieved by maintaining a constant rpm, even in treadmill studies. The normal physiologic response gives us some hope that an individualized flow might be assessed for each patient, without the need for a complex, integrated system. Additional programs have also begun to investigate the use of rotary blood pumps as CRDs. The Cleveland Clinic is developing a valveless, sensorless, pulsatile, continuous-flow double-pump TAH. The impellers of both pumps are attached to a single axial rotor. Atrial pressures are balanced by way of passive self-regulation (the rotor axial position and subsequent position of the impeller with regard to the right pump inflow). If the atrial pressure on either the left or right sides is increased, the rotor is pushed to the contralateral side, forcing the impeller either to block or open the right inflow opening, thus increasing or decreasing right pump flow. A group at Queensland University of Technology is also working on a single-unit rotary device CRD that uses a shared rotating hub with two impellers. This device employs an interpump shunt (similar to a ventricular septal defect) for left-to-right atrial pressure balance, allowing blood to flow between the two pumps and beneath the pump impellers. At Tohoku University, two centrifugal pumps, each with independent control algorithms, are being tested as a CRD.

11.8.2 Comment

Since the 1950s, when the heart-lung bypass machine was developed, many advances have been made in the surgical treatment of HF. Many surgical procedures considered impossible just 40 years ago are today considered routine. A classic example is heart transplantation. However, TAH technology has not evolved at the same pace. Although the first human heart transplantation and first human TAH implantation occurred within 2 years of each other, TAH implantation is still neither routine nor widely available. However, two TAHs undergoing clinical trials are present in the United States. The CardioWest (formerly the Jarvik-7) TAH-t, which is the more widely used of the two, is now approved for use as a bridge to transplantation. The AbioCor TAH is still in the early stages of clinical testing and is likely years away from approval as an alternative to heart transplantation. However, should its unique TET system and flow-balancing mechanism prove to be reliable for extended periods of time, the AbioCor may become a widely used alternative to heart transplantation for those patients who have no other treatment options.

There are many obstacles to overcome before any TAH is widely accepted. Infection, bleeding, thromboembolism, and biocompatibility issues are serious problems that affect nearly all implantable cardiovascular devices including TAHs. Improved biomaterials, better prevention, and more effective antibiotic and anticoagulant medications may help overcome these problems. Acceptance by the public, by some critics in the health professions, and by third-party payors may be slow. Quality in manufacturing is needed to ensure reliability of TAH components. Addressing these problems will help bring the TAH more quickly into routine clinical use.

Key Points

- Careful patient selection is key to long-term, successful outcomes. An institution's medical review board should approve potential candidates to ensure that only appropriate patients are selected for CRD implantation. Biventricular function should be carefully assessed in patients to rule out if they are better candidates for LVAD implantation.
- Bleeding complications can be multifactorial. Patients with poor platelet and liver function, those who undergo prolonged cardiopulmonary bypass times during the implant procedure, or those who have preoperative coagulopathies will likely have bleeding complications. Attention to detail during the TAH implant procedure can minimize bleeding complications and ensure faster recovery times.
- Infection prevention measures must be used in TAH patients. Should large, blood-contacting foreign bodies in the chest become infected, they are difficult to treat. All indwelling catheters should be replaced often to minimize the risk of infection. Cardiac replacement devices with external drivelines must be meticulously cared for to reduce the risk of infection.
- Device size and portability are important factors. The AbioCor Implantable Replacement Heart can only be used in larger patients with the appropriate thoracic dimensions. Imaging studies must be performed to ensure that the thoracic anatomy will allow for optimal device fit. The CardioWest TAH-t employs a pneumatic external driver, which may limit patient mobility and normal activity.

References

1. American Heart Association. Heart and stroke statistical update. 2002. Available at http://www.americanheart.org. Accessed Oct 2005.
2. Frazier OH, Myers TJ. Surgical therapy for severe heart failure [review]. Curr Probl Cardiol. 1998;23:721.
3. United Network for Organ Sharing. Critical data: U.S. Facts about transplantation. Available at http://www.unos.org. Accessed Apr 2002.

4. Kirklin JW, DuShane JW, Patrick RT, et al. Intracardiac surgery with the aid of a mechanical pump-oxygenator system (Gibbon type): report of eight cases. Staff Meet Mayo Clin. 1955;30: 201.
5. Frazier OH, Rose EA, Oz MC, et al. Multicenter clinical evaluation of the HeartMate vented electric left ventricular assist system in patients awaiting heart transplantation. J Thorac Cardiovasc Surg. 2001;122:1186.
6. Rose EA, Gelijns AC, Moskowitz AJ, et al. Long-term mechanical left ventricular assistance for end-stage heart failure. N Engl J Med. 2001;345:1435.
7. Mueller J, Weng Y, Dandel M. Long-term results of weaning from LVAD: it does work. ASAIO J. 1999;45:153.
8. Frazier OH, Myers TJ. Left ventricular assist system as a bridge to myocardial recovery. Ann Thorac Surg. 1999;68(2 Suppl):734.
9. LeGallois CJJ. Experience on the Principle of Life. Philadelphia, Thomas, 1813. Translation of LeGallois CJJ: Experience sur la principe de la vie. Paris. 1812.
10. Miller GW. King of hearts: the true story of the maverick who pioneered open heart surgery. New York: Random House; 2000.
11. Dennis C. A heart-lung machine for open-heart operations: how it came about. Trans Am Soc Artif Intern Organs. 1989;35:767.
12. Gibbon JH. Application of a heart and lung apparatus to cardiac surgery. Minn Med. 1954;37: 171.
13. Lillehei CW, Cohen M, Warden HE, Varco RL. The direct-vision intracardiac correction of congenital anomalies by controlled cross circulation. Surgery. 1955;38:11.
14. Lillehei CW, DeWall RA, Read R, et al. Direct-vision intracardiac surgery in man using a simple, disposable artificial oxygenator. Dis Chest. 1956;29:1.
15. Akutsu T, Kolff WJ. Permanent substitute for valves and hearts. Trans Am Soc Artif Intern Organs. 1958;4:230.
16. DeBakey ME. Left ventricular bypass for cardiac assistance: clinical experience. Am J Cardiol. 1971;27:3.
17. Cooley DA, Liotta D, Hallman GL, et al. Orthotopic cardiac prosthesis for two-staged cardiac replacement. Am J Cardiol. 1969;24:723.
18. DeVries WC. The permanent artificial heart: four case reports. JAMA. 1988;259:849.
19. DeVries WC, Anderson JL, Joyce LD, et al. Clinical use of the total artificial heart. N Engl J Med. 1984;310:273.
20. Copeland CG, Smith RG, Icenogle TB, Ott RA. Early experience with the total artificial heart as a bridge to cardiac transplantation. Surg Clin North Am. 1988;68:621.
21. Johnson KE, Prieto M, Joyce LD, et al. Summary of the clinical use of the Symbion total artificial heart: a registry report. Ann Surg. 1995;222:327.
22. Copeland JG. Current status and future directions for a total artificial heart with a past. Artif Organs. 1998;22:998.
23. Copeland JG, Smith RG, Arabia FA, et al. and the CardioWest Total Artificial Heart Investigators: cardiac replacement with a total artificial heart as a bridge to transplantation. N Engl J Med. 2004;351:859.
24. SoRelle R. Cardiovascular news. Totally contained AbioCor artificial heart implanted July 3, 2001. Circulation. 2001;104:E9005.
25. Kung RTV, Yu LS, Ochs BD, et al. Progress in the development of the ABIOMED total artificial heart. ASAIO J. 1995;41:M245.
26. Parnis SM, Yu LS, Ochs BD, et al. Chronic in vivo evaluation of an electrohydraulic total artificial heart. ASAIO J. 1994;40:M489.
27. Kung RTV, Yu LS, Ochs BD, et al. An artificial hydraulic shunt in a total artificial heart: a balance mechanism for the bronchial shunt. ASAIO J. 1993;39:M213.
28. Yu LS, Finnegan M, Vaughan S, et al. A compact and noise-free electrohydraulic total artificial heart. ASAIO J. 1993;39:M386.

29. Quaini E, Pavie A, Chieco S, Mambrito B. The Concerted Action 'Heart' European registry on clinical application of mechanical circulatory support systems: bridge to transplant. The Registry Scientific Committee. Eur J Cardiothorac Surg. 1997;11:182.
30. Mehta SM, Aufiero TX, Pae WE Jr, et al. Combined registry for the clinical use of mechanical ventricular assist pumps and the total artificial heart in conjunction with heart transplantation: sixth official report—1994. J Heart Lung Transplant. 1995;14:585.
31. Myers TJ, Khan T, Frazier OH. Infectious complications associated with ventricular assist systems. ASAIO J. 2000;46:S28.
32. Conger JL, Inman RW, Tamez D, et al. Infection and thrombosis in total artificial heart technology: past and future challenges—a historical review. ASAIO J. 2000;46:S22.
33. Arabia FA, Copeland JG, Smith RG, et al. International experience with the CardioWest total artificial heart as a bridge to heart transplantation. Eur J Cardiothorac Surg. 1997;11(Suppl):S5.
34. Gristina AG, Dobbins JJ, Giammara B, et al. Biomaterial-centered sepsis and the total artificial heart: microbial adhesion versus tissue integration. JAMA. 1988;259:870.
35. Joyce LD, DeVries WC, Hastings WL, et al. Response of the human body to the first permanent implant of the Jarvik-7 total artificial heart. Trans Am Soc Artif Intern Organs. 1983;29:81.
36. DeVries WC. Surgical technique for implantation of the Jarvik-7-100 total artificial heart. JAMA. 1988;259:875.
37. Copeland JG 3rd, Smith RG, Arabia FA, et al. Comparison of the CardioWest total artificial heart, the Novacor left ventricular assist system, and the Thoratec ventricular assist system in bridge to transplantation. Ann Thorac Surg. 2001;71(3 Suppl):S92.
38. Copeland JG, Smith RG, Arabia FA, et al. The CardioWest total artificial heart as a bridge to transplantation. Semin Thorac Cardiovasc Surg. 2000;12:238.
39. Copeland JG, Pavie A, Duveau D, et al. Bridge to transplantation with the CardioWest total artificial heart: the international experience 1993 to 1995. J Heart Lung Transplant. 1995;15 (1 Pt1):94.

Chapter 12
Cardiac Pacemaker and Defibrillator

Yida Tang

12.1 Introduction

Cardiac implantable electronic device (CIED) is a mainstay of contemporary cardiology practice, which provides lifesaving support and improves survival in a variety of cardiac disorders [1]. Since the first pacemaker prototype was invented in 1950 by John Hopps [2] and the first implantable pacemaker was implanted in 1958 [3], there had been many milestones in the evolutionary history of CIED therapy. There have been three primary stages of CIEDs over the past few decades: permanent pacemakers (PPMs), implantable cardioverter-defibrillators (ICDs), and cardiac resynchronization therapy (CRT) devices (PPM or ICD).

Rapid advances and transitions in pacing and defibrillator technology supported by findings of clinical trials have improved the versatility and function of implantable devices for the treatment of cardiac disorders. Since the inception of the original implantable pacemakers to provide lifesaving pacing with fixed rate in patients with bradycardia, modern pacemakers with longer battery longevity can simulate the heart's normal automaticity and atrioventricular (AV) activation patterns, additionally with the adaption of pacing to physical activity.

However, rapid advancements have created new challenges. In late 1950s, right ventricular pacing was the original breakthrough [4]. With the risk of pacing-induced dyssynchrony by right ventricular pacing, biventricular pacing provided a viable way to mitigate dyssynchrony, but only certain patients (primarily left bundle branch block and QRS duration ≥150 ms with decreased left ventricular ejection fraction, LVEF) respond. Recent advances have made His bundle and/or left bundle branch pacing an optimal option which may provide physiological pacing for a

Y. Tang (✉)
Department of Cardiology and Institute of Vascular Medicine, Peking University Third Hospital, Beijing, China
e-mail: tangyida@bjmu.edu.cn

Y.-D. Tang et al. (eds.), *Application of Biomaterials in the Treatment of Cardiovascular Diseases*, https://doi.org/10.1007/978-981-19-7712-1_12

wider patient population. Additionally, transvenous endocardial leads of traditional CIED systems have been challenged by subcutaneous ICDs and leadless pacing, which have been proposed as optimal for certain patients. In recent years, surgical implantation of pacemakers and ICDs can be performed on an outpatient basis, with low risk and minimal morbidity, which allows patients to return to social and psychological functional capacity quickly.

As mentioned above, rapid technological evolution and effectiveness promise to make the future of CIEDs exciting.

12.2 Brief History of Cardiac Pacemakers

12.2.1 Traditional Pacemakers

With the aging of population, the population of patients requiring implantation of pacemakers is expected to continue increasing [5]. Traditional pacemaker includes single-chamber pacing and dual-chamber pacing. Single-chamber pacing is usually indicated for patients with only one lead either in the atrium or ventricle, which is parallel to pacing and sensing in the respective chamber. Unfortunately, the primary drawback of single ventricular pacing is lack of AV synchrony. The advantages of dual-chamber AV pacing have been demonstrated in previous researches, including AV synchrony and arrhythmia recognizing and mode switching during an atrial tachyarrhythmia. However, the disadvantages of dual-chamber AV pacing involve the failure of "true" AV synchrony and the possibility of pacemaker-mediated tachycardia (PMT).

Fortunately, a large growth in the sophistication of current generation pacemakers has been manifested in last decades. As newer PPMs have acquired increasing functionality, the interpretation of CIEDs has become progressively complex. Enhancing physiological capabilities to sense, pace, and modulate heart rate in response to patient activity involves pacing from multiple locations within a single chamber and pacing selectively when required. In addition, CIEDs are not only for therapy but also for diagnostic purpose.

12.2.2 Biventricular Pacemakers

Researches have demonstrated the role of intraventricular conduction delay related to bundle branch block or right ventricular pacing and its translation to inefficient cardiac function [6–8]. In patients with severe systolic heart failure, the disorganized ventricular contraction may result in significant decrement in cardiac pumping efficiency and progressive mitral regurgitation.

Biventricular pacing has emerged as a significant therapeutic modality for managing impaired cardiac function, with its development driven by the concept of "dyssynchrony" [9].

12.2.2.1 Cardiac Resynchronization Therapy

The effect of electrical dyssynchrony on systolic heart failure heralded in the development of cardiac resynchronization therapy (CRT) [9]. CRT which is characterized by biventricular pacing consists of simultaneous pacing from the right and left ventricles. As "resynchronization" implies, the goal of CRT is to electrically resynchronize the ventricular septum and the LV lateral wall of the heart to improve mechanical dyssynchrony. CRT also can be incorporated into either a PPM or an ICD: a pacing-ICD lead is placed in the right ventricle, and an additional pacing lead is placed in the lateral wall of the left ventricle.

12.2.2.2 Multipoint Pacing

Although the benefits of CRT in systolic heart failure have been demonstrated, about 25% of patients who underwent CRT do not respond positively to therapy [10]. The multipoint pacing (MPP) has offered a new choice for patients who failed traditional CRT. In MPP, more myocardium of the left ventricle can be recruited to participate in resynchronization by pacing from two sites on LV lead synchronously.

12.2.2.3 His Bundle and Left Bundle Branch Pacing

The drawbacks of traditional RV pacing have been described as worsening heart failure symptoms and the progression of a pacing-mediated cardiomyopathy in patients with a higher RV pace proportion (greater than 40%) [11–13]. Therefore, physiological pacing has been pursued for the last decade in patients with a bradycardia or CRT indication for pacing. As an alternative to RV pacing, His bundle pacing (HBP) delivers physiological ventricular activation that may attenuate symptoms and reduce readmissions due to heart failure [14–22].

Data from observational studies suggests HBP can also be used to correct left bundle branch block (LBBB) and may deliver more effective ventricular resynchronization than biventricular pacing in patients with RV pacing-induced cardiomyopathy [14, 20, 23].

However, HBP which utilizes the native cardiac conduction system has inherent limitations. Conduction system capture often fails to provide adequate pacing in patients with infra Hisian block or proximal left bundle branch block (LBBB); therefore, a more distal conduction system pacing is required [24, 25]. Increased pacing thresholds, low R-wave amplitudes, and re-intervention rates of 2.7–7.5% with His pacing techniques continue to be problems in lead [26–30]. In the largest observational study, 844 patients who underwent HBP lead implantation had a mean pacing capture threshold of 1.6 V at implant and 2 V at median follow-up of 3 years [30].

At median follow-up, 28% of patients suffered increased pacing thresholds of 2.5 V at 1 ms. Nine percent of patients were indeed septal pacing in follow-up electrocardiography, and 7.6% of patients required lead revision procedures. Median time of battery exhausted was 5.8 years [30]. Additionally, the operational difficulty with relatively long fluoroscopic exposure times and the longer learning curve have been the barriers for the widespread application of HBP technique [31–33].

Since the first case of successful cardiac resynchronization with left bundle branch pacing (LBBP) was reported by Dr. Huang [34], LBBP has emerged as a promising alternative to HBP [35–37]. An optimal pacing site is characterized by the pacing lead implanted deep in the RV basal septum to capture the left bundle branch (LBB) area in a patient with heart failure and left bundle branch block (LBBB). Data from observational studies and case reports suggest LBBP is characterized with a wide target of anatomy, a low and stable pacing capture threshold related to histology, and relatively narrow QRS duration (QRSd) due to direct excitation of LBB [34, 38–43].

However, previous researches have been limited by relatively small size and short follow-up and lack of prospective recruitment and well-defined LBB capture.

12.2.3 Leadless Pacemaker

Electrophysiologists commonly use transvenous leads in pacemaker implantation due to convenience and clinical experience. However, there are some important complications of transvenous device implantation [44]. Technological advances such as innovation in battery design, miniaturization of components, and optimization of catheter-based cardiac therapies have made the application of leadless cardiac pacemakers (LCP) possible [45]. There are two available LCP devices including the Micra Transcatheter Pacing System (Medtronic, Minneapolis, MN, USA) and the Nanostim device (St. Jude Medical/Abbott Laboratories, Chicago, IL, USA).

As a unique technology, LCP is minimal invasive and free of subcutaneous generator pocket and any intravascular pacemaker leads, both of which are essential for conventional transvenous pacemakers but are associated with various device-related complications [46]. Favorable data (the high implant success rate of 99.2%, low 1-year major complication rate of 4%, and 93.3% of 6-month freedom from serious adverse events) reported in LEADLESS II trial have propelled the spreading use of LCP [47–49]. Further, exciting outcomes from the post-approval multicountry registry were similar (99.6% of implant success and 1.5% of 30-day major complications) [50].

12.3 Indications for Cardiac Pacemaker

The primary indications were mainly based on strong expert consensus before the era of the randomized controlled trial (RCT). A large number of well-designed, randomized clinical trials have supported precise indications which were published in Guidelines over past few decades.

Pacemakers were originally developed for symptomatic bradyarrhythmia or asymptomatic bradyarrhythmia which is apt to progress to a symptomatic bradyarrhythmia [51]. Symptoms of bradyarrhythmia may be subtle (dizziness, fatigue) or impressive (syncope or cardiac arrest). Generally, the most common indications for implantation of permanent pacemakers (PPMs) are for the treatment of sinus node dysfunction, second/third degree AV block, or the dysfunction of infra nodal conduction system. Previous data reveals 50% or more of pacemakers are implanted for sinus node dysfunction [52–55]. Damage to the conduction system commonly results from cardiac ischemia or fibrosis but may be caused by numerous other etiologies, such as infection, electrolyte imbalance, pharmacological agents, or thyroid disease. The imperative procedure requires ruling out potentially reversible causes before PPM implantation.

Table 12.1 Summarizes the 2018 ACC/AHA/HRS Guideline on the Evaluation and Management of Patients with Bradycardia and Cardiac Conduction Delay: A Report of the American College of Cardiology/American Heart Association Task Force on Clinical Practice Guidelines and the Heart Rhythm Society [51, 55]

12.3.1 Biventricular Pacing

In patients with a left ventricular ejection fraction between 36 and 50% and atrioventricular block, who have an indication for permanent pacing and are expected to require ventricular pacing >40% of the time, techniques that provide more *physiologic ventricular activation* (e.g., cardiac resynchronization therapy, His bundle pacing or LBBP) are preferred to right ventricular pacing to prevent heart failure (IIb, B).

Biventricular pacing is indicated for patients with symptomatic heart failure (New York Heart Association functional class III or IV, NYHA III/IV) despite optimal medical therapy, primarily LBB and a widened QRS duration.

As a novel therapy for the management of patients with systolic heart failure, *CRT* has been associated with improved left ventricular ejection fraction (LVEF) and the NYHA class, as well as reduced cardiovascular events and mortality [51, 55]. However, the guidelines for the utilization of CRT are strict. When heart failure patients with an LVEF (≤35%) and a wide QRS (>120 ms) haven't received adequate improvement in LVEF after imperative treatment with guideline-directed medicine for 3 months, CRT is to be considered either intrinsically or due to a long-term need for pacing.

Table 12.1 Indications for permanent cardiac pacing. *SND* sinus node dysfunction. (Adapted from 2012 and 2018 ACC/AHA/HRS guidelines)

	Sinus Node Dysfunction	AV Block	Other
Class I Implant is indicated	Symptomatic SND	Symptomatic third or second-degree block	Alternating bundle branch block
		Symptomatic third or second-degree bock as a result of required drug therapy	Syncope with bundle branch block and evidence of infra nodal bock during electrophysiology study
	Symptomatic bradycardia as a result of required drug therapy	Asymptomatic third-degree, Mobitz type II or high-grade AV block	
		Neuromuscular disorders with second or third-degree AV block (symptomatic or asymptomatic)	
Class IIa Implant is reasonable	Tachy-Brady syndrome with symptomatic bradycardia	Symptomatic first-degree block	Kearns-Sayre syndrome and conduction disorders
	Symptomatic chronotropic incompetence	Symptomatic second-degree Mobitz type	Epilepsy with severe symptomatic bradycardia not effectively managed with medication
Class IIb Implant may be considered	A trial of theophylline may be considered in likely symptomatic bradycardia to help evaluate the benefit of permanent pacing	Neuromuscular disease with PR interval >240 ms, QRS duration >120 ms, or fascicular block	Anderson-Fabry disease and QRS >110 ms
Class III Implant not indicated	Asymptomatic SND	Asymptomatic first-degree block	Bifascicular block without AV block or symptoms
	Symptoms do not correlate with SND and are present in the absence of bradycardia	Asymptomatic second-degree type I block	Asymptomatic fascicular block with first-degree AV block
	Sleep-related bradycardia or transient sinus pauses during sleep with no other indications for pacing	If the block is expected to resolve and not recur	Hypersensitive cardioinhibitory response to carotid sinus stimulation with no or minimal symptoms
			Situational vasovagal syncope that can be managed with avoidance behavior

Table 12.2 Indications for CRT. To qualify for CRT, a patient must be on GDMT for greater than 90 days and be more than 40 days post-MI (Adapted from 2012 and 2018 ACCF/AHA/HRS guidelines)

	NHYA Class I	NYHA Class II	NYHA Class III and Ambulatory Class IV	Other
Class I Implant is indicated		LVEF < 35%	LVEF < 35%	
		QRS >150 ms LBBB	QRS >150 ms LBBB	
		Sinus rhythm	Sinus rhythm	
Class IIa Implant is reasonable		LVEF < 35%	LVEF <35%	• Expected to require > 40% ventricular pacing
		QRS 120–149 ms LBBB	QRS 120–149 ms with LBBB or QRS >150 ms with no LBBB Sinus rhythm	• Atrial fibrillation and pacing are indicated with rate control resulting in near 100% pacing with CRT
		Sinus rhythm		
Class IIb Implant may be considered		LVEF <30%	LVEF <35%	LVEF <35%
		QRS >150 ms LBBB	QRS >150 mg No LBBB	QRS 120–149 ms No LBBB
		Ischemic cardiomyopathy	Sinus Rhythm	Sinus Rhythm
Class III Implant not indicated	QRS <150 ms No LBBB	QRS<150ms No LBBB		

Table 12.2 summarizes the guidelines from ACC/AHA/HRS for the indication of CRT [51, 55]

Compared with traditional CRT, the greater benefits of *MPP* in LVEF and NYHA class have been demonstrated in recent researches, particularly for ischemic cardiomyopathy patients who failed to propagate an electrical impulse due to scar myocardium [56].

In patients with atrioventricular block at the level of the atrioventricular node who have an indication for permanent pacing, *His bundle pacing* may be considered to maintain physiologic ventricular activation [55]. (IIb B) LBBP, a novel technique, presents an optimal alternative for cases of failed HBP or patients with heart failure accompanied by LBBB. Furthermore, LBBP can be considered as the primary strategy for achieving optimal pacing.

12.3.2 LCP

Given the overwhelming outcomes from clinical trials, LCP might be preferred for all patients suitable for VVI mode pacemakers based on its single-chamber ventricular pacing[57]. However, an optimal strategy on which system to use, a LCP or a conventional one, remains unclear, particularly for those with anatomical anomaly [58]. Heretofore, the most common implant indication for LCP was permanent atrial fibrillation (AF) with slow ventricular response, followed by planned atrioventricular (AV) node ablation. In addition, LCP is suitable for senility or young populations who have an indication for permanent pacing. Leadless pacemakers are inappropriate for patients who require dual-chamber pacing or those who frequently require pacing.

12.4 Cardiac Defibrillator

12.4.1 Brief History of the Implantable Cardioverter-Defibrillator

Sudden cardiac death (SCD) which is mainly caused by cardiac arrhythmias especially ventricular tachycardia (VT) or ventricular fibrillation (VF) has emerged as a major public health issue worldwide, accounting for almost 50% of cardiovascular deaths [59]. The implantable cardioverter-defibrillator (ICD) marks an important milestone for the prevention of SCD in high-risk patients, addressing a major unmet medical need for these patients. Claude Beck was first reported to save a patient with VF by defibrillation in 1947 [60]. Direct defibrillation of the heart from VT or VF to sinus rhythm was proposed at the initial stage. ICD was then developed by Dr. Michael Mirowski to extend the possibility to save more lives. Through several decades of efforts and evolution, the recent defibrillation system is quite distinct from the initial implanted automatic defibrillator that was performed by at Johns Hopkins in 1980 [61]. Because of its high success rate in terminating VT or VF rapidly, along with accumulating evidence of multiple clinical trials demonstrating the benefits on mortality reduction, ICD implantation is generally considered the first-line treatment option for the secondary prevention of SCD including post-myocardial infarction and congestive heart failure and for primary prevention in selected patients at high risk of SCD resulting from VT or VF [62].

12.4.2 Indications and Current Guidelines for ICD

In adherence to the current clinical practice and guidelines, ICD is the standard of care for the prevention of SCD in high-risk populations, including the primary and secondary prevention. The 2008 ACC/AHA/HRS guideline statement which was updated in 2012 and 2017 has been a very important guide for clinical practice [63, 64].

As per the results of AVID, CIDS, and CASH, ICDs become a class I indication for secondary prevention in survivors of life-threatening VF or hemodynamically unstable VT without a reversible cause. Other class I indications of ICD in secondary prevention include structural heart disease and sustained VT or syncope of unknown origin and VT or VF induced at EP study.

ICD is also indicated for the primary prevention of SCD due to VT or VF but not experienced SCD yet. For ischemic heart disease, an ICD is indicated in patients with a prior myocardial infarction (at least 40 days ago) and left ventricular ejection fraction (LVEF) $\leq$ 30%. Patients with a nonischemic cardiomyopathy are also recommended for ICD primary prevention if LVEF is $\leq$ 35% and New York Heart Association (NYHA) functional class II to III and already under optimal medical management (including use of beta blockers and angiotensin converting enzyme [ACE] inhibitors) yet at high risk of SCD. This guideline is in accordance with the inclusion criteria of SCD-HeFT and is also inclusive of the outcomes of DINAMIT and IRIS in post-MI patients [64]. Some patients with heart failure who are candidates for an ICD also have intraventricular conduction delay ($\geq$120 ms) and are candidates for CRT with a biventricular pacemaker. Such patients could be treated with a device with combined ICD and biventricular pacing functions.

Several class IIa indications exist for commonly encountered situations in which the evidence is not as robust. ICD therapy is reasonable in those with unexplained syncope, LV dysfunction, and nonischemic dilated cardiomyopathy (DCM) or those with sustained VT and near-normal LV function. Class IIa indications also include inherited abnormalities of ion channels and structural heart disease such as patients with congenital long QT syndrome who have recurrent symptoms and/or torsades de pointes despite therapy with beta blockers or other high-risk patients, high-risk patients with hypertrophic cardiomyopathy or arrhythmogenic right ventricular (RV) cardiomyopathy, and high-risk patients with Brugada syndrome, catecholaminergic polymorphic VT, and other channelopathies.

ICD therapy is not recommended in certain populations (class III indications), including patients with ventricular tachyarrhythmias due to a completely reversible disorder in the absence of structural heart disease (e.g., electrolyte imbalance, drugs, or trauma) and those with life expectancy less than 1 year, NYHA class IV heart failure, significant psychiatric illness that would be exacerbated by device implantation or prevent routine follow-up, VF or VT due to a reversible disorder, or ventricular arrhythmias that are amenable to surgical or catheter ablation.

12.5 Pacing System

A pacemaker consists of a pulse generator and leads capable of sensing and pacing. The pulse generator controls the analysis of sensed activity and battery status by its microprocessor, and pacemaker lead typically has two electrodes (bipolar) in contact with the atrial or ventricular myocardium. Impulses delivered by the pulse generator through these electrodes pace the heart. Pacemakers can be configured as single-chamber, dual-chamber, biventricular, or tri-ventricular devices.

As with a pacemaker, an ICD also consists of a pulse generator and endocardial leads. An ICD lead is characterized by its additional high-voltage coils which are integrated into the right ventricular endocardial lead and act as shocking electrodes. Therefore, an ICD pulse generator contains not only a microprocessor but also a high-voltage capacitor for longer battery longevity.

To clarify the characteristics of pacemaker, a standardized classification code which has five positions has been developed for pacing nomenclature. The *first letter* or category of the code reflexes the chamber(s) paced, and the *second letter* indicates the chamber(s) sensed. Options for these positions include A (atrium), V (ventricle), O (none), and D (dual = A + V). *The third position* of the code describes the response of the device to a sensed event; options include O (none), T (triggered), I (inhibited), and D (dual = triggered or inhibited). For instance, pacemaker programmed to VOO mode paces the ventricle at a specified rate and will ignore any signal sensed by the ventricular lead. *The fourth letter* refers to programmability of rate modulation, and the letter R refers to a rate responsive or rate adaptive pacing, which can adjust paced heart rate in response to patient activity. "O" means rate modulation is either unavailable or disabled, which is usually omitted. For example, DDD is the same as DDDO. The fifth position which is rarely used specifies the stimulation sites in multisite pacing only.

12.6 Device Implantation Procedure

12.6.1 General Endocardial Implantation

A standardized procedure with improved delivery tools and pacing leads is required, as well as long-term efficacy and safety. The utilization of contrast venography to guide the access through the axillary vein, axillary, or cephalic vein ensures the secure and efficient placement of pacemaker and defibrillator leads, promoting their safe and effective positioning. When endocardial implantation is infeasible or an open cardiac surgical procedure is necessary, epicardial lead placement approach may be a more invasive surgical alternative.

Endocardial leads are positioned in the right atrium and/or right ventricle. From a technical perspective, a myocardial injury current signifies an excellent tissue contact, which is generally observed once the pacing lead is fixed. No injury current or low amplitude of injury current indicates that either the lead helix is not screwed into the myocardium or the myocardium contact is inappropriate. If that happens, the lead should be screwed back and relocated [38]. Once positioned, the leads are inserted into the headers of the generator, which is implanted into the subcutaneous or submuscular pocket below the clavicle.

12.6.2 *Left Ventricular Lead Placement Procedure*

In the case of traditional biventricular pacing devices, an additional left ventricular (LV) lead is on the epicardium of the lateral left ventricle guided through the coronary sinus (CS). Successful placement of LV leads depends on the anatomical complexity and the availability of coronary vein branches of the main CS. When CS anatomy is unsuitable for coronary vein branch access, LV lead can be placed on the epicardial surface of the left ventricle using a minimally invasive lateral thoracotomy. The improvement of quadripolar LV leads and catheters has impelled the technology development. Quadripolar LV lead aids in programming to optimize the pacing threshold and QRS narrowing. Generally, by positioning pacemaker leads in the right ventricle and LV lead attaching to the region of left ventricle, simultaneous pacing of both walls of the left ventricle improves ventricular synchrony.

12.6.3 *LBBP Implantation Procedure*

12.6.3.1 Implant Tools and Equipment Requirements [38]

1. A SelectSecure pacing lead (Model 3830 69 cm; Medtronic, Inc., Minneapolis, MN) with an exposed 1.8mm active helix and a Select Site fixed curve sheath (Model C315 HIS; Medtronic, Inc.)
2. Electrophysiological multichannel recorder to record the intracardiac EGMs and 12-lead ECG simultaneously
3. Pacing system analyzer (PSA) to test the pacing parameters and record intracardiac EGM via the pacing lead

12.6.3.2 Implant Procedure and Methods

Recent studies have demonstrated the safety and clinical feasibility of LBBP via a transventricular-septal approach [38–40, 64]. For patients with LBBB, a temporary pacing lead is necessary to avoid inadvertent complete heart block caused by the RBBB due to temporary mechanical injury before LBBP lead achievement. A unipolar configuration is used for pacing and recording during procedure. There are two methods to identify the location for LBBP lead placement: the single-lead method (SLM) and the dual-lead method (DLM).

The Details of LBBP Procedure as Follows Under a 30° right anterior oblique fluoroscopic view, the delivery sheath with the pacing lead is first placed in the His bundle region near the tricuspid valve annulus, where His bundle potential is recorded and the image of His location is saved as a fluoroscopic marker. Then the sheath with the pacing lead is advanced toward the ventricular septum inferior to the septal leaflet of the tricuspid valve, where LBBP is attempted on the right side of the interventricular septum, 10–20 mm away from the His bundle region [38]. The tricuspid annulus can be used as a backup anatomic marker to facilitate the lead implantation when the His bundle region cannot be localized in challenging situations. The sheath is placed perpendicularly against the right side of the interventricular septum by minimal counterclockwise rotation, and the helix of the pacing lead is gently advanced out of the tip of the sheath [38, 40]. Then, pacing with an output of 2.0 V/0.5 ms or 5 V/0.5 ms is applied. The paced QRS morphology transition and pacing parameter changes are important for guiding and identifying the location of LBBP lead. The paced QRS morphology shows the LBBB morphology, usually characterized with an ECG QRSd less than 150 ms, a notch named "W" waveform at the nadir of the QRS waveform in ECG-lead V1, a positive QRS in lead II, and biphasic QRS in lead III. The initial lead fixation at this location is conducted with four to five clockwise turns of the pacing lead. The desired pacing site is confirmed when the tip of the pacing lead is perpendicular to the ventricular septum under a 30° left anterior oblique (LAO) view. Then, further clockwise rotate the pacing lead for penetrating its tip into septum. Once the paced QRS morphology shows RBBD in lead V1/V2 on the surface ECG due to delayed activation of the RBB, which satisfies the LBBP criteria as described before, the lead placement should be ceased.

12.6.3.3 Definition of LBBP

LBBP is defined as follows: (a) a paced QRS morphology on 12-lead ECG shows the pattern of right bundle branch conduction delay (RBBD) or improving the LBB conduction in patients with LBBB; (b) a left bundle branch potential can be recorded; and (c) a fast peak left ventricular activation time (LVAT) less than 90 ms in V5/V6 on 12-lead ECG is present, and the paced QRSd is usually less than 120 ms [38, 40].

After the pacing lead is successfully implanted, multiple parameters should be assessed, including pacing electrical parameters (capture threshold, lead impedance, and R-wave amplitude). In LBBP cases, the paced QRSd and fast peak left ventricular activation time (LVAT) need to be measured additionally.

12.6.4 LCP Implantation Procedure

As a miniaturized device, LCP is implanted percutaneously through femoral venous access and attaches to the right ventricular myocardium [45, 65, 66]. Femoral venous access was obtained with ultrasound guidance using the modified Seldinger technique. For Micra implantation, a 27-French (outer diameter) introducer sheath was inserted in the femoral vein, and for the Nanostim devices, an 18-French deflectable femoral venous sheath was advanced to the right atrium. The delivery catheter loaded with LCP was advanced to a suitable location in the right ventricle which fixes the device to the myocardium. After confirming that the tip of the catheter is in a suitable position by contrast injection, the device was rotated to advance the helix and deployed. To demonstrated the stability of LCP, A "pull and hold" test was performed, and imaging was employed to confirm the engagement of two or more tines. Then the electrical parameters were checked. If the measurements were unsatisfactory, the device was unscrewed and redeployed. In LEADLESS, the Nanostim device was successfully implanted in 32 of 33 patients with first targeted location of 70% [67].

12.7 Post-procedure Care and Follow-Up

12.7.1 Postoperative Care

Instructions for patients who underwent CIED implantation are as follows:

1. Keep the surgical incision clean and dry for approximately 10–14 days; sponge bathe but not shower; change dressing daily; it is recommended to put steri-strips, sutures, or staples in the region of incision for 1–2 weeks.
2. Notify their provider of any evidence of infection or swelling, drainage, or bleeding at the insertion site.
3. Limit ipsilateral arm activities to below head level and to avoid heavy lifting for 6–8 weeks to promote wound healing and prevent lead dislodgment.
4. Patients with recent permanent pacemaker implantation may drive 1 week after procedure. However, commercial drivers must wait 1 month. Driving restrictions are typically imposed in patients who underwent an ICD implantation for secondary prevention of SCD for approximately 6 months. Fortunately, patients who undergo ICD implantation for primary prevention of SCD are free of specific driving restrictions. However, commercial driving should be permanently prohibited for every patient who underwent ICD implantation [68, 69]. Patients can usually take a flight 1 day after implantation [68, 70]. Pilots must wait at least 3 months after successful implantation [68, 71].
5. Follow up regularly according to the doctor's instruction.

12.7.2 Electromagnetic Interference

Electromagnetic interference may interfere with the proper function of CIEDs. There is no restriction on cell phone and household items such as microwaves, televisions, radios, or electric blankets because these are free of electromagnetic interference. It is advised that patients with CIEDs keep away from handheld metal detectors or scanning wands during security check; a hand search is recommended instead.

MRI scanners, radiation therapy, transthoracic cardioversion, and electrocautery are all medical sources of potential electromagnetic interference. Strong magnetic field can affect CIEDs including possible movement of the device, programming changes, inhibition of pacing, asynchronous pacing (i.e., VOO), heating, and/or cardiac stimulation from induced currents in lead wires. It is necessary to consult with an electrophysiologist prior to undergoing any medical examination or therapy. Selecting MRI compatible CIEDs that are shielded from direct radiation in the generator pocket region, positioning external electrodes as far away as possible (>5 cm) from the implanted CIEDs during cardioversion, and programming the CIEDs in asynchronous mode or deactivating the ICD to avoid interference with the pacemaker or inappropriate delivery of therapy from the ICD [4, 51].

12.7.3 Follow-Up Schedule

The aims of follow-up include normal device functions and appropriate programming.

1. The recommended frequency is 1 month, 3–6 months, and 12 months, followed by once per year.
2. Adverse events, malignant arrhythmias, and device malfunction are recorded.
3. Battery voltage, lead integrity testing, and electrographic evaluation of arrhythmic events are evaluated.
4. Optimal monitoring is monitoring remotely through wireless technology over the Internet by a central device clinic. This method allows for nearcontinuous monitoring. More extensive can be done remotely on a quarterly basis or for urgent evaluation of patient symptoms. Remote monitoring has been demonstrated to reduce both morbidity and mortality in long-term follow-up.
5. Regarding arrhythmia detection, ICDs were initially programed to detect tachycardia on the basis of heart rate [71]. Although appropriate and inappropriate ICD shocks have been strongly associated with psychological effects and reduced quality of life [72], a single ICD shock is referred for evaluation and reassurance within the week and does not require an emergent emergency department visit. However, if a patient experiences the shock accompanied with worrisome symptoms such as syncope, dyspnea, persistent palpitations, or chest pain, or multiple shocks over a short period of time, an emergency department visit is required. Besides pacing for bradycardia, ICD is capable of detecting ventricular tachyarrhythmias based on a preprogrammed rate cutoff. Once a tachyarrhythmia was confirmed, an ICD may tailor the therapy according to multiple rate tiers by anti-tachyarrhythmic pacing, low-energy cardioversion, or high-energy defibrillation.

12.8 Future Directions

Advances in pacemaker and ICD technology have substantially improved survival and quality of life for patients with cardiac arrhythmias. As newer CIEDs have acquired functionality, interpretation of modern cardiac pacemaker and ICDs has become progressively complex.

Clinical development of physiological pacing is at an early but encouraging stage with great progress. In a few decades, the ICD has evolved from a hypothesis to a well-proven means of improving survival in patients who are at risk for SCD. While remote monitoring is an evolving strategy in management of patients with complex cardiac conditions and implantable devices, there are multiple areas of ongoing investigation in improving on this technology. The subcutaneous ICD (S-ICD) has been an area of significant interest and the focus of analyses that specifically address ways to optimize the efficacy of that type of device-based therapy [73]. In the future, indications for ICD therapy will probably expand as proper identification of patients at risk for future ventricular tachyarrhythmic events improves. Although LCP represents a significant technological advancement over traditional transvenous pacemakers, whether the advancements translate into a lower risk of technical and clinical adverse events remains unclear. Data comparing leadless to transvenous pacemakers with regard to electrical performance, complications, or clinical outcomes are scant [45, 74]. The implantation of LCP is rising worldwide, and there is a call for additional data on the efficacy and safety of these devices, to guide clinicians and patients in medical choice [75, 76].

Much work is still required to better define and understand the patient groups most likely to benefit from CIEDs, both to reduce the number of devices implanted that never deliver appropriate therapy and to increase the number of lives saved.

References

1. Epstein AE, DiMarco JP, Ellenbogen KA, et al. ACC/AHA/HRS 2008 guidelines for device-based therapy of cardiac rhythm abnormalities: a report of the American College of Cardiology/American Heart Association Task Force on Practice Guidelines (Writing Committee to Revise the ACC/AHA/NASPE 2002 Guideline Update for Implantation of Cardiac Pacemakers and Antiarrhythmia Devices): developed in collabo- ration with the American Association for Thoracic Surgery and Society of Thoracic Surgeons. Circulation. 2008;117:e350–408.
2. Hopps J. The development of the pace-maker. Pacing Clin Electrophysiol. 1981;4:106–8.
3. Mittal T. Pacemakers-a journey through the years. Indian J Thorac Cardiovasc Surg. 2005;21: 236–49.
4. Callaghan J, Bigelow W. An electrical artificial pacemaker for standstill of the heart. Ann Surg. 1951;1:8–17.
5. Wilkoff BL, Byrd CL, Love CJ, et al. Pacemaker lead extraction with the laser sheath: results of the pacing lead extraction with the excimer sheath (PLEXES) trial. J Am Coll Cardiol. 1999;33: 1671–6.
6. Parkash R, Philippon F, Shanks M, et al. Canadian Cardiovascular Society guide- lines on the use of cardiac resynchronization therapy: implementation. Can J Cardiol. 2013;29:1346–60.

7. Healey J, Toff W, Lamas G, et al. Cardio-vascular outcomes with atrial-based pacing compared with ventricular pacing: meta-analysis of randomzied trials, using individual patient data. Circulation. 2006;114:11–7.
8. Pakarinen S, Oikarinen L, Toivonen L. Short-term implantation-related complications of cardiac rhythm management device therapy: a retrospective single-centre 1-year study. Europace. 2010;12:103–8.
9. Tobin K, Stewart J, Westveer D, Frumin H. Acute complications of permanent pacemaker implantation: their financial implication and relation to volume and operator experience. Am J Cardiol. 2000;85:774–6.
10. Bigelow W. The pacemaker story: a cold heart spinoff. Pacing Clin Electrophysiol. 1987;10: 142–50.
11. Sweeney MO, Hellkamp AS, Ellenbogen KA, et al. Adverse effect of ventricular pacing on heart failure and atrial fibrillation among patients with normal baseline QRS duration in a clinical trial of pacemaker therapy for sinus node dysfunction. Circulation. 2003;107:2932–7.
12. Nielsen JC, Kristensen L, Andersen HR, Mortensen PT, Pedersen OL, Pedersen AK. A randomized comparison of atrial and dual-chamber pacing in 177 consecutive patients with sick sinus syndrome. J Am Coll Cardiol. 2003;42:614–23.
13. Hussain MA, Furuya-Kanamori L, Kaye G, Clark J, Doi SA. The effect of right ventricular apical and nonapical pacing on the short- and long-term changes in left ventricular ejection fraction: a systematic review and meta-analysis of randomized-controlled trials. Pacing Clin Electrophysiol. 2015;38:1121–36.
14. Arnold AD, Shun-Shin MJ, Keene D, Howard JP, Sohaib SMA, Wright IJ, Cole GD, Qureshi NA, Lefroy DC, Koa-Wing M, et al. His resynchronization versus biventricular pacing in patients with heart failure and left bundle branch block. J Am Coll Cardiol. 2018;72:3112–22.
15. Lustgarten DL, Crespo EM, Arkhipova-Jenkins I, Lobel R, Winget J, Koehler J, Liberman E, Sheldon T. His-bundle pacing versus biventricular pacing in cardiac resynchronization therapy patients: a crossover design comparison. Heart Rhythm. 2015;12:1548–57.
16. Huang W, Su L, Wu S, Xu L, Xiao F, Zhou X, Ellenbogen KA. Benefits of permanent his bundle pacing combined with atrioventricular node ablation in atrial fibrillation patients with heart failure with both preserved and reduced left ventricular ejection fraction. J Am Heart Assoc. 2017;6:e005309.
17. Abdelrahman M, Subzposh FA, Beer D, Durr B, Naperkowski A, Sun H, Oren JW, Dandamudi G, Vijayaraman P. Clinical outcomes of his bundle pacing compared to right ventricular pacing. J Am Coll Cardiol. 2018;71:2319–30.
18. Sharma PS, Naperkowski A, Bauch TD, Chan JYS, Arnold AD, Whinnett ZI, Ellenbogen KA, Vijayaraman P. Permanent His bundle pacing for cardiac resynchronization therapy in patients with heart failure and right bundle branch block. Circ Arrhythm Electrophysiol. 2018;11: e006613.
19. Abdelrahman M, Subzposh FA, Beer D, et al. Clinical outcomes of his bundle pacing compared to right ventricular pacing. J Am Coll Cardiol. 2018;71:2319–30.
20. Huang W, Su L, Wu S, et al. Long-term outcomes of his bundle pacing in patients with heart failure with left bundle branch block. Heart. 2019;105:137–43.
21. Huang W, Su L, Wu S, et al. Benefits of permanent his bundle pacing combined with atrioventricular node ablation in atrial fibrillation patients with heart failure with both preserved and reduced left ventricular ejection fraction. J Am Heart Assoc. 2017;6:1–11.
22. Wilson W, Taubert K, Gewitz M, et al. Prevention of infective endocarditis: guide-lines from the American Heart Association. Circulation. 2007;116:1736–54.
23. American Society of Anesthesiologists Task Force on Perioperative Management. Practice advisory for the perioperative management of patients with cardiac rhythm management devices: Pacemakers and implantable cardioverter-defibrillators. Anesthesiology. 2005;103: 186–98.

24. Vijayaraman P, Naperkowski A, Ellenbogen KA, Dandamudi G. Electrophysiologic insights into site of atrioventricular block: lessons from permanent his bundle pacing. JACC Clin Electrophysiol. 2015;1:571–81.
25. Upadhyay GA, Cherian T, Shatz DY, Beaser AD, Aziz Z, Ozcan C, Broman MT, Nayak HM, Tung R. Intracardiac delineation of septal conduction in left bundle-branch block patterns. Circulation. 2019;139:1876–88.
26. Jastrzębski M, Moskal P, Bednarek A, Kiełbasa G, Czarnecka D. His-bundle pacing as a standard approach in patients with permanent atrial fibrillation and bradycardia. Pacing Clin Electrophysiol. 2018;41:1508–12.
27. Keene D, Arnold AD, Jastrzebski M, Burri H, Zweibel S, Crespo E, Chandrasekaran B, Bassi S, Joghetaei N, Swift M, et al. His bundle pacing, learning curve, procedure characteristics, safety, and feasibility: insights from a large international observational study. J Cardiovasc Electrophysiol. 2019;30:1984–93.
28. Zanon F, Ellenbogen KA, Dandamudi G, Sharma PS, Huang WJ, Lustgarten DL, Tung R, Tada H, Koneru JN, Bergemann T, et al. Permanent His-bundle pacing: a systematic literature review and meta-analysis. Europace. 2018;20:1819–26.
29. Bhatt AG, Musat DL, Milstein N, et al. The efficacy of His bundle pacing: lessons learned from implementation for the first time at an experienced electrophysi- ology center. JACC Clin Electrophysiol. 2018;4:1397–406.
30. Zanon F, Abdelrahman M, Marcantoni L, et al. Long term performance and safety of His bundle pacing: a multicenter experience. J Cardiovasc Electrophysiol. 2019;30:1594–601.
31. Vijayaraman P, Chung MK, Dandamudi G, et al. His bundle pacing. J Am Coll Cardiol. 2018;72:927–47.
32. Lustgarten DL, Crespo EM, Arkhipova-Jenkins I, et al. His-bundle pacing versus biventricular pacing in cardiac resynchronization therapy patients: a crossover design comparison. Heart Rhythm. 2015;12:1548–57.
33. Sharma PS, Dandamudi G, Herweg B, et al. Permanent His-bundle pacing as an alternative to biventricular pacing for cardiac resynchronization therapy: a multicenter experience. Heart Rhythm. 2018;15:413–20.
34. Huang W, Su L, Wu S, Xu L, Xiao F, Zhou X, Ellenbogen KA. A novel pacing strategy with low and stable output: pacing the left bundle branch immediately beyond the conduction block. Can J Cardiol. 2017;33:1736.e1–3.
35. Hou X, Qian Z, Wang Y, Qiu Y, Chen X, Jiang H, Jiang Z, Wu H, Zhao Z, Zhou W, Zou J. Feasibility and cardiac synchrony of permanent left bundle branch pacing through the interventricular septum. Europace. 2019;21:1694–702.
36. Vijayaraman P, Subzposh FA, Naperkowski A, Panikkath R, John K, Mascarenhas V, Bauch TD, Huang W. Prospective evaluation of feasibility and electrophysiologic and echocardiographic characteristics of left bundle branch area pacing. Heart Rhythm. 2019;16:1774–82.
37. Padala S, Ellenbogen K. Left bundle branch pacing is the best approach to physiological pacing. Heart Rhythm. 2020;O2(1):59–67.
38. Chen K, Li Y. How to implant left bundle branch pacing lead in routine clinical practice. J Cardiovasc Electrophysiol. 2019;30:2569–77.
39. Chen K, Li Y, Dai Y, et al. Comparison of electrocardiogram characteristics and pacing parameters between left bundle branch pacing and right ventricular pacing in patients receiving pacemaker therapy. Europace. 2019;21:673–80.
40. Li Y, Chen K, Dai Y, et al. Left bundle branch pacing for symptomatic bradycardia: implant success rate, safety and pacing characteristics. Heart Rhythm. 2019;16(12):1758–65. https://doi.org/10.1016/j.hrthm.2019.05.014.
41. Vijayaraman P, Huang W. Atrioventricular block at the distal His bundle: electrophysiological insights from left bundle branch pacing. Heart Rhythm Case Rep. 2019;5:233–6. https://doi.org/10.1016/j.hrcr.2019.01.006072-14.

42. Vijayaraman P, Subzposh FA, Naperkowski A, et al. Prospective evaluation of feasibility, electrophysiologic and echocardio-graphic characteristics of left bundle branch area pacing. Heart Rhythm. 2019:30442–4. https://doi.org/10.1016/j.hrthm.2019.05.011.
43. Li X, Li H, Ma W, et al. Permanent left bundle branch area pacing for atrioventricular block: feasibility, safety, and acute effect. Heart Rhythm. 2019;16(12):1766–73. https://doi.org/10.1016/j.hrthm.2019.04.043.
44. Mulpuru SK, Madhavan M, McLeod CJ, Cha YM, Friedman PA. Cardiac pacemakers: function, troubleshooting, and management: part 1 of a 2-part series. J Am Coll Cardiol. 2017;69: 189–210.
45. Reynolds D, Duray GZ, Omar R, et al. A leadless intracardiac transcatheter pacing system. N Engl J Med. 2016;374:533–41.
46. Kirkfeldt RE, Johansen JB, Nohr EA, Jorgensen OD, Nielsen JC. Complications after cardiac implantable electronic device implantations: an analysis of a complete, nationwide cohort in Denmark. Eur Heart J. 2014;35:1186–94.
47. Duray GZ, Ritter P, El-Chami M, Narasimhan C, Omar R, Tolosana JM, Zhang S, Soejima K, Steinwender C, Rapallini L, Cicic A, Fagan DH, Liu S, Reynolds D. Micra transcatheter pacing study group. Long-term performance of a transcatheter pacing system: 12-month results from the Micra transcatheter pacing study. Heart Rhythm. 2017;14:702–9.
48. Reynolds D, Duray GZ, Omar R, Soejima K, Neuzil P, Zhang S, Narasimhan C, Steinwender C, Brugada J, Lloyd M, Roberts PR, Sagi V, Hummel J, Bon-Giorni MG, Knops RE, et al. A leadless intracardiac transcatheter pacing system. N Engl J Med. 2016;374:533–41.
49. Reddy VY 3rd, Exner DV, Cantillon DJ, Doshi R, Bunch TJ, Tomassoni GF, Fried-Man PA, NAM E, Ip J, Niazi I, Plunkitt K, Banker R, Porterfield J, Ip JE, Dukkipati SR, et al. Percutaneous implantation of an entirely intracardiac lead-less pacemaker. N Engl J Med. 2015;373:1125–35.
50. Roberts PR, Clementy N, Al Samadi F, Garweg C, Martinez-Sande JL, Iacopino S, Johansen JB, Prat XV, Kowal RC, Klug D, Mont L, Steffel J, Li S, Van Osch D, El-Chami MF. A leadless pacemaker in the real-world setting: the Micra transcatheter pacing system post-approval registry. Heart Rhythm. 2017;14:1375–9.
51. Tracy CM, Epstein AE, Darbar D, Dimarco JP, Dunbar SB, Estes NA 3rd, Ferguson TB Jr, Hammill SC, Karasik PE, Link MS, Marine JE, Schoenfeld MH, Shanker AJ, Silka MJ, Stevenson LW, Stevenson WG, Varosy PD. 2012 ACCF/AHA/HRS focused update of the 2008 guidelines for device-based therapy of cardiac rhythm abnormalities: a report of the American College of Cardiology Foundation/American Heart Association Task Force on Practice Guidelines. J Am Coll Cardiol. 2012;60:1297–313.
52. Byrd CL, Wilkoff BL, Love CJ, et al. Clinical study of the laser sheath for lead extraction: the total experience in the United States. Pacing Clin Electrophysiol. 2002;25:804–8.
53. Klug D, Jarwe M, Messaoudene SA, et al. Pacemaker lead extraction with the needle's eye snare for countertraction via a femoral approach. Pacing Clin Electrophysiol. 2002;25:1023–8.
54. Ellison K, Sharma PS, Trohman R. Advances in cardiac pacing and defibrillation. Expert Rev Cardiovasc. 2017;15(6):429–40.
55. Kusumoto FM, Schoenfeld MH, Barrett C, Edgerton JR, Ellenbogen KA, et al. 2018 ACC/AHA/HRS Guideline on the evaluation and management of patients with bradycardia and cardiac conduction delay: a report of the American College of Cardiology/American Heart Association Task Force on Clinical Practice Guidelines and the Heart Rhythm Society. Circulation. 2019;140(8):e382–482.
56. Baddour L, Epstein A, Erickson C, et al. Update on cardiovascular implantable electronic device infections and their management: a scientific statement from the American Heart Association. Circulation. 2010;121:458–77.
57. Da Costa A, Lelievre H, Kirkorian G, et al. Role of the preaxillary flora in pacemaker infections: a prospective study. Circulation. 1998;97:1791–5.
58. Alyesh D, Cunnane R. Successful leadless pacemaker deployment in a patient with challenging right heart anatomy using a double snare technique. HeartRhythm Case Rep. 2019;5:399–401.

59. Myerburg RJ, Junttila MJ. Sudden cardiac death caused by coronary heart disease. Circulation. 2012;125:1043–52.
60. Beck CS, Pritchard WH, Feil HS. Ventricular fibrillation of long duration abolished by electric shock. J Am Med Assoc. 1947;135:985.
61. Mirowski M, Reid PR, Mower MM, Watkins L, Gott VL, Schauble JF, Langer A, Heilman MS, Kolenik SA, Fischell RE, Weisfeldt ML. Termination of malignant ventricular arrhythmias with an implanted automatic defibrillator in human beings. N Engl J Med. 1980;303:322–4.
62. Bardy GH, Lee KL, Mark DB, Poole JE, Packer DL, Boineau R, Domanski M, Troutman C, Anderson J, Johnson G, McNulty SE, Clapp-Channing N, Davidson-Ray LD, Fraulo ES, Fishbein DP, Luceri RM, Ip JH. Amiodarone or an implantable cardioverter-defibrillator for congestive heart failure. N Engl J Med. 2005;352:225–37.
63. Epstein AE, DiMarco JP, Ellenbogen KA, Estes NA 3rd, Freedman RA, Gettes LS, Gillinov AM, Gregoratos G, Hammill SC, Hayes DL, Hlatky MA, Newby LK, Page RL, Schoenfeld MH, Silka MJ, Stevenson LW, Sweeney MO, Smith SC Jr, Jacobs AK, Adams CD, Anderson JL, Buller CE, Creager MA, Ettinger SM, Faxon DP, Halperin JL, Hiratzka LF, Hunt SA, Krumholz HM, Kushner FG, Lytle BW, Nishimura RA, Ornato JP, Page RL, Riegel B, Tarkington LG, Yancy CW. ACC/AHA/HRS 2008 Guidelines for Device-based therapy of cardiac rhythm abnormalities: a report of the American College of Cardiology/American Heart Association Task Force on Practice Guidelines (Writing Committee to Revise the ACC/AHA/NASPE 2002 Guideline Update for Implantation of Cardiac Pacemakers and Antiarrhythmia Devices): developed in collaboration with the American Association for Thoracic Surgery and Society of Thoracic Surgeons. Circulation. 2008;117:e350–408.
64. Al-Khatib SM, Stevenson WG, Ackerman MJ, Bryant WJ, Callans DJ, Curtis AB, Deal BJ, Dickfeld T, Field ME, Fonarow GC, Gillis AM, Granger CB, Hammill SC, Hlatky MA, Joglar JA, Kay GN, Matlock DD, Myerburg RJ, Page RL. 2017 AHA/ACC/HRS Guideline for Management of patients with ventricular arrhythmias and the prevention of sudden cardiac death: a report of the American College of Cardiology/American Heart Association Task Force on Clinical Practice Guidelines and the Heart Rhythm Society. J Am Coll Cardiol. 2018;72: e91–e220.
65. Reddy VY, Knops RE, Sperzel J, et al. Permanent leadless cardiac pacing: results of the LEADLESS trial. Circulation. 2014;129:1466–71.
66. Ritter P, Duray GZ, Steinwender C, et al. Early performance of a miniaturized leadless cardiac pacemaker: the Micra Transcatheter Pacing Study. Eur Heart J. 2015;36:2510–9.
67. Padala SK, Ellenbogen KA. Left bundle branch pacing is the best approach to physiological pacing. Heart Rhythm O2. 2020;1(1):59–67.
68. Bains P, Chatur S, Ignaszewski M, Ladhar S, Bennett M. John Hopps and the pacemaker: a history and detailed overview of devices, indications, and complications. BC Med J. 2017;59 (1):29–37. bcmj.org
69. BC Ministry of Public Safety, Solicitor General, Office of the Superintendent of Motor Vehicles. 2010 BC guide in determining fitness to drive. http://www2.gov.bc.ca/assets/gov/driving-and-transportation/driving/ publications/2010-guide-in-determining-fitness-to-drive.pdf. Accessed 28 Oct 2016.
70. Wielgosz AT. Guidelines for the assessment of cardiovascular fitness in licensed aviation personnel 2012. Ottawa: Transport Canada Civil Aviation Medicine; 2012. http://www.tc.gc.ca/media/documents/caopssvs/cardio vascular-2012.pdf. Accessed 31 Oct 2016.
71. Miller JD, Yousuf O, Berger RD. The implantable cardioverter-defibrillator: an update. Trends Cardiovasc Med. 2015;25:606–11.
72. Passman R, Kadish A. Sudden death prevention with implantable devices. Circulation. 2007;116:561–71.
73. Amin AK, Gold MR, Burke MC, Knight BP, Rajjoub MR, Duffy E, Husby M, Stahl WK, Weiss R. Factors associated with high-voltage impedance and subcutaneous implantable defibrillator ventricular fibrillation conversion success. Circ Arrhythm Electrophysiol. 2019;12:e006665.

74. Tjong FV, Reddy VY. Permanent leadless cardiac pacemaker therapy: a comprehensive review. Circulation. 2017;135:1458–70.
75. Cano Perez O, Pombo Jimenez M, Fidalgo Andres ML, Lorente Carreno D, Coma Samartin R. Spanish pacemaker registry. 14th official report of the Spanish Society of Cardiology working group on cardiac pacing (2016). Rev Esp Cardiol. 2017;70:1083–97.
76. Preminger MW, Mittal S. Leadless pacing meets the real world: the maturation of clinical evidence behind a miniaturized pacemaker. Heart Rhythm. 2017;14:1380–1.

Chapter 13
Future Challenges and Perspective

Yi-Da Tang

13.1 Coronary Artery Interventional Devices

13.1.1 Coronary Artery Imaging Technique

Coronary imaging and functional assessment are critical to the development of coronary interventional devices. Our understanding of coronary arteries has long evolved from 2D coronary angiography to the era of intravascular imaging. The major intravascular imaging assessment techniques currently available include IVUS, OCT, and NIRS, with different methods having their own characteristics and application scenarios. Future research directions include clearer and faster image acquisition, integration of multiple techniques, and artificial intelligence interpretation.

The latest Artificial Intelligence Optical Coherence Tomography (AI-OCT) imaging platform Ultreon™ 1.0 Software from Abbott was licensed by the US FDA [1]. This imaging platform combines OCT with artificial intelligence (AI) to provide physicians with an enhanced and integrated view of coronary blood flow and blockages to aid in decision-making. The software is integrated with Abbott's new Dragonfly OpStar imaging catheter and PressureWire X guidewire, providing physicians with multiple tools to access coronary flow obstructions, and the catheter can navigate through tortuous anatomy to access distal lesions to provide brighter, better quality images.

The first intravascular photoacoustic-ultrasound-OCT multimodal imaging system integrates ultrasound, OCT, and the latest photoacoustic spectroscopy imaging technologies in a precision imaging catheter with an outer diameter of only 0.9 mm.

Y.-D. Tang (✉)
Department of Cardiology and Institute of Vascular Medicine, Peking University Third Hospital, Beijing, China
e-mail: tangyida@bjmu.edu.cn

Y.-D. Tang et al. (eds.), *Application of Biomaterials in the Treatment of Cardiovascular Diseases*, https://doi.org/10.1007/978-981-19-7712-1_13

The system not only observes the overall morphology and microstructure of the vessel wall but also simultaneously identifies the lipid component of the plaque with 3D imaging and provides its quantified concentration information, allowing simultaneous acquisition of multidimensional information on the overall morphology, microstructure, lipid identification, and its concentration quantification of the vessel wall [2]. This work is of great importance to promote the in-depth study of the physiopathological evolution and rupture mechanism of vulnerable plaques, as well as the early prevention and intervention of acute cardiovascular and cerebrovascular diseases.

13.1.2 Balloon Catheter

In recent years, the balloon has gradually returned to center stage from a supporting role. PTCA balloon technological advances are most evident in drug-coated balloons (DCB). DCBs use the same antiproliferative drugs as drug-eluting stents (DES) to prevent the growth of a new endothelium. DCBs are mainly used to treat in-stent restenosis and de novo lesions in small vessels. Screening for more effective drugs, longer duration of action in the vasculature, and improving drug delivery systems to reduce drug loss during delivery and improve drug balloon passage performance will be major directions for future research and development.

Another revolutionary innovation is undoubtedly Shockwave Medical's Shockwave C2 balloon catheter, which uses ultrasound waves to generate a reflected wave on the surface of the calcification, which is compressed and ruptured. When the intensity of the wave is greater than the expansion intensity of the stone, the calcification ruptures. The ultrasound waves continue to propagate and gradually attenuate in a liquid or liquid-like soft tissue medium without causing damage to the soft tissue. In February 2021, the Shockwave C2 balloon catheter received FDA approval for the treatment of severely calcified coronary arteries. As its clinical use increases, it will certainly have a significant impact on current interventional strategies [3].

13.1.3 Coronary Stents

Continuous optimization of conventional metallic stents remains an important direction in the development of coronary stents, including thinner struts, lower struts-drug ratios, better passing ability and radial support. With innovations in conventional metal stent science, particularly the continued development of DES, innovations in antiplatelet agents, and the addition of intravascular imaging during implantation, TLFs, including stent thrombosis and restenosis, have declined to extremely low levels. However, we still retain anxiety about permanent implants.

Biodegradable stents are another direction of development for coronary stents, mainly including bioresorbable polymeric vascular scaffold (BPVS) and biodegradable metallic alloy scaffold (BMAS). Among them, BPVS has a significantly thicker stent struts in order to achieve higher radial support and needs to face the problem of uneven degradation rate, which leads to a higher incidence of in-stent thrombosis and limits its application scenarios; BMAS can overcome some of the problems of BPVS but still needs to be validated in more complex clinical situations. New skeletal materials are also being developed. STENTiT Stent is a new type of stent made of biodegradable electrospun polymer, which differs from conventional stents in three ways: prepared from biodegradable electrospun fibers with internal microstructures that provide an environment for the patient's own cell growth; the combination of bioresorbable material and graft design helps it to be tightly anchored to the vessel wall, which has a positive impact on tissue remodeling. STENTiT Stent is currently undergoing animal testing, and it is hoped that new materials will expand the development of coronary stents [4].

13.1.4 Devices for CTO

Chronic total occlusion (CTO) is a complete occlusion of a coronary artery for more than 3 months. The low success rate and high complication rate have made CTO the last bastion of PCI. Current CTO-specific devices include specialized CTO guidewires and devices such as the Stingray balloon catheter, which are designed to achieve specific ADR procedures. Making CTO-PCI free from complex inverse strategies and making forward PCI easier is an important mission for CTO devices to accomplish. The SoundBite Crossing System utilizes Soundbite's patented technology to produce a shock wave through the source console, which continuously impacts the calcified lesion in front of it. The SoundBite Crossing System is available in 0.018"and 0.014" guidewire sizes. It is able to pass through the hardest lesions in CTO, calcium plaques, and fibrotic lesions, while, keeping healthy tissues unharmed, allows control of the amount and shape of the shock wave released by the guidewire, improves the torque response of the guidewire, and the bendable tip with better steering capability makes CTO PCI easy and fast, allowing general PCI performers to accomplish CTO PCI [5].

13.2 Structural Heart Devices

13.2.1 Devices for Aortic Valve Valves

Even after more than 10 years of development, the TAVR device remains imperfect, and its research remains on the fast track in the field of interventional cardiac device research. After the first generation of transcatheter prosthetic valves derived many problems in their application, such as excessive PPM implantation rates, coronary risk, and PVL. At the same time, VIV remains difficult to avoid in pure AR patients. In addition to the current mainstream valves such as SAPIEN 3 and Evolute, several major medical device companies have launched new generation valves to address the above aspects.

In 2021, the Abbott Portico valve and the FlexNav delivery system were jointly approved by the FDA for use in patients with severe AS who are at excessive surgical risk. Portico is designed with intra-annular leaflet position in a self-expanding valve provides early valve function and hemodynamic stability throughout the procedure and with large-cell geometry which preserves coronary access for further coronary intervention. The Portico is available in four sizes – 23, 25, 27, and 29 mm – with covering an annular range from 19 to 27 mm [6].

Boston Scientific's first-generation Acurate neo valve performed poorly compared to both the current generation SAPIEN and CoreValve. Boston Scientific's second-generation Acurate neo2 continues the generation's larger effective orifice area, supra-annular valve design, open frame construction, and stable arch with automatic coaxial adjustment, a special "top-down" release mechanism for more controlled operation, and a flexible delivery system, with a sealed skirt. The ACURATE neo AS study support the 30-day and 12-month safety and performance of TAVR with the ACURATE neo2 valve, and further data is still needed [7].

JENAVALVE Technologies' Trilogy™ System was the first transfemoral device to receive CE for the treatment of both severe symptomatic AR and AS. While TAVI devices continue to appear on the market, they have limited ability to treat aortic valve disease beyond aortic stenosis, which is characterized by severe calcification and the need to adequately immobilize the valve. The Trilogy system uses its proprietary positioner technology that does not rely on calcification to immobilize the valve. Instead, the positioner clips onto the patient's local anatomy to provide valve security. The valve's unique design also enables anatomic valve positioning, which facilitates future access to the coronary arteries and beneficial hemodynamics, both of which are significant clinical advantages in the treatment of both AR and AS [8].

Aortic valve calcification increases the risk of PVL, annular rupture, and stroke and can limit the open area of the implant valve and shorten the life of the prosthetic valve. Aortic valve calcification modification techniques are also a direction of development for aortic valve interventional devices. Technologies that have been proposed include AorticLab's GEMINI, a unique Double-Frequency Ultrasound Ablator. This allows the calcium to be removed more effectively from the surface of the leaflets. According to AorticLab, the whole GEMINI procedure takes only 30 min, which is very efficient, and the GEMINI procedure needs to be used in conjunction with AorticLab's embolic protector EMBRACE, which can effectively collect the calcium shed during the procedure and prevent the calcium from flowing into the whole body with blood flow [9].

In addition, the Leaflex™ Performer transcatheter aortic valve repair system, designed to etch the calcified portion of the aortic valve leaflet (i.e., create an indentation line in the calcified portion of the aortic valve leaflet without tearing the ventricular tissue), completely separates the area of calcified deposits on either side of the etch, allowing the leaflet to regain elasticity and restore mobility, thereby increasing the open area of the valve. All these techniques are currently in the clinical research phase [10].

13.2.2 Devices for the Treatment of Mitral Valve

The incidence of severe mitral regurgitation (MR) is second only to that of aortic valve disease. However, due to its complex anatomy, the design of transcatheter mitral valve replacement (TMVR) systems is much more complex than that of TAVR. Transcatheter mitral valve repair (TMVr) remains the dominant clinical intervention, particularly edge-to-edge repair. There are many more novel designs about TMVr devices.

The tendon cord repair technique is another idea for mitral valve repair. Most of the transcatheter tendon repair techniques that are currently available or under development are basically similar to the NeoChord DS1000 and Edwards Harpoon, both of which have a transapical approach. The transfemoral ChordArt consists of a delivery system and an implant to replace a ruptured or stretched tendon cord and reconnect the valve leaflet to the papillary muscle to restore normal mitral valve function. The implant allows preloading within the delivery system and consists of three parts: proximal anchorage (for leaflet anchorage), cords (ePTFE sutures), and distal anchorage (papillary muscle anchorage). The ChordArt implant thereby follows nature, anchoring the new chord at the same places as natural chords, resulting in minimal interference with the biological valve structure [11].

In addition, Vesalius Cardiovascular is also developing a new interventional artificial tendon repair technology, the Calla TMVr. Calla TMVr currently has less published data. It consists of a delivery system and implantable components. Calla TMVr is a clinical procedure in which an ePTFE is sutured to the free edge of the anterior mitral leaflet and a Vesalius anchor is attached to the papillary muscle. Thus, the ePTFE suture and Vesalius replace the original tendon cords and carry the majority of the anterior mitral leaflet, serving to repair the mitral valve and avoid mitral regurgitation [12].

For TMVR, the product requires attention to structures such as left ventricular left atrial size, leaflet annular calcification, location of papillary tendon attachment, posterior flap and tendon structure function, location of anterior-posterior junction sutures, and adjacent tissue around the annulus. It is the relatively complex constraints that have led to the absence of a unified gold standard and design philosophy for mitral valve replacement in clinical practice. In addition to addressing these anatomic constraints, the design of stent patterns, leaflet attachment, sealing and anchoring mechanisms, access, and dimensions of the valve design pose significant challenges to the design of mitral valve replacement products.

HighLife is a transcatheter mitral valve replacement prosthesis that can be matched to either a transatrial septal approach or a transapical approach, depending on its unique design. The valve is designed to be self-coaxial and adaptive, and the valve is anchored by a valve in ring strategy. There is no need for rotational positioning or consideration of coaxiality, and the device's self-release ring can be used as a landing mark for valve release. So we can clarify the position of the valve by X-ray fluoroscopy and echocardiography [13].

Foldax Tria valve features the patented biopolymer LifePolymer and an innovative valve design designed to resist calcification, withstand stress and not fail strain, and restore patients' quality of life without the need for lifelong anticoagulants. Tria is the first and only heart valve manufactured using automated technology due to its polymeric leaflets that can be manufactured at a precise thickness in a consistent manner and is designed to be a valve with predictable performance that lasts a lifetime [14].

13.2.3 Devices for Pulmonary Valve

TPVR products currently available worldwide are primarily balloon-expandable valves, such as Medtronic's Melody and Edwards' SAPIEN, which are limited in use due to the small size of the balloon-expandable valve, which is not suitable for the anatomic characteristics of the RVOT, which is a tumor-like dilatation due to prolonged PR. The small size of the balloon-expandable valve is expected to be addressed with the FDA approval of Medtronic's self-expanding valve, Harmony, which is dumbbell-shaped with large diameters proximal and distal to the porcine pericardial valve located in the central stenotic portion and relies on the shape memory properties of nitinol to self-expand at the lesion site. The porcine pericardial tissue is sterilized with an alpha-amino-oleic acid anticalcification process and 0.2% glutaraldehyde. Harmony is available in only one model with distal corona, waist, and proximal corona diameters of 34 mm, 23.5 mm, and 42 mm, respectively, and a total stent length of 55 mm, whereas Medtronic's balloon-expandable valve Melody is available in only two models with maximum internal diameters of 20 mm and 22 mm. 3-year results from the Native TPV EFS revealed stable Harmony TPV device position, good valve function in most, and the absence of moderate/severe PVL and significant late frame fractures [15].

13.2.4 Devices for Tricuspid Valve

The transcatheter therapeutic device design concept for flow is similar to that of the mitral valve in that it includes both valve repair and valve replacement, but the proximity to the upper and lower vena cava structures allows for more options in the design of the device.

OrbusNeich P&F recently announced CE approval for its transcatheter dual-chamber valve system, TricValve®, a system consisting of two self-expanding bioprosthetic valves for the treatment of patients with tricuspid valve insufficiency and endoluminal regurgitation [16]. The valve is available in two models, SVC and IVC, both involving different diameters to best match the anatomical characteristics of the superior and inferior vena cava. It offers a new treatment option for patients who are not surgical candidates and for patients with a large degree of annular dilatation or even with a pacemaker.

TRISOL VALVE is a transcatheter tricuspid valve replacement valve with a very simple and intuitive implantation method, stent, and a specially designed dome-shaped leaflet. The most important feature is that the valve leaflet structure is completely different from the published leaflets of other companies, with a domed design and a completed leaflet and a traditional leaflet suture formation [17].

The Cuspa Medical Co. came up with a solution to the problem of inadequate closure that occurs in all valves of the heart. The Cusper is an artificial valve cusp with a grasper that attaches to a native valve cusp. The cusp is designed to compensate for the inadequate fit of the native cusp, effectively eliminating the regurgitant gap orifice while maintaining normal valve function. The Cusper moves with the valve leaflets, so that when the valve is open, the Cusper maintains the valve opening at its original size, thus ensuring proper blood flow through. When the valve is closed, the Cusper ensures that there are no openings between the valve leaflets, thus preventing backflow of blood. The relatively small size of the Cusper implant allows for a transvascular approach to each valve, and the small diameter of the delivery system causes less trauma and allows for rapid recovery [18].

In addition to biological tissues such as pericardium as a material for valve leaflets, new composite materials have entered the circuit. Since currently approved heart valves are for elderly patients, they are not applicable for pediatric patients. In addition to the limited life span of biologic valves (too short), in addition to the fact that the physiology of children is an unsettled one relative to adults and will continue to change as children develop, the heart valves that are already on the market are set in stone and cannot be selected to change as the physiology of the patient changes. Pediatric patients can only undergo multiple valve surgeries to match the patient's physiological changes.

Xeltis PV is an absorbable implant (pulmonary valve) that utilizes endogenous tissue repair (ETR) technology, allowing the patient's own tissue to colonize the implant and evolve into a fully functional natural heart valve over time, while the absorbable implant (valve) is gradually degraded and absorbed over time. ETR technology, Xeltis is not limited to Xeltis PV but has now also been expanded to include aortic valve replacement [19].

13.3 Congenital Heart Disease Devices

Because of the elevated risk of AF after PFO and ASD occlusion, preservation of puncturable septal structures for future interventions such as radiofrequency ablation is an issue that needs to be addressed. The main body of PFO and ASD occluders currently in clinical use is a four-layer metal mesh structure densely woven with nickel-titanium alloy wire and the multi-layer flow-blocking membrane of the double disc produces hypertrophic fibrous hyperplasia tissue between them, making it almost impossible to puncture through the device from the center of the occluder and lose the atrial septum as an important interventional channel. ReAces is the world's first punctable occluder to be tested in humans. The main body of the device maintains the basic structure and working principle of the traditional occluder "double disc and one waist" to ensure safe and effective occluding. The central zone of the occluder has a metal-free, thin flow-blocking membrane design that effectively blocks blood flow through the occluder and makes it easy to puncture and deliver the sheath from the blocker. In addition, the metal content is significantly reduced compared to conventional occluders, which reduces metal-related side effects. While safely and effectively treating atrial septal defects, ReAces preserve access for subsequent interventional procedures performed through the atrial septum puncture, preserving the opportunity for minimally invasive interventional procedures for patients [20].

13.4 Devices for Pacing and Ablation

Transvenous catheter ablation for intractable tachyarrhythmias is one of the advances in cardiovascular pathology in the past two decades. In 2020, the total number of tachyarrhythmias in China exceeds 30 million, with atrial fibrillation and supraventricular tachycardia predominating; more than 212,000 of these patients undergo electrophysiological procedures. Ablation therapy remains the best initial treatment option for the majority of refractory atrial tachycardias and junctional tachycardias. The elucidation of arrhythmia mechanisms has driven the development of catheter ablation for the radical treatment of arrhythmias, and the application of new energy, new devices, and new markers has helped to improve the success rate of catheter ablation of arrhythmias. Several clinical studies, such as CABANA, have validated that electrophysiological ablation procedures are significantly superior to pharmacological therapy, significantly reducing the risk of recurrence and improving the quality of patient survival.

13.4.1 New Technologies for Atrial Fibrillation Ablation

The safety and efficacy of conventional transcatheter ablation of AF cannot be achieved simultaneously, and optimization and improvement based on existing techniques and devices have not yet broken this bottleneck. Tumor ablation is often performed using an electroporation technique (Pulse Field), which allows complete ablation of local lesions without affecting peripheral vascular blood flow, while having no major impact on peripheral circulation, nerves, or adjacent tissues.

Recently, the results of the first human trial of pulsed electric field ablation of AF based on this principle show a 100% rate of pulmonary vein isolation immediately after the procedure, no stroke, phrenic nerve injury, pulmonary vein stenosis, or left AV esophageal fistula during follow-up and a sinus rhythm maintenance rate of 87.4% [21]. Recent international multicenter studies have shown that high-power (45–50 W), short-duration (2–15 s) discharges (Kill Dot) are expected to replace conventional RF ablation modalities, reduce the risk of perioperative complications, shorten procedure time and discharge time, and produce more limited and durable transmural injury.

13.4.2 Ventricular Arrhythmia Ablation

Ventricular arrhythmias are a heterogeneous group of disorders that can be one of the concomitant manifestations of certain diseases or can be the earliest or only manifestation of cardiac abnormalities. Catheter ablation is an important treatment modality for this group of diseases, and its status has been rising in recent years, with more than 20 years of exploration and refinement to reach maturity.

Advances in instrumentation and technology have contributed to the development of catheter ablation of ventricular arrhythmias. The advent of three-dimensional electroanatomical landmarking systems, high-precision landmarking techniques, image fusion techniques, and radiofrequency ablation needles has helped to identify the underlying mechanisms of arrhythmias and improve the success rate of the procedure [22]. In idiopathic ventricular arrhythmias, the success rate of its catheter ablation procedure has reached more than 80%, and it can already be used as the first-line treatment for ventricular premature right ventricular outflow tract (Class I recommendation, Class B evidence). In organic ventricular tachycardia, catheter ablation has been shown to reduce arrhythmia burden and the number of ICD discharges, but there is still no conclusive evidence of improvement in clinical "hard endpoints". The use of radiation therapy for refractory ventricular tachycardia is an exciting innovation.

13.4.3 Radiofrequency Ablation of Hypertrophic Obstructive Cardiomyopathy

The treatment of hypertrophic obstructive cardiomyopathy (HOCM) is one of the clinical challenges that cannot be ignored. Traditional non-pharmacological treatments include surgical septal resection and chemical ablation of the septum. The former is surgically invasive, while the latter causes an uncontrollable extent of myocardial infarction, a high incidence of atrioventricular block, and coronary conditions unsuitable for chemical ablation in 5% to 15% of patients. In recent years, transcatheter septal radiofrequency ablation, assisted by intracardiac ultrasound and three-dimensional electroanatomical landmarking system, has been performed by releasing radiofrequency energy through the catheter to cause local coagulation and necrosis of septal tissue, and outflow tract obstruction is effectively relieved. Because this procedure is less destructive to the myocardial cells around the ablation target, it can avoid conduction bundle injury and reduce the risk of pacemaker implantation. 2018 Journal of the American College of Cardiology (JACC) published the remarkable Liwen procedure from the First Affiliated Hospital of Air Force Military Medical University (Xijing Hospital) in China, which is based on the principle of implanting a radiofrequency needle into the septum through the apical part of the epicardium via percutaneous intercostal puncture. The ablation of HOCM was performed. Patients with HOCM treated had a significant reduction in resting/excited LV outflow tract pressure gradient and septal thickness with no perioperative complications, and all physiological parameters improved at 6 months of follow-up.

13.4.4 Ablation-Assisted Calibrations

Three-dimensional (3D) imaging is also a very active field in arrhythmias, and the images show a great development in the application and integration of various new technologies, both in 3D imaging and in high-precision electrophysiologic measurements. The new 3D mapping technique ripple mapping has been successfully applied to persistent AF, atrial tachycardia, and ARVC, while an observational trial showed that ablation of persistent AF guided by noncontact balloon mapping had a good prognosis (59%) at 12 months. In addition, a clinical trial of the feasibility of the STAR scaler system for AF ablation has been conducted.

13.5 Devices for Heart Failure Treatment

Interventional treatment of heart failure has evolved rapidly in recent years, and the main design ideas include left ventricular assist devices for hemodynamic support and improved ventricular remodeling from different perspectives such as morphology/electrophysiology. Let us first introduce a device for pulmonary artery pressure detection.

Pulmonary artery pressure-guided heart failure management is rapidly becoming the standard of care for NYHA Class III patients. The Cordella pulmonary artery pressure sensor enables clinicians to remotely monitor and proactively adjust therapy and medication without the need for patients to come to the hospital. The First-in-Hand (FIH) trial is currently in the follow-up phase to demonstrate the safety and accuracy of the Cordella sensor. The study met key accuracy and safety nodes, and patients demonstrated high compliance with daily measurements [23].

The ventricular assistant device (VAD) is another important area of interventional devices for heart failure. Abiomed has introduced a new generation of Impella, the Impella ECP, the world's smallest hemodynamically supported heart pump, with a diameter of only 9 F (3 mm). Despite its small size its flow rate is not small, reaching flows of more than 3.5 L/min. Once the sheath is removed, it can be expanded to approximately 18 Fr. Postoperative removal is also very simple, as the pump is simply retracted into the 9 Fr sheath and withdrawn via the same route. The Impella ECP uses a specially designed pigtail tube, prepared with a low stiffness polyurethane, so that the pigtail tube is very flexible and can be used like a guidewire to guide the heart pump through the aortic valve and into the left ventricle without damage. Delivery system to the heart, whereas the Impella ECP delivery sheath is only 9F), which means that the smaller puncture holes will help to address the current need for large-diameter vascular access and the associated bleeding complications, which may lead to an increase in the use of heart pumps for high-risk PCI [24].

In addition to the classic Impella design, which places the VAD pump deep into the left ventricle, the placement of a flow pump into the descending aorta is a relatively well-established option. The representative device is the Aortix from Procyrion, a very small device with a diameter of only 6 mm and a length of 6.5 cm. The Aortix is placed in the descending aorta for up to 7 days through a percutaneous catheter intervention, reducing the cardiac load and thus allowing the heart to rest and increasing the renal blood supply. It has already had great success in the field of heart failure treatment. Its intended use is now being expanded into the field of cardiorenal syndrome. To provide new treatment options for patients with cardiorenal syndrome [25].

Another way in which VAD is achieved is through physical means, by giving assisted contraction, or by reducing the internal diameter of the left ventricle and delaying left ventricular remodeling. The concept of extraventricular mechanical circulatory support (MCS) is to deliver the required MCS by squeezing the heart from its outer surface (epicardium), thus avoiding blood contact. Thus, it also completely avoids the blood contact complications associated with VADs. The BEAT system is a new generation of extracardiac MCS devices developed by AdjuCor Germany and consists of an implant attached to a portable driver with a pneumatic air cushion that mechanically assists the heart by inflating and deflating it. Supply lines are used to exchange compressed air between the driver and electrodes and to transmit ECG data from the heart to the driver. The portable drive unit is responsible for generating compressed air, interpreting the ECG data, and controlling the inflation/deflation of the implant pneumatic cushion. The BEAT system offers the following advantages: (1) completely avoids contact with blood, (2) provides single and/or dual ventricular support by adjusting the left and right ventricular support separately on the BEATview monitor, (3) allows for minimally invasive implantation without any cardiac anastomosis or sutures implantation, (4) provides true pulsatile support as it is synchronized in real time with the local cardiac ECG, and (5) is a non-compulsory system, meaning that in the unlikely event of a technical failure, the patient may be weakened by lack of support but will not die from thrombosis due to blood quiescence, as the heart will continue to beat even if the BEAT is not actively assisting the heartbeat [26].

There are also devices that physically reduce the internal diameter of the left ventricle to reduce cardiac load and improve cardiac remodeling. V-sling is a transcatheter ventricular repair device designed and developed by Cardiac Success, to simultaneously reduce the left ventricle and correct papillary muscle misalignment to restore the conical shape of the left ventricle, thereby improving the ejection fraction of patients with heart failure and addressing the heart failure problem at its source. The success of the V-sling concept will need to be in future clinical trials [27].

The AccuCinch from Ancora Heart offers an option that works by anchoring a flexible implant to the inner wall of the left ventricle, thereby shrinking the left ventricle, reducing ventricular wall stress, and supporting and strengthening the heart wall. AccuCinch treats mitral regurgitation in addition to heart failure [28].

Heart failure is the final "battlefield" of cardiovascular disease and the area where the structure of the heart is most closely linked to electrophysiology. Cardiac resynchronization therapy (CRT) is a relatively well-established treatment for heart failure and is recognized by guidelines. However, 30% of patients do not respond to CRT treatment, which is mainly due to the placement of the pacing lead in the epicardium (outside), making the pacing results inconsistent with the actual heart rhythm. Endocardial stimulation of the left ventricle is preferable to epicardial pacing, but pacing leads placed in the left ventricle can lead to complications such as thrombosis and stroke. In addition, pacing leads may break, displace, or otherwise fail, resulting in adverse events in approximately 12% of cases. These reasons ultimately lead to the inability of conventional CRTs to meet clinical needs.

EBR Systems develops the world's only wireless cardiac pacing product for heart failure—WiSE CRT System. The WiSE CRT System uses a new delivery system that implants electrodes in the left bundle branch region and uses wireless technology to deliver the pacing stimulus directly to the inside of the left ventricle of the heart, acting as left bundle branch area pacing (LBBAP). This approach is designed to overcome the limitations of conventional CRTs, where pacing from the inside of the left ventricle is considered more physiological. The electrodes of the WiSE CRT System are only the size of a grain of rice. The WiSE CRT System has electrodes the size of a grain of rice, thus completely eliminating the need for a conventional CRT that requires a pacing lead on the outside of the left ventricle of the heart and the associated complications. The design of the WiSE CRT System also allows for flexibility in the choice of pacing site, allowing for customization of the treatment for the patient [29].

13.6 Conclusion

Clinical needs and physicians' experiences are crucial to the development of devices. It can be said that clinical demand is the source of vitality of medical devices. If a device is not clinically required or is only clinically pseudo-demanded, then the device, no matter how powerful and perfectly designed it is, will be in a situation where no one will use it. Medical devices generally need to be used by doctors, and some devices are used by patients themselves. As with other industrial products, user experience is very important. If a product is cumbersome to use and has a long learning curve, it will limit the promotion of the product. Physicians have a natural understanding of clinical needs and user experience.

The pursuit of minimally invasive and noninvasive is the eternal theme of medical technology development. Reviewing the development of medical devices, stent implantation instead of coronary artery bypass, large intracavitary stents instead of open vascular surgery, the rise of percutaneous subpenetrating treatment and so on, all of them are performing this law. Currently, structural heart disease is the biggest boom in the development of interventional cardiology, and it is in the development of something that can be done with transcatheter minimally invasive methods to replace traditional surgical procedures. However, minimally invasive treatment may be somewhat different or even slightly inferior to traditional open surgery in terms of results. Therefore, the re-interventional nature of the technique becomes important. For example, TAVR valve failure occurs and then TAVR, arrhythmia radiofrequency ablation recurs and then ablation, and new lesions appear after coronary stenting and then intervention. If it is feasible and effective to intervene minimally in these patients again, then such technology and medical devices are perfect.

An ideal medical device must be a perfect combination of science and art and must be scientifically valuable and artistic at the same time. The shape of each medical device implanted into the body must be streamlined to avoid fracture of the device after implantation due to uneven local force and to improve its fatigue durability. The streamlined design of in vitro operating systems such as delivery systems can also improve user experience and meet ergonomic standards, while the beautification and artistic design can increase the ornamental nature of the product and improve users' favorable impression of the product. The future process of medical device research, development, and innovation is bound to be the pursuit of the ultimate fusion of device performance and artistry!

References

1. Chu M, Jia H, Gutiérrez-Chico JL, Maehara A, Ali ZA, Zeng X, He L, Zhao C, Matsumura M, Peng W, Zeng M, Kubo T, Bo X, Chen L, Bo Y, Mintz GS, Wijns W, Holm NR, Shengxian T. Artificial intelligence and optical coherence tomography for the automatic characterisation of human atherosclerotic plaques. EuroIntervention. 2021;17(1):41–50.
2. Leng J, Zhang J, Li C, Shu C, Wang B, Lin R, Liang Y, Wang K, Shen L, Lam K-h, Xie Z, Gong X, Ge J, Song L. Multi-spectral intravascular photoacoustic/ultrasound/optical coherence tomography tri-modality system with a fully-integrated 0.9-mm full field-of-view catheter for plaque vulnerability imaging. Biomed. Opt Express. 2021;12(4):1934–46.

3. Umapathy S, Keh YS, Wong N, Ho KW, Tan JWC, Wong ASL, Lim ST, Goh YS, Yeo KK. Real-world experience of coronary intravascular lithotripsy in an asian population: a retrospective, observational, single-center, all-comers registry. J Invasive Cardiol. 2021;33(6):E417–24.
4. Duijvelshoff R, Cabrera MS, Sanders B, Dekker S, Smits AIPM, Baaijens FPT, Bouten CVC. Transcatheter-delivered expandable bioresorbable polymeric graft with stenting capacity induces vascular regeneration. JACC Basic Transl Sci. 2020;5(11):1095–110.
5. Benko A, Bérubé S, Buller CE, Dion S, Riel L-P, Brouillette M, Généreux P. Novel crossing system for chronic total occlusion recanalization: first-in-man experience with the soundbite crossing system. J Invasive Cardiol. 2017;29(2):E17–20.
6. Taramasso M, Miura M, Gavazzoni M, Andreas M, Saccocci M, Gülmez G, Puri R, Maisano F. The Portico transcatheter aortic valve for the treatment of severe aortic stenosis. Future Cardiol. 2019;15(1):31–7.
7. Reardon MJ. The ACURATE neo™ and neo2™ Valve Systems. Heart Int. 2021;15(1):37–41.
8. Hamid N, Ranard LS, Khalique OK, Hahn RT, Nazif TM, George I, Ng V, Leon MB, Kodali SK, Vahl TP. Commissural alignment after transfemoral transcatheter aortic valve replacement with the jenavalve trilogy system. JACC Cardiovasc Interv. 2021;14(18):2079–81.
9. Fermi E, Benettin D, Bernava G, Pesce M, Pasquino E. Trans-catheter double-frequency ultrasound ablator for the treatment of aortic valve leaflets calcification. BJSTR. 2021;33 (4):25952–7.
10. https://www.venusmedtech.com/index.php/list-7
11. https://www.coremedic.de/chordart
12. https://vesaliuscardio.com/calla-tmvr-system/
13. Lange R, Piazza N. HighLife transcatheter mitral valve replacement. EuroIntervention. 2016;12 (Y):Y81–3.
14. https://www.foldax.com/lifepolymer
15. Bergersen L, Benson LN, Gillespie MJ, Cheatham SL, Crean AM, Hor KN, Horlick EM, Lung T-H, McHenry BT, Osten MD, Powell AJ, Cheatham JP. Harmony feasibility trial: acute and short-term outcomes with a self-expanding transcatheter pulmonary valve. JACC Cardiovasc Interv. 2017;10(17):1763–73.
16. Donà C, Goliasch G, Schneider M, Hengstenberg C, Mascherbauer J. Transcatheter TricValve implantation for the treatment of severe tricuspid regurgitation. Eur Heart J Cardiovasc Imaging. 2021;22(7):e92.
17. https://trisol-medical.com/
18. https://cuspamedical.com/the-product/
19. Morales DL, Herrington C, Bacha EA, Morell VO, Prodán Z, Mroczek T, Sivalingam S, Cox M, Bennink G, Asch FM. A novel restorative pulmonary valve conduit: early outcomes of two clinical trials. Front Cardiovasc Med. 2020;7:83360.
20. Burning No Bridges. Atrial Septal Defect Closure With the ReAces Device. JACC Basic Transl Sci. 2022;7(12):1211–3.
21. Reddy VY, Anic A, Koruth J, Petru J, Funasako M, Minami K, Breskovic T, Sikiric I, Dukkipati SR, Kawamura I, Neuzil P. Pulsed field ablation in patients with persistent atrial fibrillation. J Am Coll Cardiol. 2020;76(9):1068–80.
22. Nakagawa H, Kuroda S, Hussein AA. Radiofrequency ablation using a needle electrode combined with heated saline injection: three different mechanisms of tissue heating. Heart Rhythm. 2021;18(3):453–4.
23. Mullens W, Sharif F, Dupont M, Rothman AMK, Wijns W. Digital health care solution for proactive heart failure management with the cordella heart failure system: results of the SIRONA first-in-human study. Eur J Heart Fail. 2020;22(10):1912–9.
24. https://www.abiomed.com/about-us/news-and-media/press-releases/fda-approves-impella-ecp-pivotal-heart-pump-and-first-patients-enrolled-in-pivotal-clinical-trial
25. Kapur NK, Esposito ML, Whitehead E. Aortix™: a novel intra-aortic entrainment pump. Future Cardiol. 2021;17(2):283–91.
26. https://www.adjucor.com/

27. https://www.cardiacsuccess.com/
28. Gooley RP, Meredith IT. The Accucinch transcatheter direct mitral valve annuloplasty system. EuroIntervention. 2015;11(Suppl W):W60–1.
29. Montemerlo E, Pozzi M, De Ceglia S, Santini F, Piazzi E, Rovaris G. First-in-man fully leadless transvenous CRT-P with a transseptal implant of WISE-CRT® system and Micra® PM. Pacing Clin Electrophysiol. 2019;42(11):1489–92.

MIX
Papier aus verantwortungsvollen Quellen
Paper from responsible sources
FSC® C105338

If you have any concerns about our products,
you can contact us on
ProductSafety@springernature.com

In case Publisher is established outside the EU,
the EU authorized representative is:
Springer Nature Customer Service Center GmbH
Europaplatz 3, 69115 Heidelberg, Germany

Printed by Libri Plureos GmbH
in Hamburg, Germany